KV-019-531

N-Heterocyclic Carbenes

From Laboratory Curiosities to Efficient Synthetic Tools

RSC Catalysis Series

Series Editor:
Professor James J Spivey, *Louisiana State University, Baton Rouge, USA*

Advisory Board:
Krijn P de Jong, *University of Utrecht, The Netherlands*, James A Dumesic, *University of Wisconsin-Madison, USA*, Chris Hardacre, *Queen's University Belfast, Northern Ireland*, Enrique Iglesia, *University of California at Berkeley, USA*, Zinfer Ismagilov, *Boreskov Institute of Catalysis, Novosibirsk, Russia*, Johannes Lercher, *TU München, Germany*, Umit Ozkan, *Ohio State University, USA*, Chunshan Song, *Penn State University, USA*

Titles in the Series:
1: Carbons and Carbon Supported Catalysts in Hydroprocessing
2: Chiral Sulfur Ligands: Asymmetric Catalysis
3: Recent Developments in Asymmetric Organocatalysis
4: Catalysis in the Refining of Fischer-Tropsch Syncrude
5: Organocatalytic Enantioselective Conjugate Addition Reactions: A Powerful Tool for the Stereocontrolled Synthesis of Complex Molecules
6: N-Heterocyclic Carbenes: From Laboratory Curiosities to Efficient Synthetic Tools

How to obtain future titles on publication:
A standing order plan is available for this series. A standing order will bring delivery of each new volume immediately on publication.

For further information please contact:
Book Sales Department, Royal Society of Chemistry, Thomas Graham House, Science Park, Milton Road, Cambridge, CB4 0WF, UK
Telephone: +44 (0)1223 420066, Fax: +44 (0)1223 420247, Email: books@rsc.org
Visit our website at http://www.rsc.org/Shop/Books/

N-Heterocyclic Carbenes
From Laboratory Curiosities to Efficient Synthetic Tools

Edited by

Silvia Díez-González
Department of Chemistry, Imperial College London, London, UK
Institute of Chemical Research of Catalonia (ICIQ), Tarragona, Spain

GEORGE GREEN LIBRARY OF
SCIENCE AND ENGINEERING

RSC Publishing

1006340155

RSC Catalysis Series No. 6

ISBN: 978-1-84973-042-6
ISSN: 1757-6725

A catalogue record for this book is available from the British Library

© Royal Society of Chemistry 2011

All rights reserved

Apart from fair dealing for the purposes of research for non-commercial purposes or for private study, criticism or review, as permitted under the Copyright, Designs and Patents Act 1988 and the Copyright and Related Rights Regulations 2003, this publication may not be reproduced, stored or transmitted, in any form or by any means, without the prior permission in writing of The Royal Society of Chemistry or the copyright owner, or in the case of reproduction in accordance with the terms of licences issued by the Copyright Licensing Agency in the UK, or in accordance with the terms of the licences issued by the appropriate Reproduction Rights Organization outside the UK. Enquiries concerning reproduction outside the terms stated here should be sent to The Royal Society of Chemistry at the address printed on this page.

The RSC is not responsible for individual opinions expressed in this work.

Published by The Royal Society of Chemistry,
Thomas Graham House, Science Park, Milton Road,
Cambridge CB4 0WF, UK

Registered Charity Number 207890

For further information see our website at www.rsc.org

Printed and bound in Great Britain by CPI Antony Rowe, Chippenham and Eastbourne

Foreword

A journalist recently asked me if I foresaw, when my group prepared the first stable carbene, that this topic would become a field in its own right. My answer was clear-cut: certainly not! Indeed, in 1988 and even for a few years after, I believed that these species were too fragile to be really useful, and that they would remain laboratory curiosities. I was not totally wrong as far as our group's first stable carbene is concerned, and it is obvious that Arduengo's carbenes, the famous NHCs, have been responsible for the fantastic development of this field of chemistry.

As can be seen in this Book, during the last fifteen years or so, following the pioneering work by Herrmann *et al.*, carbenes have mostly been used as ancillary ligands for the preparation of transition metal-based catalysts. Compared to phosphorus ligands, carbenes tend to bind more strongly to metal centres, avoiding the necessity for the use of excess ligand in catalytic reactions. The corresponding complexes are often less sensitive to air and moisture, and have proven remarkably resistant to oxidation. It is noteworthy that, although the first carbene–transition metal complexes were prepared as early as 1915 by Chugaev (Fischer and Maasböl were the first to fully characterize a carbene–metal species), the recent developments in their application in catalysis have considerably been facilitated by the availability of carbenes stable enough to be bottled. Moreover, the existence of metal-free carbenes has allowed their use as organocatalysts.

It is amazing to realize that the first method of taming a carbene was by attaching it to a transition metal, whereas nowadays carbenes can be used to stabilize transition metal centres that otherwise are not accessible. Similarly striking are some recent developments, which show that carbenes, which were considered for years as the prototypes of reactive intermediates, can be used for stabilizing highly reactive main group species.

For the future, since the robustness of carbene complexes is largely due to the presence of strong carbon–metal bonds, other types of stable low-valent carbon species are highly desirable, and I believe that a second generation of carbon-based L ligands will soon appear.

Lastly, I wish to say here that, for Bo Arduengo and myself, the enormous amount of results summarized in this excellent Book is a wonderful gift.

Professor Guy Bertrand
Distinguished Professor of Chemistry
UCR-CNRS Joint Research Chemistry Laboratory,
Department of Chemistry, University of California, USA

Preface

N-Heterocyclic carbenes (NHCs) are certainly peculiar chemical entities. Having been considered as transient species, at the most, for many years, they are currently at the heart of numerous advances in different chemical fields. One of the most conspicuous peculiarities of NHCs is the ongoing debate about their Lewis representation, particularly when used as ligands in metal complexes. A number of representations can be found indistinguishably in the literature and, just as in fashion, the representation of choice is normally dependent on the most popular version at the time, and/or the research group implicated (Figure 1, five-membered ring diaminocarbenes are represented as examples). Representation **A** was soon abandoned with early data showing the poor π back-donating ability of NHCs, making NHC–M bond single in nature (at least with cyclic diaminocarbenes).[1] Although the current picture of NHCs is far more complex and it is now known that π back-donation can represent over 30% of the bonding (hardly negligible!),[2] the analytical data stills point relentlessly towards NHC–M single bonds. Similarly, the representation of unsaturated NHCs as aromatic derivatives (**B**) has gradually decreased following reports on the predominance of the carbenic form over ylidic resonance structures.[3] Even if subsequent studies pointed towards a cyclic electron stabilization, the resulting aromatic character of imidazol-2-ylidenes would be substantially smaller than in benzene or imidazolium salts.[4] Most probably, structure **E** might be

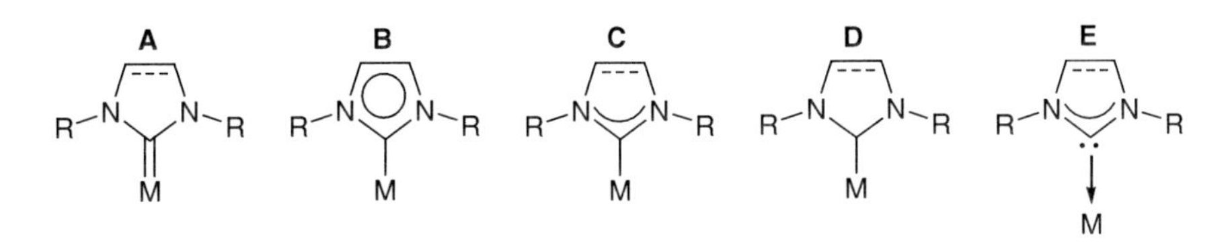

Figure 1 Possible representations of [(NHC)M] complexes.

RSC Catalysis Series No. 6
N-Heterocyclic Carbenes: From Laboratory Curiosities to Efficient Synthetic Tools
Edited by Silvia Díez-González

© Royal Society of Chemistry 2011

Published by the Royal Society of Chemistry, www.rsc.org

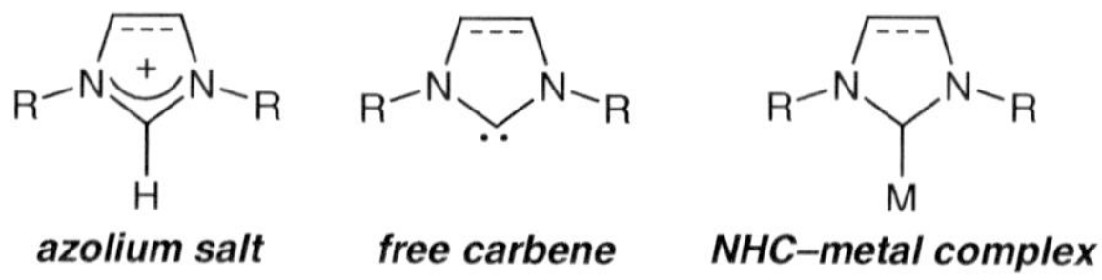

azolium salt *free carbene* *NHC–metal complex*

Figure 2 Chosen representations for the present book.

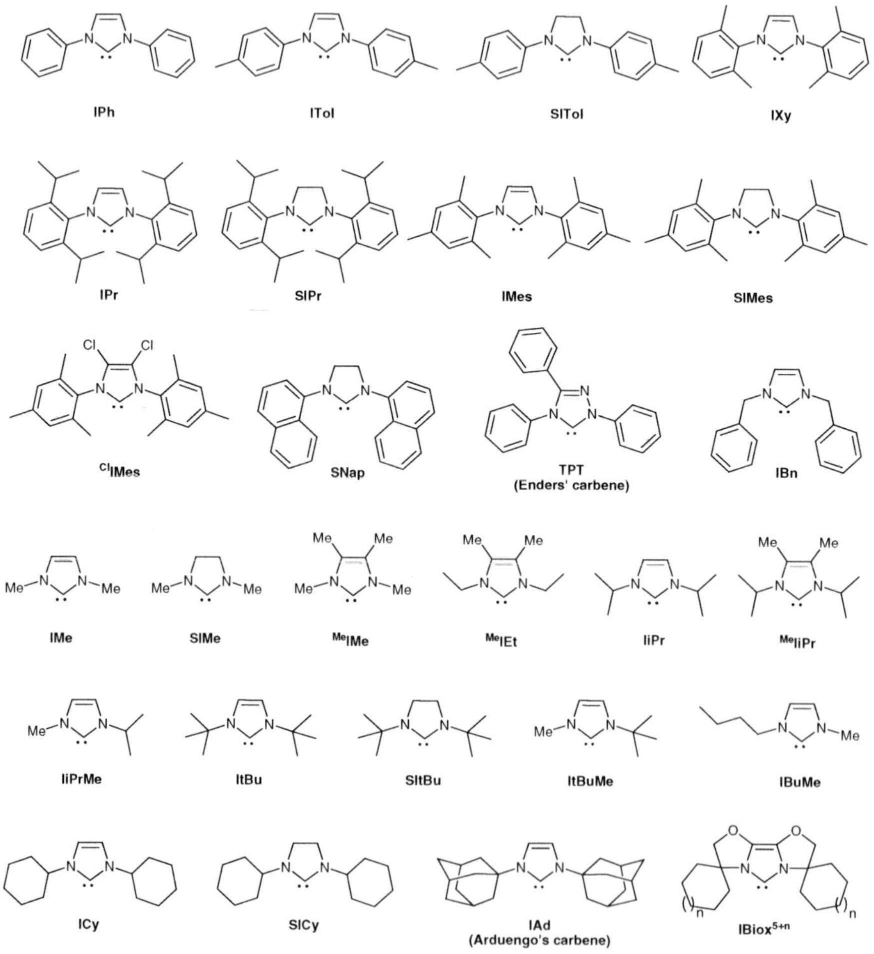

IPh ITol SITol IXy

IPr SIPr IMes SIMes

ClIMes SNap TPT (Enders' carbene) IBn

IMe SIMe MeIMe MeIEt IiPr MeIiPr

IiPrMe ItBu SItBu ItBuMe IBuMe

ICy SICy IAd (Arduengo's carbene) IBiox5+n

Figure 3 Structures and acronyms of NHCs.

the one closer to the real bonding situation in these species; however, it is also the least straightforward to use. Actually, **C** and **D** are the most popular versions nowadays, despite the fact of symbolizing a trivalent carbon bond (**D**) in a confusing manner for any chemist unfamiliar with the field.

In the present book, version **D** has been consistently used. This is unarguably an arbitrary decision and does not pretend to make any statement in this debate. For the reader's convenience, the general representations used for azolium salts, free NHCs and NHC/metal complexes are depicted in Figure 2. Also, the structures of all NHC acronyms used throughout the different chapters can be found in Figure 3.

Regardless of their importance, NHCs are not the only peculiarity of this book. When I was first contacted by the RSC about a book project, I was so astonished I concluded it had to be an internet scam. How could I expect to receive such an invitation with my incipient career? But once it was clear that no one would be asking for my credit card information, I was set to make the best book on NHCs to date . . . or at least to try! My relative 'juniority' (which will be inexorably cured over time) makes me even prouder of having gathered fourteen researchers I knew and respected to join me in this exciting project. From here I would like to thank and congratulate them all for the outstanding result of the joint efforts.

One last peculiarity I would like to share is the wish I cherish for this book. In spite of being a long and demanding journey, I actually hope this book will be soon outdated! Despite their high popularity, the chemistry of NHCs is far from mature and their reactivity, even beyond catalysis, will certainly provide more mind-blowing advances. Personally, I am looking forward to knowing what will happen next.

Dr Silvia Díez-González

References

1. (a) K. Öfele and C. G. Kreiter, *Chem. Ber.*, 1972, **105**, 529–540; (b) K. Öfele and M. Herberhold, *Z. Naturforsch B.*, 1973, **28**, 306–309.
2. X. Hu, Y. Tang, P. Gantzel and K. Meyer, *Organometallics*, 2003, **22**, 612–614.
3. For instance, see: (a) D. A. Dixon and A. J. Arduengo III, *J. Phys. Chem.*, 1991, **95**, 4180–4182; (b) A. J. Arduengo III, H. Bock, H. Chen, M. Denk, D. A. Dixon, J. C. Green, W. A. Herrmann, N. L. Jones, M. Wagner and R. West, *J. Am. Chem. Soc.*, 1994, **116**, 6641–6649.
4. J. F. Lehmann, S. G. Urquhart, L. E. Ennios, A. P. Hitchcock, K. Hatano, S. Gupta and M. K. Denk, *Organometallics*, 1999, **18**, 1862–1872.

"Big thoughts and small words
How could I possibly solve such a problem?
A desert's mistery ..."

Contents

RSC Catalysis Series No. 6
N-Heterocyclic Carbenes: From Laboratory Curiosities to Efficient Synthetic Tools
Edited by Silvia Díez-González
© Royal Society of Chemistry 2011
Published by the Royal Society of Chemistry, www.rsc.org

Abbreviations

2M2BN	2-methyl-2-butenenitrile
2M3BN	2-methyl-3-butenenitrile
3PN	3-pentenenitrile
AB	ammonia–borane
Ac	acyl
acac	acetylacetonato
Ad	adamantyl [tricyclo[3.3.1.1^{3,7}]decyl]
ADC	acyclic diaminocarbene
ADMET	acyclic diene metathesis polymerisation
AdN	adiponitrile
Am	amyl
An	*para*-anisyl
aq	aqueous
Ar	aryl
ATH	asymmetric transfer hydrogenation
ATRP	atom transfer radical polymerisation
AYC	amino(ylidene)carbene
BARF	tetrakis[3,5-bis(trifluoromethyl)phenyl]borate
BET	back-electron transfer
BIMY	benzimidazolinylidene
BINAM	1,1′-bi(2-naphthylamine)
BMIM	1-butyl-3-methylimidazolium
Bn	benzyl
Boc	*tert*-butyloxycarbonyl
bpy	2,2′-bipyridine
Bu	butyl
Bz	benzoyl
CAAC	cyclic alkyl(amino)carbene
cat	catecholato
CF	cystic fibrosis
CHM	Chalk–Harrod mechanism
CM	cross-metathesis
COD	1,5-cyclooctadiene
COE	cyclooctene
^{Cl}IMes	4,5-dichloro-1,3-bis(2,4,6-trimethylphenyl)imidazol-2-ylidene

coMAO	co-polymeric isobutyl methylaluminoxane
conv.	conversion
cot	cyclooctatetraene
Cp	cyclopentadienyl
Cp*	1,2,3,4,5-pentamethylcyclopentadienyl
Cy	cyclohexyl
Cyp	cyclopentyl
DAC	diaminocarbene
dba	dibenzylideneacetone
DBU	1,8-diazabicyclo[5.4.0]undec-7-ene
DCC	*N*,*N*′-dicyclohexylcarbodiimide
DCE	1,2-dichloroethane
DCM	dichloromethane
de	diastereomeric excess
DFT	density functional theory
DIPEA	*N*,*N*-diisopropylethylamine
Dipp	2,6-diisopropylphenyl
DMA	*N*,*N*-dimethylacetamide
DMAO	dry methylaluminoxane
DMAP	4-(dimethyl)pyridine
DME	1,2-dimethoxyethane
DMF	dimethylformamide
DMSO	dimethylsulfoxide
dpe	1,2-diphosphinoethane
dppe	1,2-bis(diphenylphosphino)ethane
dppp	1,3-bis(diphenylphosphino)propane
dr	diastereoisomeric ratio
ee	enantiomeric excess
equiv	equivalent
Et	ethyl
ETM	early transition metal
EWG	electron-withdrawing group
Fc	ferrocene
FLP	frustrated Lewis pair
GC	gas chromatography
GGA	generalised gradient approximation
Hept	heptyl
Hex	hexyl
HIV	human immunodeficiency virus
HMDS	hexamethyldisilazide
HOAT	1-hydroxy-7-azabenzotriazole
HOMO	highest occupied molecular orbital
i	iso
IAd	1,3-diadamantylimidazol-2-ylidene
IBn	1,3-dibenzylimidazol-2-ylidene
IBox	bisoxazoline-based N-heterocyclic carbene

IBuMe	1-butyl-3-methylimidazol-2-ylidene
IC_{50}	half maximal inhibitory concentration
ICy	1,3-dicyclohexylimidazol-2-ylidene
IH	imidazol-2-ylidene
IiPr	1,3-diisopropylimidazol-2-ylidene
IiPrMe	1-methyl-3-isopropylimidazol-2-ylidene
IL	ionic liquid
IMe	1,3-dimethylimidazol-2-ylidene
IMes	1,3-bis(2,4,6-trimethylphenyl)imidazol-2-ylidene
IPh	1,3-diphenylimidazol-2-ylidene
IPr	1,3-bis(2,6-diisopropylphenyl)imidazol-2-ylidene
IR	infrared
ItBu	1,3-di-*tert*-butylimidazol-2-ylidene
ItBuMe	1-*tert*-butyl-3-methylimidazol-2-ylidene
ITol	1,3-bis(4-methylphenyl)imidazol-2-ylidene
IXy	1,3-bis(2,6-dimethylphenyl)imidazol-2-ylidene
KIE	kinetic isotope effect
LB	lysogeny broth
LDA	lithium diisopropylamide
LHMDS	lithium hexamethyldisilazane
LTM	late transition metal
m	*meta*
MAO	methylaluminoxane
Me	methyl
MECP	minimum energy crossing point
^{Me}IEt	1,3-diethyl-4,5-dimethyl-imidazol-2-ylidene
^{Me}IiPr	1,3-diisopropyl-4,5-dimethyl-imidazol-2-ylidene
^{Me}IMe	1,3,4,5-tetramethylimidazol-2-ylidene
Mes	mesityl [1,3,5-trimethylphenyl]
MIC	minimum inhibitory concentrations
MMAO	modified methylaluminoxane
MMP	mitochondrial membrane permeabilisation
M_n	number-average molar mass
MO	molecular orbital
MOM	methoxymethyl ether
MS	molecular sieves
Ms	mesyl [methanesulfonyl]
MTT	3-(4,5-dimethylthiazol-2-yl)-2,5-diphenyltetrazolium bromide
MW	microwave
Napht	1-naphthyl
NBS	*N*-bromosuccinimide
NHC	N-heterocyclic carbene
NMP	*N*-methylpyrrolidinone
NMR	nuclear magnetic resonance
nbd	norbornadiene

nbe	norbornene
Np	neopentyl
Nu	nucleophile
o	*ortho*
Oct	octyl
p	*para*
PDI	polydispersity index
PE	polyethene
PEG	polyethylene glycol
PEPPSI	pyridine enhanced precatalyst preparation stabilisation and initiation
Ph	phenyl
PHC	P-heterocyclic carbene
pin	pinacolato
PKR	Pauson–Khand reaction
PMHS	polymethylhydrosiloxane
Pr	propyl
PS	polystyrene
PTC	phase transfer catalyst
pyr	pyridine
rfx	reflux
RCM	ring-closing metathesis
RIM	reaction injection moulding
ROCM	ring-opening cross metathesis
ROMP	ring-opening metathesis polymerisation
RT	room temperature
s	*sec*
SET	single electron transfer
SICy	1,3-dicyclohexylimidazolin-2-ylidene
SIH	imidazolin-2-ylidene
SIMe	1,3-dimethylimidazolin-2-ylidene
SIMes	1,3-bis(2,4,6-trimethylphenyl)imidazolin-2-ylidene
SINap	1,3-di(1-naphthyl)imidazolin-2-ylidene
SIPr	1,3-bis(2,6-diisopropylphenyl)imidazolin-2-ylidene
SItBu	1,3-di(*tert*-butyl)imidazolin-2-ylidene
SITol	1,3-bis(4-methylphenyl)imidazolin-2-ylidene
SOMO	single occupied molecular orbital
t	*tert*
tacn	1,4,7-triazacyclononane
TB	*Mycobacterium tuberculosis*
TBDMS	*tert*-butyldimethylsilane
TBHP	*tert*-butyl hydrogen peroxide
TEM	transmission electron microscopy
TEMPO	2,2,6,6-tetramethylpiperidine 1-oxyl
TF	transfer hydrogenation
Tf	triflyl [trifluoromethylsulfonyl]

TFA	trifluoroacetic acid
TFAA	trifluoroacetic acid anhydride
THF	tetrahydrofuran
TIBAL	triisobutylaluminium
$TIMEN^{Ar}$	tris[2-(3-arylimidazol-2-ylidene)ethyl]amine
TIPS	triisopropylsilyl
thd	tris(2,2,6,6-tetramethylheptane-3,5-dionato)-ligand
TM	transition metal
tmeda	*N*,*N*,*N′*,*N′*-tetramethylethylenediamine
TMS	trimethylsilyl
TOF	turnover frequency
Tol	tolyl [methylphenyl]
TON	turnover number
TPT	1,3,4-triphenyl-1,2,4-triazol-5-ylidene
Tr	trityl [triphenylmethyl]
TrxR	thioredoxin reductase
Ts	tosyl [4-methylphenylsulfonyl]
TS	transition state
Xyl	xylyl [dimethylphenyl]

CHAPTER 1

Introduction to N-Heterocyclic Carbenes: Synthesis and Stereoelectronic Parameters

MAREIKE C. JAHNKE AND F. EKKEHARDT HAHN*

Institut für Anorganische und Analytische Chemie, Westfälische Wilhelms-Universität Münster, Corrensstrasse 30, D-48149 Münster, Germany

1.1 Introduction

Chemists have been fascinated with carbenes for more than 150 years.[1,2] The simplest member of this class of compounds, possessing a neutral divalent carbon atom and six electrons in its valence shell, is methylene, CH_2. Numerous attempts to isolate methylene or related compounds failed,[3] although 'carbenic reactivity' of methylene derivatives was described[4,5] in connection with cyclopropanation reactions[5–8] as early as 1953.

Even if free carbenes could not be isolated, carbene complexes have been known for a long time. The first complex with a heteroatom-stabilized carbene ligand, most likely unrecognized as such, was prepared as early as 1925 by Tschugajeff (English transcription, Chugaev). Tschugajeff's 'red salt' **1** was obtained by the reaction of the tetrakis(methylisocyanide) platinum(II) cation with hydrazine. The 'yellow salt' **2** was formed upon treatment of **1** with HCl in a reversible reaction.[9] The determination of the molecular structures of **1** and **2** in 1970 demonstrated that these compounds were correctly described as diaminocarbene complexes (Scheme 1.1).[10]

In 1964, Fischer prepared and characterized unambiguously the first metal carbene complex **3** obtained by nucleophilic attack of phenyl lithium at

RSC Catalysis Series No. 6
N-Heterocyclic Carbenes: From Laboratory Curiosities to Efficient Synthetic Tools
Edited by Silvia Díez-González
© Royal Society of Chemistry 2011
Published by the Royal Society of Chemistry, www.rsc.org

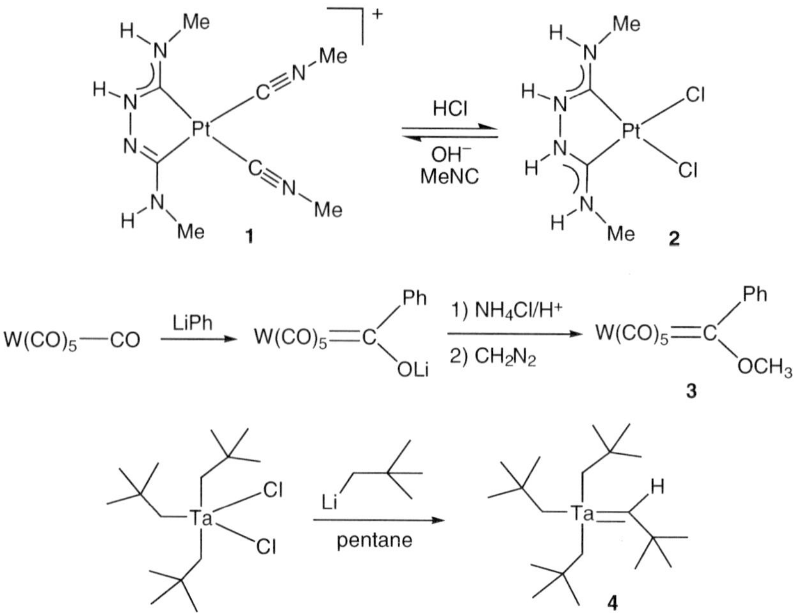

Scheme 1.1 Early syntheses of carbene complexes.

tungsten hexacarbonyl followed by *O*-alkylation.[11] This was followed by Schrock's synthesis of a high oxidation state metal alkylidene complex **4** obtained by α-hydrogen abstraction from tris(neopentyl) tantalum(V) dichloride (Scheme 1.1).[12]

Parallel to these efforts, Wanzlick tried to prepare a stable N-heterocyclic carbene by α-elimination of chloroform from **5**.[13] The free carbene, however, could not be isolated and instead its dimer, the entetraamine **6=6**, was always obtained (Scheme 1.2). Wanzlick's initially postulated cleavage of the entetraamine according to **6=6**→2×**6** could not be demonstrated conclusively. Cross-metathesis experiments with differently *N,N′*-substituted entetraamines failed, excluding an equilibrium between the monomer **6** and the dimer **6=6** (Scheme 1.2).[14]

Around 1960, it was already known that unsaturated heterocyclic azolium cations reacted in a base-catalyzed H,D-exchange reaction.[15] Hoping that the delocalization of the six π-electrons in such derivatives might stabilize the intermediately formed carbene species, Wanzlick attempted to prepare the free carbene **7** by deprotonation of tetraphenylimidazolium perchlorate with KO*t*-Bu (Equation (1.1)). Again, he could not isolate free **7** but its intermediate formation was demonstrated indirectly by identification of some of its reaction products with water or with $[Hg(OAc)_2]$.[16] Almost three decades later, Arduengo succeeded in the preparation of free **7** by the deprotonation method originally suggested by Wanzlick.[17]

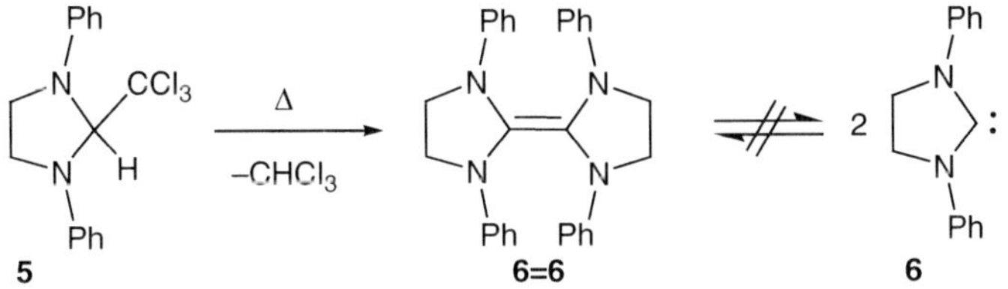

Scheme 1.2 Wanzlick's attempt to synthesize carbene **6** by α-elimination from **5**.

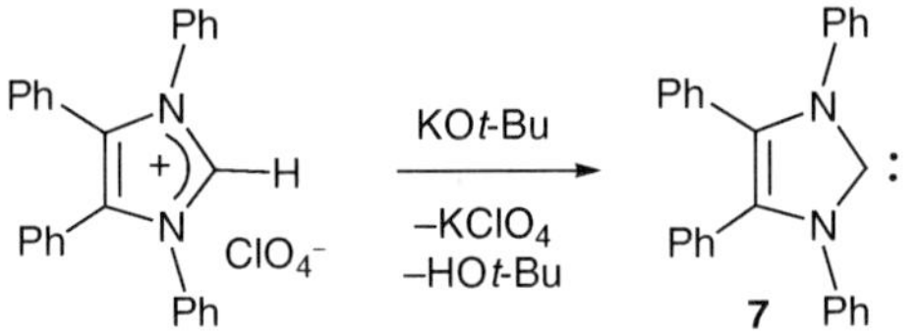

(1.1)

While up to 1990 all attempts to isolate a stable N-heterocyclic carbene failed, metal complexes of unsaturated imidazol-2-ylidenes were known as early as 1968. The first complexes of this type were obtained by *in situ* deprotonation of imidazolium salts using mercury(II) acetate[18] or dimethylimidazolium hydridopentacarbonylchromate(–II)[19] followed by coordination of the carbene to the metal center (Scheme 1.3). Shortly thereafter, the stabilization of the saturated imidazolin-2-ylidene in a metal complex was described by Lappert who treated electron-rich entetraamines of type **6=6** with coordinatively unsaturated transition metal complexes to obtain complexes with imidazolin-2-ylidene ligands (Scheme 1.3).[20]

While Bertrand and co-workers described in 1988 the stable λ^3-phosphinocarbene **8**, which did not act as a ligand,[21] Arduengo *et al.* prepared in 1991 the first free and stable 'bottleable' N-heterocyclic carbene **9** by deprotonation of the corresponding imidazolium salt (Scheme 1.4).[22] This deprotonation method was later supplemented by Kuhn, who introduced the reductive desulfurization of thiones for the preparation of stable imidazol-2-ylidenes.[23]

The isolation of compound **9** demonstrated that free carbenes were not invariably unstable intermediates. Its isolation initiated an intensive search for additional stable N-heterocyclic carbenes (NHCs) leading to the isolation of derivatives with different heteroatoms in the carbene ring and different N-heterocyclic ring sizes. General aspects of the synthesis of NHCs[24–26] and their coordination chemistry with transition metals,[26–30] coinage metals,[31,32] f-block metals[33] and main group elements[34] have already been reviewed. Additional reviews deal with selected classes of polydentate ligands containing NHC donor functions[35–37] and chiral NHC ligands[38,39] or NHCs as organocatalysts.[40,41] Important aspects of carbene dimerization[42] and catalytic applications of NHC complexes have also been reviewed.[43–45]

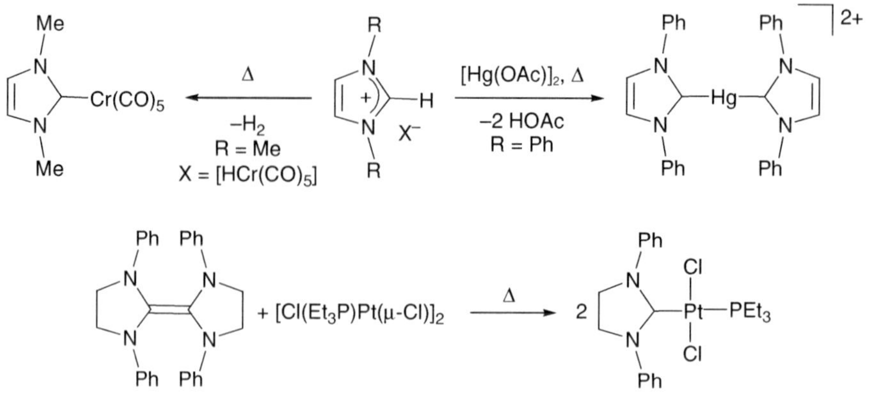

Scheme 1.3 Synthesis of complexes with imidazol-2-ylidene and imidazolin-2-ylidene ligands.

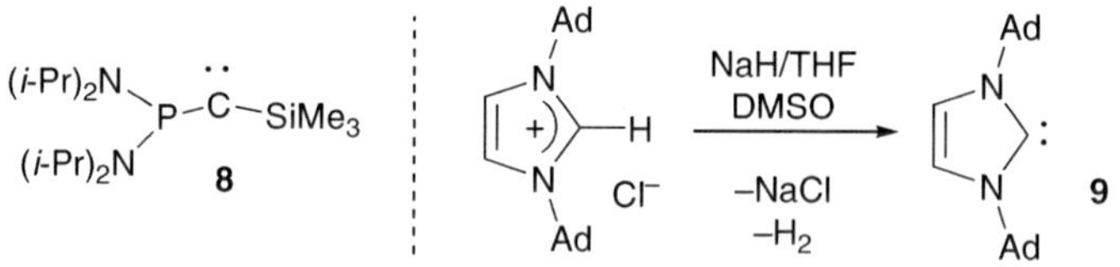

Scheme 1.4 Bertrand's carbene **8** and Arduengo's synthesis of the first stable NHC **9** (Ad = adamantyl).

1.2 Electronic Structure and Stabilization of N-Heterocyclic Carbenes

Carbenes are defined as neutral compounds of divalent carbon where the carbon atom possesses only six valence electrons. If methylene, CH_2, is considered the simplest carbene, a linear or bent geometry at the carbene carbon atom can be considered. The linear geometry is based on a sp-hybridized carbene carbon atom leading to two energetically degenerated p orbitals (p_x, p_y). This geometry constitutes an extreme and most carbenes contain a sp^2-hybridized carbon atom with a non-linear geometry at this atom. The energy of the non-bonding p orbital (p_y), conventionally called p_π after sp^2-hybridization, does practically not change upon the $sp \rightarrow sp^2$ transition. The sp^2-hybrid orbital, normally described as the σ orbital, possesses partial s-character and is thus energetically stabilized relative to the original p_x orbital (Figure 1.1).

The two non-bonding electrons at the sp^2-hybridized carbene carbon atom can occupy the two empty orbitals (p_π and σ) with a parallel spin orientation leading to a triplet ground state ($\sigma^1 p_\pi^1$, $^3B_1$ state, Figure 1.1). Alternatively, the

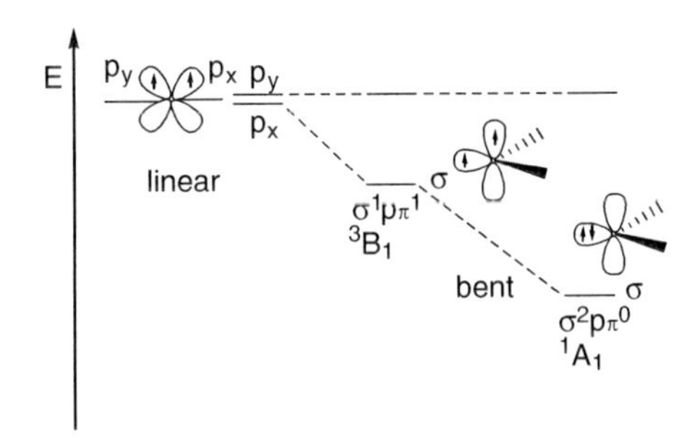

Figure 1.1 Frontier orbitals and possible electron configurations for carbene carbon atoms.

two electrons occupy the σ orbital with an antiparallel spin orientation ($\sigma^2 p_\pi^0$, $^1A_1$ state). An additional, generally less stable singlet state ($\sigma^0 p_\pi^2$, $^1A_1$ state) and an excited singlet state with an antiparallel occupation of the p_π and σ orbitals ($\sigma^1 p_\pi^1$, $^1B_1$ state) are conceivable but of no relevance for the present discussion.

The multiplicity of the ground state determines the properties and the reactivity of a carbene.[46] Singlet carbenes possessing a filled σ and an empty p_π orbital exhibit an ambiphilic behavior, while triplet carbenes can be considered as diradicals. The multiplicity of the ground state is determined by the relative energies of the σ and p_π orbitals as depicted in Figure 1.1. Quantum chemical calculations showed that an energy difference of about 2 eV is required for the stabilization of the singlet ground state ($^1A_1$), while an energy difference of less than 1.5 eV between the relative energies of the σ and p_π orbitals favored the triplet ground state ($^3B_1$).[47]

Steric and electronic effects of the α substituents at the carbene carbon atom control the multiplicity of the ground state. It is generally accepted that the singlet ground state is stabilized by σ-electron withdrawing, generally more electronegative substituents.[48] This negative inductive effect causes a lowering of the relative energy of the non-bonding σ orbital, while the relative energy of the p_π orbital remains essentially unchanged. Substituents with σ-electron donating properties decrease the energy gap between the σ and the p_π orbital and thus stabilize the triplet ground state.

In addition, mesomeric effects play a crucial role.[47,49] The substituents at the carbene carbon atom can be classified into different categories depending on their π donor/π acceptor properties.[24,27] Singlet carbenes of type X_2C:, substituted with two π donors X, are strongly bent at the carbene carbon atom. The interaction of the π-electron pairs at the α substituents with the p_π orbital at the carbene carbon atom raises the relative energy of this orbital. The relative energy of the σ orbital at the carbene carbon atom is not affected by the π-interaction. Consequently, the σ–p_π energy gap becomes larger leading to a further stabilization of the bent singlet ground state. The interaction of the π electrons of the substituents with the p_π orbital at the carbene carbon atom

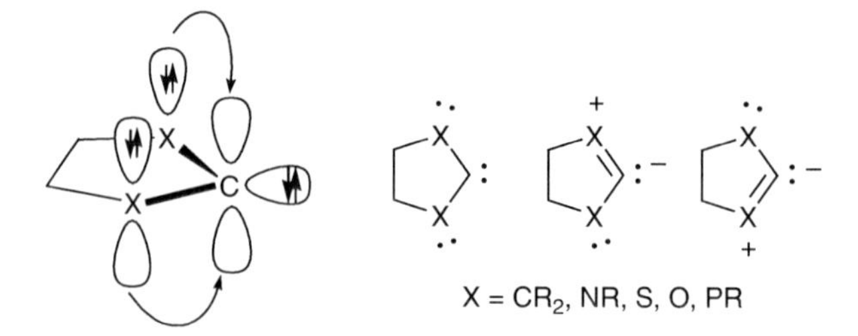

Figure 1.2 Electronic configuration and resonance structures of carbenes derived from a five-membered heterocycle and containing an X_2C: carbene centre.

leads to some extent to the formation of a four-electron three-centre π system where the X–C bonds acquire partial double-bond character (Figure 1.2). Important members of this class of compounds are the dimethoxycarbenes[50] and the dihalocarbenes.[51] The most important singlet carbenes stabilized by two π donors are the N-heterocyclic carbenes. The bonding situation in N-heterocyclic carbenes derived from five-membered heterocycles has been discussed in detail.[52]

1.3 N-Heterocyclic Carbene Ligands

The isolation of the first stable N-heterocyclic carbenes and their successful use as ancillary ligands for the preparation of various metal complexes initiated an intensive search for new NHC ligands by variation of the size of the heterocycle, the heteroatoms within the cycle and the substituents at the nitrogen atoms of the heterocycle and the heterocycle itself. Access to N-heterocyclic carbenes is largely controlled by the availability of suitable NHC precursors. Most NHCs are prepared by deprotonation of azolium cations found in imidazolium, triazolium, benzimidazolium, imidazolinium or thiazolium salts or by reductive desulfurization of imidazol-, benzimidazol- and imidazolin-2-thiones. In addition, stable NHCs have also been obtained from various imidazolidines by thermally induced α-elimination reactions (Scheme 1.5). The preparation of suitable azolium salts, 2-thiones and imidazolines is presented in this section, followed by the description of methods for the liberation of the free NHCs from these compounds. Imidazol-2-ylidenes are the most widely used NHC ligands and therefore special emphasis is placed on synthetic procedures leading to these unsaturated NHC ligands featuring a five-membered diaminoheterocycle.

1.3.1 Synthesis of NHC Precursors

Several methods for the preparation of imidazolium salts **10** (Scheme 1.6) have been described. The two most common routes are the alkylation of the nitrogen atoms of imidazole[53] and the multi-component reactions of primary amines,

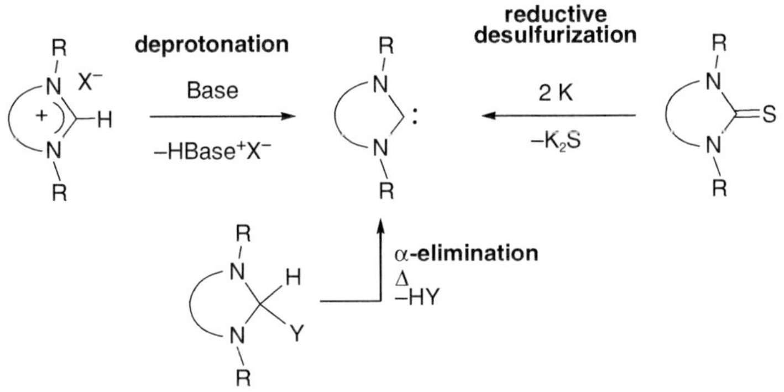

Scheme 1.5 Preparation of NHCs from cyclic precursors.

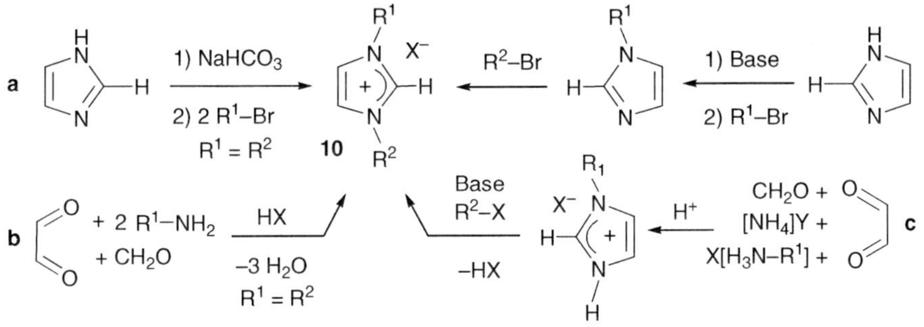

Scheme 1.6 Synthesis of symmetrically und unsymmetrically substituted imidazolium salts.

glyoxal and formaldehyde giving symmetrical *N,N′*-substituted azolium salts (Scheme 1.6a and b).[54] The second method is particularly useful for the synthesis of imidazolium salts bearing aromatic, very bulky or functionalized *N,N′*-substituents.[54,55] Unsymmetrically substituted imidazolium salts can be obtained by either stepwise alkylation of imidazole (Scheme 1.6a) or by combination of a multi-component cyclization[56] with a subsequent *N*-alkylation reaction (Scheme 1.6c).[57] *N*-Arylated imidazole derivatives can also be prepared from imidazole *via* a copper-catalyzed Ullman-coupling reaction.[58] Imidazolium salts bearing two different aryl substituents at the nitrogen atoms[59] or bisoxazoline-derived imidazolium salts leading to NHC ligands with a flexible steric bulk have also been described.[60]

The saturated imidazolinium salts of type **11** (Scheme 1.7) can be obtained by alkylation of dihydroimidazole or by the cyclization reactions between *N,N′*-dialkyl-α,β-ethyldiamines with orthoesters (Scheme 1.7a).[61] A multi-component reaction leading to unsymmetrical derivatives of type **11**, which is particularly interesting for the synthesis of imidazolinium salts with substituents at the C4 and C5 positions of the heterocycle, was reported by Orru

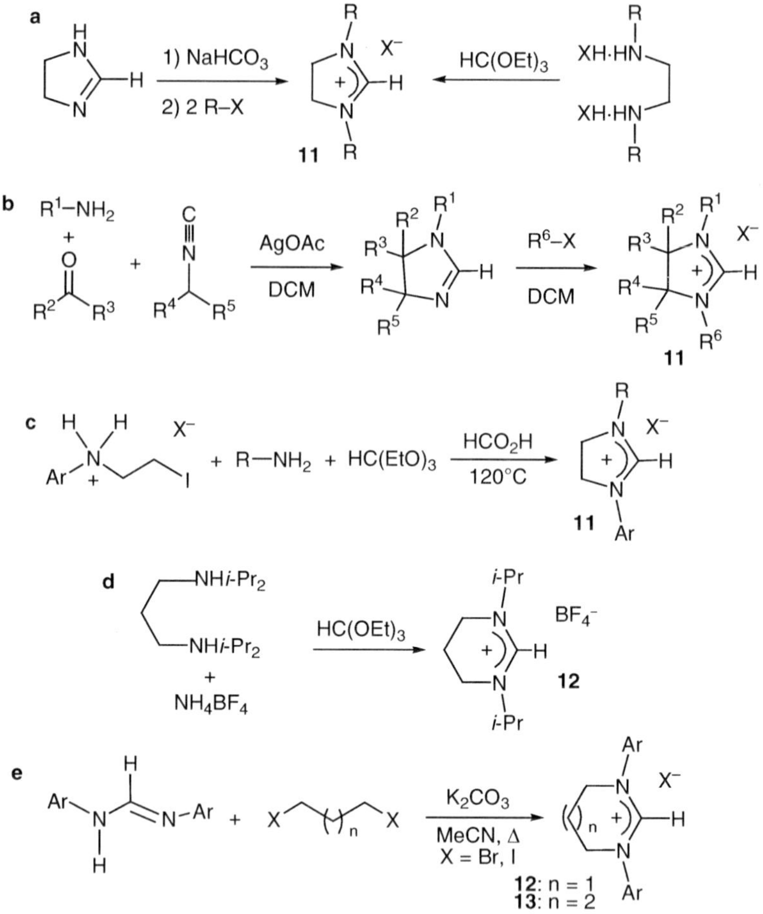

Scheme 1.7 Synthetic methods for the preparation of imidazolinium and amidinium salts.

and co-workers (Scheme 1.7b).[62] Also, the reaction of stable *N*-(2-iodoethyl)-arylammonium salts with an amine and triethylorthoformiate was reported to yield various imidazolinium salts of type **11** (Scheme 1.7c).[63]

Precursor **12** for a N-heterocyclic diaminocarbene possessing a six-membered heterocycle was obtained from the reaction of a suitable 1,3-diaminopropane with triethyl orthoformiate (Scheme 1.7d).[64] Related compounds with an aromatic backbone were obtained from diaminonaphthalene.[65] While these methods are based on the ring closure by introduction of a CH^+ fragment,[66] Bertrand and co-workers presented a different approach based on the reaction of a 1,3-diazaallyl anion with compounds featuring two leaving groups.[67] Amidinium salts with both six- or seven-membered heterocycles (**12** or **13**) were obtained by this method (Scheme 1.7e) which was further developed by Cavell[68]

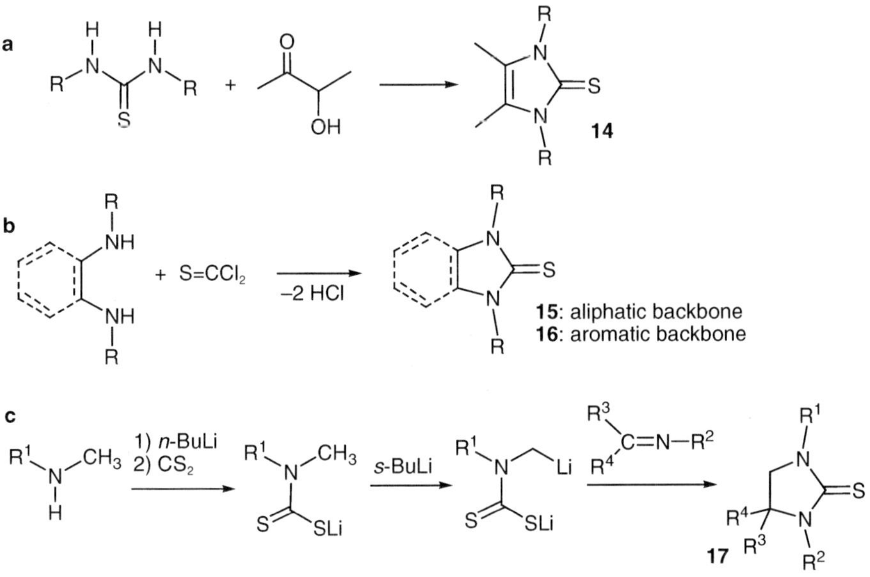

Scheme 1.8 Preparation of cyclic 2-thione precursors for NHCs.

and others[69] to yield a variety of derivatives with different ring sizes and substituents at the nitrogen atoms.

Alternative precursors for the synthesis of NHCs are thiourea derivatives of type **14**. Kuhn and Kratz first reported a facile method for the synthesis of symmetrically substituted imidazol-2-thiones by the reaction of 3-hydroxy-2-butanone with suitable thiourea derivatives (Scheme 1.8a).[23] Related saturated imidazolin-2-thiones[70] **15** or benzimidazol-2-thiones[71] **16** were obtained by reaction of aliphatic or aromatic 1,2-diamino compounds with thiophosgene (Scheme 1.8b). Unsymmetrically substituted imidazolin-2-thiones **17** were obtained by the reaction of lithium-*N*-lithiomethyldithiocarbamates with aldimines and ketimines, respectively (Scheme 1.8c).[72]

N-Heterocyclic carbenes can also be prepared by α-elimination from imidazolidines (see Scheme 1.5). Differently substituted imidazolidines or benzimidazolines were prepared by addition of alkali metal alkoxides to azolium salts[73] or by condensation of suitable diamines with benzaldehydes bearing fluorinated aromatic rings.[74] More exotic carbene precursors like 1,3-dimethyltetrahydropyrimidin-2-ium chloride were described by Bertrand and co-workers.[75]

Reports on *N,N'*-donor-functionalized, chiral and polydentate N-heterocyclic carbenes appeared almost immediately after the reports on the preparation of the first stable NHCs. Herrmann's report on the first donor-functionalized imidazolium cations **18** and **19** (Figure 1.3) from 1996[54a] was followed by a large number of publications dealing with imidazolium

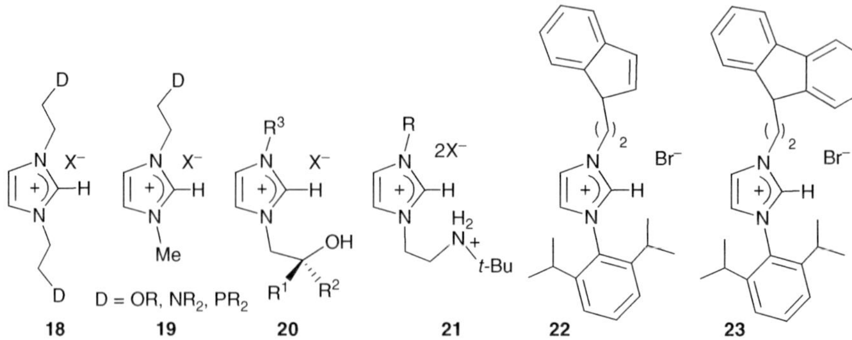

D = OR, NR_2, PR_2

18 19 20 21 22 23

Figure 1.3 Donor-functionalized imidazolium salts.

precursors for donor-functionalized carbene ligands.[36,76] The preparation of NHCs from imidazolium precursors bearing acidic *N*-substituents like alcohols or secondary amines was of special interest. Studies by Arnold and co-workers demonstrated that alcohol functionalized imidazolium salts **20** (Figure 1.3) were readily accessible by nucleophilic opening of epoxides.[77] Imidazolium salts **21** bearing *N*-*s*-amine substituents were also prepared.[77] Fryzuk and co-workers succeeded with the synthesis of an *N*,*N′*-di(*s*-amine) substituted imidazolium salt.[78] Indenyl-[79,80] (**22**) and fluorenyl-[80] (**23**) substituted imidazolium cations were also described (Figure 1.3).

A large number of differently bridged diazolium salts of type **24** (Figure 1.4) have been described.[81] Compound **25**, the precursor for an interesting bis-carbene containing carbene and alcoholato donor functions, was prepared by Arnold *et al.* by the reaction of two equivalents of an *N*-alkylimidazole with a functionalized epichlorhydrin (Figure 1.4).[82] Dibenzimidazolium salts[83] are also known. Trisimidazolium salt **26** was first described by Dias and Jin,[84] shortly thereafter followed by the description of trisimidazolium salts of type **27**.[85] Tripodal tris-imidazolium salts of types **26**, **28** and **29** were prepared by Hu and Meyer.[86]

The successful development of pincer ligands containing phosphine or amine donor groups by Milstein,[87] and van Koten[88] led to attempts to transfer this useful rigid ligand topology to tridentate ligands with NHC donor groups.[37] Today, pyridine- (**30**),[89] lutidine- (**31**),[90] and phenylene-bridged **32**[91] and **33**[90] bis-imidazolium salts are known in addition to the diethyl amine-bridged derivative **36**.[92] The lutidine- and phenylene-bridged benzimidazolium salts **34**[93] and **35**[94] were also described (Figure 1.5).

Precursors for pincer ligands containing only one NHC and two additional donor groups are also known. Phosphines (**37**),[95] *s*-amines (**38**),[78] pyridyl (**39**),[96] or phenoxy groups (**40**)[97] can function as additional donors. A double phosphine-substituted benzimidazolium salt **41**[98a] (Figure 1.5) and a similar imidazolinium derivative[98b] were reported in addition to a *N*,*N′*-diallyl substituted derivative.[99]

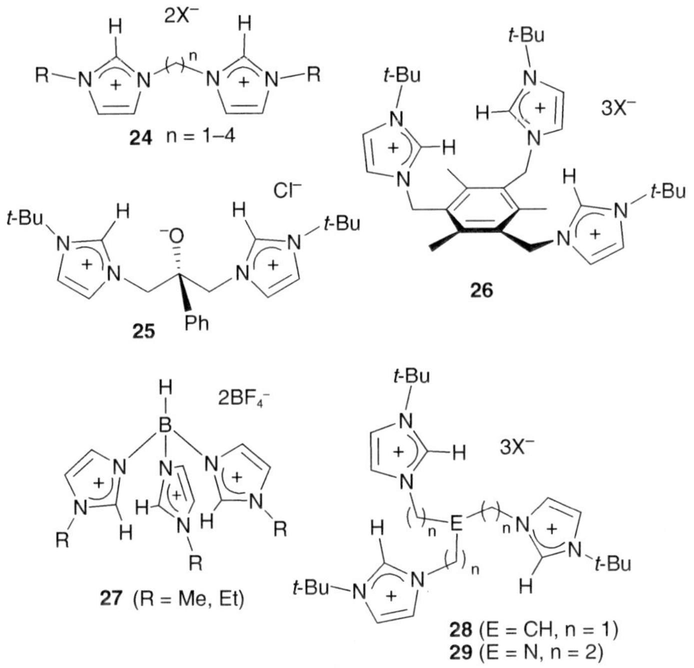

Figure 1.4 Bis- and tris-imidazolium salts.

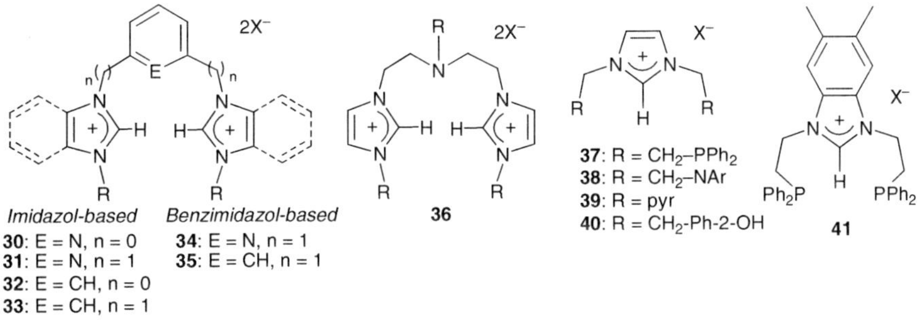

Figure 1.5 Precursors for pincer ligands with one or two NHC donor groups.

Among the polydentate carbene ligands, particular interest has recently been placed on cyclic polycarbenes. The synthesis of such ligands required the preparation of suitable cyclic polyazolium salts and Thummel reported the first cyclic dibenzimidazolium salts **42** (Figure 1.6).[100] Related diimidazolium salts **43**[101] were subsequently prepared. Cyclic tetrabenzimidazolium[100c,102] and tetrakisimidazolium salts[102] like **44** and **45** were synthesized during the search for new anion receptors. Bisimidazolium salt **46**, prepared by Baker *et al.*,[103] possesses, after C2 deprotonation, both NHC and pyridine donor functions, as tetrakisimidazolium salt **47**[104] which after four-fold C2 deprotonation gave a

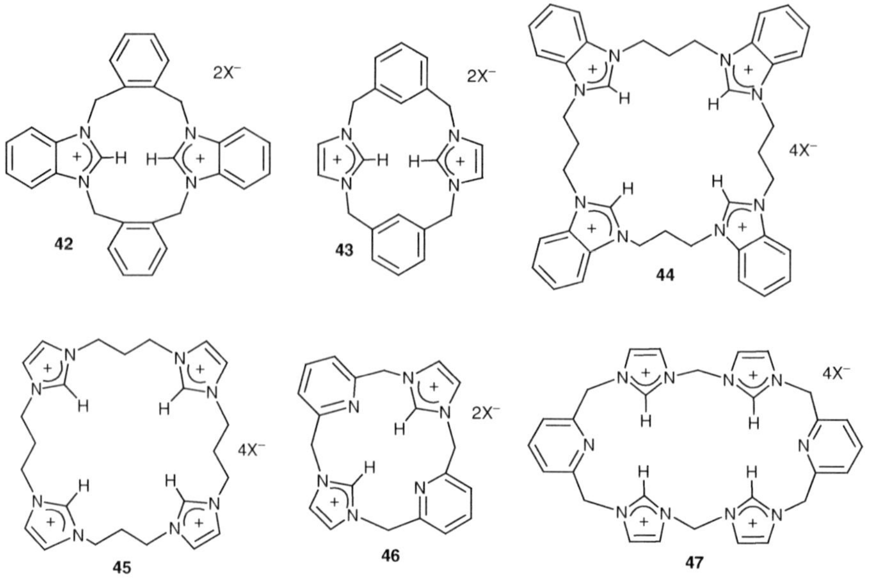

Figure 1.6 Cyclic polyazolium salts.

ligand containing two endocyclic pincer subunits.[105] Even a cyclic triply lutidine-bridged hexaimidazolium salt was described.[105]

Following the classification by Gade and Bellemin-Laponnaz[106] precursors for chiral N-heterocyclic carbenes can be subdivided into five groups (Figure 1.7). The first group comprises imidazolium salts like **48**, with a center of chirality within the *N*-substituents.[107] Additional imidazolium salts with chiral *N*,*N'*-substituents like **49**,[108] **50**[109] and **51**[110] were subsequently synthesized. Triazolium salts **52**,[111] **53**[112] and related compounds[113] containing chiral *N*-substituents were also described.

The second type of chiral azolium precursors possesses a chiral centre within the N-heterocyclic ring, which is normally substituted at the C4 and C5 positions, like in compounds **54**[114] or **55**.[115]

Bridging of two carbene donor groups with the 1,1'-binaphthyl moiety led to ligands with axial chirality. Bis-imidazolium salt **56**[116] and the analogous benzimidazolium derivative[117] belong to these ligand precursors as well as the hydroxyl substituted mono-imidazolium derivative **57**.[118] Additional examples for carbenes and carbenes precursors with chiral modified N-heterocycles are presented in Schemes 1.7 and 1.8.

The fourth group of chiral imidazolium precursors is made up from planar–chiral derivatives. The first representative of this type, **58**, was prepared by Bolm *et al.*[119] Togni and co-workers described the C_2-symmetrical ferrocenyl-substituted imidazolium salt **59**[120] and related derivatives are also known.[121,122] The planar–chiral imidazolium salt with *N*-paracyclophane substituents **60**[123] and some derivatives[124] are known. The enantiomerically pure *trans*-1,2-diaminocyclohexane was an useful starting material for the generation of the chiral diimidazolium salts **61**[125] and **62**[126] and of mono-imidazolium salts.[127]

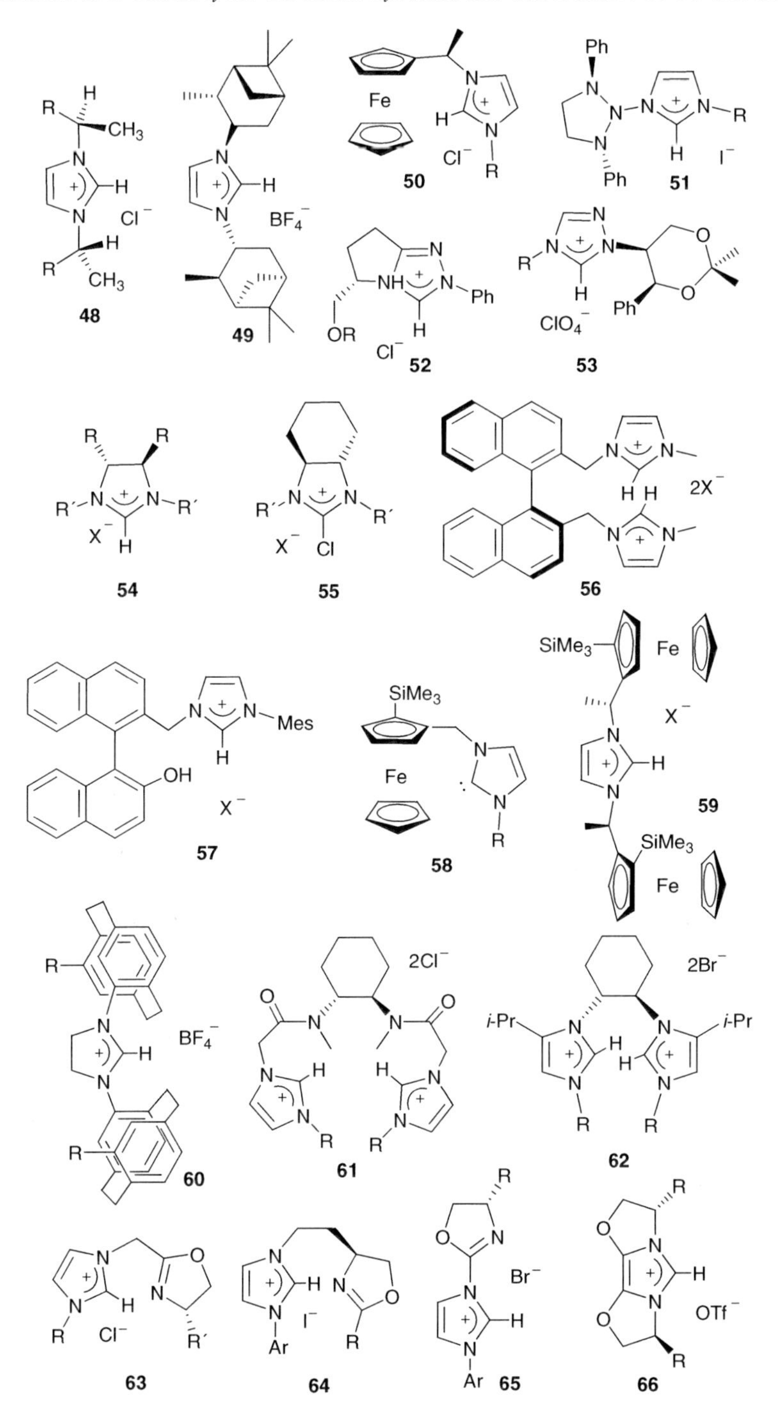

Figure 1.7 Chiral azolium salts.

The last group of chiral NHC ligand precursors is comprised of oxazoline-substituted imidazolium salts. The first derivative of this type **63** was reported by Herrmann *et al.* in 1998.[128] A slight modification of the ligand backbone and use of the oxazoline C4 atom for bridging gave the imidazolium salt **64**.[129] The direct connection of the two heterocycles led to **65**.[130] Imidazolium salts built from oxazoline heterocycles like **66** were prepared by Glorius *et al.*[131]

1.3.2 Preparation of Free N-Heterocyclic Carbenes

Azolium salts are easily accessible and different types of imidazolium, imidazolinium and benzimidazolium salts can be deprotonated at the C2 position to yield the free NHCs. The first successful deprotonation of an imidazolium salt to the free NHC was achieved with NaH in THF/DMSO (see Scheme 1.4).[22] Subsequently, other bases like KO*t*-Bu or NaH in THF[52g,61b] or NaH in liquid ammonia[132] were used (Scheme 1.9). For azolium salts with acidic substituents or in case of the formamidinium salts derived from six- or seven-membered heterocycles,[68,69] the use of a sterically demanding base like MHMDS (M = Li, Na, K; HMDS = hexamethyldisilazide) was required for the selective deprotonation at C2.

The reductive desulfurization of cyclic thiones (Scheme 1.9) has become an alternative method for the preparation of saturated,[70,72,133] unsaturated[23] and benzannulated NHCs.[71,134] It is, however, less frequently used than the deprotonation of azolium salts. The choice of reducing agent depends on the starting thione. Imidazol-2-thiones were reduced to the free carbenes with potassium in boiling THF within 4 h,[23] while benzimidazol-2-thiones were reduced with a Na/K alloy in toluene[71] requiring a reaction time of up to 3 weeks at ambient temperature.

Certain imidazolinium salts, particularly derivatives with bulky *N,N'*-substituents and some triazolium salts could not be deprotonated with strong bases like NaH, but a thermally induced α-elimination reaction proved successful in these cases. This approach was attempted as early as 1960 by Wanzlick and Schikora who heated C2-trichloromethyl substituted imidazolidine but could only isolate the carbene dimer (see Scheme 1.2).[13] Enders *et al.* successfully heated triazoline **67** under elimination of methanol from the C5 carbon atom and formation of triazol-5-ylidene **68** (Scheme 1.10a).[135]

The corresponding alcohol elimination from 2-alkoxyimidazolidines **69** to give imidazolin-2-ylidenes of type **70** was reported by Grubbs and co-workers[73] after

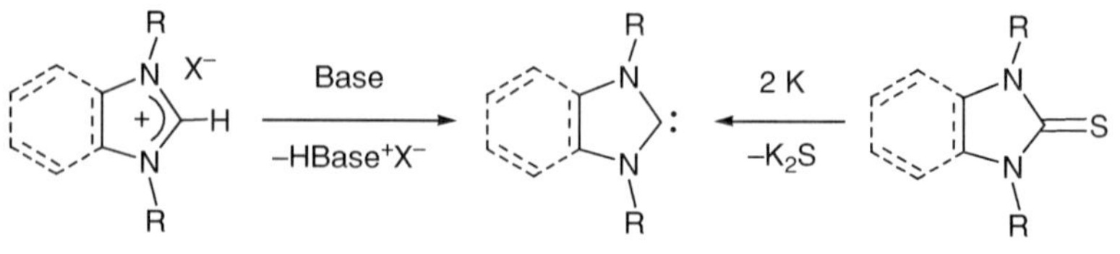

R N X⁻ +)—H N R —Base / −HBase⁺X⁻→ R N): N R ←2 K / −K$_2$S— R N =S N R

Scheme 1.9 Free N-heterocyclic carbenes *via* deprotonation or by reductive desulfurization.

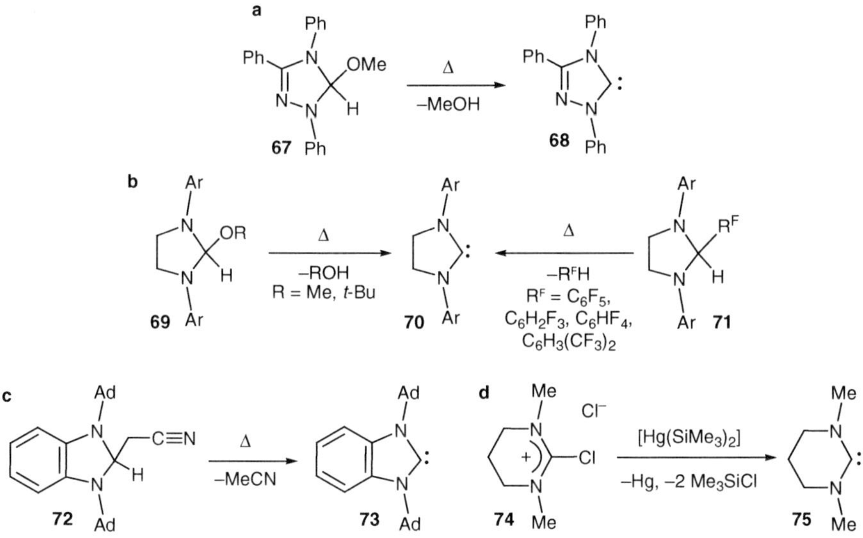

Scheme 1.10 Preparation of NHCs by α-elimination or dehalogenation.

early unsuccessful attempts by Wanzlick and Kleiner.[136] Imidazolin-2-ylidenes **70** were also accessible by α-elimination of fluorinated aryls from 2-(fluorophenyl) imidazolines **71** (Scheme 1.10b).[74] The α-elimination of acetonitrile from **72** to yield the benzimidazol-2-ylidene **73** has also been described (Scheme 1.10c).[137] Bertrand and co-workers reported the interesting dechlorination reaction between tetrahydropyrimidinium chloride **74** and bis(trimethylsilyl)mercury leading to NHC **75** in about 50% yield.[75] This reaction is of general applicability for the dechlorination of various chloroiminium and chloroamidinium salts. The electrochemical reduction of *N*,*N*-dimesitylimidazolium chloride was also reported.[138a]

Air and moisture stable imidazolium-2-carboxylates can act as a carbene transfer agent for various metals like Rh, Ru, Ir and Pd,[138b] but access to such derivatives is limited. Therefore, readily available *N*,*N*′-dimesitylenimidazolium-2-isobutylester was used as an alternative carbene source.[138c] Exclusion of air and moisture is not necessary in this carbene transfer making free carbenes unlikely intermediates in these reactions. In general, carbene transfer from NHC-2-carboxylates constitutes an interesting high yield procedure for the preparation of NHC complexes under mild reaction conditions.

1.4 Comparison of Different Types of N-Heterocyclic Carbenes

1.4.1 Carbenes Derived from Four-membered Heterocycles

The simplest stable singlet carbene derived from a cyclic precursor is cyclopropylidene.[139] Like other carbocyclic carbenes[140] it does not contain a

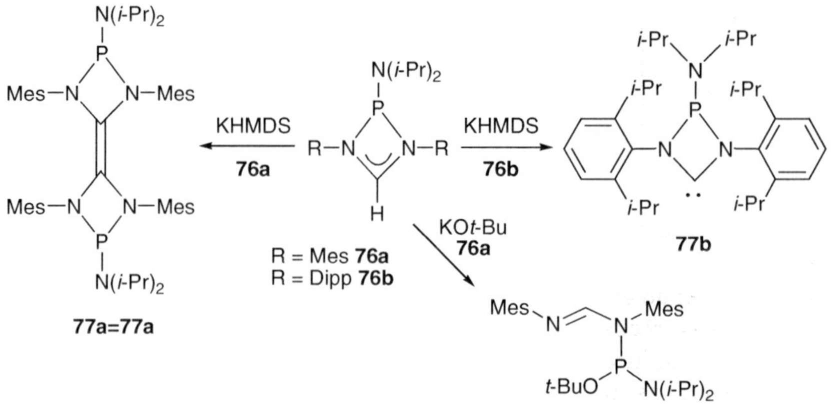

Scheme 1.11 Reactions of four-membered iminimum salts (Mes = mesityl; Dipp = 2,6-diisopropylphenyl).

heteroatom within the ring and is therefore not further discussed here. The first cyclic carbene based on a four-membered heterocycle was described in 2004. Deprotonation of iminium salt **76a** with KHMDS gave carbene dimer **77a**=**77a** while the same reaction with **76b**, bearing even bulkier *N*,*N'*-substituents, led to the isolation of **77b** (Scheme 1.11).[141] Attempts to deprotonate **76a** with KO*t*-Bu resulted in the opening of an endocyclic P–N bond demonstrating the electrophilic character of the phosphorus atom in this compound. Carbene **77b** exhibited a remarkable downfield shift for the resonance of the carbene carbon atom ($\delta = 285$ ppm) in the ^{13}C NMR spectrum. Structure analysis revealed that carbene **77b** did not possess C_2 symmetry in the solid state and lacks perfectly planarized endocyclic nitrogen atoms resulting in a suboptimal stabilization of the carbene centre which was most likely responsible for the formation of the carbene dimer **77a**=**77a** bearing the sterically less demanding *N*,*N'*-substituents. The N–C–N bond angle in **77b** was close to rectangular measuring 96.72(13)°. Ruthenium and rhodium complexes bearing carbene ligand **77b** have been prepared.[142]

1.4.2 Carbenes Derived from Five-membered Heterocycles

Most of the known N-heterocyclic carbenes are derived from five-membered heterocycles with additional oxygen, sulfur or phosphorus heteroatoms. Stable NHCs with up to four heteroatoms and saturated or unsaturated five-membered heterocycles are discussed in the following Section.

1.4.2.1 Imidazol-2-ylidenes

Unsaturated imidazol-2-ylidenes **78** (Scheme 1.12) form the largest group of stable heterocyclic diaminocarbenes.[24] Following the isolation of the first stable

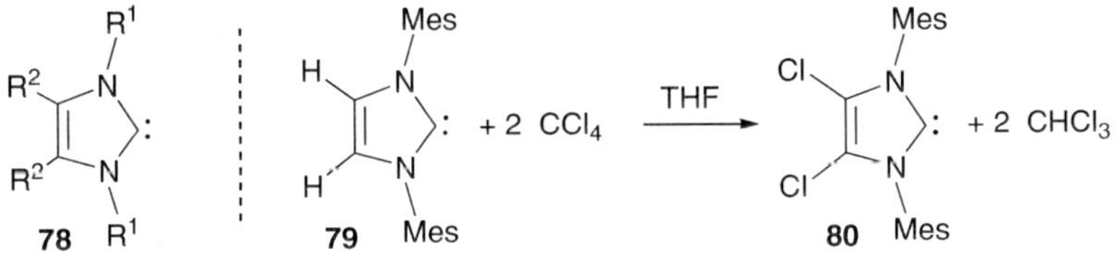

Scheme 1.12 Imidazol-2-ylidenes **78** and chlorination of **79**.

NHC by Arduengo *et al.* in 1991,[22] a large number of differently substituted **78** derivatives was prepared by deprotonation of azolium salts or reductive desulfurization of imidazol-2-thiones (see section 1.3.2). Carbenes **78** are normally obtained as colorless, diamagnetic, crystalline solids which show remarkably high melting points with the exception of *N*,*N′*-dimethylimidazol-2-ylidene which was isolated as a colorless liquid.[52g] Imidazol-2-ylidenes are stable under an inert gas atmosphere for months. ^{Cl}IMes **80**, which was obtained by chlorination of IMes **79** with CCl_4, was even stable in air at ambient temperature for some days (Scheme 1.12).[143]

Formation of an imidazol-2-ylidene **78** from an imidazolium salt (**10**, see Scheme 1.6) or an imidazol-2-thione (**14**, see Scheme 1.8) is detected by the characteristic downfield shift of the resonance of the C2 carbon atom from about $\delta = 140–150$ ppm (**10**) or $\delta = 161–162$ ppm (**14**) to $\delta = 211–220$ ppm (**78**) in the ^{13}C NMR spectra. The N–C–N bond angles in imidazol-2-ylidenes (101–102°) are significantly smaller than the equivalent angles in the parent imidazolium salts. In addition, a lengthening of the N–$C_{carbene}$ bond distances of approximately 0.05 Å was observed upon NHC formation by deprotonation of imidazolium salts, indicative of a reduced π-delocalization in the imidazol-2-ylidenes **78** in comparison to the precursor salts **10**.

Regardless of the substitution pattern, no dimerization to the electron-rich tetraazafulvalenes was observed for unsaturated **78**. The $^1A_1$ singlet ground state (see Figure 1.1) is sufficiently stabilized by inductive and mesomeric effects. Quantum chemical calculations[52g,144] demonstrated that the energy gap between the singlet and triplet states amounted to about 85 kcal mol^{-1}. Based on these calculations the strengths of the C=C double bond which would result from the dimerization of two unsaturated imidazol-2-ylidenes was estimated to be approximately 2 kcal mol^{-1}.[144a] According to the Carter and Goddard estimate,[145] the strength of the central C=C double bond in a dimer of two imidazol-2-ylidenes should correspond to that of a canonical C=C bond (normally that of ethane, 172 kcal mol^{-1}) minus the sum of the singlet–triplet energy difference for the two carbenes involved. For the dimer of two imidazol-2-ylidenes one would thus expect a bond strengths of $[172 - (2\times85)] \approx 2$ kcal mol^{-1}. Experimental evidence for the weakness of the C=C double bond in tetraazafulvalenes was presented by Taton and Chen (Scheme 1.13).[144a] Both the deprotonation of the doubly propylene-bridged bis-imidazolium salt **81** as well as the two-electron reduction of derivate **82** gave the doubly bridged tetraazafulvalene **83** with a true C=C double bond (δ(C=C) = 1.337(5) Å),

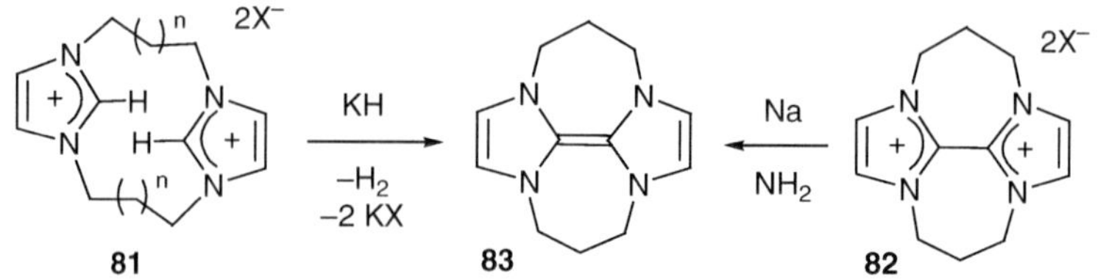

Scheme 1.13 Synthesis of tetraazafulvalenes and bis-NHCs from bridged diimidazolium salts.

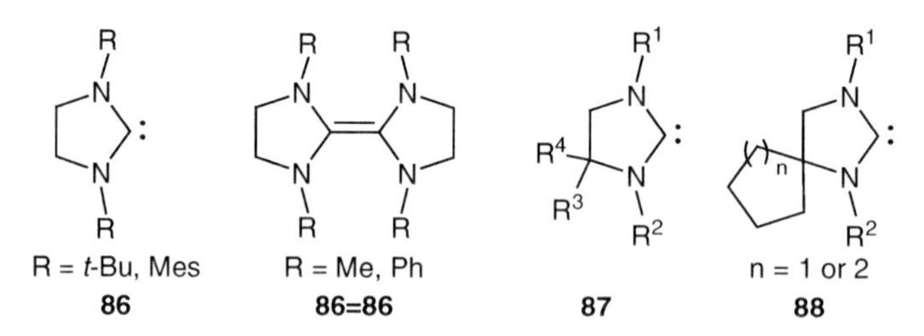

Figure 1.8 Stable imidazolin-2-ylidenes **86**, *rac*-**87** and **88** and the entetraamines **86=86**.

while the deprotonation of the more flexible doubly butylene-bridged analogue **84** always led to the bis(imidazol-2-ylidene) **85** and no formation of a C=C double bond was observed.

1.4.2.2 Imidazolin-2-ylidenes

Wanzlick and Schikora were the first to attempt the preparation of saturated imidazolin-2-ylidenes by an α-elimination of chloroform from imidazoline derivatives but could only obtain the electron-rich entetraamines (see Scheme 1.2).[13] It was Arduengo *et al.* who presented the first stable crystalline imidazolin-2-ylidene **86** (Figure 1.8) obtained by deprotonation of the corresponding imidazolinium salt.[146] Shortly thereafter, Denk *et al.* used the reductive desulfurization of imidazolin-2-thiones to obtain the *N,N′*-di(*tert*-butyl) substituted stable carbene of type **86** and the entetraamine **86=86** with the sterically less demanding *N,N′*-dimethyl substituents (Figure 1.8).[70] Additional imidazolin-2-ylidenes with varying *N,N′*-substituents were obtained by α-elimination reactions from imidazolidines.[73,74] Unsymmetrically substituted derivatives like **87** were obtained as racemic mixtures of the stereoisomers[72a] and spirocyclic derivatives **88** were also reported.[72b]

The isolation of imidazolin-2-ylidenes **86** convincingly showed that a supposedly aromatic 6π-electron system within the heterocycle is not required for the stabilization of an N-heterocyclic diaminocarbene. The singlet carbene center is sufficiently stabilized by inductive and mesomeric effects caused by the nitrogen substituents leading to N1–C2–N3 π-delocalization. The smaller energy gap between the singlet and triplet states in carbenes **86**,[42,144] however, causes a rapid dimerization of derivatives with sterically less demanding *N,N′*-substituents to yield entetraamines **86=86**. Kinetic stabilization of the imidazolin-2-ylidenes by sterically demanding *N,N′*-substituents is much more important for **86** than for imidazol-2-ylidenes **78** where even the *N,N′*-dimethyl substituted derivative was stable towards dimerization.[23] Alder *et al.*[42] and Graham *et al.*[144b] studied the dimerization reaction of imidazolin-2-ylidenes in detail. From these studies, it was concluded that dimerization of imidazolin-2-ylidenes was most likely a proton-catalyzed process, which in exceptional cases could also be catalyzed by other Lewis acids.[147] In addition, the dimerization reaction depends on factors like the N–$C_{carbene}$–N bond angle and the basicity of the carbene carbon atom.

Molecular structure data of saturated imidazolin-2-ylidenes **86** revealed N–$C_{carbene}$–N angles in the range of 104.7(3)–106.4(1)°,[24] significantly larger than the equivalent angles observed for the unsaturated analogues **78** (101–102°). As was observed for **78**, the N–$C_{carbene}$–N angles were smaller and the N–$C_{carbene}$ bond distances were slightly longer in **86** in comparison to the equivalent values found in the parent azolium salts. The nitrogen atoms within the heterocycle in **86** are planarized, but the heterocycle itself was not always planar which contrasted the situation found in imidazol-2-ylidenes **78**. Dimerization of NHCs **86** to entetraamines **86=86** caused an expansion of the N–C2 distances (1.418(1)–1.443(1) Å) and the N–C2–N bond angles (108.7(1)–111.3(1)°).[24] Actually, the nitrogen atoms in the entetraamines are significantly pyramidalized.

Imidazolin-2-ylidenes can be identified by the characteristic downfield resonance for the carbene carbon atom at $\delta \approx 240$ ppm in the ^{13}C NMR spectra relative to imidazolinium salts **11** ($\delta = 157.2$–160.9 ppm) or imidazolin-2-thiones of type **17** ($\delta = 180.8$–183.1 ppm).[24] The downfield shift of the carbene carbon resonance is more pronounced for saturated imidazolin-2-ylidenes than for unsaturated imidazol-2-ylidenes ($\Delta\delta \approx 25$ ppm). Entetraamines **86=86** exhibit a ^{13}C NMR resonance in the range $\delta = 124.3$–130.4 ppm.

1.4.2.3 Benzimidazol-2-ylidenes and Related Benzannulated NHCs

Deprotonation of benzimidazolium salts[137] or reduction of benzimidazol-2-thiones[71] yielded benzimidazol-2-ylidenes **89**[71,137,148] or dibenzotetraazafulvalenes **89=89**[71,149] depending on the steric demand of the *N,N′*-substituents (Figure 1.9).

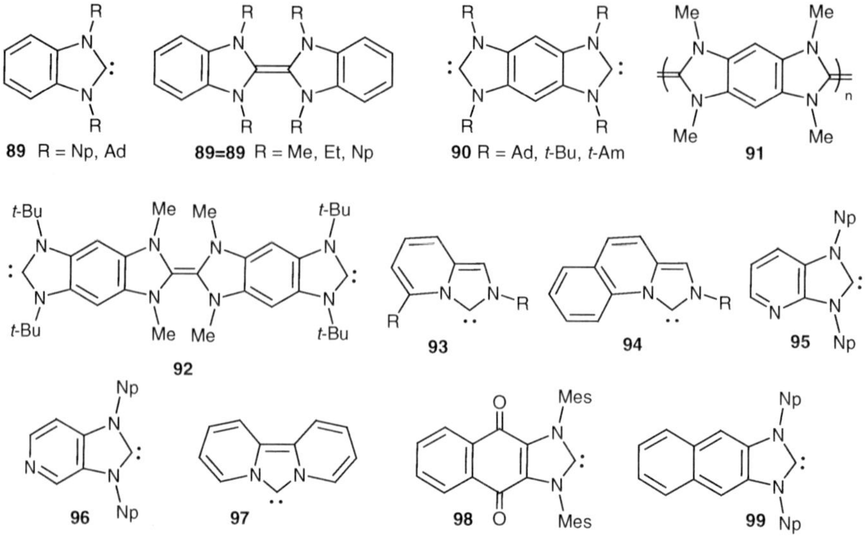

Figure 1.9 Benzannulated and heterocycle-annulated NHCs (Np = neopentyl; Am = amyl).

Bielawski and co-workers described 'Janus-head' like benzobis(imidazol-2-ylidenes) **90**,[150] which were obtained by deprotonation of the corresponding benzobis(imidazolium) salts.[151] Depending on the steric demand of the *N,N'*-substituents, these dicarbenes formed monomers **90**,[150] dimers **92**[150] or polymers **91**[152] (Figure 1.9). These rigid carbenes were also used for the generation of organometallic polymers[153] and for the synthesis of metalosupramolecular assemblies featuring NHC donors.[154]

To vary the electronic properties of the carbene carbon atom a number of carbo- and heterocycle-annulated NHCs were prepared and studied. Single pyrido-annulated NHCs (**93**–**96**)[155] as well as the doubly pyrido-annulated carbene **97**[156] were obtained by deprotonation of the corresponding azolium salts. NHCs such as the quinone-annulated carbene **98**,[157] the naphtho-annulated carbene **99**[155d] and related polycyclic compounds have also been obtained from more extended ring systems.[158]

DFT studies showed that annulation of imidazol-2-ylidenes generally destabilizes the NHC.[159] This becomes also apparent when analogies between benzannulated N-heterocyclic carbenes and saturated imidazolin-2-ylidenes **86** (see Figure 1.8) are considered. Both types of NHC dimerize rapidly if the nitrogen atoms of the heterocycle are substituted with sterically less demanding substituents. No such dimerization is observed with imidazol-2-ylidenes **78**, which possess an unsaturated heterocycle like the benzimidazol-2-ylidenes.

A detailed inspection of structural and ^{13}C NMR spectroscopic parameters provided additional evidence for some special properties of benzimidazol-2-ylidenes **89** and their dibenzotetraazafulvalene dimers **89**=**89**. The ^{13}C NMR resonance for the $C_{carbene}$ carbon atom in compounds **89**, for example, is observed at a chemical shift which lies in between the typical values for the $C_{carbene}$ resonance of saturated imidazolin-2-ylidenes **86** and unsaturated imidazol-2-ylidenes **78**. In spite of the unsaturated nature of the five-membered carbene ring in **89**, N1–C2–N3 angles which are typical for saturated imidazolin-2-ylidenes **86**, were determined crystallographically.[71] It appears that the five-membered ring in carbenes **89** possesses the topology of an unsaturated N-heterocyclic carbene but exhibits spectroscopic and structural parameters similar to those observed for the saturated imidazolin-2-ylidenes **86**. This intermediate position of the benzimidazol-2-ylidenes between unsaturated and saturated N-heterocyclic carbenes indicates, even in the absence of experimental data, that the energy gap between the singlet and triplet states for compounds **89** must also assume an intermediate position between that of the stable imidazol-2-ylidenes **78** and the rapidly dimerizing imidazolin-2-ylidenes **86**.

The dimerization behaviour of benzimidazol-2-ylidenes is, as expected, determined by kinetic factors (steric demand of the *N,N′*-substituents). *N,N′*-Dimethylbenzimidazol-2-ylidene dimerized to give the dibenzotetraazafulvalene **89**=**89** (Figure 1.9).[71] Sterically more demanding *N,N′*-substituents (neopentyl,[71] adamantyl[137]) enabled the isolation of the monomeric benzimidazol-2-ylidenes **89**. Variation of the steric demand of the *N,N′*-substituents between those extremes led to the synthesis of *N,N′*-di(isobutyl)benzimidazol-2-ylidene which in solution co-existed together with its dibenzotetraazafulvalene dimer.[134] The thermally induced cleavage of a dibenzotetraazafulvalene **89**=**89** (R = Me) at 110–140 °C and of the *N,N′,N″,N‴*-tetraethyl substituted derivative into two benzimidazol-2-ylidenes was also reported.[160] No conclusive mechanistic information about the cleavage of dibenzotetraazafulvalenes into benzannulated N-heterocyclic carbenes is available at this time. Both a monomolecular cleavage as well as an electrophile-catalyzed reaction,[42] leading to the monomer–dimer mixture are conceivable. The rapid dimerization of benzimidazol-2-ylidenes with sterically less demanding *N,N′*-substituents led to an interesting observation when the nitrogen atoms bore allyl groups, and after C2-deprotonation of the *N,N′*-diallylbenzimidazolium salt a dibenzotetraazafulvalene was obtained which rearranged under cleavage of one or two N–C_{allyl} bonds.[161]

The ^{13}C NMR resonances for the carbene carbon atom in monomeric benzimidazol-2-ylidenes **89** fall in the range of $\delta = 223.0$–231.5 ppm while the C2 resonance for dibenzotetraazafulvalenes **89**=**89** are observed at higher field at $\delta \approx 119.0$ ppm. The geometry of the nitrogen atoms in carbenes **89** is trigonal–planar, while pyramidalized nitrogen atoms and a non-planar $N_2C{=}CN_2$ moiety are found in the dibenzotetraazafulvalenes **89**=**89**.

1.4.2.4 Triazol-5-ylidenes

Enders *et al.* isolated the first triazol-5-ylidene **68** by thermal methanol elimination from 5-methoxytriazoline in 1995 (see Scheme 1.10a).[135] **68** was stable up

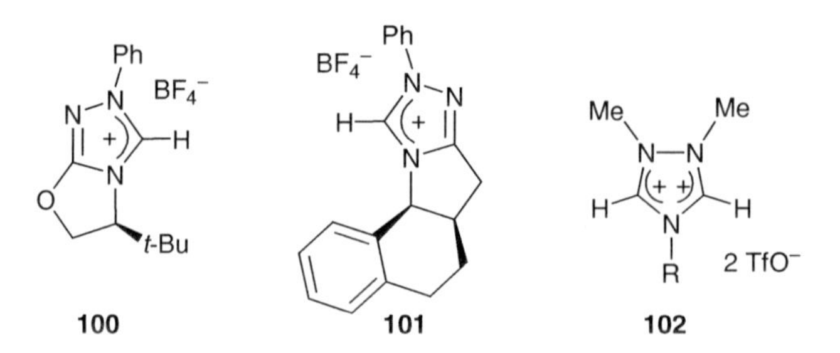

Figure 1.10 Chiral triazolium salts **100–101** and the dicationic triazolium salt **102**.

to a temperature of 150 °C. ^{13}C NMR spectroscopic (δ(C5) = 214.6 ppm) and structural parameters such as the short $C_{carbene}$–N bond lengths (1.351(3) Å and 1.373(4) Å) and the small N–$C_{carbene}$–N bond angle (100.6(2)°) confirmed similar properties of **68** and the unsaturated imidazol-2-ylidenes **78**. NHC **68** showed no tendency to dimerize to the entetraamine and thus behaved also chemically like an unsaturated imidazol-2-ylidene. Different chiral triazolium salts like **100–101** (Figure 1.10) were used after C5 deprotonation as nucleophiles in organocatalysis.[40,162]

The 1,2,4-triazolium salt **102** (Figure 1.10) was prepared by Bertrand and co-workers.[163] Even if all attempts to isolate the free dicarbene by deprotonation of the dication failed thus far, it was possible to obtain the mono- and disilver complexes by deprotonation and metalation of the carbon atoms of the heterocycle.[164a] Peris and co-workers obtained the diiridium complex by reaction of **102** (R = Me) with two equivalents of $[IrCl(COD)]_2$ in the presence of KO*t*-Bu. A heterobimetallic Ir^I/Rh^I complex was obtained by successive reaction of **102** with $[IrCl(COD)]_2$ and $[RhCl(COD)]_2$.[164b]

1.4.2.5 Thiazol-2-ylidenes and Benzothiazol-2-ylidenes

Arduengo *et al.* obtained the first stable thiazol-2-ylidene **104** by deprotonation of the corresponding thiazolium salt **103** with potassium hydride in THF (Scheme 1.14).[165] In the presence of protons, NHC **104** dimerized to olefin **104=104** and an equilibrium between monomer and dimer was observed. Both **104** and **104=104** were the first and only carbene/olefin pair where both components could be characterized crystallographically. Thiazol-2-ylidenes with sterically less demanding *N*-substituents like mesityl or methyl, however, dimerized rapidly at room temperature and only their dimers can be isolated.[165,166] None of the analogous benzothiazol-2-ylidenes could be isolated up to now, although benzothiazolium salts **105** and complexes of benzothiazol-2-ylidenes are readily accessible.[167]

Thiazol-2-ylidenes exhibit spectroscopic and structural properties similar to those of saturated imidazolin-2-ylidenes **86**. Upon deprotonation of the thiazolium salts to thiazol-2-ylidenes the N–C–S angle becomes smaller and the endocyclic N–C2 bond length increases slightly within an essentially planar

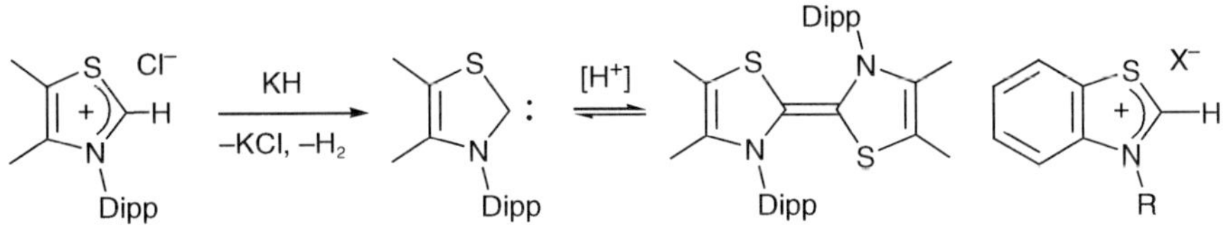

Scheme 1.14 Synthesis and dimerization of thiazol-2-ylidene and benzothiazolium salt **105**.

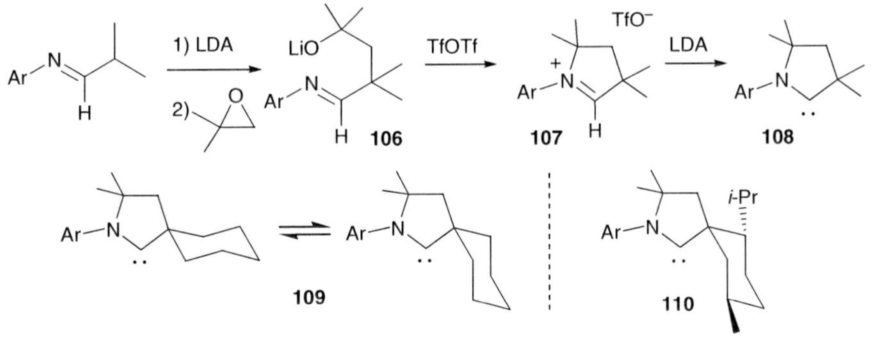

Scheme 1.15 Synthesis of the cyclic alkyl(amino)carbenes **108**–**110**.

heterocycle.[165] The ^{13}C NMR spectra of thiazol-2-ylidenes exhibit a significant downfield shift of the resonance for the C2 atom ($\delta = 252$–254 ppm) compared to the parent thiazolium salts ($\delta = 155$–160 ppm) or the (N,S)C=C(N,S) carbene dimers ($\delta = 110$–120 ppm).[165,166] Thiazolium cations such as thiamin (vitamin B1) play an important role in various enzymatic C–C coupling reactions.[168] Breslow[169] and others[170] used model compounds to demonstrate that this enzymatic activity is caused by deprotonation of the C2 atom of the azolium salt under formation of a nucleophilic carbene.

1.4.2.6 Cyclic Alkyl(amino)carbenes and Related Compounds

The isolation of the non-cyclic amino(aryl)carbenes[171] and amino(alkyl)carbenes[172] demonstrated that singlet carbene centres can be sufficiently stabilized by only one α-nitrogen atom. In 2005, Bertrand and co-workers succeeded in preparing the first cyclic alkyl(amino)carbenes (CAACs; Scheme 1.15).[173] The precursor for CAAC **108** was obtained from an imine by deprotonation with LDA (LDA = lithium diisopropylamide) and subsequent reaction with 1,2-epoxy-2-methylpropane to give **106**, which was converted into cyclic aldiminium salt **107** by reaction with trifluoromethanesulfonic acid anhydride. Deprotonation of **107** with LDA afforded CAAC **108** as a colorless solid (Scheme 1.15). The presence of a quaternary carbon atom in the α-position to the carbene centre offered the possibility to construct ligands with a large variety of steric environments. CAAC **109**, for example, contained a spiro-carbon atom next to the carbene centre leading to a flexible steric demand caused by conformational changes of the cyclohexyl ring. Similar behaviour was observed for related bisoxazoline derived NHC ligands.[60] The influence of the 'flexible cyclohexyl wing' reached a maximum in **110**, where an intelligent substitution of the cyclohexyl ring fixes the 'wing' in one conformation giving maximum steric protection to the carbene carbon atom. Coordination of CAAC **110** to transition metals led to complexes with low coordination number and strong steric protection of the coordinated metal centre.[173b,174] CAACs with other bulky substituents at the carbon atom in the α-position to the carbene centre were also prepared by Bertrand and co-workers.[175]

The substitution of one electronegative amine substituent in classical NHCs for the strong σ-donor carbon makes CAACs particularly electron-rich. Consequently, CAACs are excellent donor ligands with a donor strength, which is often superior to that of phosphines or classical NHCs. The N–$C_{carbene}$ bond length (1.315(3) Å)[173a] is comparable to the value found in the saturated imidazolin-2-ylidenes **86**, while the $C_{carbene}$–C bond length is typical for a C–C single bond. The similarities between CAACs and imidazolin-2-ylidenes is also apparent from the N–$C_{carbene}$–C bond angle in CAAC **110** (106.5(2)°)[173a] which is significantly larger than the equivalent N–C–N bond angle in imidazol-2-ylidenes **78**, but similar to the N–$C_{carbene}$–N angles observed in imidazolin-2-ylidenes **86**. The carbene carbon atom in CAACs is strongly deshielded leading to downfield ^{13}C NMR resonances (**108** $\delta = 304.2$ ppm, **109** $\delta = 309.4$ ppm, **110** $\delta = 319.0$ ppm).[173a]

The preparation of CAAC precursors depicted in Scheme 1.15 is tedious and time consuming. To overcome this problem Bertrand and co-workers developed the 'hydroiminiumation' reaction for the preparation of the cyclic aldiminium salts of type **107**.[176] The cyclization step in this reaction is based on the addition of an imine to a double bond and allows also the preparation of larger heterocycles. An asymmetric version of the reaction was described, generating a stereocentre at an atom of the heterocyclic backbone.[177]

Bertrand also demonstrated that CAAC **110** could activate dihydrogen to give compound **111** (Scheme 1.16). Contrary to the activation of dihydrogen by electrophilic transition metals, the carbene acted primarily as a nucleophile creating a hydride-like hydrogen, which subsequently reacted with the vacant p orbital of the positively polarized carbene carbon atom.[178] Importantly, this activation was not observed with the slightly less nucleophilic imidazol-2-ylidenes and imidazolin-2-ylidenes.

Pairs of Lewis acids and bases normally form stable donor–acceptor adducts. When steric effects prevent this adduct formation, Frustrated Lewis Pairs (FLPs) are obtained instead. Such FLPs are promising reagents for the activation of small molecules.[179] Tamm[180] and Stephan[181] combined an NHC with sterically demanding *N,N′*-substituents and tris(pentafluorphenyl)borane

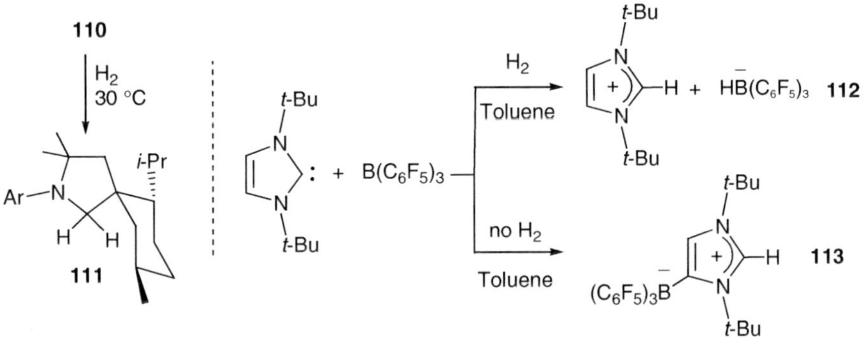

Scheme 1.16 Activation of dihydrogen by CAAC **110** and FLPs.

to form a non-quenching carbene-based FLP (Scheme 1.16). In toluene purged with dihydrogen the FLP reacted with the dihydrogen to form the imidazolium borate salt **112** in quantitative yield. The limitations of this FLP system became apparent when, in the absence of dihydrogen, the resulting FLP/toluene solution lost its reactivity towards dihydrogen within 1 h and the 1:1 zwitterionic carbene–borane adduct **113** was isolated instead.

The electron-donating ability of NHCs is determined by the mesomeric and inductive effects of the α-substituents (Section 1.2).[52] While the electronegativity of the α-carbon atom in CAACs is lower than that of nitrogen in NHCs, the sp^3-hybridized α-carbon atom in CAACs is certainly no π donor. To enhance the π-donating ability of carbon atoms in α-position to the carbene center, amino(ylide)carbenes (AYCs) were developed. Similar compounds had been known for some time[182] but up to then had little impact compared to their diamino-stabilized relatives.

Kawashima and co-workers prepared the phosphorus ylide-stabilized AYC ligand **114** which could not be isolated due to an intramolecular rearrangement to **115**. The intermediate formation of **114** was confirmed by its reaction with sulfur, and the formation of the rhodium complexes **116** and **117** (Scheme 1.17).[183] IR spectroscopy with the carbonyl complex **117** confirmed the superb donor properties of AYC **114**. Fürstner *et al.* prepared the sulfur ylide-stabilized AYC **118** which again could not be directly observed but was stabilized in the rhodium complex **119**.[184] Moving away from the indole scaffold, AYC **120** was prepared and characterized in solution ($\delta C_{carbene} = 218$ ppm, $J_{C,P} = 51.2$ Hz). Finally, it was tested if polarized C=C bonds also lend themselves to the stabilization of a vicinal carbene. The 'carbon ylide' stabilized AYC **121** was characterized by its reaction with sulfur. Compound **122** reacted with [Pd(acac)$_2$] (acac = acetylacetonato) under formation of a complex bearing a 'carbon ylide' stabilized carbene ligand coordinated in a chelating fashion (Scheme 1.17).[184]

Yet another ylide/carbene combination exists in bidentate CAAC–phosphonium ylide and diaminocarbene–phosphonium ylide ligands prepared by Bertrand[185] and by Canac *et al.*[186] While the free carbenes were not isolated, they could be stabilized in palladium complexes **123–125** (Scheme 1.17).

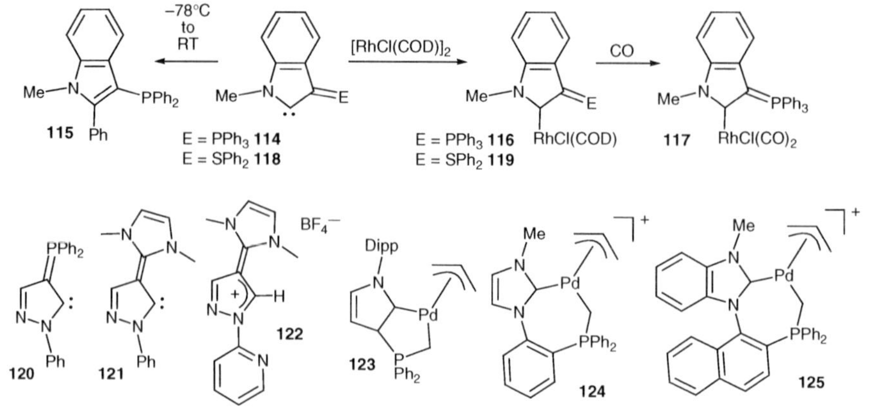

Scheme 1.17 Ylide-stabilized CAAC ligands and complexes with bidentate carbene–phosphonium ylide ligands.

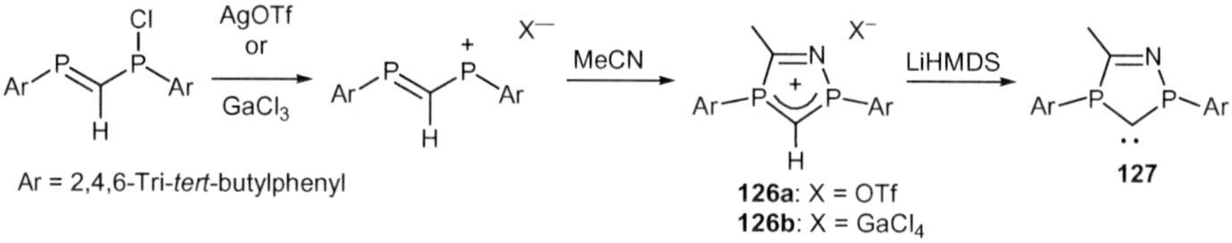

Scheme 1.18 Synthesis of PHC **127**.

1.4.2.7 P-Heterocyclic Carbenes

Since the stability of N-heterocyclic carbenes is based on the stabilizing effect of the α-nitrogen atoms, it is surprising that the heavier analogues of nitrogen, such as phosphorus, were initially disregarded for the stabilization of cyclic carbenes. Acyclic phosphorus-stabilized carbenes have been known since 1988.[21] Experimental and theoretical studies revealed that the π-donor capabilities of the third-row elements were as large or even larger than those of their second-row counterparts.[187] Calculations also showed that the phosphorus atoms in a hypothetical P-heterocyclic carbene were expected to be pyramidalized[188] and thus incapable of stabilizing the carbene centre, while the nitrogen atoms in NHCs are trigonal–planar.[52]

The first complex with a P-heterocyclic carbene ligand, analogue of an imidazol-2-ylidene, was described by Le Floch and co-workers in 2004.[189] The free P-heterocyclic carbene (PHC) ligand, however, could not be isolated. Avoiding the pyramidalization of the phosphorus atoms by bulky *P,P′*-substituents, Bertrand isolated the first stable P-heterocyclic carbene **127** in 2005 (Scheme 1.18).[190] Since the phosphorus analogues of azolium salts were unknown, a new protocol for the preparation of cationic PHC precursors had to be developed. This involved the dehalogenation of a phosphaalkene with

AgOTf or $GaCl_3$ followed by a [3+2] cycloaddition of the intermediate diphosphaallylic cation with acetonitrile as dipolarophile. Compounds **126** were then deprotonated with LiHMDS to give the stable PHC **127**.

The molecular structures of **126b** and **127** provided evidence that the use of the sterically demanding 2,4,6-tri-*tert*-butylphenyl substituents led to the desired planarization of the phosphorus atoms. The slight deviation from planarity and the resulting *trans*-arrangement of the aryl substituents at the phosphorus atoms led to chiral compounds in the solid state. The ^{1}H and ^{13}C NMR spectra in solution exhibited only one resonance for the diastereotopic groups, even at low temperature (−100 °C), giving evidence for a rapid interconversion of the enantiomers in solution and a low inversion barrier at the phosphorus atoms. Additionally, the ^{13}C NMR resonance for the carbene carbon atom in PHC **127** ($\delta = 184$ ppm) was highfield shifted relative to the chemical shift observed for NHC analogues.

The structure analysis of **127** confirmed the donor character of the phosphorus atoms. The P–$C_{carbene}$ distances (1.673(6) and 1.710(6) Å) were significantly shorter than P–C single bonds. Deprotonation of the PHC precursor caused the P–C–P angle to become more acute from 106.2(5)° in **126b** to 98.2(3)° in **127**. This behavior is analogous to the azolium salts and the NHCs obtained from these by deprotonation. *Ab initio* calculations demonstrated that a decrease in the steric bulk of the *P,P′*-substituents led to a pyramidalization of the phosphorus atoms and a decrease of the singlet–triplet energy gap thereby destabilizing the *P,P′*-heterocyclic carbene.[191] Some derivatives of PHC **127** were prepared[190b] as well as metal complexes.[190] Moreover, a report on the first persistent P,N-stabilized heterocyclic carbene also appeared.[192]

1.4.3 Heterocyclic Carbenes Containing Boron within the Heterocycle

Diverse N-heterocyclic carbenes containing Lewis-acidic boron atoms within the carbene ring have been described in the literature. Carbene **128** derived from a four-membered heterocycle (Figure 1.11)[193] showed several similarities with the previously discussed carbene **77b** (see Scheme 1.11). Substitution of the

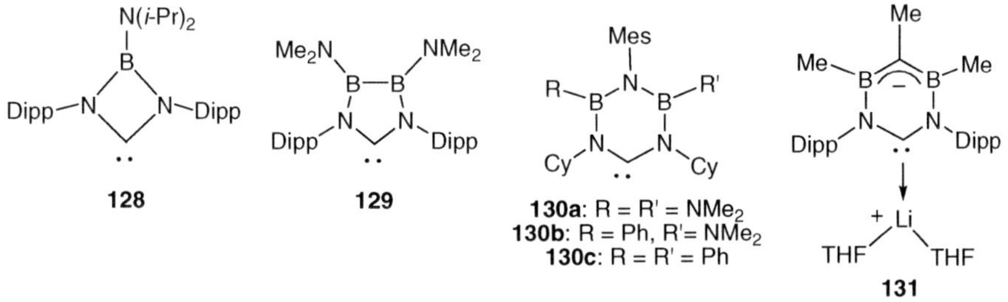

128 **129**

130a: R = R′ = NMe_2
130b: R = Ph, R′= NMe_2
130c: R = R′ = Ph

131

Figure 1.11 N-Heterocyclic carbenes containing boron atoms within the heterocycle.

phosphorus atom in **77b** with a boron atom in **128** resulted in a shift of the ^{13}C NMR resonance for the carbene carbon atom down to the lowest value observed so far for cyclic diaminocarbenes ($\delta = 312.6$ ppm in C_6D_6). The four-membered carbene ring in **128** is planar and the exocyclic nitrogen atom is surrounded in a trigonal–planar fashion. The lengths of the endocyclic C–N and B–N bonds indicated an efficient π-electron donation from the nitrogen atoms to the electron-poor carbon and boron atoms. The endocyclic N–C–N angle in **128** (94.0(2)°) represents the smallest value observed so far for N-heterocyclic carbenes.

The five-membered carbene ring in **129** (Figure 1.11) is nearly planar.[194] The resonance for the carbene carbon atom appeared downfield at $\delta = 304$ ppm in the ^{13}C NMR spectrum. A comparison of the geometrical parameters of **129** with imidazol-2-ylidenes **78** or imidazolin-2-ylidenes **86** revealed a relatively long B–B bond (1.731(2) Å) in **129** in addition to an enlarged value for the N–C–N angle (108.45(8)°). The endocyclic B–N bonds were about 0.1 Å longer than the exocyclic ones, indicating that N→B π-delocalization is weak within the heterocycle and stronger involving the exocyclic amine substituents. A significant N→B π-interaction to the exocyclic amine groups was also confirmed by the observation of a hindered rotation around the B–NMe_2 bond in both the ^{1}H and ^{13}C NMR spectra at room temperature. First experiments indicated that **129** might be a better σ-donor than its analogues **78** and **86** with a carbon backbone.[194]

Heterocycles of type **130** are isoelectronic with the stable borazine. The carbene ring in **130c** is nearly planar and the cyclohexyl groups are oriented to achieve a maximal steric protection of the carbene centre.[195] The N–C–N angle in **130c** is enlarged to 114.5(1)° in accordance with expectations for a six-membered ring. The resonances for the carbene carbon atoms in **130a–c** fall in the narrow range of $\delta = 281.5$–282.9 ppm in the ^{13}C NMR spectra. The IR spectra of complexes [RhCl$(CO)_2$(**130**)] were used to demonstrate that the σ-donor strength of the carbenes **130** could be modulated by the exocyclic substituents at the boron atoms. The σ-donor strength of the carbene ligand decreased together with the donor strength of the exocyclic substituents at the boron atoms in the order **130a** > **130b** > **130c**.[195]

The anionic carbene **131** was isolated and characterized as a lithium adduct.[196] Crystallographically determined molecular parameters of **131** are very similar to those found for the isoelectronic lithium terphenyl derivatives.[197] DFT calculations suggested that **131** exhibited a σ-donor strength in between imidazol-2-ylidenes and terphenyl anions.[196]

1.4.4 N-Heterocyclic Carbenes Derived from Six- or Seven-membered Heterocycles

Aside from the ubiquitous NHCs featuring a five-membered heterocycle, significant advances have been made regarding the preparation of NHCs derived from larger heterocycles. The first stable NHC featuring a six-membered aliphatic heterocycle **132a** was prepared in 1999[64] followed by a number of differently *N,N′*-substituted derivatives.[68c,69] Aromatic derivatives **133a** are also

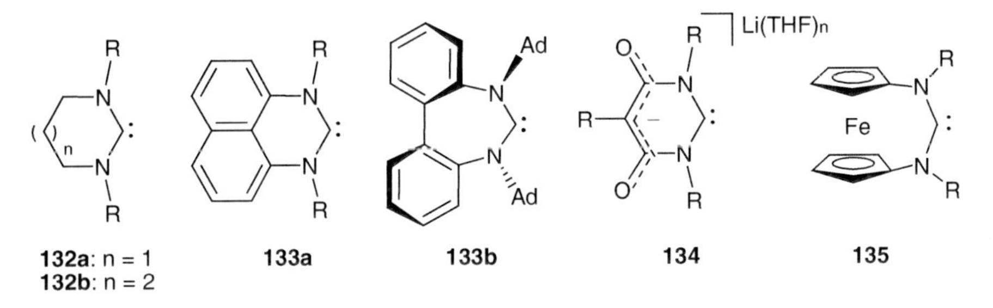

Figure 1.12 Stable NHCs with six- and seven-membered heterocycles.

known (Figure 1.12).[65] The NHCs derived from six-membered heterocycles were obtained by deprotonation of the corresponding cyclic amidinium salts with sterically demanding bases since the use of alkali *tert*-butoxides often led to the formation of alcohol adducts. Carbenes with a saturated six-membered scaffold, such as **132a**, showed no tendency to dimerize to the entetraamines. However, compounds of type **133a** rapidly dimerized to the entetraamines if bearing aromatic substituents or if unsymmetrically *N,N′*-substituted.[65]

NHCs featuring an aliphatic seven-membered ring scaffold of type **132b**[68c,69,198] were also obtained by deprotonation of amidinium salts with sterically demanding bases. In this case, only the *in situ* deprotonation and immediate coordination of the generated NHC to a metal centre was successful.[68,198] Very recently, Stahl and co-workers were able to synthesize carbene **133b** by base-induced α-elimination from the phenol adduct of the carbene.[66,199,200] The seven-membered NHCs **132b** and **133b** are not planar but adopted a twisted conformation instead. The anionic N-heterocyclic carbene **134** and its zwitterionic complexes were also prepared.[201] In their search for NHCs, which allow tuning of the electronic properties by a redox reaction, Bielawski[202] and Siemeling[203] prepared the diaminocarbene[3]ferrocenophanes **135** (Figure 1.12) and their metal complexes.

The characteristic ^{13}C NMR resonances for the carbene carbon atoms of the NHCs featuring six-membered heterocycles (**132a**, **133a**) fall in the range $\delta = 235–245$ ppm,[65,68c,69] similarly to imidazolin-2-ylidenes **86**, whereas the carbene carbon resonances for **132b** and **133b** appeared more downfield ($\delta = 258–260$ ppm).[68c]

The N–C–N angles in NHCs derived from six- and seven-membered heterocycles were larger that in their five-membered analogues ($\approx 115°$ in **132a** and **133a**, 116.6° in **132b**, 113.4° in **133b**).

1.5 Conclusions and Outlook

At this time a large number of NHCs featuring different ring sizes, heteroatoms and substitution patterns are accessible. Still, reports on new NHCs appear regularly in the literature. As diverse as the topology of NHCs is their stereochemical and electronic properties.[204] Different, sometimes surprisingly facile routes leading to stable NHCs or directly to their metal complexes have been developed.

Particularly interesting new work includes the development of 'remote'[205] and 'abnormal'[206] NHCs (see Chapter 5 for further details). Also, NHCs have not only been used as ligands in transition metal chemistry but also as nucleophilic organocatalysts.[40,41] In an extension of this concept, Lavallo and Grubbs presented recently the first organometallic transformation catalyzed by NHCs.[207] This illustrate nicely that almost 20 years after Arduengo's first report on an N-heterocyclic carbene, new NHCs as well as new applications for these interesting and versatile molecules are still and will be for some time the subject of intensive research.

References

1. M. Hermann, *Justus Liebigs Ann. Chem.*, 1855, **95**, 211–255.
2. A. Geuther, *Justus Liebigs Ann. Chem.*, 1862, **123**, 121–122.
3. J. U. Nef, *Justus Liebigs Ann. Chem.*, 1897, **298**, 202–374.
4. W. von E. Doering and L. H. Knox, *J. Am. Chem. Soc.*, 1953, **75**, 297–303.
5. W. von E. Doering and A. K. Hoffmann, *J. Am. Chem. Soc.*, 1954, **76**, 6162–6165.
6. R. A. Moss, *Acc. Chem. Res.*, 1980, **13**, 58–64.
7. R. A. Moss, *Acc. Chem. Res.*, 1989, **22**, 15–21.
8. P. S. Skell and S. R. Sandler, *J. Am. Chem. Soc.*, 1958, **80**, 2024–2025.
9. L. Tschugajeff, M. Skanawy-Grigorjewa and A. Posnjak, *Z. Anorg. Allg. Chem.*, 1925, **148**, 37–42.
10. (a) G. Rouschias and B. L. Shaw, *Chem. Commun.*, 1970, 183; (b) A. Burke, A. L. Balch and J. H. Enemark, *J. Am. Chem. Soc.*, 1970, **92**, 2555–2557; (c) W. M. Butler, J. H. Enemark, J. Parks and A. L. Balch, *Inorg. Chem.*, 1973, **12**, 451–457.
11. E. O. Fischer and A. Maasböl, *Angew. Chem., Int. Ed.*, 1964, **3**, 580–581.
12. R. R. Schrock, *J. Am. Chem. Soc.*, 1974, **96**, 6796–6797.
13. (a) H.-W. Wanzlick and E. Schikora, *Angew. Chem.*, 1960, **72**, 494; (b) H.-W. Wanzlick and E. Schikora, *Chem. Ber.*, 1961, **94**, 2389–2393; (c) H.-W. Wanzlick, *Angew. Chem., Int. Ed. Engl.*, 1962, **1**, 75–80.
14. (a) D. M. Lemal, R. A. Lovald and K. I. Kawano, *J. Am. Chem. Soc.*, 1964, **86**, 2518–2519; (b) H. E. Winberg, J. E. Carnahan, D. D. Coffman and M. Brown, *J. Am. Chem. Soc.*, 1965, **87**, 2055–2056.
15. (a) H. Quast and S. Hünig, *Angew. Chem., Int. Ed. Engl.*, 1964, **3**, 800–801; (b) R. A. Olofson, W. R. Thompson and J. S. Michelman, *J. Am. Chem. Soc.*, 1964, **86**, 1865–1866.
16. H.-J. Schönherr and H.-W. Wanzlick, *Justus Liebigs Ann. Chem.*, 1970, **731**, 176–179.
17. A. J. Arduengo III, J. R. Goerlich, R. Krafczyk and W. J. Marshall, *Angew. Chem., Int. Ed.*, 1998, **37**, 1963–1965.

18. H.-W. Wanzlick and H.-J. Schönherr, *Angew. Chem., Int. Ed. Engl.*, 1968, **7**, 141–142.
19. K. Öfele, *J. Organomet. Chem.*, 1968, **12**, P42–P43.
20. M. F. Lappert, *J. Organomet. Chem.*, 2005, **690**, 5467–5473.
21. A. Igau, H. Grützmacher, A. Baceiredo and G. Bertrand, *J. Am. Chem. Soc.*, 1988, **110**, 6463–6466.
22. A. J. Arduengo III, R. L. Harlow and M. Kline, *J. Am. Chem. Soc.*, 1991, **113**, 361–363.
23. N. Kuhn and T. Kratz, *Synthesis*, 1993, 561–562.
24. F. E. Hahn and M. C. Jahnke, *Angew. Chem., Int. Ed.*, 2008, **47**, 3122–3172.
25. D. Bourissou, O. Guerret, F. P. Gabbaï and G. Bertrand, *Chem. Rev.*, 2000, **100**, 39–91.
26. W. A. Herrmann and C. Köcher, *Angew. Chem., Int. Ed. Engl.*, 1997, **36**, 2162–2187.
27. P. de Frémont, N. Marion and S. P. Nolan, *Coord. Chem. Rev.*, 2009, **253**, 862–892.
28. H. Braband, T. I. Kückmann and U. Abram, *J. Organomet. Chem.*, 2005, **690**, 5421–5429.
29. K. J. Cavell and D. S. McGuinness, *Coord. Chem. Rev.*, 2004, **248**, 671–681.
30. C. M. Crudden and D. P. Allen, *Coord. Chem. Rev.*, 2004, **248**, 2247–2273.
31. J. C. Y. Lin, R. T. W. Huang, C. S. Lee, M. Bhattacharyya, W. S. Hwang and I. J. B. Lin, *Chem. Rev.*, 2009, **109**, 3561–3598.
32. K. M. Hindi, M. J. Panzner, C. A. Tessier, C. L. Cannon and W. J. Youngs, *Chem. Rev.*, 2009, **109**, 3859–3884.
33. P. L. Arnold and I. J. Casely, *Chem. Rev.*, 2009, **109**, 3599–3611.
34. N. Kuhn and A. Al-Sheik, *Coord. Chem. Rev.*, 2005, **249**, 829–857.
35. M. Poyatos, J. A. Mata and E. Peris, *Chem. Rev.*, 2009, **109**, 3677–3707.
36. A. T. Normand and K. J. Cavell, *Eur. J. Inorg. Chem.*, 2008, 2781–2800.
37. D. Pugh and A. A. Danopoulos, *Coord. Chem. Rev.*, 2007, **251**, 610–641.
38. L. H. Gade and S. Bellemin-Laponnaz, *Coord. Chem. Rev.*, 2007, **251**, 718–725.
39. V. César, S. Bellemin-Laponnaz and L. H. Gade, *Chem. Soc. Rev.*, 2004, **33**, 619–636.
40. D. Enders, O. Niemeier and A. Henseler, *Chem. Rev.*, 2007, **107**, 5606–5655.
41. N. Marion, S. Díez-González and S. P. Nolan, *Angew. Chem., Int. Ed.*, 2007, **46**, 2988–3000.
42. R. W. Alder, M. E. Blake, L. Chaker, J. N. Harvey, F. Paolini and J. Schütz, *Angew. Chem., Int. Ed.*, 2004, **43**, 5896–5911.
43. S. Díez-González, N. Marion and S. P. Nolan, *Chem. Rev.*, 2009, **109**, 3612–3676.
44. W. A. Herrmann, *Angew. Chem., Int. Ed.*, 2002, **41**, 1290–1309.
45. T. M. Trnka and R. H. Grubbs, *Acc. Chem. Res.*, 2001, **34**, 18–29.

46. G. B. Schuster, *Adv. Phys. Org. Chem.*, 1987, **22**, 311–361.
47. R. Hoffmann, G. D. Zeiss and G. W. Van Dine, *J. Am. Chem. Soc.*, 1968, **90**, 1485–1499.
48. (a) C. W. Bauschlicher Jr, H. F. Schaefer III and P. S. Bagus, *J. Am. Chem. Soc.*, 1977, **99**, 7106–7110; (b) J. F. Harrison, R. C. Liedtke and J. F. Liebman, *J. Am. Chem. Soc.*, 1979, **101**, 7162–7168; (c) D. Feller, W. T. Borden and E. R. Davidson, *Chem. Phys. Lett.*, 1980, **71**, 22–26.
49. N. C. Baird and K. F. Taylor, *J. Am. Chem. Soc.*, 1978, **100**, 1333–1338.
50. (a) R. A. Moss, M. Włostowski, S. Shen, K. Krogh-Jespersen and A. Matro, *J. Am. Chem. Soc.*, 1988, **110**, 4443–4444; (b) X.-M. Du, H. Fan, J. L. Goodman, M. A. Kesselmayer, K. Krogh-Jespersen, J. A. LaVilla, R. A. Moss, S. Shen and R. S. Sheridan, *J. Am. Chem. Soc.*, 1990, **112**, 1920–1926.
51. R. A. Moss and C. B. Mallon, *J. Am. Chem. Soc.*, 1975, **97**, 344–347.
52. (a) D. Nemcsok, K. Wichmann and G. Frenking, *Organometallics*, 2004, **23**, 3640–3646; (b) M. Tafipolski, W. Scherer, K. Öfele, G. Artus, B. Pedersen, W. A. Herrmann and G. S. McGrady, *J. Am. Chem. Soc.*, 2002, **124**, 5865–5880; (c) C. Boehme and G. Frenking, *J. Am. Chem. Soc.*, 1996, **118**, 2039–2046; (d) C. Heinemann, T. Müller, Y. Apeloig and H. Schwarz, *J. Am. Chem. Soc.*, 1996, **118**, 2023–2038; (e) A. J. Arduengo III, H. V. R. Dias, D. A. Dixon, R. L. Harlow, W. T. Kloster and T. F. Koetzle, *J. Am. Chem. Soc.*, 1994, **116**, 6812–6822; (f) C. Heinemann and W. Thiel, *Chem. Phys. Lett.*, 1994, **217**, 11–16; (g) A. J. Arduengo III, H. V. R. Dias, R. L. Harlow and M. Kline, *J. Am. Chem. Soc.*, 1992, **114**, 5530–5534; (h) D. A. Dixon and A. J. Arduengo III, *J. Phys. Chem.*, 1991, **95**, 4180–4182.
53. (a) P. Fournari, P. de Cointet and E. Laviron, *Bull. Soc. Chim. Fr.*, 1968, 2438–2446; (b) B. K. M. Chan, N.-H. Chang and M. R. Grimmett, *Aust. J. Chem.*, 1977, **30**, 2005–2013.
54. (a) W. A. Herrmann, C. Köcher, L. J. Gooßen and G. R. J. Artus, *Chem.–Eur. J.*, 1996, **2**, 1627–1636; (b) V. P. W. Böhm, T. Weskamp, C. W. K. Gstöttmayr and W. A. Herrmann, *Angew. Chem., Int. Ed.*, 2000, **39**, 1602–1604.
55. B. Bildstein, M. Malaun, H. Kopacka, K. Wurst, M. Mitterböck, K.-H. Ongania, G. Opromolla and P. Zanello, *Organometallics*, 1999, **18**, 4325–4336.
56. (a) A. A. Gridnev and I. M. Mihaltseva, *Synth. Commun.*, 1994, **24**, 1547–1555; (b) J. Liu, J. Chen, J. Zhao, Y. Zhao, L. Li and H. Zhang, *Synthesis*, 2003, 2661–2666.
57. W. A. Herrmann, L. J. Gooßen and M. Spiegler, *J. Organomet. Chem.*, 1997, **547**, 357–366.
58. J.-H. Cristau, P. P. Cellier, J.-F. Spindler and M. Taillefer, *Chem.–Eur. J.*, 2004, **10**, 5607–5622.
59. A. Fürstner, M. Alcarazo, V. César and C. W. Lehmann, *Chem. Commun.*, 2006, 2176–2178.

60. (a) G. Altenhoff, R. Goddard, C. W. Lehmann and F. Glorius, *Angew. Chem., Int. Ed.*, 2003, **42**, 3690–3693; (b) G. Altenhoff, R. Goddard, C. W. Lehmann and F. Glorius, *J. Am. Chem. Soc.*, 2004, **126**, 15195–15201.
61. (a) S. Saba, A. Brescia and M. K. Kaloustian, *Tetrahedron Lett.*, 1991, **32**, 5031–5034; (b) A. J. Arduengo III, R. Krafczyk, R. Schmutzler, H. A. Craig, J. R. Goerlich, W. J. Marshall and M. Unverzagt, *Tetrahedron*, 1999, **55**, 14523–14534; (c) A. Paczal, A. C. Bényei and A. Kotschy, *J. Org. Chem.*, 2006, **71**, 5969–5979.
62. (a) R. S. Bon, C. Hong, M. J. Bouma, R. F. Schmitz, F. J. J. de Kanter, M. Lutz, A. L. Spek and R. V. A. Orru, *Org. Lett.*, 2003, **5**, 3759–3762; (b) R. S. Bon, B. van Vliet, N. E. Sprenkels, R. F. Schmitz, F. J. J. de Kanter, C. V. Stevens, M. Swart, F. M. Bickelhaupt, M. B. Groen and R. V. A. Orru, *J. Org. Chem.*, 2005, **70**, 3542–3553; (c) R. S. Bon, F. J. J. de Kanter, M. Lutz, A. L. Spek, M. C. Jahnke, F. E. Hahn, M. B. Groen and R. V. A. Orru, *Organometallics*, 2007, **26**, 3639–3650.
63. B. A. B. Prasad and S. R. Gilbertson, *Org. Lett.*, 2009, **11**, 3710–3713.
64. R. W. Alder, M. E. Blake, C. Bortolotti, S. Bufali, C. P. Butts, E. Linehan, J. M. Oliva, A. G. Orpen and M. J. Quayle, *Chem. Commun.*, 1999, 241–242.
65. P. Bazinet, T.-G. Ong, J. S. O'Brien, N. Lavoie, E. Bell, G. P. A. Yap, I. Korobkov and D. S. Richeson, *Organometallics*, 2007, **26**, 2885–2895.
66. C. C. Scarborough, M. J. W. Grady, I. A. Guzei, B. A. Gandhi, E. E. Bunel and S. S. Stahl, *Angew. Chem., Int. Ed.*, 2005, **44**, 5269–5272.
67. R. Jazzar, H. Liang, B. Donnadieu and G. Bertrand, *J. Organomet. Chem.*, 2006, **691**, 3201–3205.
68. (a) M. Iglesias, D. J. Beetstra, B. Kariuki, K. J. Cavell, A. Dervisi and I. A. Fallis, *Eur. J. Inorg. Chem.*, 2009, 1913–1919; (b) A. Binobaid, M. Iglesias, D. J. Beetstra, B. Kariuki, A. Dervisi, I. A. Fallis and K. J. Cavell, *Dalton Trans.*, 2009, 7099–7112; (c) M. Iglesias, D. J. Beetstra, J. C. Knight, L.-L. Ooi, A. Stasch, S. Coles, L. Male, M. B. Hursthouse, K. J. Cavell, A. Dervisi and I. A. Fallis, *Organometallics*, 2008, **27**, 3279–3289.
69. E. L. Kolychev, I. A. Portnyagin, V. V. Shuntikov, V. N. Khrustalev and M. S. Nechaev, *J. Organomet. Chem.*, 2009, **694**, 2454–2462.
70. M. K. Denk, A. Thadani, K. Hatano and A. J. Lough, *Angew. Chem., Int. Ed. Engl.*, 1997, **36**, 2607–2609.
71. F. E. Hahn, L. Wittenbecher, R. Boese and D. Bläser, *Chem.–Eur. J.*, 1999, **5**, 1931–1935.
72. (a) F. E. Hahn, M. Paas, D. Le Van and T. Lügger, *Angew. Chem., Int. Ed.*, 2003, **42**, 5243–5246; (b) F. E. Hahn, M. Paas, D. Le Van and R. Fröhlich, *Chem.–Eur. J.*, 2005, **11**, 5080–5085.
73. M. Scholl, S. Ding, C. W. Lee and R. H. Grubbs, *Org. Lett.*, 1999, **1**, 953–956.
74. (a) G. W. Nyce, S. Csihony, R. W. Waymouth and J. L. Hedrick, *Chem.–Eur. J.*, 2004, **10**, 4073–4079; (b) A. P. Blum, T. Ritter and R. H. Grubbs, *Organometallics*, 2007, **26**, 2122–2124.

75. M. Otto, S. Conejero, Y. Canac, V. D. Romanenko, V. Rudzevitch and G. Bertrand, *J. Am. Chem. Soc.*, 2004, **126**, 1016–1017.
76. (a) D. S. McGuinness and K. J. Cavell, *Organometallics*, 2000, **19**, 741–748; (b) A. A. Danopoulos, S. Winston, T. Gelbrich, M. B. Hursthouse and R. P. Tooze, *Chem. Commun.*, 2002, 482–483.
77. I. S. Edworthy, M. Rodden, S. A. Mungur, K. M. Davis, A. J. Blake, C. Wilson, M. Schröder and P. L. Arnold, *J. Organomet. Chem.*, 2005, **690**, 5710–5719.
78. L. P. Spencer and M. D. Fryzuk, *J. Organomet. Chem.*, 2005, **690**, 5788–5803.
79. B. Wang, D. Wang, D. Cui, W. Gao, T. Tang, X. Chen and X. Jing, *Organometallics*, 2007, **26**, 3167–3172.
80. S. P. Downing, S. C. Guadaño, D. Pugh, A. A. Danopoulos, R. M. Bellabarba, M. Hanton, D. Smith and R. P. Tooze, *Organometallics*, 2007, **26**, 3762–3770.
81. (a) W. A. Herrmann, J. Schwarz and M. G. Gardiner, *Organometallics*, 1999, **18**, 4082–4089; (b) J. A. Mata, A. R. Chianese, J. R. Miecznikowski, M. Poyatos, E. Peris, J. W. Faller and R. H. Crabtree, *Organometallics*, 2004, **23**, 1253–1263; (c) S. Ahrens, E. Herdtweck, S. Goutal and T. Strassner, *Eur. J. Inorg. Chem.*, 2006, 1268–1274; (d) J.-W. Wang, Q.-S. Li, F.-B. Xu, H.-B. Song and Z.-Z. Zhang, *Eur. J. Org. Chem.*, 2006, 1310–1316.
82. P. L. Arnold, A. C. Scarisbrick, A. J. Blake and C. Wilson, *Chem. Commun.*, 2001, 2340–2341.
83. F. E. Hahn and M. Foth, *J. Organomet. Chem.*, 1999, **585**, 241–245.
84. H. V. R. Dias and W. Jin, *Tetrahedron Lett.*, 1994, **35**, 1365–1366.
85. (a) U. Kernbach, M. Ramm, P. Luger and W. P. Fehlhammer, *Angew. Chem., Int. Ed. Engl.*, 1996, **35**, 310–312; (b) R. Fränkel, U. Kernbach, M. Bakola-Christianopoulou, U. Plaia, M. Suter, W. Ponikwar, H. Nöth, C. Moinet and W. P. Fehlhammer, *J. Organomet. Chem.*, 2001, **617–618**, 530–545.
86. X. Hu and K. Meyer, *J. Organomet. Chem.*, 2005, **690**, 5474–5484.
87. M. E. van der Boom and D. Milstein, *Chem. Rev.*, 2003, **103**, 1759–1792.
88. M. Albrecht and G. van Koten, *Angew. Chem., Int. Ed.*, 2001, **40**, 3750–3781.
89. (a) J. C. C. Chen and I. J. B. Lin, *J. Chem. Soc., Dalton Trans.*, 2000, 839–840; (b) E. Peris, J. A. Loch, J. Mata and R. H. Crabtree, *Chem. Commun.*, 2001, 201–202.
90. (a) S. Gründemann, M. Albrecht, J. A. Loch, J. W. Faller and R. H. Crabtree, *Organometallics*, 2001, **20**, 5485–5488; (b) A. A. Danopoulos, A. A. D. Tulloch, S. Winston, G. Eastham and M. B. Hursthouse, *Dalton Trans.*, 2003, 1009–1015.
91. G. T. S. Andavan, E. B. Bauer, C. S. Letko, T. K. Hollis and F. S. Tham, *J. Organomet. Chem.*, 2005, **690**, 5938–5947.

92. (a) R. E. Douthwaite, J. Houghton and B. M. Kariuki, *Chem. Commun.*, 2004, 698–699; (b) I. S. Edworthy, A. J. Blake, C. Wilson and P. L. Arnold, *Organometallics*, 2007, **26**, 3684–3689.
93. F. E. Hahn, M. C. Jahnke, V. Gomez-Benitez, D. Morales-Morales and T. Pape, *Organometallics*, 2005, **24**, 6458–6463.
94. F. E. Hahn, M. C. Jahnke and T. Pape, *Organometallics*, 2007, **26**, 150–154.
95. H. M. Lee, J. Y. Zeng, C.-H. Hu and M.-T. Lee, *Inorg. Chem.*, 2004, **43**, 6822–6829.
96. V. J. Catalano, M. A. Malwitz and A. O. Etogo, *Inorg. Chem.*, 2004, **43**, 5714–5724.
97. H. Aihara, T. Matsuo and H. Kawaguchi, *Chem. Commun.*, 2003, 2204–2205.
98. (a) F. E. Hahn, M. C. Jahnke and T. Pape, *Organometallics*, 2006, **26**, 5927–5936; (b) T. Steinke, B. K. Shaw, H. Jong, B. O. Patrick and M. D. Fryzuk, *Organometallics*, 2009, **28**, 2830–2836.
99. F. E. Hahn, C. Holtgrewe, T. Pape, M. Martin, E. Sola and L. A. Oro, *Organometallics*, 2005, **24**, 2203–2209.
100. (a) Z. Shi and R. P. Thummel, *Tetrahedron Lett.*, 1994, **35**, 33–36; (b) Z. Shi and R. P. Thummel, *Tetrahedron Lett.*, 1995, **36**, 2741–2744; (c) Z. Shi and R. P. Thummel, *J. Org. Chem.*, 1995, **60**, 5935–5945.
101. (a) E. Alcalde, C. Alvarez-Rúa, S. García-Granda, E. García-Rodriguez, N. Mesquida and L. Pérez-García, *Chem. Commun.*, 1999, 295–296; (b) S. Ramos, E. Alcalde, G. Doddi, P. Mencarelli and L. Pérez-García, *J. Org. Chem.*, 2002, **67**, 8463–8468; (c) A. M. Magill, D. S. McGuinness, K. J. Cavell, G. J. P. Britovsek, V. C. Gibson, A. J. P. White, D. J. Williams, A. H. White and B. W. Skelton, *J. Organomet. Chem.*, 2001, **617–618**, 546–560; (d) M. V. Baker, S. K. Brayshaw, B. W. Skelton, A. H. White and C. C. Williams, *J. Organomet. Chem.*, 2005, **690**, 2312–2322; (e) M. V. Baker, D. H. Brown, P. V. Simpson, B. W. Skelton, A. H. White and C. C. Williams, *J. Organomet. Chem.*, 2006, **691**, 5845–5855; (f) M. V. Baker, B. W. Skelton, A. H. White and C. C. Williams, *J. Chem. Soc., Dalton Trans.*, 2001, 111–120; (g) P. J. Barnard, L. E. Wedlock, M. V. Baker, S. J. Berners-Price, D. A. Joyce, B. W. Skelton and J. H. Steer, *Angew. Chem., Int. Ed.*, 2006, **45**, 5966–5970.
102. W. W. H. Wong, M. S. Vickers, A. R. Cowley, R. L. Paul and P. D. Beer, *Org. Biomol. Chem.*, 2005, **3**, 4201–4208.
103. M. V. Baker, B. W. Skelton, A. H. White and C. C. Williams, *Organometallics*, 2002, **21**, 2674–2678.
104. K. Chellappan, N. J. Singh, I.-C. Hwang, J. W. Lee and K. S. Kim, *Angew. Chem., Int. Ed.*, 2005, **44**, 2899–2903.
105. F. E. Hahn, C. Radloff, T. Pape and A. Hepp, *Chem.–Eur. J.*, 2008, **14**, 10900–10904.
106. L. H. Gade and S. Bellemin-Laponnaz, *Top. Organomet. Chem.*, 2007, **21**, 117–157.

107. (a) W. A. Herrmann, L. J. Goossen, C. Köcher and G. R. J. Artus, *Angew. Chem., Int. Ed. Engl.*, 1996, **35**, 2805–2807; (b) S. K. Schneider, J. Schwarz, G. D. Frey, E. Herdtweck and W. A. Herrmann, *J. Organomet. Chem.*, 2007, **692**, 4560–4568.
108. S. Lee and J. F. Hartwig, *J. Org. Chem.*, 2001, **66**, 3402–3415.
109. H. Seo, B. Y. Kim, J. H. Lee, H.-J. Park, S. U. Son and Y. K. Chung, *Organometallics*, 2003, **22**, 4783–4791.
110. A. Ros, D. Monge, M. Alcarazo, E. Álvarez, J. M. Lassaletta and R. Fernández, *Organometallics*, 2006, **25**, 6039–6046.
111. R. L. Knight and F. J. Leeper, *J. Chem. Soc., Perkin Trans. 1*, 1998, 1891–1893.
112. D. Enders and H. Gielen, *J. Organomet. Chem.*, 2001, **617–618**, 70–80.
113. J. Read de Alaniz and T. Rovis, *J. Am. Chem. Soc.*, 2005, **127**, 6284–6289.
114. A. Alexakis, C. L. Winn, F. Guillen, J. Pytkowicz, S. Roland and P. Mangeney, *Adv. Synth. Catal.*, 2003, **345**, 345–348.
115. D. Kremzow, G. Seidel, C. W. Lehmann and A. Fürstner, *Chem.–Eur. J.*, 2005, **11**, 1833–1853.
116. D. S. Clyne, J. Jin, E. Genest, J. C. Gallucci and T. V. RajanBabu, *Org. Lett.*, 2000, **2**, 1125–1128.
117. W.-L. Duan, M. Shi and G.-B. Rong, *Chem. Commun.*, 2003, 2916–2917.
118. J. J. Van Veldhuizen, S. B. Garber, J. S. Kingsbury and A. H. Hoveyda, *J. Am. Chem. Soc.*, 2002, **124**, 4954–4955.
119. C. Bolm, M. Kesselgruber and G. Raabe, *Organometallics*, 2002, **21**, 707–710.
120. (a) D. Broggini and A. Togni, *Helv. Chim. Acta*, 2002, **85**, 2518–2522; (b) F. Visentin and A. Togni, *Organometallics*, 2007, **26**, 3746–3754.
121. H. Seo, H.-J. Park, B. Y. Kim, J. H. Lee, S. U. Son and Y. K. Chung, *Organometallics*, 2003, **22**, 618–620.
122. (a) S. Gischig and A. Togni, *Organometallics*, 2004, **23**, 2479–2487; (b) S. Gischig and A. Togni, *Eur. J. Inorg. Chem.*, 2005, 4745–4754.
123. Y. Ma, C. Song, C. Ma, Z. Sun, Q. Chai and M. B. Andrus, *Angew. Chem., Int. Ed.*, 2003, **42**, 5871–5874.
124. T. Focken, G. Raabe and C. Bolm, *Tetrahedron: Asymmetry*, 2004, **15**, 1693–1706.
125. M. C. Perry, X. Cui and K. Burgess, *Tetrahedron: Asymmetry*, 2002, **13**, 1969–1972.
126. (a) L. G. Bonnet, R. E. Douthwaite and R. Hodgson, *Organometallics*, 2003, **22**, 4384–4386; (b) G. Dyson, J.-C. Frison, A. C. Whitwood and R. E. Douthwaite, *Dalton Trans.*, 2009, 7141–7151.
127. (a) L. G. Bonnet, R. E. Douthwaite and B. M. Kariuki, *Organometallics*, 2003, **22**, 4187–4189; (b) R. Hodgson and R. E. Douthwaite, *J. Organomet. Chem.*, 2005, **690**, 5822–5831.
128. W. A. Herrmann, L. J. Goossen and M. Spiegler, *Organometallics*, 1998, **17**, 2162–2168.
129. M. C. Perry, X. Cui, M. T. Powell, D.-R. Hou, J. H. Reibenspies and K. Burgess, *J. Am. Chem. Soc.*, 2003, **125**, 113–123.

130. V. César, S. Bellemin-Laponnaz and L. H. Gade, *Organometallics*, 2002, **21**, 5204–5208.
131. F. Glorius, G. Altenhoff, R. Goddard and C. Lehmann, *Chem. Commun.*, 2002, **2**, 2704–2705.
132. W. A. Herrmann, M. Elison, J. Fischer, C. Köcher and G. R. J. Artus, *Chem.–Eur. J.*, 1996, **2**, 772–780.
133. M. K. Denk, K. Hatano and M. Ma, *Tetrahedron Lett.*, 1999, **40**, 2057–2060.
134. F. E. Hahn, L. Wittenbecher, D. Le Van and R. Fröhlich, *Angew. Chem., Int. Ed.*, 2000, **39**, 541–544.
135. D. Enders, K. Breuer, G. Raabe, J. Runsink, J. H. Teles, J.-P. Melder, K. Ebel and S. Brode, *Angew. Chem., Int. Ed. Engl.*, 1995, **34**, 1021–1023.
136. W.-H. Wanzlick and H.-J. Kleiner, *Angew. Chem.*, 1961, **73**, 493.
137. N. I. Korotkikh, G. F. Raenko, T. M. Pekhtereva, O. P. Shvaika, A. H. Cowley and J. N. Jones, *Russ. J. Org. Chem.*, 2006, **42**, 1822–1833.
138. (a) B. Gorodetsky, T. Ramnial, N. R. Branda and J. A. C. Clyburne, *Chem. Commun.*, 2004, 1972–1973; (b) A. M. Voutchkova, L. N. Appelhans, A. R. Chianese and R. H. Crabtree, *J. Am. Chem. Soc.*, 2005, **127**, 17624–17625; (c) A. M. Voutchkova, M. Feliz, E. Clot, O. Eisenstein and R. H. Crabtree, *J. Am. Chem. Soc.*, 2007, **129**, 12834–12846.
139. (a) V. Lavallo, Y. Canac, B. Donnadieu, W. W. Schoeller and G. Bertrand, *Science*, 2006, **312**, 722–724; (b) D. Holschumacher, C. G. Hrib, P. G. Jones and M. Tamm, *Chem. Commun.*, 2007, 3661–3663; (c) G. Kuchenbeiser, B. Donnadieu and G. Bertrand, *J. Organomet. Chem.*, 2008, **693**, 899–904.
140. K. Öfele, C. Taubmann and W. A. Herrmann, *Chem. Rev.*, 2009, **109**, 3408–3444.
141. E. Despagnet-Ayoub and R. H. Grubbs, *J. Am. Chem. Soc.*, 2004, **126**, 10198–10199.
142. E. Despagnet-Ayoub and R. H. Grubbs, *Organometallics*, 2005, **24**, 338–340.
143. A. J. Arduengo III, F. Davidson, H. V. R. Dias, J. R. Goerlich, D. Khasnis, W. J. Marshall and T. Prakasha, *J. Am. Chem. Soc.*, 1997, **119**, 12742–12749.
144. (a) T. A. Taton and P. Chen, *Angew. Chem., Int. Ed. Engl.*, 1996, **35**, 1011–1013; (b) D. C. Graham, K. J. Cavell and B. F. Yates, *J. Phys. Org. Chem.*, 2005, **18**, 298–309.
145. (a) E. A. Carter and W. A. Goddard III, *J. Phys. Chem.*, 1986, **90**, 998–1001; (b) J. A. Blush, H. Clauberg, D. W. Kohn, D. W. Minsek, X. Zhang and P. Chen, *Acc. Chem. Res.*, 1992, **25**, 385–392.
146. A. J. Arduengo III, J. R. Goerlich and W. J. Marshall, *J. Am. Chem. Soc.*, 1995, **117**, 11027–11028.
147. Y. Liu and D. M. Lemal, *Tetrahedron Lett.*, 2000, **41**, 599–602.
148. B. Gehrhus, P. B. Hitchcock and M. F. Lappert, *J. Chem. Soc., Dalton Trans.*, 2000, 3094–3099.
149. E. Çetinkaya, P. B. Hitchcock, H. Küçükbay, M. F. Lappert and S. Al-Juaid, *J. Organomet. Chem.*, 1994, **481**, 89–95.

150. D. M. Khramov, A. J. Boydston and C. W. Bielawski, *Angew. Chem., Int. Ed.*, 2006, **45**, 6186–6189.
151. (a) D. M. Khramov, A. J. Boydston and C. W. Bielawski, *Org. Lett.*, 2006, **8**, 1831–1834; (b) A. J. Boydston, K. A. Williams and C. W. Bielawski, *J. Am. Chem. Soc.*, 2005, **127**, 12496–12497.
152. J. W. Kamplain and C. W. Bielawski, *Chem. Commun.*, 2006, 1727–1729.
153. (a) A. J. Boydston and C. W. Bielawski, *Dalton Trans.*, 2006, 4073–4077; (b) A. J. Boydston, J. D. Rice, M. D. Sanderson, O. L. Dykhno and C. W. Bielawski, *Organometallics*, 2006, **25**, 6087–6098.
154. (a) F. E. Hahn, C. Radloff, T. Pape and A. Hepp, *Organometallics*, 2008, **27**, 6408–6410; (b) C. Radloff, F. E. Hahn, T. Pape and R. Fröhlich, *Dalton Trans.*, 2009, 7215–7222; (c) C. Radloff, J. J. Weigand and F. E. Hahn, *Dalton Trans.*, 2009, 9392–9394.
155. (a) M. Alcarazo, S. J. Roseblade, A. R. Cowley, R. Fernández, J. M. Brown and J. M. Lassaletta, *J. Am. Chem. Soc.*, 2005, **127**, 3290–3291; (b) S. J. Roseblade, A. Ros, D. Monge, M. Alcarazo, E. Álvarez, J. M. Lassaletta and R. Fernández, *Organometallics*, 2007, **26**, 2570–2578; (c) C. Burstein, C. W. Lehmann and F. Glorius, *Tetrahedron*, 2005, **61**, 6207–6217; (d) S. Saravanakumar, A. I. Oprea, M. K. Kindermann, P. G. Jones and J. Heinicke, *Chem.–Eur. J.*, 2006, **12**, 3143–3154; (e) F. Ullah, G. Bajor, T. Veszprémi, P. G. Jones and J. W. Heinicke, *Angew. Chem., Int. Ed.*, 2007, **46**, 2697–2700.
156. (a) R. Weiss, S. Reichel, M. Handtke and F. Hampel, *Angew. Chem., Int. Ed.*, 1998, **37**, 344–347; (b) M. Nonnenmacher, D. Kunz, F. Romminger and T. Oeser, *J. Organomet. Chem.*, 2005, **690**, 5647–5653; (c) M. Nonnenmacher, D. Kunz, F. Rominger and T. Oeser, *Chem. Commun.*, 2006, 1378–1380.
157. M. D. Sanderson, J. W. Kamplain and C. W. Bielawski, *J. Am. Chem. Soc.*, 2006, **128**, 16514–16515.
158. (a) S. Saravanakumar, M. K. Kindermann, J. Heinicke and M. Köckerling, *Chem. Commun.*, 2006, 640–642; (b) F. Ullah, M. K. Kindermann, P. G. Jones and J. Heinicke, *Organometallics*, 2009, **28**, 2441–2449; (c) D. Tapu, C. Owens, D. VanDerveer and K. Gwaltney, *Organometallics*, 2009, **28**, 270–276.
159. L. Pause, M. Robert, J. Heinicke and O. Kühl, *J. Chem. Soc., Perkin Trans. 2*, 2001, 1383–1388.
160. Y. Liu, P. E. Lindner and D. M. Lemal, *J. Am. Chem. Soc.*, 1999, **121**, 10626–10627.
161. (a) C. Holtgrewe, C. Diedrich, T. Pape, S. Grimme and F. E. Hahn, *Eur. J. Org. Chem.*, 2006, 3116–3124; (b) J. A. Chamizo and M. F. Lappert, *J. Org. Chem.*, 1989, **54**, 4684–4686; (c) B. Çetinkaya, E. Çetinkaya, J. A. Chamizo, P. B. Hitchcock, H. A. Jasim, H. Küçükbay and M. F. Lappert, *J. Chem. Soc., Perkin Trans. 1*, 1998, 2047–2054.
162. (a) D. Enders and T. Balensiefer, *Acc. Chem. Res.*, 2004, **37**, 534–541; (b) D. Enders and A. A. Narine, *J. Org. Chem.*, 2008, **73**, 7857–7870; (c) D. Enders, O. Niemeier and T. Balensiefer, *Angew. Chem., Int. Ed.*, 2006, **45**,

1463–1467; (d) D. Enders and U. Kallfass, *Angew. Chem., Int. Ed.*, 2002, **41**, 1743–1745; (e) D. Enders, K. Breuer and J. H. Teles, *Helv. Chim. Acta*, 1996, **79**, 1217–1221.
163. O. Guerret, S. Solé, H. Gornitzka, M. Teichert, G. Trinquier and G. Bertrand, *J. Am. Chem. Soc.*, 1997, **119**, 6668–6669.
164. (a) O. Guerret, S. Solé, H. Gornitzka, G. Trinquier and G. Bertrand, *J. Organomet. Chem.*, 2000, **600**, 112–117; (b) E. Mas-Marzá, J. A. Mata and E. Peris, *Angew. Chem., Int. Ed.*, 2007, **46**, 3729–3731.
165. A. J. Arduengo III, H. R. Goerlich and W. J. Marshall, *Liebigs Ann./Recueil*, 1997, 365–374.
166. G. Morel, G. Gachot and D. Lorcy, *Synlett*, 2005, 1117–1120.
167. (a) H. V. Huynh, N. Meier, T. Pape and F. E. Hahn, *Organometallics*, 2006, **25**, 3012–3018; (b) F. E. Hahn, N. Meier and T. Pape, *Z. Naturforsch.*, 2006, **61b**, 820–824; (c) S. K. Yen, L. L. Koh, F. E. Hahn, H. V. Huynh and T. S. A. Hor, *Organometallics*, 2006, **25**, 5105–5112; (d) V. Caló, R. Del Sole, A. Nacci, E. Schingaro and F. Scordari, *Eur. J. Org. Chem.*, 2000, 869–871; (e) S. K. Yen, L. L. Koh, H. V. Huynh and T. S. A. Hor, *J. Organomet. Chem.*, 2009, **694**, 332–338.
168. R. Kluger, *Chem. Rev.*, 1987, **87**, 863–876.
169. R. Breslow, *J. Am. Chem. Soc.*, 1958, **80**, 3719–3726.
170. (a) Y.-T. Chen and F. Jordan, *J. Org. Chem.*, 1991, **56**, 5029–5038; (b) F. J. Leeper and D. H. C. Smith, *J. Chem. Soc., Perkin Trans. 1*, 1995, 861–873.
171. (a) S. Solé, H. Gornitzka, W. W. Schoeller, D. Bourissou and G. Bertrand, *Science*, 2001, **292**, 1901–1903; (b) Y. Canac, M. Soleilhavoup, S. Conejero and G. Bertrand, *J. Organomet. Chem.*, 2004, **689**, 3857–3865.
172. V. Lavallo, J. Mafhouz, Y. Canac, B. Donnadieu, W. W. Schoeller and G. Bertrand, *J. Am. Chem. Soc.*, 2004, **126**, 8670–8671.
173. (a) V. Lavallo, Y. Canac, C. Präsang, B. Donnadieu and G. Bertrand, *Angew. Chem., Int. Ed.*, 2005, **44**, 5705–5709; (b) V. Lavallo, Y. Canac, A. DeHope, B. Donnadieu and G. Bertrand, *Angew. Chem., Int. Ed.*, 2005, **44**, 7236–7239.
174. G. D. Frey, R. D. Dewhurst, S. Kousar, B. Donnadieu and G. Bertrand, *J. Organomet. Chem.*, 2008, **693**, 1674–1682.
175. (a) X. Zeng, G. D. Frey, S. Kousar and G. Bertrand, *Chem.–Eur. J.*, 2009, **15**, 3056–3060; (b) X. Zeng, G. D. Frey, R. Kinjo, B. Donnadieu and G. Bertrand, *J. Am. Chem. Soc.*, 2009, **131**, 8690–8696.
176. R. Jazzar, R. D. Dewhurst, J.-B. Bourg, B. Donnadieu, Y. Canac and G. Bertrand, *Angew. Chem., Int. Ed.*, 2007, **46**, 2899–2902.
177. R. Jazzar, J.-B. Bourg, R. D. Dewhurst, B. Donnadieu and G. Bertrand, *J. Org. Chem.*, 2007, **72**, 3492–3499.
178. G. Frey, V. Lavallo, B. Donnadieu, W. W. Schoeller and G. Bertrand, *Science*, 2007, **316**, 439–441.
179. A. L. Kenward and W. E. Piers, *Angew. Chem., Int. Ed.*, 2008, **47**, 38–41.
180. D. Holschumacher, T. Bannenberg, C. G. Hrib, P. G. Jones and M. Tamm, *Angew. Chem., Int. Ed.*, 2008, **47**, 7428–7432.

181. P. A. Chase and D. W. Stephan, *Angew. Chem., Int. Ed.*, 2008, **47**, 7433–7437.
182. (a) G. Facchin, M. Mozzon, R. A. Michelin, M. T. A. Ribeiro and A. J. L. Pombeiro, *J. Chem. Soc., Dalton Trans.*, 1992, 2827–2835; (b) A. J. L. Pombeiro, *J. Organomet. Chem.*, 2005, **690**, 6021–6040.
183. S.-y. Nakafuji, J. Kobayashi and T. Kawashima, *Angew. Chem., Int. Ed.*, 2008, **47**, 1141–1144.
184. A. Fürstner, M. Alcarazo, K. Radkowski and C. W. Lehmann, *Angew. Chem., Int. Ed.*, 2008, **47**, 8302–8306.
185. M. Asay, B. Donnadieu, B. Baceiredo, M. Soleilhavoup and G. Bertrand, *Inorg. Chem.*, 2008, **47**, 3949–3951.
186. (a) Y. Canac, C. Duhayon and R. Chauvin, *Angew. Chem., Int. Ed.*, 2007, **46**, 6313–6315; (b) Y. Canac, C. Lepetit, M. Abdalilah, C. Duhayon and R. Chauvin, *J. Am. Chem. Soc.*, 2008, **130**, 8406–8413; (c) I. Abdellah, N. Debono, Y. Canac, C. Duhayon and R. Chauvin, *Dalton Trans.*, 2009, 7196–7202.
187. J. Kapp, C. Schade, A. M. El-Nahasa and P. von Ragué Schleyer, *Angew. Chem., Int. Ed. Engl.*, 1996, **35**, 2236–2238.
188. (a) A. Fekete and L. Nyulászi, *J. Organomet. Chem.*, 2002, **643–644**, 278–284; (b) W. W. Schoeller and D. Eisner, *Inorg. Chem.*, 2004, **43**, 2585–2589.
189. T. Cantat, N. Mézailles, N. Maigrot, L. Ricard and P. Le Floch, *Chem. Commun.*, 2004, 1274–1275.
190. (a) D. Martin, A. Baceiredo, H. Gornitzka, W. W. Schoeller and G. Bertrand, *Angew. Chem., Int. Ed.*, 2005, **44**, 1700–1703; (b) J. D. Masuda, D. Martin, C. Lyon-Saunier, A. Baceiredo, H. Gornitzka, B. Donnadieu and G. Bertrand, *Chem.–Asian J.*, 2007, **2**, 178–187.
191. L. Nyulászi, *Tetrahedron*, 2000, **56**, 79–84.
192. G. D. Frey, M. Song, J.-B. Bourg, B. Donnadieu, M. Soleilhavoup and G. Bertrand, *Chem. Commun.*, 2008, 4711–4713.
193. Y. Ishida, B. Donnadieu and G. Bertrand, *Proc. Natl. Acad. Sci. U. S. A*, 2006, **103**, 13585–13588.
194. K. E. Krahulic, G. D. Enright, M. Parvez and R. Roesler, *J. Am. Chem. Soc.*, 2005, **127**, 4142–4143.
195. C. Präsang, B. Donnadieu and G. Bertrand, *J. Am. Chem. Soc.*, 2005, **127**, 10182–10183.
196. T. D. Forster, K. E. Krahulic, H. M. Tuononen, R. McDonald, M. Parvez and R. Roesler, *Angew. Chem., Int. Ed.*, 2006, **45**, 6356–6359.
197. P. P. Power, *Appl. Organomet. Chem.*, 2005, **19**, 488–493.
198. M. Iglesias, D. J. Beetstra, A. Stasch, P. N. Horton, M. B. Hursthouse, S. J. Coles, K. J. Cavell, A. Dervisi and I. A. Fallis, *Organometallics*, 2007, **26**, 4800–4809.
199. C. C. Scarborough, B. V. Popp, I. A. Guzei and S. S. Stahl, *J. Organomet. Chem.*, 2005, **690**, 6143–6155.
200. C. C. Scarborough, I. A. Guzei and S. S. Stahl, *Dalton Trans.*, 2009, 2284–2284.

201. V. César, N. Lugan and G. Lavigne, *J. Am. Chem. Soc.*, 2008, **130**, 11286–11287.
202. D. M. Kramov, E. L. Rosen, V. M. Lynch and C. W. Bielawski, *Angew. Chem., Int. Ed.*, 2008, **47**, 2267–2270.
203. (a) U. Siemeling, C. Färber and C. Bruhn, *Chem. Commun.*, 2009, 98–100; (b) U. Siemeling, C. Färber, M. Leibold, C. Bruhn, P. Mücke, R. F. Winter, B. Sarkar, M. von Hopffgarten and G. Frenking, *Eur. J. Inorg. Chem.*, 2009, 4607–4612.
204. (a) H. Jacobsen, A. Correa, C. Constabile and L. Cavallo, *Coord. Chem. Rev.*, 2009, **253**, 687–703; (b) S. Díez-González and S. P. Nolan, *Coord. Chem. Rev.*, 2007, **251**, 874–883.
205. H. G. Raubenheimer and S. Cronje, *Dalton. Trans.*, 2008, 1265–1272.
206. (a) P. L. Arnold and S. Pearson, *Coord. Chem. Rev.*, 2007, **251**, 596–609; (b) M. Albrecht, *Chem. Commun.*, 2008, 3601–3610; (c) E. Aldeco-Perez, A. J. Rosenthal, B. Donnadieu, P. Parameswaran, G. Frenking and G. Bertrand, *Science*, 2009, **326**, 556–559.
207. V. Lavallo and R. H. Grubbs, *Science*, 2009, **326**, 559–562.

CHAPTER 2

Computational Studies on the Reactivity of Transition Metal Complexes Featuring N-Heterocyclic Carbene Ligands

L. JONAS L. HÄLLER, STUART A. MACGREGOR* AND JULIEN A. PANETIER

School of Engineering and Physical Sciences, Perkin Building, Heriot–Watt University, Edinburgh EH14 4AS, UK

2.1 Introduction

The isolation and characterisation of the first stable N-heterocyclic carbene (NHC) by Arduengo *et al.* in 1991[1] sparked an explosion in the use of these species as ligands for transition metals (TMs) and in the following 20 years NHCs have become ubiquitous in organometallic chemistry and homogeneous catalysis.[2] Interestingly, this period has coincided with the rapid growth in the use of computational methods to study the reactivity of transition metal systems. This is largely due to the development of density functional theory (DFT) as a practical tool for the study of reaction mechanisms,[3] coupled with improvements in computational algorithms for the location and characterisation of reactive species (in particular transition states) and the availability of ever more powerful computing hardware. A symbiosis now exists, with computational chemistry being used routinely to characterise and understand the structure and reactivity of TM systems, while the possibility of modelling larger and more intricate experimental problems is constantly provoking theory to rise to the challenge of these more technically demanding scenarios.

RSC Catalysis Series No. 6
N-Heterocyclic Carbenes: From Laboratory Curiosities to Efficient Synthetic Tools
Edited by Silvia Díez-González
© Royal Society of Chemistry 2011
Published by the Royal Society of Chemistry, www.rsc.org

This Chapter will focus on computational studies on the reactivity of NHC–TM systems and aspects of the bonding in NHC–TM complexes will only be discussed when relevant to reactivity. As mentioned above, the use of DFT dominates this field and, unless directly pertinent for the discussion, no mention of the specific methodology employed will be given. The original papers should be consulted for full details. Equally, the energies reported here (*i.e.* simple electronic energies, zero-point energy corrected enthalpies or free energies) reflect the choice made by the authors of the original studies. In general, only the model systems employed in the computational studies will be discussed; the original papers should again be consulted for the full experimental details.

2.2 Non-innocent Behaviour of NHCs at TM Centres

The popularity of NHCs as ligands in transition metal catalysis is founded on their ability to form strong bonds to metal centres, a feature attributed to powerful σ-donation from the C2 position. In addition, NHCs offer variable steric and electronic properties that promise control of the metal coordination environment and in this respect they are similar to phosphines. These aspects have been reviewed[4] and are discussed in Chapter 1. One further advantage is that NHCs were perceived to be less prone to ligand-based decomposition reactions. However, this view had to be modified to some extent by the observation of a number of processes where the NHC ligand was directly involved, including ligand loss due to unexpected lability or *via* imidazolium formation. NHC ligands may also be modified by reactions at the metal coordination sphere, such as cyclometalation or migratory insertion. Computational studies on these NHC-based reactions are reviewed in this Section.

2.2.1 Reductive Elimination/Oxidative Addition

The reductive elimination of imidazolium species represents a major catalyst deactivation pathway for NHC–TM complexes. This phenomenon (along with its reverse process, oxidative addition) was first reported by Cavell[5] who, in collaboration with Yates, published a series of computational studies on this subject. The first of these considered loss of 1,2,3-trimethylimidazolium from $[(IMe)Pd(Me)(PR_3)_2]^+$ (R = H, OPh, Scheme 2.1), with calculations indicating that this proceeded *via* a concerted three-centred transition state consistent with a reductive elimination step (as opposed to an alternative process involving migration of the Me ligand onto the NHC ligand).[6] With R = H, the computed barrier was $+22.3\,\text{kcal mol}^{-1}$ and reductive elimination was exothermic by $3.7\,\text{kcal mol}^{-1}$. The process became even more favourable with the larger model (R = OPh: $\Delta E^{\ddagger} = +14.1\,\text{kcal mol}^{-1}$, $\Delta E = -9.2\,\text{kcal mol}^{-1}$). The calculated barrier in this instance was somewhat lower than the observed activation enthalpy of $+23.5 \pm 1.3\,\text{kcal mol}^{-1}$, although the experimental systems did employ a ^{Me}IMe ligand. Alternative mechanisms based on reductive

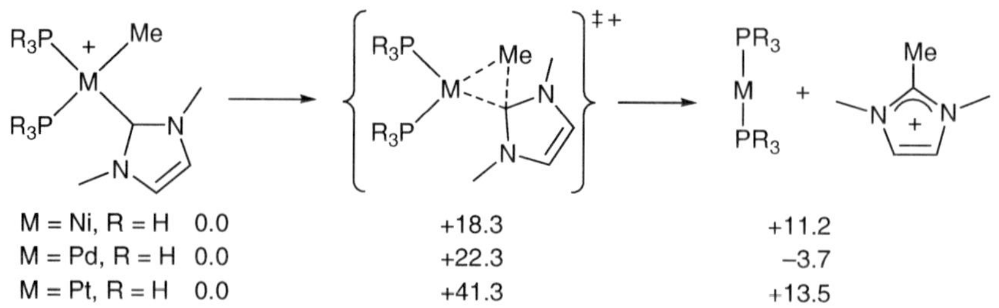

M = Ni, R = H 0.0	+18.3	+11.2
M = Pd, R = H 0.0	+22.3	−3.7
M = Pt, R = H 0.0	+41.3	+13.5

Scheme 2.1 Reductive elimination of 1,2,3-trimethylimidazolium from $[(IMe)M(Me)(PR_3)_2]^+$. Computed energetics indicated in kcal mol^{-1}.

elimination from 3-coordinate intermediates formed *via* initial phosphine dissociation were much less accessible ($\Delta E^{\ddagger} = +35.1$ kcal mol^{-1}), consistent with the first-order kinetics determined experimentally.

Further calculations on a $[(IMe)Pd(Me)(PMe_3)_2]^+$ model considered the factors controlling reductive elimination in more detail. Increased bite angles between the spectator PMe_3 ligands promoted the reaction and gave the lowest activation barriers and the greatest exothermicity.[7] The use of small bite angle chelate ligands may therefore confer greater stability. The orientation of the NHC ligand relative to the metal coordination plane had less effect on the activation barrier, although the 4-coordinate reactant was strongly destabilized when the NHC lay near to the coordination plane, such that reductive elimination became strongly favoured thermodynamically. *N*-Substituents also affected the kinetic stability and for these systems the trend with different substituents on the nitrogen was R = Cl < H < Ph < Me < Cy < *i*-Pr < CH_2*t*-Bu < *t*-Bu.[8] The larger barriers obtained with bulky alkyl substituents reflected a greater electron donating capacity, rather than steric effects. The latter were less important due to the perpendicular orientation of the NHC in the transition state. Conversely, electron-withdrawing substituents promoted lower barriers. An interesting effect was observed with aryl substituents. When R = Ph, delocalization of electron density away from the NHC pπ orbital reduced the activation barrier. However, this effect was lost with *ortho*-substituted aryls, as these were forced to lie roughly perpendicular to the imidazol-2-ylidene ring, disrupting the electron delocalization effect.

The use of pincer ligands could also suppress imidazolium loss, as in $[(CNC)Pd(Me)]^+$ complexes **1** and **2** which required prolonged heating at 140 °C before undergoing decomposition to Pd black (Figure 2.1).[9] Calculations confirmed that imidazolium formation in these systems entailed high activation energies of +44.2 and +35.0 kcal mol^{-1}, respectively. The more sterically encumbered species **3** had a lower computed barrier of +30.0 kcal mol^{-1}, consistent with its instantaneous decomposition at 140 °C.

The realisation that imidazolium loss involved reductive elimination suggested ways to manipulate this process to promote the reverse reaction, the oxidative addition of imidazolium salts. Platinum(0) is well known to promote oxidative addition compared to palladium(0) and, accordingly, $[(IMe)Pt(Me)(PH_3)_2]^+$ was found to be +13.5 kcal mol^{-1} more stable than $[Pt(PH_3)_2]$ and 1,2,3-trimethylimidazolium as well as very kinetically stable, the barrier for reductive elimination

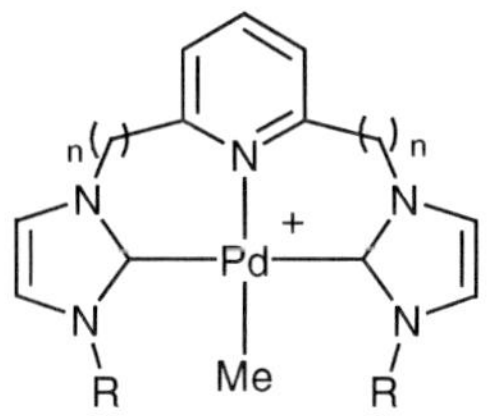

1 ($n = 0$, R = Me)
2 ($n = 1$, R = Mc)
3 ($n = 1$, R = *t*-Bu)

Figure 2.1 Cationic Pd–CNC pincer complexes studied by Nielsen *et al.*[9]

being $+41.3\,\text{kcal mol}^{-1}$ (see Scheme 2.1).[10] Further calculations showed that the analogous nickel(II) species was almost as stable thermodynamically although kinetically less so.[11] As suggested by earlier studies, the use of small bite angle chelate ligands also favoured the metal(II) species. Thus, C–C oxidative addition could be rendered favourable even with palladium(0), as in the reaction of 1,2,3-trimethylimidazolium with [Pd(dpe)] (dpe = 1,2-diphosphinoethane: $\Delta E^{\ddagger} = +7.3\,\text{kcal mol}^{-1}$, $\Delta E = -15.8\,\text{kcal mol}^{-1}$). As expected, oxidative addition was easier with C–H rather than C–C bonds and activation of $[\text{IMe·H}]^+$ at [Pd(dpe)] occurred without any energy barrier to give a very stable palladium(II) hydride ($\Delta E = -27.5\,\text{kcal mol}^{-1}$). C–Br activation was also found to be barrierless and even more exothermic ($\Delta E = -65.1\,\text{kcal mol}^{-1}$). The C–H activation of $[\text{IMe·H}]^+$ could be further promoted by the use of electron donating substituents on the phosphine.

The ease of imidazolium C2–H activation and, under favourable circumstances, C2–Me activation suggested that this may be exploited as a general route to the synthesis of NHC complexes and a number of computational studies considered such processes. One early example was the C–H activation of $[\text{IMe·H}]^+$ at models of Wilkinson's catalyst, $[\text{RhCl(PR}_3)_3]$ (R = H, Me), although in this case the pseudo-octahedral hydrido-NHC products were thermodynamically unstable.[12] With Ir, however, this activation was feasible. The reactions of bis-imidazolium salts with $[\text{IrCl}_2(\text{COD})]_2$ dimers (COD = cyclooctadiene) required the use of weak bases, due to the presence of vulnerable acidic sites in the linker groups.[13] NHC–Ir bond formation may occur either *via* initial deprotonation of imidazolium or *via* oxidative addition at the [IrCl(COD)] moiety to form an iridium hydride which would be subsequently deprotonated by an external base (Scheme 2.2a). These two processes were computed to be close in energy when simple amines were employed as the external base. Activation of the second imidazolium presented a similar issue and in this case the oxidative addition/deprotonation route was favoured. A double deprotonation of the bis-imidazolium species prior to coordination was computed to be too high in energy. The final products observed experimentally depended on the linker group and only with $n = 3$ was reductive elimination of HCl kinetically accessible to yield the 4-coordinate product **5** (Scheme 2.2b). A related study considered the reactivity of C2–Me substituted bis-imidazolium salts at an $[\text{IrCl}_2(\text{Cp})]$ fragment.[14] In this case the first metalation occurred at the sp^3 C–H bond of one of the 2-Me groups. For the second metalation, activation at the C5 position was kinetically preferred to give a product with one abnormal carbene ligand.

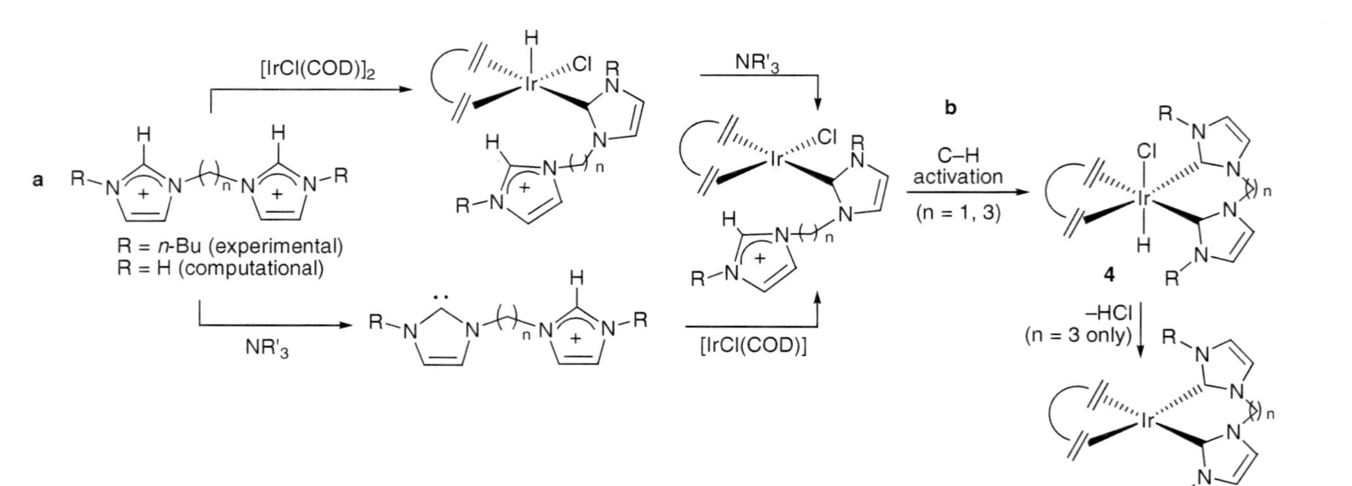

Scheme 2.2 Formation of the **4** and **5** in the presence of amines.

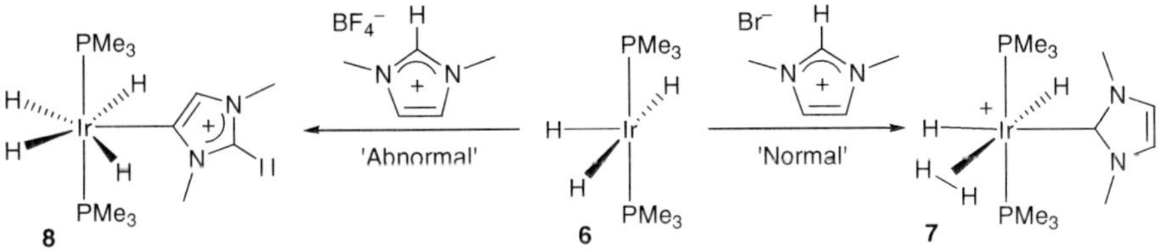

Scheme 2.3 Anion-dependent 'normal' and 'abnormal' activation of imidazolium salts with **6**.

The reactions between 2-pyridylmethyl imidazolium salts and $[Ir(H)_5(PPh_3)_2]$ offered further possibilities that were dependent on the anion employed. With Br^-, activation at the 'normal' 2-position was favoured, but metallation of the 'abnormal' 5-position was observed with BF_4^-.[15] Model calculations for the reaction of $[IMe{\cdot}H]^+$ at the $[Ir(H)_3(PMe_3)_2]$ intermediate indicated that activation at the C2 position proceeded *via* C–H bond heterolysis to give a η^2-H_2 intermediate **7** (Scheme 2.3). In contrast, C5–H activation produced an iridium(V) tetrahydride **8**, possibly due to the C5 position being a stronger donor and so able to stabilise this higher oxidation state. The observed preference for C2 activation with Br^- salts arose from the high positive charge associated with the transferring hydrogen and the ability of the smaller Br^- anion to access this site and stabilise the transition state. Additional calculations on the full product geometries revealed an intrinsic preference for the 'normal' C2 NHC, although this difference was reduced in the ion pairs due to particularly strong H-bonding to the exposed C2–H bond in the 'abnormal' C5 form. Nevertheless, the C2 NHC complexes remained thermodynamically more stable and therefore the observation of the C5-bound species must be kinetic in origin.

A novel synthetic route to NHC complexes species involving C–C activation of imidazolium carboxylates was also characterised computationally.[16] Reaction with the [RhCl(COD)] fragment required access to two vacant sites at the metal and this could be achieved by displacement of either one alkene moiety or Cl^- dissociation. Alkene dissociation was a very high-energy process (+47 kcal mol^{-1}), whereas in more polar solvents Cl^- loss to give the *O,O*-bound adduct **9** was shown to be feasible (Equation (2.1)). This route was also consistent with experiments showing that added Cl^- retards the reaction. The subsequent C–C cleavage had an activation barrier of +25.5 kcal mol^{-1}, with the overall reaction being driven by the ultimate release of a CO_2 molecule.

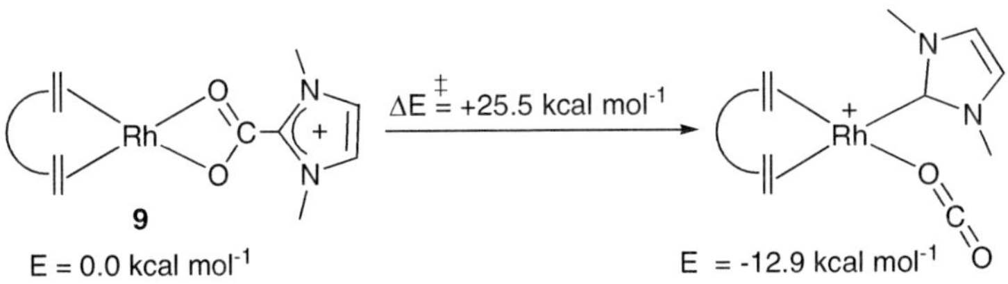

$\Delta E^{\ddagger}$ = +25.5 kcal mol^{-1}

9 E = 0.0 kcal mol^{-1} → E = -12.9 kcal mol^{-1} (2.1)

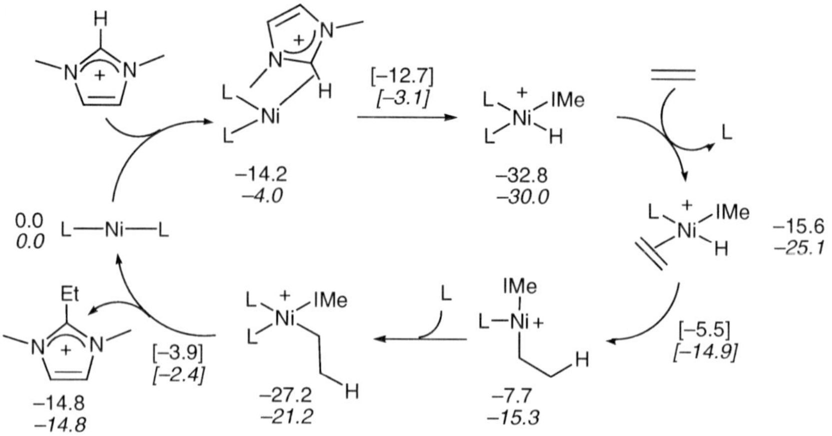

Scheme 2.4 Computed catalytic cycle for the coupling of $[IMe{\cdot}H]^+$ with ethene. Energies (kcal mol^{-1}) are relative to the reactants set to zero in each case with values in plain text for L = IMe and in italics for L = PMe_3. Values in square brackets refer to computed transition state energies.

Cavell exploited this imidazolium C–H activation/C–C reductive elimination chemistry for the coupling of imidazolium salts with ethene (Scheme 2.4).[17] Calculations on the reaction of $[IMe{\cdot}H]^+$ with $[NiL_2]$ species allowed for direct comparison for L = IMe cf. PMe_3. C–H activation in these systems was highly exothermic ($\Delta E > 30$ kcal mol^{-1}) and proceeded *via* an initial σ-bound precursor with minimal barriers. Ethene insertion then occurred through a 4-coordinate species, substitution reaction being far more accessible when L = PMe_3. The prior L/C_2H_2 barriers for migratory insertion were very similar for both models and, after re-coordination of L (now favoured for IMe), reductive elimination released the 2-ethylimidazolium product. The barrier for this final step was larger for the small, highly electron donating IMe ligand. The concepts underpinning these processes also play a central role in the direct functionalisation of nitrogen heterocycles (see Section 2.5). Similar behaviour has been observed in a $[(PCP)Ni(H)]^+$ pincer system, which upon exposure to ethene forms a nickel(0) species in which the final imidazolium moiety is held in close contact with the metal centre.[18]

A related process was characterised for the exchange of an allyl with an aryl group at the C2 position of an imidazolium cation and involved the reaction of $[(^{Me}IMe)Pd(\eta^3\text{-}C_3H_5)(PMe_3)]^+$ with phenyl iodide. In this case, transmetalation between a NHC–Pd and a Pd-iodo intermediate and the C–C reductive elimination both had similar activation energies of around +20 kcal mol^{-1}.[19]

2.2.2 Migratory Insertion

Studies on [(CNC)Pd] systems revealed another class of NHC-based reactivity. Mixing the free CNC ligand with $[Pd(Me)_2(tmeda)]$ (tmeda = *N*,*N*,*N*′,*N*′-tetramethylethylenediamine) at −78 °C yielded **11** in which one methyl group had migrated onto an adjacent NHC C2 (Equation (2.2)).[20] Calculations suggested that **11** was formed from an intermediate **10** *via* transfer of a Me group from Pd to the NHC C2 position. This step should be feasible as it entailed a computed barrier of only $+7.8\,\text{kcal mol}^{-1}$ and was extremely exothermic ($\Delta E = -32.2\,\text{kcal mol}^{-1}$). Unlike the earlier examples where C–C bond formation involved reductive elimination, in this case, no change in the formal metal oxidation state is implied. Such processes have therefore been described as a migratory insertion.

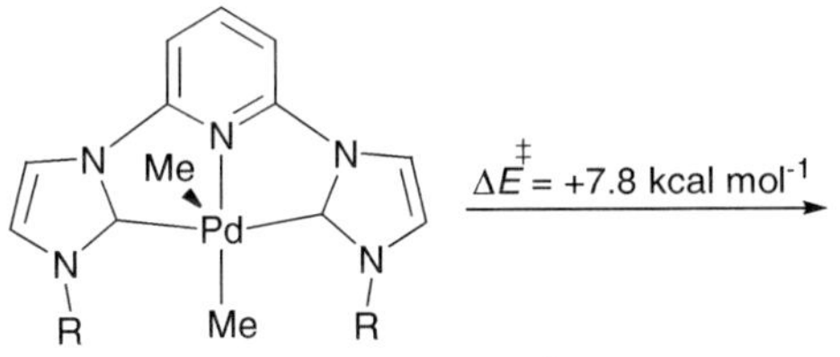

$\Delta E^{\ddagger} = +7.8$ kcal mol^{-1}

10 (E = 0.0 kcal mol^{-1})
R = 2,6-(i-Pr)$_2$C$_6$H$_3$ (experimental)
R = H (computational)

11 (E = −32.2 kcal mol^{-1})

(2.2)

Further examples of such migratory insertion processes were subsequently studied computationally. Kirchner showed that the formation of a stable allyl-alkylidene **13** derived from the metalacyclopentatriene species **12** *via* intramolecular attack of the NHC ligand (Scheme 2.5).[21] The migratory insertion of an alkyne into a NHC–Ru bond was also suggested to account for the formation of **16**.[22,23] This reaction, for which there was no precedent in phosphine chemistry, involved addition of alkyne to **12** to give **14** and the subsequent facile insertion to give a metalacyclopropene **15**. The alkyne addition step dominated the energetics of this process and was controlled by steric bulk of the *N*-substituent on the NHC ligand: $\Delta G^{\ddagger} = +18.7\,\text{kcal mol}^{-1}$

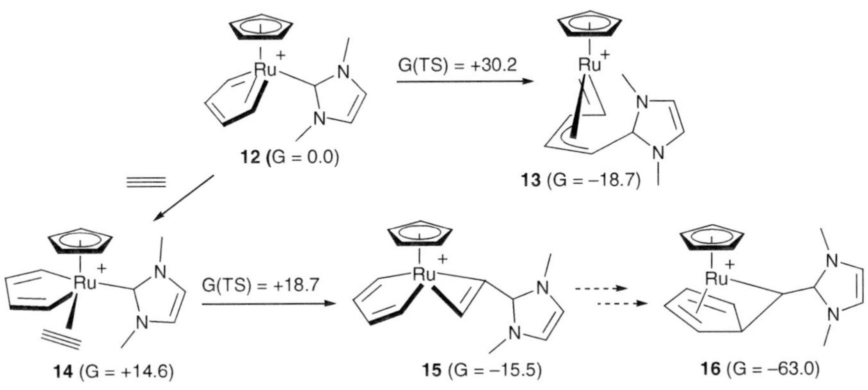

12 (G = 0.0) — G(TS) = +30.2 → **13** (G = −18.7)

14 (G = +14.6) — G(TS) = +18.7 → **15** (G = −15.5) ⇢ **16** (G = −63.0)

Scheme 2.5 Reactions of metalacyclopentatriene complex **12**. Relative energies in kcal mol^{-1}.

for IMe and +35.8 kcal mol^{-1} for IPh. C–C bond coupling in **15** and subsequent insertion of the alkene moiety into the metal-vinyl unit yielded the very stable final observed product ($G = -63.0$ kcal mol^{-1}).

Nolan and co-workers also showed that alkenes could react with NHCs in what appeared to be a net migratory insertion, although in this case an intermolecular attack was proposed.[24] Reaction of diene adduct **17** with an excess of IPr led to **18** (Equation (2.3)). Calculations (R = H) showed that **18** formed *via* initial addition of an NHC ligand to the Pt centre followed by intermolecular attack of a second NHC moiety. The intermolecular attack occurred with an activation barrier of +7.6 kcal mol^{-1}, whereas intramolecular migratory insertion involving the Pt-bound NHC would entail a barrier of +40 kcal mol^{-1}.

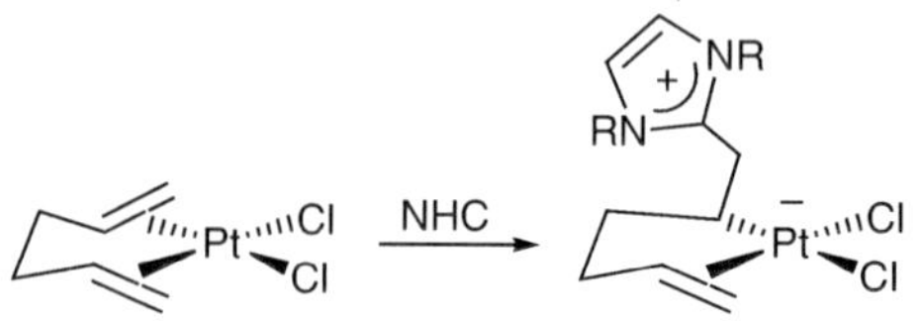

(2.3)

2.2.3 Reactivity at the *N*-Substituents

The most common process involving the *N*-substituents of NHCs is C–H activation to give cyclometalated species. Such processes may not necessarily be deleterious for catalysis, as they can be reversed by the addition of H_2. Indeed, in some cases the cyclometalated species were directly involved in mediating H_2 transfer.[25] Other reactions involving C–C and C–N activation of the *N*-substituents tend to have more drastic implications for catalyst integrity. However, in some cases these reactions involve novel processes that are of interest in their own right. A number of studies considered the decomposition reactions of Grubbs' alkene metathesis catalysts and these will be considered in more detail in Section 2.4.

One well characterised example of C–H activation was observed with [(IMes)Ru(H)$_2$(CO)(PPh$_3$)$_2$], **19**. Dehydrogenation with a sacrificial alkene at room temperature produced a ruthenium(0) intermediate (**20**) which underwent C–H activation of one *ortho*-Me group of the IMes (**21**) (Scheme 2.6).[26] The C–H activation step had a computed barrier of +12.1 kcal mol^{-1} and was exothermic by 11.2 kcal mol^{-1}.[27] In contrast, prolonged heating of the bis-IMes species, **22**, at 110 °C led to a highly unusual cleavage of an unstrained C(aryl)–C(Me) bond to form **23**.

Calculations showed that formation of the bis-IMes system was necessary for this C–C activation to take place, as the increased steric encumbrance allowed the formation of the highly reactive 14e intermediate [(IMes)$_2$Ru(CO)] **24**. This species could undergo reversible C–H activation to **25**, but C–C activation was also accessible under the reaction conditions to form **26**. This species was then trapped out by H_2 and PPh$_3$ to produce the thermodynamically stable **23** (Scheme 2.7). The activation of other *ortho*-C–X bonds (X = H, Me, F, OH,

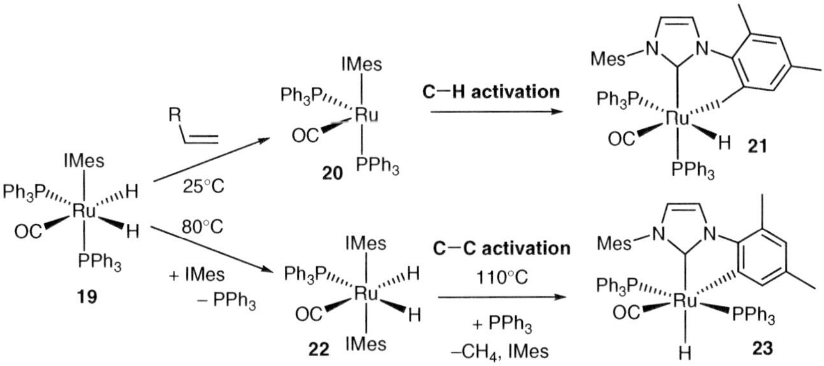

Scheme 2.6 C–H and C–C activation reactions of $[(IMes)Ru(H)_2(CO)(PPh_3)_2]$ **19**.

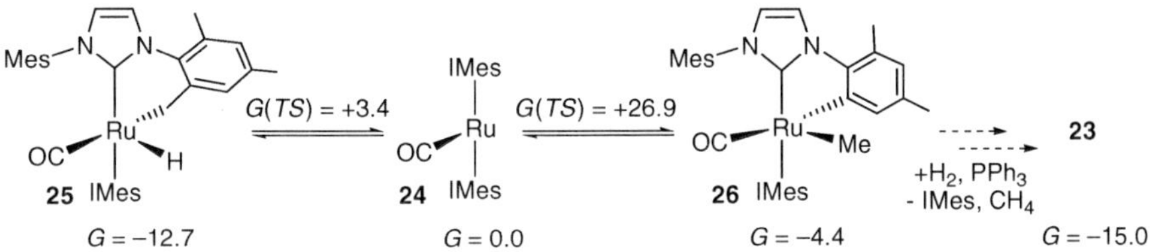

Scheme 2.7 Competing C–H and C–C activation in **24**. Relative free energies in $kcal\,mol^{-1}$.

NH_2, OMe, CF_3) at model $[(NHC)Ru(CO)(PH_3)_2]$ species was also considered computationally.[28] The highest activation barrier was computed for X = Me, suggesting that activation of all these C–X bonds might be feasible if suitable reactive ruthenium(0) species could be accessed. Generally, the computed barriers and reaction exothermicities correlated with the strength of the Ru–X bond formed in the product, although the ability of certain X groups (NH_2, OR) to stabilise the ruthenium(0) species produced less favourable energetics.

C–H activation of *N*-alkyl substituents has also been studied computationally, an early example being an assessment of the electronic structure of the double C–H activated species formed with ItBu at Rh and Ir metal centres. These 14e species have novel bent 'see-saw' geometries and it has been proposed that the NHC ligands are acting as π-donors in these systems.[29] Additionally, C–H activation in agostically stabilised **27** occurred in the presence of strong bases to give cyclometalated **28** (Equation (2.4)). Labelling studies showed that C–H activation must involve deprotonation of the methyl group of the *iso*-propyl substituent. This was thought to most likely involve deprotonation of the agostic C–H bond; however calculation showed that the proton geminal to the agostic interaction in fact was much more acidic and that deprotonation of this site by a strong base can readily occur.[30] This result is significant as many NHC–TM complexes are prepared by phosphine substitution. As this often requires excess NHC, such systems may be vulnerable to cyclometalation.

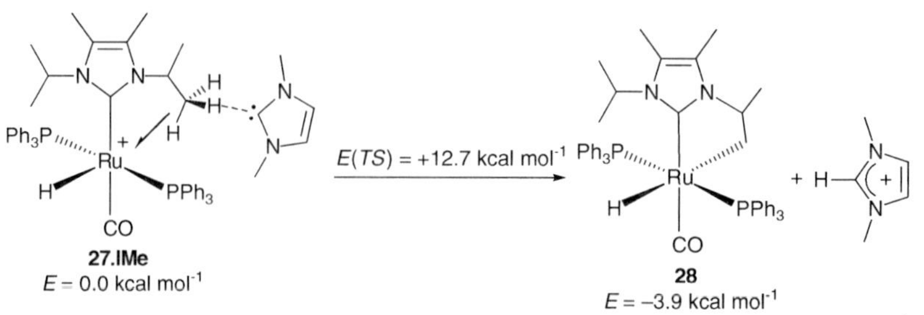

(2.4)

The double C–H activation of a single *N*-Me substituent was also observed in the thermolysis of [(IMe)$M_3(CO)_{11}$] clusters **29**.[31] After initial CO loss, a first C–H bond activation proceeded through a transition state where the transferring hydrogen has contacts with two metal centres. For M = Ru, this process ($\Delta E^{\ddagger} = +19.6\,\text{kcal mol}^{-1}$) produced a new intermediate in which one Ru–Ru connectivity was bridged by a hydrogen and a cyclometalated IMe ligand. After further loss of CO, a second C–H activation occurred *via* a more conventional single-centred oxidative addition transition state ($\Delta E^{\ddagger} = +30.3\,\text{kcal mol}^{-1}$) to give the final observed product **30** (Equation (2.5)). The Os analogue behaved similarly, albeit with slightly higher barriers consistent with the more forcing conditions required experimentally.

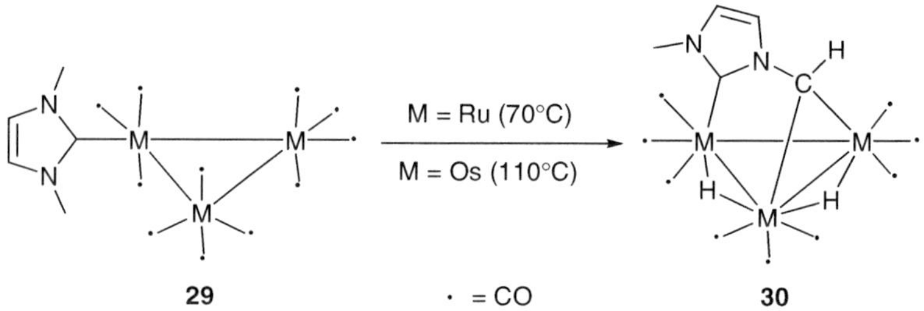

(2.5)

Such cyclometalation reactions are not confined to the late transition metals. For example, reaction of a tridentate NCN ligand with [$TaCl_2(CH_2Ph)_3$] gave an unexpected metallaaziridine **33** (Scheme 2.8).[32] Starting from a model intermediate **31**, calculations assessed two possible pathways: (i) a concerted σ-bond metathesis involving direct loss of toluene and; (ii) α-H migration from a benzyl group resulting in loss of toluene to form an alkylidene intermediate **32**, followed by H transfer from the chelate arm onto the alkylidene ligand. This latter pathway was found to be more accessible, although it had a rather high activation barrier of $+36.4\,\text{kcal mol}^{-1}$, given that the reaction occured at $-30\,^{\circ}$C. However, this may reflect the greatly reduced steric bulk of the model used in the calculations and the proposed mechanism was certainly consistent with labelling studies using a [$TaCl_2(CD_2Ph)_3$] precursor.

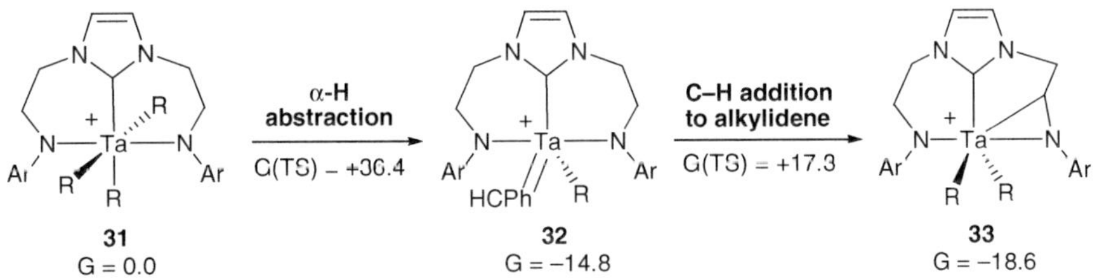

α-H abstraction
G(TS) = +36.4
C–H addition to alkylidene
G(TS) = +17.3

31
G = 0.0

32
G = −14.8

33
G = −18.6

Scheme 2.8 C–H activation in **31** to give metalaaziridine **33** experimental: R = CH_2Ph, Ar = Mes; computational: R = Me, Ar = H. Computed relative free energies given in kcal mol^{-1}.

2.3 Binding and Activation of Small Molecules at NHC–TM Complexes

The oxidative stability of NHCs means that they present certain advantages over phosphines and may therefore play an important role in the development of catalysis incorporating the environmentally friendly O_2 as oxidant. In this context, palladium complexes have been particularly closely studied.

The addition of triplet O_2 to $[PdL_2]$ species was calculated to be favourable for L = IMe ($\Delta G = -5.7$ kcal mol^{-1}), but endergonic with L = PMe_3 ($\Delta G = +2.3$ kcal mol^{-1}).[33] This difference was thought to reflect stronger donation from the NHC ligand, which enhanced charge transfer to the η^2-O_2 moiety. Solvent effects were also important in stabilising the adduct, which had a significant dipole moment. O_2 could be removed from this system, but only under rather forcing conditions (80 °C under vacuum for 1 week). The mechanism of O_2 addition involved two sequential single electron transfer steps: an η^1-O_2 adduct was first formed bound to a *trans*-L–Pd–L unit ($G = +0.7$ kcal mol^{-1}) which then isomerised to a *cis* form which was formulated as containing a superoxide ligand, *i.e.* $[L_2Pd^I$-$(\eta^1$-$O_2^-)]$ ($G = +6.7$ kcal mol^{-1}) (Scheme 2.9). The reaction was completed *via* a readily accessible triplet–singlet minimum energy crossing point (MECP) to give the symmetrically bound singlet η^2-O_2 adduct. The O–O distance of 1.43 Å in this species was consistent with the presence of a peroxide ligand.

In *trans*-$[(NHC)_2RhCl(\eta^2$-$O_2)]$ (NHC = IMes, IPr) species the O_2 ligand exhibited very short O–O distances in the range 1.267(13)–1.315(3) Å and a ν_{O-O} stretching frequency above 1000 cm^{-1}.[34] Analysis of the computed electronic structure, coupled with X-ray absorption spectroscopy suggested Rh was present in the +1 oxidation state in these species, which were interpreted as adducts of a *singlet* O_2 ligand. In a separate study, the $[(^{Me}IiPr)_4Ru(H)(\eta^2$-$O_2)]^+$ cation ^{1}**35** was shown to have similar properties. In this case, however, the mechanism of O_2 addition was computed and revealed that O_2 addition to $[(^{Me}IiPr)_4Ru(H)]^+$ **34**, formed a ruthenium(III)-(η^1-O_2^-) triplet intermediate ^{3}**35a** (Figure 2.2).[35] A second triplet form of this η^1-O_2 adduct was characterised ^{3}**35b** from which the triplet-singlet MECP could be accessed to give the final singlet η^2-O_2 adduct, ^{1}**35**. Importantly, a natural population analysis revealed the dioxygen ligand was *more* reduced in ^{1}**35** than in the superoxo intermediates

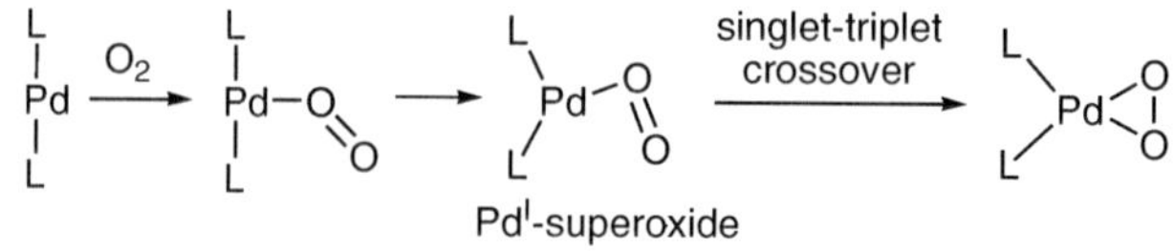

Scheme 2.9 Addition of dioxygen to PdL_2 species (L = PMe_3, IMe).

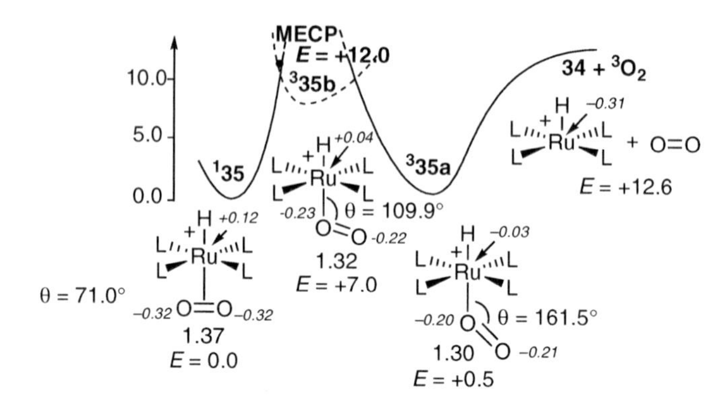

Figure 2.2 Computed energy profile (kcal mol^{-1}) for the addition of $^3O_2$ to **34** with potential energy curves for variation of the Ru–O–O angle, θ, in ^{3}**35a**, ^{3}**35b** and ^{1}**35** (L = ^{Me}IiPr). Values for the Ru–O–O angle and the O–O distances (Å, degrees, plain text) and natural atomic charges at Ru and O (italics) also shown. Reproduced, with permission, from Burling *et al.*[36]

^{3}**35a** and ^{3}**35b**. In this case, the description of the O_2 moiety as a singlet O_2 ligand did not seem appropriate. Instead the quandary presented by the range of $L_nM(\eta^2\text{-}O_2)$ geometries in the literature appears analogous to that seen in metal–alkene ↔ metallacyclopropane spectrum.

The free energy of O_2 binding in ^{1}**35** was close to zero in these Ru systems and experimentally O_2 binding was truly reversible, the adduct readily forming at − 80 °C with O_2 being removed by simple freeze–pump–thaw cycles. The size of the NHC ligand was important in defining the correct pocket for O_2 binding in these systems: further experimental data and calculations showed that small molecule binding energies (O_2, CO, H_2 and N_2) for $[(NHC)_4Ru(H)]^+$ cations followed the trend NHC = ^{Me}IMe ≫ ^{Me}IEt > ^{Me}IiPr.[36]

The design of new ligands for the catalytic reduction of N_2 to NH_3 in Mo complexes was considered by Schenk and Rieher by computing the energetics of each reduction step in the cycle.[37] Replacing the tris-amido chelate of the original Schrock system with a related tris-NHC ligand as in **36** gave promising thermodynamics for these steps (Figure 2.3). However, the crucial NH_3/N_2 substitution reaction was very endothermic, mainly as the neutral NHC-based ligand resulted in a 3 + charge on the complex. The strongly Lewis acidic metal centre therefore favoured NH_3 binding, but was unable to support the N_2 ligand through π back-donation. A triply reduced tris-NHC analogue,

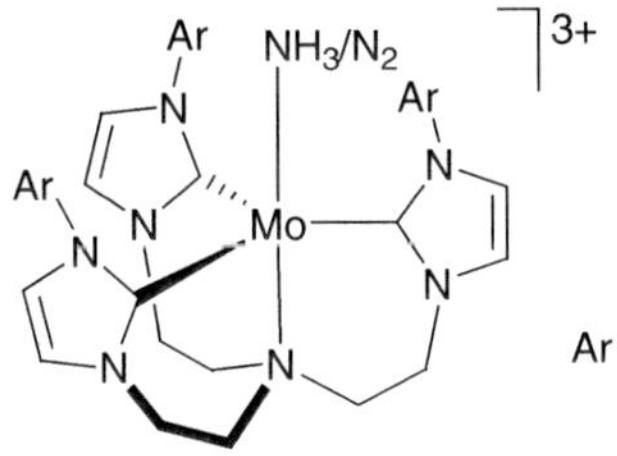

Figure 2.3 A molybdenum(III) tris-NHC chelate system for N_2 reduction.

however, showed promise as the NH_3/N_2 substitution step was calculated to become exothermic.

Reactions involving the cleavage of small molecules have also been studied. Calculations on H_2, CH_4 and C_2H_6 activation at $[(NHC)_{3-n}Ru(CO)(PH_3)_n]$ ($n = 1$–3) species showed that PH_3/IH substitution (IH = imidazol-2-ylidene) at the axial sites of these species had a minimal effect on the reaction energetics.[38] PH_3/IH substitution did create a more electron-rich metal centre, as gauged by computed ν_{CO} frequencies. However, analysis showed that the metal-based σ-acceptor orbital was also pushed up in energy, reducing Lewis acidity. PH_3/IMe substitution actually reduced the ability of the metal centre to effect oxidative addition, an observation attributed to steric effects. A full treatment of the IMes ligand also proved important in obtaining realistic energetics for H–X activation (X = OH, F, SH) at $[(IMes)_2RuH_2(CO)]$ species.[39] These reactions were initiated at a $[(IMes)_2RuH_2(HX)(CO)]$ adduct *via* proton transfer directly onto a *cis*-hydride ligand. H_2 loss and isomerisation yields the 16e $[(IMes)_2RuH(X)(CO)]$ products observed experimentally. Calculations with a range of HX species showed the ease of H transfer reflected the pK_a of the HX moiety.

2.4 Ru-catalysed Alkene Metathesis

Alkene metathesis promoted by the second generation Grubbs' catalyst was among the first transition metal NHC systems to be addressed using computational methods[40] and remains a popular topic for modelling. This reflects the importance of this process in organic and materials chemistry, although interest in this system also derives from it being the most palpable example of enhanced performance upon substitution of a phosphine in **37** by an NHC in **38** (Scheme 2.10). From a computational point of view, obtaining a correct description of the initial phosphine dissociation step has also presented specific technical challenges. Early computational studies on small model systems (L = IH, SIH (imidazolin-2-ylidene), $PR_3 = PH_3$, PMe_3; R, R′ = H, Ph)[40,41] confirmed the general mechanism of alkene metathesis in Scheme 2.10 that had been derived from experimental studies,[42] including the dissociative nature of the initial phosphine/alkene substitution step. However, such models gave a poor description of the energetics of phosphine dissociation, and in particular failed to reproduce the reduced lability of the phosphine ligand observed experimentally

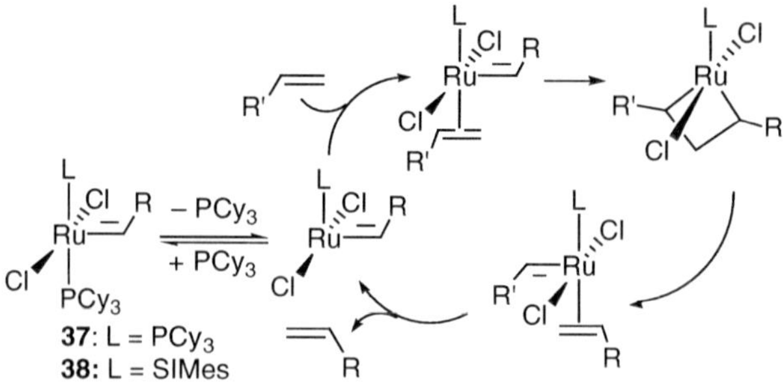

Scheme 2.10 Alkene metathesis catalysed by Grubbs' first- and second-generation catalysts.

in the second generation catalyst **38**. This reduce lability was puzzling, as the *stronger* σ-donor NHC ligands might be expected to weaken the *trans*-Ru–PR_3 bond. Several studies on the full experimental systems were attempted to address this situation, but, once the correct conformational minima had been located, these discrepancies remained.[41c]

The situation was ultimately resolved with the advent of the M06-L functional, developed by Zhao and Truhlar.[43] As in the earlier studies, calculations with a range of generalised gradient approximation (GGA), meta-GGA and hybrid functionals underestimated the PCy_3 dissociation energies and gave larger values for **37**. A key difference with the M06-L functional was that it included a treatment of medium range electron correlation, allowing dispersion effects to be taken into account. With this approach the experimental gas-phase phosphine dissociation enthalpies (+33.4 kcal mol^{-1} for **37** and +36.9 kcal mol^{-1} for **38**)[44] were reproduced to within experimental error, with the Ru–PCy_3 dissociation energy being +3.4 kcal mol^{-1} higher in **38**.[45] This difference could be linked directly to dispersion effects between the SIMes, PCy_3 and alkylidene ligands, derived from calculations on the optimised catalyst structure where the Ru and two Cl ligands had been removed. With the M06 functional this ligand interaction term was found to be stabilising, by 9.9 kcal mol^{-1} and 14.4 kcal mol^{-1} in **37** and **38**, respectively. In contrast, calculations with B3LYP resulted in repulsive interactions of 4.5 and 8.0 kcal mol^{-1}. Hence, weak inter-ligand attractive forces were responsible for the differences between the different generations of Grubbs' catalysts. Later studies, however, suggested that B3LYP calculated geometries in better agreement with experiment than the M06-L functional[46] and that this may be due to overestimation of attractive non-covalent interactions in M06-L. Calculating the geometry with B3LYP and the energy with an M06 functional could therefore be a good approach.

These series of investigations highlight several issues that need to be considered when undertaking computational work, especially when this involves modelling quantitative experimental data. As well as a consideration of a variety of different functional approaches, the large systems employed here presented significant conformational flexibility and steps must be taken to ensure the lowest energy forms are located. Geometry optimizations are run in the gas phase and so solvation effects are not directly taken into account. The impact of entropic terms is also likely to be overestimated in the gas phase.[47] For this reason the availability of experimental gas-phase bond dissociation enthalpies was particularly useful. It is likely that the issue of ligand dissociation in Grubbs' catalysts will remain a benchmark process against which new computational methodologies will be tested.[48]

Functional choice was also important is the alkene binding step.[49] M06 predicted alkene binding to the $[(SIMes)Ru(=CHPh)Cl_2]$ fragment to be highly exothermic, while B3LYP calculated it to be endothermic. Highly exothermic alkene binding was in agreement with experiment[44] and it was therefore concluded that the non-covalent attractive interactions described by M06 were essential.

Computational studies of the subsequent steps in the catalytic cycle showed metathesis to proceed through a migratory insertion to give a metalacyclobutane intermediate. Cleavage of the C–C bond associated with the original alkene formed the new alkylidene and alkene ligands, and the latter dissociated in the final step. These processes were computed to have low reaction barriers, especially for the second-generation catalysts and the origins of this were considered by Adlhart and Chen[50] and by Straub.[51] The migratory insertion step involved a planar 4-centred transition state and this required the alkylidene ligand to rotate such that it lay in the Cl–Ru–Cl plane and the alkene to be parallel to the Ru=CHR direction. The transition state for this process involved a decrease in the NHC–Ru–alkene angle, which produced an antibonding interaction between the NHC σ-bond and an occupied metal d-orbital (Figure 2.4). This interaction was alleviated by 4d/5p orbital mixing that served to enhance π back-bonding to the empty p-orbital on the alkylidene, thus stabilising the transition state. As phosphine ligands are weaker σ-donors but better π-acceptors than NHCs, the stabilising π-bonding effect should be greater in the second-generation species.

Lord *et al.*[52] studied *N*-substituent effects (R = H, Me, F) on the barriers for migratory insertion with both saturated and unsaturated NHCs. For the saturated species low barriers were computed to follow the trend R = F < Me < H. In contrast, the unsaturated species were both more sensitive to the *N*-substituent and showed the reverse trend (R = Me: $\Delta E^{\ddagger} = +2.4$ kcal mol^{-1}; R = F: $\Delta E^{\ddagger} = +7.6$ kcal mol^{-1}).

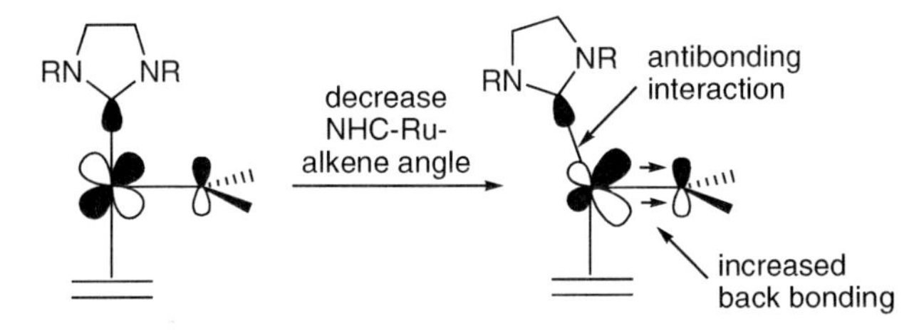

Figure 2.4 Key orbital interactions in the plane of migratory insertion during alkene metathesis with Grubbs' second-generation catalyst. Adapted from Straub.[51a]

A number of studies compared the reactivity of the standard form of Grubbs' second-generation catalyst with *trans* Cl ligands with that of the *cis* isomer.[49] Calculations by Goddard on the pyridyl-containing species **39** (Figure 2.5) showed that while the *trans* catalyst was more stable in the gas phase by $+6.8\,\mathrm{kcal\,mol^{-1}}$, the introduction of dichloromethane solvent favoured the *cis* form by $+0.7\,\mathrm{kcal\,mol^{-1}}$.[53] This result corresponded to a ratio of the intermediates of 76:24, in very good agreement with experiment (78:22). Correa and Cavallo[54] also showed that the *cis* form of the 5-coordinate catalysts could become favoured in dichloromethane. Increased steric bulk, however, acted against this trend and so the enantioselective catalyst **40** (Figure 2.5) was more stable as the *trans* isomer. Solvent effects also lowered the barriers for migratory insertion in the *cis* isomers; however, in this case the transition states with *trans* Cl ligands were always more accessible. Ultimately, the preferred reaction pathway will depend on the balance between solvent, electronic and steric effects in each case.

Costabile and Cavallo also addressed the origin of the enantioselectivity arising from the use of catalyst **40**.[55] Calculations located two diastereomeric intermediates, the *re* and *si* forms (Figure 2.5), and concluded that the rotation of the *N*-substituent away from the phenyl group on the backbone increased steric repulsion between the *N*-substituent and the *re* diastereomer. This was therefore destabilised relative to the *si* diastereomer and accounted for the high

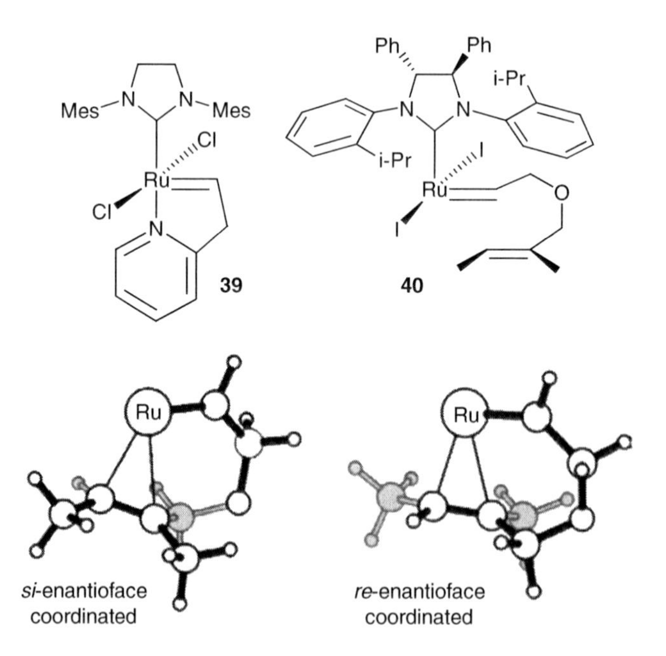

Figure 2.5 Variants of Grubbs' second-generation catalyst studied by computational methods. Enantioselective alkene binding modes with achiral triene. Reproduced, with permission, from Costabile and Cavallo.[55] © 2004 Americal Chemical Society.

enantioselectivity observed experimentally. A number of other papers considered asymmetric catalysis in a range of related systems.[56]

Although most computational work in this area focused on alkene substrates, the reactions of alkynes and enynes also received attention. Lippstreu and Straub considered the reactivity of ethyne at a [(SIMe)Ru(=CH_2)Cl_2(PMe_3)] model catalyst to generate a [L_nRu=CH–C_2H_3] vinylcarbene species.[57] This could then react with ethene to form butadiene and regenerate the original catalysts. An alternative polymerisation pathway where the vinylcarbene reacted with further ethyne was computed to be kinetically inaccessible. Ethyne actually bound stronger than ethene, but the rotation of ethyne to become aligned with the Ru–methylene axis was higher in energy than for ethene. Nolan and coworkers found similar trends in the initial steps of the reactions of 1,11-dien-6-ynes at 14e [Ru(=CH_2)Cl_2L] species (L = PCy_3, SIMes and IiPr), with binding of the C≡C triple bond moiety being favoured over the C=C double bond; however, as before, the insertion of the latter remained more accessible kinetically.[58]

Although the Grubbs' catalysts are generally stable and robust, understanding their decomposition pathways is essential for further improvement. van Rensburg *et al.* considered the possibility of β-hydrogen abstraction from the ruthenacyclobutane intermediate to generate propene and ultimately the deactivation of the catalyst (Scheme 2.11).[59] Using [Ru(=CH_2)Cl_2(C_2H_4)L] systems they computed the energy barrier for this reaction to be +16.9 kcal mol^{-1} for L = PCy_3 and +24.3 kcal mol^{-1} when L = IMes. The higher barrier for the second-generation Grubbs' catalysts was ascribed to a larger C_{NHC}–Ru–C^{α} angle of 118.9° compared with a P–Ru–C^{α} angle of 96.1°, the difference being due to the greater steric effect of the NHC ligand. Experimental evidence for such processes was seen in the formation of 1-butene when propene was used as a substrate. Trace amounts of 2-butene were also observed arising from the alternative regioselectivity in the migratory insertion step.

The susceptibility of Grubbs' second-generation catalysts to decomposition *via* activation of the pendant *N*-substituents was also considered. Cavallo[60] showed that addition of a π-acid such as CO *trans* to the methylene in [(SIMes)Ru(=CH_2)Cl_2(PMe_3)] could induce the methylene ligand to attack the *ipso* carbon of the mesityl group to form a cyclopropane adduct (Scheme 2.12). Ring expansion to give a seven-membered product then occurred with a minimal activation barrier, with product **44** being trapped by a second carbonyl ligand. Indeed, this process was observed experimentally when CO was deliberately added to quench the metathesis reaction.[61] The presence of other π acids could induce similar reactivity. For example, Webster showed that binding of a

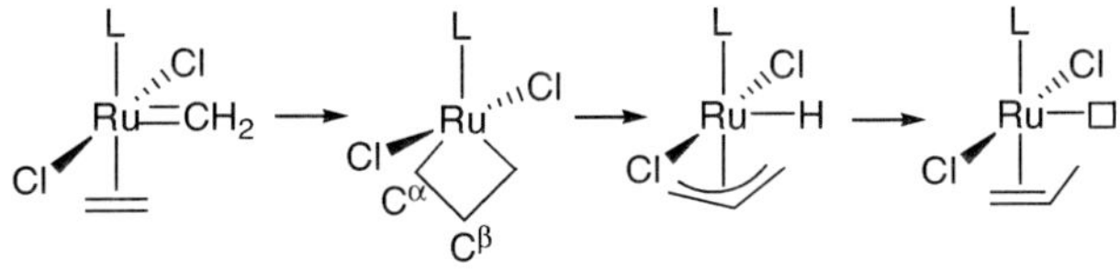

Scheme 2.11 Deactivation of Grubbs' catalysts *via* β-hydrogen abstraction.

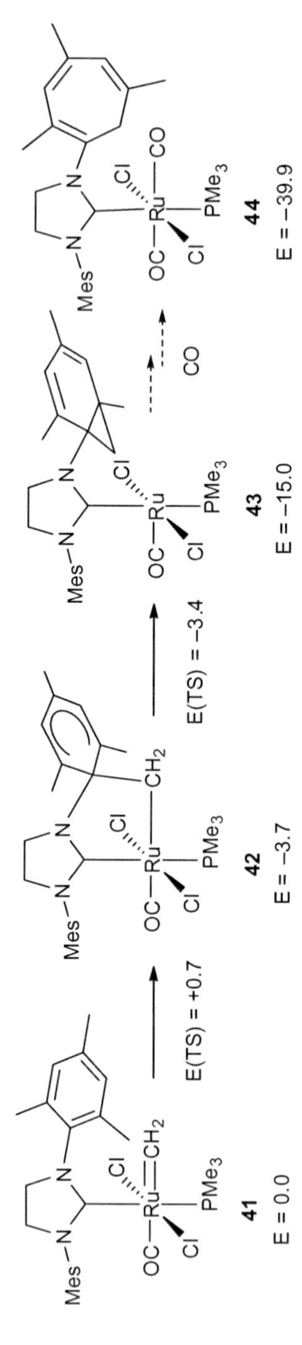

Scheme 2.12 CO-induced decomposition of the model catalyst **41**. Relative energies given in kcal mol^{-1}.

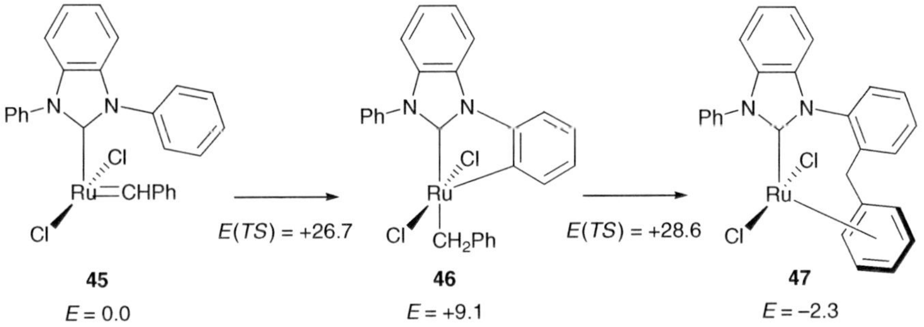

Scheme 2.13 Key intermediates in the computed profile for decomposition of 14e $[(NHC)Ru(=CHPh)Cl_2]$, with the energy of the highest-lying TS between each intermediate indicated. Relative energies given in $kcal\,mol^{-1}$.

second alkene may be feasible and although this was computed to raise the transition state energy for migratory insertion by about 5–6 $kcal\,mol^{-1}$ such species may play a role in inducing decomposition.[62]

Mathew *et al.*[63] also considered C–H activation of an *N*-phenyl substituent in the 14e species $[(NHC)Ru(=CHPh)Cl_2]$ **45** (Scheme 2.13). In this case, upon phosphine dissociation, isomerisation could occur such that the methylene ligand moved *trans* to the NHC, allowing the phenyl substituent to engage in an agostic interaction with Ru. This activated C–H bond then underwent σ-bond metathesis to form species **46** which featured a benzyl ligand and a cyclometalated phenyl group. These two species could then undergo reductive coupling to give **47**.

2.5 Group 9

Jensen published a detailed study on the role of NHCs as possible co-ligands in the catalytic hydroformylation of alkenes by $[RhH(CO)_3L]$ systems (where L = CO, phosphites, phosphines or NHCs).[64] When L = CO or electron-withdrawing phosphites, the rate determining step was associated with the hydrogenolysis of the acyl intermediate. For NHC ligands (and electron-rich phosphines) this changed to the initial alkene coordination/insertion steps. In general, the barriers of the rate-controlling steps were higher with electron donating ligands, making standard NHCs a poor choice for this process. However, electron-withdrawing NHCs based on a tetrazole architecture were proposed and computed to have similar energetics to phosphites. Further computational studies were performed on other catalytic cycles involving NHC–Rh systems, including the hydrosilylation of ketones[65] and the cross-coupling of boronic acids and aldehydes at Rh_2 dimers.[66]

An interesting application of the non-innocent behaviour of NHCs at transition metal centres was developed by Tan *et al.*[67] They found that cyclisation of substituted benzimidazoles was efficiently catalysed by $[RhCl(COE)_2]_2/PCy_3$ at 135 °C (COE = cyclooctene). Mechanistic studies isolated a square planar NHC–rhodium(I)/alkene complex **48**, as the resting state of the catalyst.

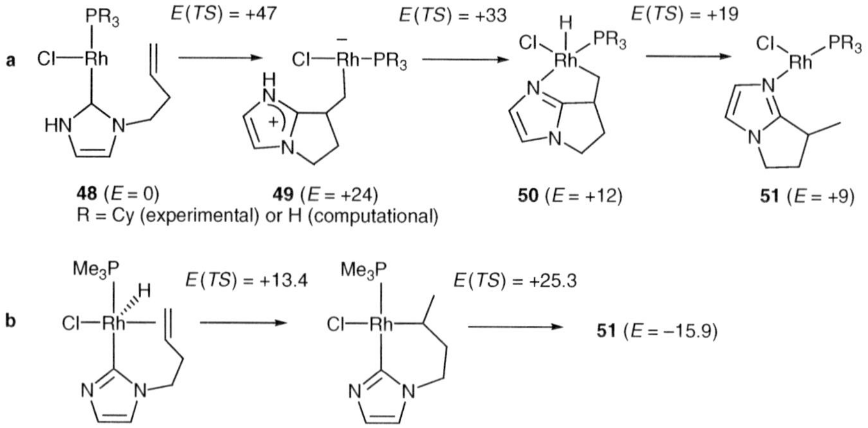

48 ($E = 0$) **49** ($E = +24$) **50** ($E = +12$) **51** ($E = +9$)
R = Cy (experimental) or H (computational)

52 ($E = 0.0$) **53** ($E = -1.9$)

Scheme 2.14 Rh-catalysed C–H activation and cyclisation of substituted heterocycles. Relative energies given in kcal mol^{-1}.

Calculations proposed the insertion of the alkene moiety into the NHC–Rh bond to form a zwitterionic intermediate **49**, to be the rate-determining step with a computed activation barrier of +47 kcal mol^{-1} (Scheme 2.14a). N to Rh proton transfer in **49** formed **50** from which reductive elimination of the cyclized product could occur, this being N-bound to Rh in **51**. The mechanism of N- to C-tautomerism implicit in these reactions has been studied for a hydropyrimidine-based system.[68]

Hawkes *et al.* also considered the mechanism of Rh-catalysed C–H activation/C–C coupling reactions and proposed that may be initiated via a C–H activation of the heterocycle to generate a hydride **52** followed by alkene insertion into the Rh–H bond to give **53** (Scheme 2.14b).[69] C–C bond coupling then produced the final product and was the overall rate determining step. This mechanism had a lower barrier than the one involving hydride transfer to the nitrogen from the heterocycle.[67] Also, acid catalysis facilitated this route. The factors that control the relative energies of the N- and C-bound tautomers of NHCs were also considered computationally.[70]

2.6 Group 10

One of the first mechanistic computational studies on NHC–TM chemistry was published in 1998 by Rösch on the Heck reaction.[71] This study employed a simple diaminocarbene (DAC) model ligand and characterised the alkene insertion and β-H elimination steps of the cationic pathway, derived from a *cis*-[$(DAC)_2PdBr(Ph)$] intermediate *via* Br^-/C_2H_4 substitution. The neutral pathway (*via* initial DAC/C_2H_4 substitution) was discounted on the basis of the non-observation of free NHC experimentally and the fact that cleavage of the NHC–Pd bond would be prohibitively high in energy. A mixed phosphine/NHC chelate was also considered and may allow access to the neutral pathway due to more facile phosphine dissociation. This topic was revisited by Lee and co-workers who compared the behaviour of the [Pd(PMe_3)$_2$], [(IMe)$_2$Pd] and [(IMe)(IMe′)Pd] systems, where the last species features one IMe ligand bound through the abnormal C5 position (IMe′). [(IMe)$_2$Pd] was computed to be +16.6 kcal mol^{-1} more stable than [(IMe)(IMe′)Pd], although the higher dipole of the latter reduced this difference to +8.7 kcal mol^{-1} in DMSO solvent.[72] While the oxidative addition of phenyl bromide was found to be much more exothermic for the NHC systems, it was the [Pd(PMe_3)$_2$] model that displayed the lowest activation barrier for this process. The subsequent steps, whether *via* the neutral or the cationic mechanisms, showed only subtle variations between the ligand sets. The only exceptions arose from the higher NHC dissociation energies that again tended to disfavour the neutral mechanism for the NHC systems. The computational results indicated no significant advantage in the use of the abnormal NHC ligand, in contrast with experimental reports where a species featuring an abnormal NHC was proposed, although in that case the much bulkier IMes ligand was involved.[73]

A number of computational studies addressed the effect of NHC sterics on the fundamental properties of their Group 10 metal complexes. The stability of *fac*-[(NHC)$_3$PtMe$_3$]$^+$ species towards the reductive elimination of ethane was studied.[74] Calculations showed that even with the relatively small IMe ligand the binding energy of a third NHC ligand was very low. The tris-IMe species could therefore readily access a reactive 5-coordinate intermediate from which reductive elimination occured with a free energy barrier of only +8.8 kcal mol^{-1}. In general, changing the nature of the NHC had much more effect on the ligand binding energies rather than on the barrier to reductive elimination.

Steric factors were also important in understanding the reactivity of [(P∼C)PdMe$_2$] **54**, featuring a mixed NHC–phosphine ligand.[75] Protonation of this species in the presence of pyridine led to substitution *trans* to phosphine and not *trans* to the NHC, an unexpected result given the anticipated higher *trans* effect and influence of the NHC ligand. Calculations on a small model system reaffirmed this expectation and the experimental observations could only be reproduced by considering the full model (Equation (2.6)). This result was linked to steric strain between the large 2,6-diisopropylphenyl substituent and the *cis* Me group, which was relieved upon protonation by the formation of a much more flexible σ-CH_4 ligand, allowing for subsequent substitution by pyridine at that site.

54 —— 1) H$^+$ 2) pyridine ——→ Favoured by Model 1 (R′ = R = H) or Favoured by Model 2 and observed experimentally (R′ = Ph, R = 2,6-(i-Pr)$_2$C$_6$H$_3$)

(2.6)

Rotation barriers around the NHC–Ni bond in [(NHC)Ni(η^3-allyl)(X)] species also reflected the bulk of the NHC, with increased barriers arising from a steric clash in the transition state where the NHC substituents must pass through the Ni coordination plane.[76] Computed barriers were in good agreement with experimental trends and it was interesting to note the increased barrier for the ^{Me}IMe system ($\Delta G^{\ddagger} = +17.8\,\text{kcal mol}^{-1}$) compared to the IMe analogue ($\Delta G^{\ddagger} = +15.2\,\text{kcal mol}^{-1}$) even though the extra steric bulk was imposed at the remote 4 and 5 positions.

In their study of Ar–Cl activation by [(NHC)$_2$Pd] complexes (NHC = IH, IMe, ItBu), Green and co-workers showed that increased steric bulk promotes both NHC dissociation and the subsequent C–Cl oxidative addition.[77] ItBu also produced reduced activation barriers for the C–Cl cleavage step, but this effect was far less significant than those responsible for accessing the reactive [(NHC)Pd(η^2-C_6H_5Cl)] precursor. In the context of aryl amination, this work showed that the initial C–Cl activation was the overall rate determining step with ItBu.

The activation of C–F bonds at [(IiPr)$_2$Ni] was also studied computationally for a range of fluoroaromatics.[78] Evidence for 3-coordinate η^2-arene adducts was obtained experimentally and the computed C–F activation barrier of $+24.3\,\text{kcal mol}^{-1}$ for C_6F_6 was in good agreement with the experimental values. The greater accessibility of C–F activation with NHC ligands was ascribed to the electron-rich and hence nucleophilic Ni centre. Further calculations on C–F activation of isomers of $C_6H_3F_3$ reproduced the experimental selectivities. C–F activation in these systems was very thermodynamically favourable, meaning that the reverse reaction, C–F reductive elimination was likely to be problematic. Indeed, C–F reductive elimination from [(IMe)Pd-F(aryl)] species was shown to have very high barriers; moreover, reductive elimination of aryl-imidazolium side-products was facile in comparison.[79]

Even subtle changes in the nature of a bulky NHC ligand can have significant effects on the outcome of a reaction. This was observed in alkyl–alkyl Negishi cross-coupling, where with IPr extremely efficient coupling took place at room temperature, even when the alkyl halide contained β-hydrogens.[80] In contrast, with IXy β-H elimination dominated the reaction leading to undesirable side-products. Calculations on the mono-ligated [(IPr)Pd] system suggested that a 'steric wall' was set up at Pd, in which a large number of weak Pd···H interactions were possible. Subtle entropic effects were proposed to account for the selectivity of this system. NHCs were also used as co-ligands in biaryl cross-coupling reactions between aryl halides and aryl Grignards with MF_2 catalysts (M = Fe, Co and Ni). The use of fluorides made a higher oxidation state pathway involving a metal(IV) intermediate accessible. Once such species were formed, cross-coupling of the desired product could rapidly occur.[81]

Louie and co-workers considered the mechanism of the 1,3-sigmatropic shift of vinylcyclopropanes to form cyclopentenes catalysed by [Ni(COD)$_2$]/IPr.[82] Calculations with a [(IMe)Ni] fragment showed the lowest energy pathway involved initial formation of an alkene adduct, followed by oxidative addition to form a vinylmetallacyclobutane (Scheme 2.15). This rearranged to a metalacyclohexene species *via* TS2, which then underwent reductive coupling to generate cyclopentene. All three transition states were close in energy.

Further calculations on alkyl- and aryl-substituted substrates reproduced experimental trends, although the identity of the rate determining process varied from one substrate to another. Calculations on the catalytic cycle using [Ni(PMe_3)] gave only minor changes in energy and this, along with an experimental assessment of alternative co-ligands, underlined the importance of NHC sterics in promoting the reaction. Calculations on other NHCs confirmed that electronic effects were not significant in this system and that the importance of using a [Ni($COD)_2$] precursor lay in the ability of the COD ligand to facilitate the formation of a mono-ligated [(NHC)Ni] as the active species.

Popp and Stahl extended their studies of O_2 chemistry at [PdL_2] systems (described in Section 2.3) to consider the insertion of O_2 into the Pd–H bond in *trans*-[$(IMes)_2$Pd(H)(X)] species to form *trans*-[$(IMes)_2$Pd(OOH)(X)].[83] This was a key step in aerobic oxidation reactions and calculations (NHC = IMe) indicated two possible pathways close in energy: (i) hydrogen atom abstraction by O_2 followed by radical recombination; or (ii) reductive elimination of HX followed by O_2 addition and protonolysis. Strongly donating ligands X, *trans* to hydride favoured the H abstraction mechanism by labilising the Pd–H bond. Although the two mechanisms were close in energy, on balance the second pathway was more consistent with experimental observations, including a rate enhancement in the presence of excess HOAc.

The presence of a strongly donating NHC ligand had an important effect on the mechanism of β-H elimination of ketones from Pd-alkoxide species, [(IPr)Pd(OCHRR′)(OAc)].[84] Calculations suggested that direct H transfer onto the acetate ligand occured in one step *via* a transition state **55**, featuring a short Pd···H contact of only 1.75 Å (Figure 2.6). This process was promoted by the strongly donating NHC ligand, which disfavoured a two-step mechanism *via* a hydride intermediate, and this was important in maintaining the selectivity of ketone formation.

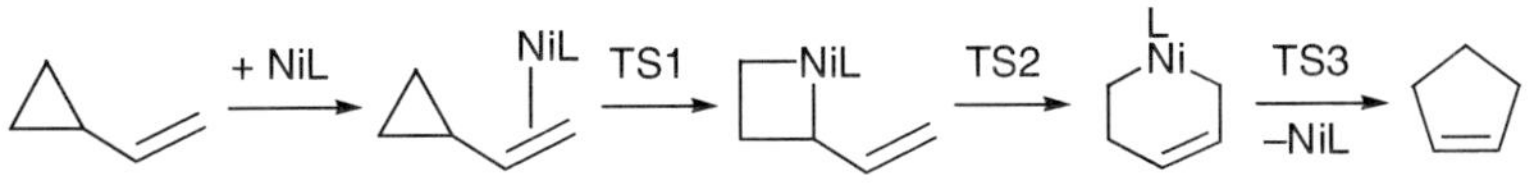

Scheme 2.15 Ni-catalysed vinylcyclopropane to cyclopentene rearrangement.

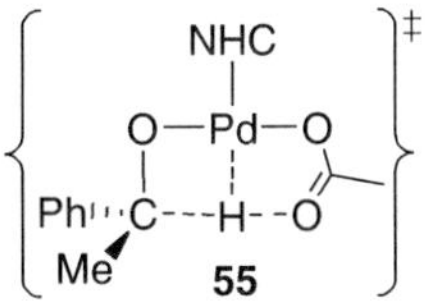

Figure 2.6 Transition state for β-H transfer in [(IPr)Pd(OCHRR′)(OAc)].

Calculations provided insight into the reduction of CO_2 to CO at a [(IPr)Ni] dimer **56**, where each Ni centre was stabilised by one σ bond to one NHC ligand and is η^6-bound to the 2,6-diisopropylphenyl substituent of the other.[85] The reaction proceeded *via* displacement of one η^6-interaction at Ni1 by CO_2 to form an η^2-(C,O)-CO_2 ligand. This then rearranged to bridge both Ni centres, which weakens the η^6-NHC interaction at Ni2, and allowed a second CO_2 molecule to bind (**57**) (Figure 2.7). Cleavage of the bridging C–O bond then occured with an activation barrier of +24.1 kcal mol^{-1}, giving a final product featuring a CO ligand at Ni2 and a carbonate ligand at Ni1 (**58**).

Catalytic dehydrogenation of amineborane, H_3BNH_3 has been an area of intense investigation in recent years and one of the most promising systems involves [(NHC)$_2$Ni] species (NHC = Enders' carbene in **59**). Experimentally, this system caused the loss of 2.5 equivalents of H_2 and k_H/k_D kinetic isotope effects (KIE) were reported for deuteration at B and N (individually and together). Calculations by Yang and Hall to address the mechanism of the loss of the first H_2 molecule highlighted a novel role for the NHC, whereby a hydrogen was transferred from N in **59** onto an NHC C2 position, generating a bound imidazolium intermediate **60** (Scheme 2.16).[86] After reorientation of the σ-borane moiety, C–H activation of the imidazolium yielded a nickel hydride intermediate. The ability of the NHC ligand to accept and donate H in this way is reminiscent of the reductive elimination chemistry described in Section 2.2.1. Dehydrogenation was then completed by facile B–H bond activation and dissociation of both H_2 and the aminoborane product.

Further studies by Yang and Hall proposed a mechanism for the loss of a second H_2 molecule. In this case the NHC ligand again played a non-innocent role, as it facilitated B–H activation in an aminoborane adduct **61**, by transferring onto the nascent BH group in **62** (Scheme 2.16). N–H bond activation then led to H_2 loss with the NHC ligand then being transferred back to the metal centre.[87] Hall also addressed the observed KIE by showing that use of appropriately labelled substrates increased the activation barrier associated with both the N–H and C–H activation steps in the first dehydrogenation as well as in the initial B–H activation in the second dehydrogenation. Thus, the observation of KIEs associated with both the borane and amine groups could

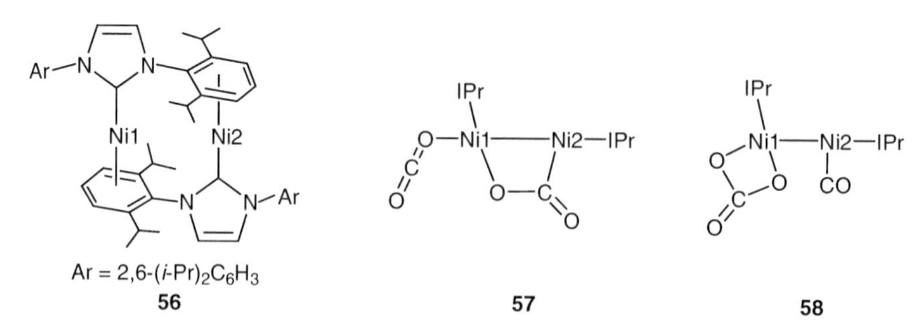

Figure 2.7 Intermediates computed for the reduction of CO_2 at [IPrNi]$_2$.

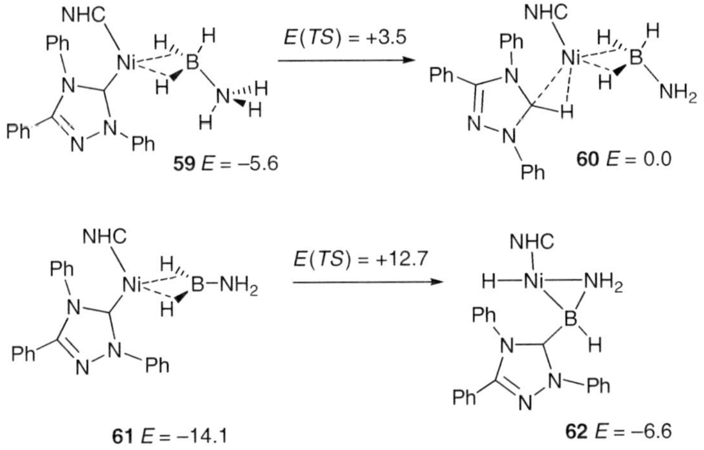

Scheme 2.16 Ni-catalysed dehydrogenation of amineborane.

be accounted for and the fact that the former was not observed experimentally in the first 20 min of catalysis may be consistent with it only appearing once the second dehydrogenation step had begun to turn over.

This interpretation of events was questioned by Musgrave, though, who proposed the importance of [(NHC)Ni(H_2BNH_2)], formed after initial dehydrogenation at the bis-NHC precursor, as the active species in catalysis.[88] The calculated activation barrier associated with this species is over $+20$ kcal mol^{-1} but this value appeared more in keeping with a reaction temperature of 60 °C. The computed N–H/N–D KIE was consistent with experimental values for the first dehydrogenation. This study also stressed the importance of delocalisation onto the phenyl substituents in Enders' carbene in facilitating dehydrogenation. In a further study, a mechanism for the second dehydrogenation was proposed in which the free ligand played the central role, with no direct role for a metal species.[89]

2.7 Group 11

The reactivity of [(NHC)Cu(BR_2)] species was extensively investigated in a series of joint experimental (NHC = IPr; BR_2 = B(cat) or B(pin), where cat = catecholato, and pin = pinacolato) and computational (NHC = IMe; BR_2 = B(–OCH_2CH_2O–)) studies by Marder and co-workers (Scheme 2.17). The reactivity in these systems depended upon a high-lying Cu–B σ bond, which acted as a nucleophile and was aided by the strong *trans* influence NHC ligand that weakened the Cu–B interaction, facilitating the transfer of the boryl group such that insertion into the Cu–B bond could occur.[90] In the case of CO_2 reduction, the process involved an O-bound intermediate, which then extruded CO to generate a [(NHC)Cu(OBR_2)] species.[91] Catalyst regeneration occurred

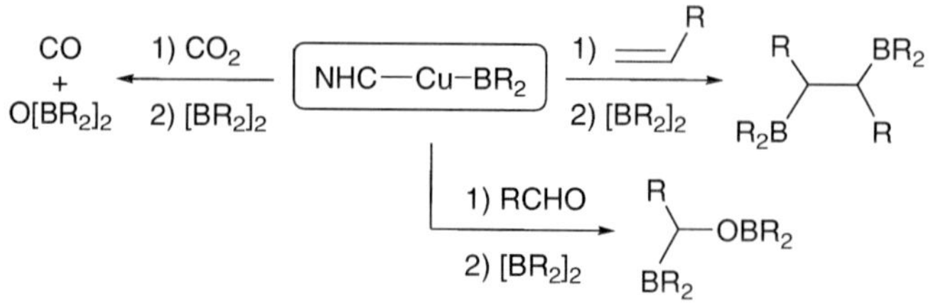

Scheme 2.17 Borylation reactions catalysed by $[(NHC)Cu(BR_2)]$ species.

via reaction with a diborane, which added over Cu–O bond *via* a σ-bond metathesis mechanism and expelled $O(BR_2)_2$. Bulky NHC ligands could be detrimental to this final step due to a sterically encumbered transition state.

Alkenes also inserted into the Cu–B bond, and this process was more accessible with alkenes bearing electron withdrawing substituents.[92] These preferentially undergo 2,1-insertion, with boryl migrating to the unsubstituted carbon. The regioselectivity of insertion for alkenes with electron donating groups was more subtle, as electronic and steric effects were counter-directing.[90] The final step of the cycle again involved transmetalation with a diborane and this was promoted by use of the more Lewis acidic $[B(cat)]_2$ rather than $[B(pin)]_2$.[92] In a separate study, Lillo *et al.* showed that use of a [(IPr)CuB(cat)] catalyst resulted in the highly selective 1,2-diboration of styrene.[93] With other catalysts hydroboration products were observed and this could be explained by β-H elimination from the initial 2-borylalkyl intermediates formed upon alkene insertion. Diborane addition over the resultant Cu–H bond generated monoboranes that then reacted with Cu-alkyl to generate the hydroboration products. Calculations showed that these process involved a number of σ-bound adducts which reacted *via* σ-bond metathesis rather than oxidative addition. Marder and Lin also studied the diboration of aldehydes.[94] Nucleophilic attack of the boryl group then occurred at the aldehyde carbon to give an O-bound intermediate. This was a kinetic product, however, and rearranged to a more stable C-bound form, which could readily undergo σ-bond metathesis with diborane reagents to generate the final products.

Other computational studies involving NHC–Cu species considered the formation of phenylisocyanates from nitrobenzene,[95] and the development of [3 + 2] cycloaddition reactions for the formation of 1,2,3-triazoles.[96] In the latter case the use of NHCs allowed the direct use of copper(I) catalysts, whereas copper(II) precursors were predominant before. With [(NHC)CuBr] the reaction could be run in water and was successful even for internal alkynes, an unusual observation as the intermediacy of Cu-acetylides had previously been assumed. Calculations showed that the [(SIMes)Cu] fragment was ideally set up to bind internal alkynes in an η^2-fashion and hence activate them towards cycloaddition. With terminal alkynes the acetylide route may still be operative.

The use of NHC complexes of silver is less well developed in catalysis; however, such species are widely employed as NHC transfer agents in the synthesis of NHC–TM complexes. Hayes *et al.* showed that Ag_2O readily deprotonated $[IMe{\cdot}H]^+$ and that facile AgI transfer resulted in the formation of a [(IMe)AgI] complex.[97] The presence of a second imidazolium strongly facilitated these processes by H-bonding through the C2–H group. The second imidazolium was then first deprotonated by AgOH and then metallated. The overall formation of two [(IMe)AgI] complexes and water was strongly exothermic and the efficacy of this process was attributed to the high basicity of the Ag_2O reagent.

Much of the computational work on catalysis by $[(NHC)Au]^+$ species is founded on the ability of this fragment to activate unsaturated organic substrates towards nucleophilic attack. In addition, intermediates are often subject to a number of rearrangements that make predicting and understanding reaction outcomes problematic. Calculations have an obvious role shedding light on such processes.

One simple example involved the Au-catalysed rearrangement of allylic acetates.[98] Initially the $[(NHC)Au]^+$ moiety (NHC = IMe, ItBu) bound to the C=C double bond (**63**) and the reaction proceeded through a six-membered acetoxonium intermediate **64** with a computed barrier for the IMe model of $+17.3\,kcal\,mol^{-1}$ in dichloromethane (Scheme 2.18). This barrier was not significantly affected by use of the bulker ItBu ligand or by including the BF_4^- counter-ion in the calculation. By comparison, the computed barrier in the gold-free reaction was $+31.7\,kcal\,mol^{-1}$ and gold was proposed to promote the rearrangement by stabilising the developing charges during this process.

Propargylic acetates can bind to a $[Au(L)]^+$ moiety through the C≡C triple bond (**66**) and this interaction can promote both 1,2- and 1,3-acetate shifts, resulting in the formation of gold vinyl carbenoid **67** or gold allene **69** intermediates, respectively (Scheme 2.19).[99] Calculations suggested that all three species were close in energy and could readily interconvert. Hence the outcome of the cycloisomerisation reactions of a complex substrate such as **70** would depend subtly on the reaction condition and the nature of the co-ligand. Calculations suggested that with L = IMe the allene form **69** was marginally favoured, while with L = PMe_3 an intermediate **67** on the pathway between **66** and **68** may accumulate. This was consistent with product distributions formed with **70** that favoured **71** when $[(NHC)Au]^+$ catalysts were employed. Calculations showed that the formation of **71** could be traced back to the gold allene intermediate.[100]

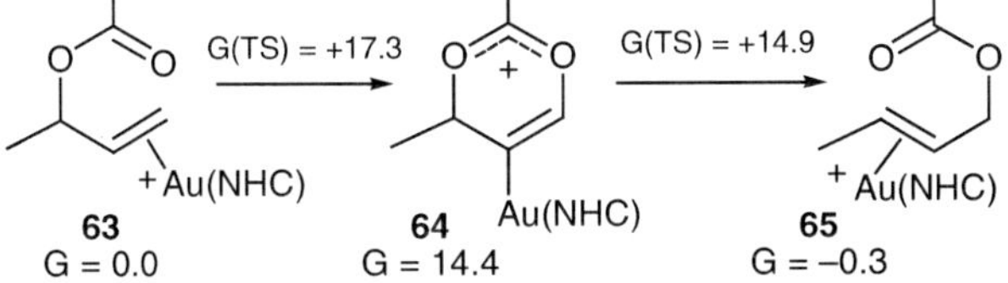

G(TS) = +17.3
G(TS) = +14.9
+Au(NHC)
63
G = 0.0
64
Au(NHC)
G = 14.4
+ Au(NHC)
65
G = −0.3

Scheme 2.18 Au-catalysed rearrangement of allylic acetates. Relative free energies given in $kcal\,mol^{-1}$.

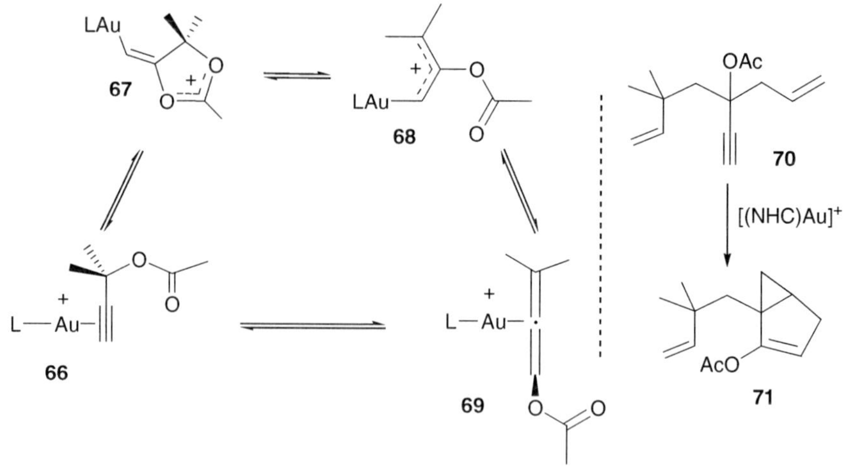

Scheme 2.19 Au-catalysed cycloisomerisation/rearrangements of propargylic acetates.

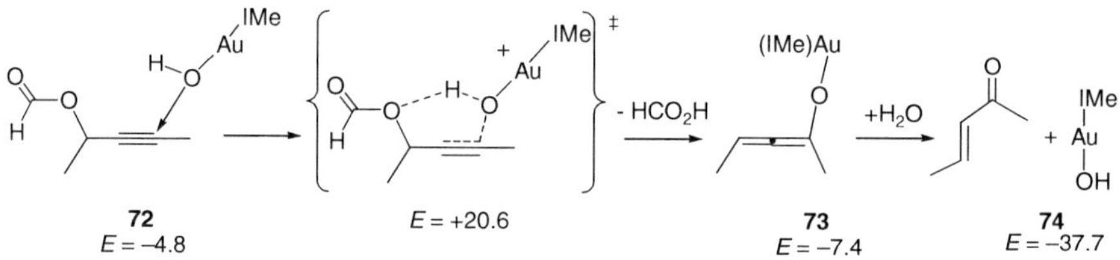

Scheme 2.20 Au-catalysed formation of enones from propargylic acetate. Relative energies given in kcal mol^{-1}.

In contrast to the above examples, calculations modelling the gold-catalysed formation of conjugated enones and enals from propargylic acetates could not define a pathway based on the initial activation of the C≡C triple bond.[101] Instead, the observation that this transformation was promoted by the addition of water led to the suggestion that [(NHC)Au(OH)] may be the active species, and act as a nucleophile at the free alkyne (as in **72**) to give a gold allenolate intermediate **73** (Scheme 2.20). The computed barrier (with NHC = IMe) for this process was +25.4 kcal mol^{-1} compared to +36.2 kcal mol^{-1} for the equivalent reaction with water. The gold centre then assisted the delivery of a proton from a second water molecule to form the final conjugated enone (or enal) product **74**, and regenerate [(NHC)Au(OH)].

2.8 Conclusions

This Chapter illustrates the important role that computational chemistry has played in understanding the reactivity of transition metal NHC complexes. In particular, the combination of experiment and computation provides a powerful tool for the elucidation of reaction mechanisms. As the transition metal chemistry of NHC systems has developed rapidly over the last 20 years, so have the types of problems that can be addressed computationally. Early calculations often employed small model systems and focused on the electronic effect of simple NHCs against, for example, small, representative phosphines. Often such studies showed that these two types of ligands were only subtly different in their effect on reactivity. Increasingly, computational work is taking into account the full size of the, often bulky, NHC ligands and in many cases the treatment of steric effects is clearly vital in order to reproduce (and ultimately predict) experimental behaviour. The work on ligand dissociation from Grubbs' alkene metathesis catalysts highlights the need for an accurate representation of such effects. Addressing these challenges will ensure that computational chemistry will become an even more influential tool in the study of the reactivity of transition metal NHC systems in the future.

References

1. A. J. Arduengo III, R. L. Harlow and M. Kline, *J. Am. Chem. Soc.*, 1991, **113**, 361–363.
2. (a) S. Díez-González, N. Marion and S. P. Nolan, *Chem. Rev.*, 2009, **109**, 3612–3676; (b) W. J. Sommer and M. Weck, *Coord. Chem. Rev.*, 2007, **251**, 860–873; (c) E. Peris and R. H. Crabtree, *Coord. Chem. Rev.*, 2004, **248**, 2239–2246; (d) W. A. Herrmann, *Angew. Chem., Int. Ed.*, 2002, **41**, 1290–1309; (e) W. A. Herrmann, T. Weskamp and V. P. M. Böhm, *Adv. Organomet. Chem.*, 2002, **48**, 1–69; (f) W. A. Herrmann and C. Köcher, *Angew. Chem., Int. Ed. Engl.*, 1997, **36**, 2162–2187.
3. (a) T. Ziegler, *Chem. Rev.*, 1991, **91**, 651–667; (b) A. Dedieu, *Chem. Rev.*, 2000, **100**, 543–600; (c) M. Torrent, M. Solá and G. Frenking, *Chem. Rev.*, 2000, **100**, 439–494; (d) S. Q. Niu and M. B. Hall, *Chem. Rev.*, 2000, **100**, 353–405; (e) W. Koch and M. C. Holthausen, *A Chemist's Guide to Density Functional Theory*, Wiley-VCH, 2001.
4. (a) L. Cavallo, A. Correa, C. Costabile and H. Jacobsen, *J. Organomet. Chem.*, 2005, **690**, 5407–5413. See also: (b) R. A. Kelly III, H. Clavier, S. Giudice, N. M. Scott, E. D. Stevens, J. Bordner, I. Samardjiev, C. D. Hoff, L. Cavallo and S. P. Nolan, *Organometallics*, 2008, **27**, 202–210.
5. D. S. McGuinness, M. J. Green, K. J. Cavell, B. W. Skelton and A. H. White, *J. Organomet. Chem.*, 1998, **565**, 165–178.
6. D. S. McGuinness, N. Saendig, B. F. Yates and K. J. Cavell, *J. Am. Chem. Soc.*, 2001, **123**, 4029–4040.
7. D. C. Graham, K. J. Cavell and B. F. Yates, *Dalton Trans.*, 2005, 1093–1100.
8. D. C. Graham, K. J. Cavell and B. F. Yates, *Dalton Trans.*, 2006, 1768–1775.

9. D. J. Nielsen, A. M. Magill, B. F. Yates, K. J. Cavell, B. W. Skelton and A. H. White, *Chem. Commun.*, 2002, 2500–2501.
10. D. S. McGuinness, K. J. Cavell and B. F. Yates, *Chem. Commun.*, 2001, 355–356.
11. (a) D. S. McGuinness, K. J. Cavell, B. F. Yates, B. W. Skelton and A. H. White, *J. Am. Chem. Soc.*, 2001, **123**, 8317–8328; (b) D. C. Graham, K. J. Cavell and B. F. Yates, *Dalton Trans.*, 2007, 4650–4658.
12. K. J. Hawkes, D. S. McGuinness, K. J. Cavell and B. F. Yates, *Dalton Trans.*, 2004, 2505–2513.
13. M. Viciano, M. Poyatos, M. Sanaú, E. Peris, A. Rossin, G. Ujaque and A. Lledós, *Organometallics*, 2006, **25**, 1120–1134.
14. M. Viciano, M. Feliz, R. Corberán, J. A. Mata, E. Clot and E. Peris, *Organometallics*, 2007, **26**, 5304–5314.
15. L. N. Appelhans, D. Zuccaccia, A. Kovacevic, A. R. Chianese, J. R. Miecznikowski, A. Macchioni, E. Clot, O. Eisenstein and R. H. Crabtree, *J. Am. Chem. Soc.*, 2005, **127**, 16299–16311.
16. A. M. Voutchkova, M. Feliz, E. Clot, O. Eisenstein and R. H. Crabtree, *J. Am. Chem. Soc.*, 2007, **129**, 12834–12846.
17. A. T. Normand, K. J. Hawkes, N. D. Clement, K. J. Cavell and B. F. Yates, *Organometallics*, 2007, **26**, 5352–5363.
18. T. Steinke, B. K. Shaw, H. Jong, B. O. Patrick, M. D. Fryzuk and J. C. Green, *J. Am. Chem. Soc.*, 2009, **131**, 10461–10466.
19. A. T. Normand, M. S. Nechaev and K. J. Cavell, *Chem.–Eur. J.*, 2009, **15**, 7063–7073.
20. A. A. Danopoulos, N. Tsoureas, J. C. Green and M. B. Hursthouse, *Chem. Commun.*, 2003, 756–757.
21. E. Becker, V. Stingl, G. Dazinger, M. Puchberger, K. Mereiter and K. Kirchner, *J. Am. Chem. Soc.*, 2006, **128**, 6572–6573.
22. E. Becker, V. Stingl, G. Dazinger, K. Mereiter and K. Kirchner, *Organometallics*, 2007, **26**, 1531–1535.
23. K. Kirchner, *Monatsh. Chem.*, 2008, **139**, 337–348.
24. S. Fantasia, H. Jacobsen, L. Cavallo and S. P. Nolan, *Organometallics*, 2007, **26**, 3286–3288.
25. S. Burling, B. M. Paine, D. Nama, V. S. Brown, M. F. Mahon, T. J. Prior, P. S. Pregosin, M. K. Whittlesey and J. M. J. Williams, *J. Am. Chem. Soc.*, 2007, **129**, 1987–1995.
26. R. F. R. Jazzar, S. A. Macgregor, M. F. Mahon, S. P. Richards and M. K. Whittlesey, *J. Am. Chem. Soc.*, 2002, **124**, 4944–4945.
27. R. A. Diggle, S. A. Macgregor and M. K. Whittlesey, *Organometallics*, 2008, **27**, 617–625.
28. R. A. Diggle, A. A. Kennedy, S. A. Macgregor and M. K. Whittlesey, *Organometallics*, 2008, **27**, 938–944.
29. N. M. Scott, R. Dorta, E. D. Stevens, A. Correa, L. Cavallo and S. P. Nolan, *J. Am. Chem. Soc.*, 2005, **127**, 3516–3526.
30. L. J. L. Häller, M. J. Page, S. A. Macgregor, M. F. Mahon and M. K. Whittlesey, *J. Am. Chem. Soc.*, 2009, **131**, 4604–4605.

31. J. A. Cabeza and E. Pérez-Carreño, *Organometallics*, 2008, **27**, 4697–4702.
32. L. P. Spencer, C. Beddie, M. B. Hall and M. D. Fryzuk, *J. Am. Chem. Soc.*, 2006, **128**, 12531–12543.
33. B. V. Popp, J. E. Wendlandt, C. R. Landis and S. S. Stahl, *Angew. Chem., Int. Ed.*, 2007, **46**, 601–604.
34. J. M. Praetorius, D. P. Allen, R. Y. Wang, J. D. Webb, F. Grein, P. Kennepohl and C. M. Crudden, *J. Am. Chem. Soc.*, 2008, **130**, 3724–3725.
35. L. J. L. Häller, E. Mas-Marzá, A. Moreno, J. P. Lowe, S. A. Macgregor, M. F. Mahon, P. S. Pregosin and M. K. Whittlesey, *J. Am. Chem. Soc.*, 2009, **131**, 9618–9619.
36. S. Burling, L. J. L. Häller, E. Mas-Marzá, A. Moreno, S. A. Macgregor, M. F. Mahon, P. S. Pregosin and M. K. Whittlesey, *Chem.–Eur. J.*, 2009, **15**, 10912–10923.
37. S. Schenk and M. Reiher, *Inorg. Chem.*, 2009, **48**, 1638–1648.
38. R. A. Diggle, S. A. Macgregor and M. K. Whittlesey, *Organometallics*, 2004, **23**, 1857–1865.
39. S. L. Chatwin, M. G. Davidson, C. Doherty, S. M. Donald, R. F. R. Jazzar, S. A. Macgregor, G. J. McIntyre, M. F. Mahon and M. K. Whittlesey, *Organometallics*, 2006, **25**, 99–110.
40. T. Weskamp, F. J. Kohl, W. Hieringer, D. Gleich and W. A. Herrmann, *Angew. Chem., Int. Ed.*, 1999, **38**, 2416–2419.
41. (a) S. F. Vyboishchikov, M. Bühl and W. Thiel, *Chem.–Eur. J.*, 2002, **8**, 3962–3975; (b) C. Adlhart and P. Chen, *J. Am. Chem. Soc.*, 2004, **126**, 3496–3510; (c) A. C. Tsipis, A. G. Orpen and J. N. Harvey, *Dalton Trans.*, 2005, 2849–2858.
42. M. S. Sanford, J. A. Love and R. H. Grubbs, *J. Am. Chem. Soc.*, 2001, **123**, 6543–6554.
43. Y. Zhao and D. G. Truhlar, *J. Chem. Theor. Comp.*, 2009, **5**, 324–333.
44. S. Torker, D. Merki and P. Chen, *J. Am. Chem. Soc.*, 2008, **130**, 4808–4814.
45. Y. Zhao and D. G. Truhlar, *Org. Lett.*, 2007, **9**, 1967–1970.
46. I. C. Stewart, D. Benitez, D. J. O'Leary, E. Tkatchouk, M. W. Day, W. A. Goddard and R. H. Grubbs, *J. Am. Chem. Soc.*, 2009, **131**, 1931–1938.
47. (a) B. O. Leung, D. L. Reid, D. A. Armstrong and A. Rauk, *J. Phys. Chem., A*, 2004, **108**, 2720–2725; (b) D. Ardura, R. López and T. L. Sordo, *J. Phys. Chem. B*, 2005, **109**, 23618–23623.
48. N. Fey, *Dalton Trans.*, 2010, **39**, 296–310.
49. (a) D. Benitez, E. Tkatchouk and W. A. Goddard, *Chem. Commun.*, 2008, 6194–6196; (b) D. Benitez, E. Tkatchouk and W. A. Goddard, *Organometallics*, 2009, **28**, 2643–2645.
50. C. Adlhart and P. Chen, *Angew. Chem., Int. Ed.*, 2002, **41**, 4484–4487.
51. (a) B. F. Straub, *Angew. Chem., Int. Ed.*, 2005, **44**, 5974–5978; (b) B. F. Straub, *Adv. Synth. Catal.*, 2007, **349**, 204–214.
52. R. L. Lord, H. Wang, M. Vieweger and M. H. Baik, *J. Organomet. Chem.*, 2006, **691**, 5505–5512.

53. D. Benitez and W. A. Goddard, *J. Am. Chem. Soc.*, 2005, **127**, 12218–12219.
54. A. Correa and L. Cavallo, *J. Am. Chem. Soc.*, 2006, **128**, 13352–13353.
55. C. Costabile and L. Cavallo, *J. Am. Chem. Soc.*, 2004, **126**, 9592–9600.
56. (a) S. F. Vyboishchikov and W. Thiel, *Chem.–Eur. J.*, 2005, **11**, 3921–3935; (b) M. R. Buchmeiser, D. Wang, Y. Zhang, S. Naumov and K. Wurst, *Eur. J. Inorg. Chem.*, 2007, 3988–4000; (c) F. Grisi, C. Costabile, E. Gallo, A. Mariconda, C. Tedesco and P. Longo, *Organometallics*, 2008, **27**, 4649–4656.
57. J. J. Lippstreu and B. F. Straub, *J. Am. Chem. Soc.*, 2005, **127**, 7444–7457.
58. H. Clavier, A. Correa, E. C. Escudero-Adán, J. Benet-Buchholz, L. Cavallo and S. P. Nolan, *Chem.–Eur. J.*, 2009, **15**, 10244–10254.
59. (a) W. J. van Rensburg, P. J. Steynberg, W. H. Meyer, M. M. Kirk and G. S. Forman, *J. Am. Chem. Soc.*, 2004, **126**, 14332–14333; (b) W. J. van Rensburg, P. J. Steynberg, M. M. Kirk, W. H. Meyer and G. S. Forman, *J. Organomet. Chem.*, 2006, **691**, 5312–5325.
60. A. Poater, F. Ragone, A. Correa and L. Cavallo, *J. Am. Chem. Soc.*, 2009, **131**, 9000–9006.
61. B. R. Galan, M. Pitak, M. Gembicky, J. B. Keister and S. T. Diver, *J. Am. Chem. Soc.*, 2009, **131**, 6822–6832.
62. C. E. Webster, *J. Am. Chem. Soc.*, 2007, **129**, 7490–7491.
63. J. Mathew, N. Koga and C. H. Suresh, *Organometallics*, 2008, **27**, 4666–4670.
64. M. Sparta, K. J. Børve and V. R. Jensen, *J. Am. Chem. Soc.*, 2007, **129**, 8487–8499.
65. N. Schneider, M. Finger, C. Haferkemper, S. Bellemin-Laponnaz, P. Hofmann and L. H. Gade, *Chem.–Eur. J.*, 2009, **15**, 11515–11529.
66. A. F. Trindade, P. M. P. Gois, L. F. Veiros, V. André, M. T. Duarte, C. A. M. Afonso, S. Caddick and F. G. N. Cloke, *J. Org. Chem.*, 2008, **73**, 4076–4086.
67. K. L. Tan, R. G. Bergman and J. A. Ellman, *J. Am. Chem. Soc.*, 2002, **124**, 3202–3203.
68. S. H. Wiedemann, J. C. Lewis, J. A. Ellman and R. G. Bergman, *J. Am. Chem. Soc.*, 2006, **128**, 2452–2462.
69. K. J. Hawkes, K. J. Cavell and B. F. Yates, *Organometallics*, 2008, **27**, 4758–4771.
70. (a) G. Sini, O. Eisenstein and R. H. Crabtree, *Inorg. Chem.*, 2002, **41**, 602–604; (b) L. J. L. Häller and S. A. Macgregor, *Eur. J. Inorg. Chem.*, 2009, 2000–2009.
71. K. Albert, P. Gisdakis and N. Rösch, *Organometallics*, 1998, **17**, 1608–1616.
72. M.-T. Lee, H. M. Lee and C.-H. Hu, *Organometallics*, 2007, **26**, 1317–1324.
73. H. Lebel, M. K. Janes, A. B. Charette and S. P. Nolan, *J. Am. Chem. Soc.*, 2004, **126**, 5046–5047.
74. R. Lindner, C. Wagner and D. Steinborn, *J. Am. Chem. Soc.*, 2009, **131**, 8861–8874.

75. A. A. Danopoulos, N. Tsoureas, S. A. Macgregor and C. Smith, *Organometallics*, 2007, **26**, 253–263.
76. L. C. Silva, P. T. Gomes, L. F. Veiros, S. I. Pascu, M. T. Duarte, S. Namorado, J. R. Ascenso and A. R. Dias, *Organometallics*, 2006, **25**, 4391–4403.
77. J. C. Green, B. J. Herbert and R. Lonsdale, *J. Organomet. Chem.*, 2005, **690**, 6054–6067.
78. T. Schaub, P. Fischer, A. Steffen, T. Braun, U. Radius and A. Mix, *J. Am. Chem. Soc.*, 2008, **130**, 9304–9317.
79. D. V. Yandulov and N. T. Tran, *J. Am. Chem. Soc.*, 2007, **129**, 1342–1358.
80. G. A. Chass, C. J. O'Brien, N. Hadei, E. A. B. Kantchev, W. H. Mu, D. C. Fang, A. C. Hopkinson, I. G. Csizmadia and M. G. Organ, *Chem.–Eur. J.*, 2009, **15**, 4281–4288.
81. T. Hatakeyama, S. Hashimoto, K. Ishizuka and M. Nakamura, *J. Am. Chem. Soc.*, 2009, **131**, 11949–11963.
82. S. C. Wang, D. M. Troast, M. Conda-Sheridan, G. Zuo, D. LaGarde, J. Louie and D. J. Tantillo, *J. Org. Chem.*, 2009, **74**, 7822–7833.
83. B. V. Popp and S. S. Stahl, *J. Am. Chem. Soc.*, 2007, **129**, 4410–4422.
84. R. J. Nielsen and W. A. Goddard, *J. Am. Chem. Soc.*, 2006, **128**, 9651–9660.
85. J. Li and Z. Y. Lin, *Organometallics*, 2009, **28**, 4231–4234.
86. X. Z. Yang and M. B. Hall, *J. Am. Chem. Soc.*, 2008, **130**, 1798–1799.
87. X. Z. Yang and M. B. Hall, *J. Organomet. Chem.*, 2009, **694**, 2831–2838.
88. P. M. Zimmerman, A. Paul and C. B. Musgrave, *Inorg. Chem.*, 2009, **48**, 5418–5433.
89. P. M. Zimmerman, A. Paul, Z. Zhang and C. B. Musgrave, *Angew. Chem., Int. Ed.*, 2009, **48**, 2201–2205.
90. L. Dang, H. T. Zhao, Z. Y. Lin and T. B. Marder, *Organometallics*, 2007, **26**, 2824–2832.
91. H. T. Zhao, Z. Y. Lin and T. B. Marder, *J. Am. Chem. Soc.*, 2006, **128**, 15637–15643.
92. L. Dang, H. T. Zhao, Z. Y. Lin and T. B. Marder, *Organometallics*, 2008, **27**, 1178–1186.
93. V. Lillo, M. R. Fructos, J. Ramírez, A. A. C. Braga, F. Maseras, M. M. Díaz-Requejo, P. J. Pérez and E. Fernández, *Chem.–Eur. J.*, 2007, **13**, 2614–2621.
94. H. T. Zhao, L. Dang, T. B. Marder and Z. Y. Lin, *J. Am. Chem. Soc.*, 2008, **130**, 5586–5594.
95. A. B. Kazi, T. R. Cundari, E. Baba, N. J. DeYonker, A. Dinescu and L. Spaine, *Organometallics*, 2007, **26**, 910–914.
96. S. Díez-González, A. Correa, L. Cavallo and S. P. Nolan, *Chem.–Eur. J.*, 2006, **12**, 7558–7564.
97. J. M. Hayes, M. Viciano, E. Peris, G. Ujaque and A. Lledós, *Organometallics*, 2007, **26**, 6170–6183.

98. C. Gourlaouen, N. Marion, S. P. Nolan and F. Maseras, *Org. Lett.*, 2009, **11**, 81–84.
99. A. Correa, N. Marion, L. Fensterbank, M. Malacria, S. P. Nolan and L. Cavallo, *Angew. Chem., Int. Ed.*, 2008, **47**, 718–721.
100. N. Marion, G. Lemière, A. Correa, C. Costabile, R. S. Ramón, X. Moreau, P. de Frémont, R. Dahmane, A. Hours, D. Lesage, J. C. Tabet, J. P. Goddard, V. Gandon, L. Cavallo, L. Fensterbank, M. Malacria and S. P. Nolan, *Chem.–Eur. J.*, 2009, **15**, 3243–3260.
101. N. Marion, P. Carlqvist, R. Gealageas, P. de Frémont, F. Maseras and S. P. Nolan, *Chem.–Eur. J.*, 2007, **13**, 6437–6451.

CHAPTER 3

Synthesis, Activation and Decomposition of N-Heterocyclic Carbene-containing Complexes

JEREMY M. PRAETORIUS AND CATHLEEN M. CRUDDEN*

Queen's University, Chernoff Hall, 90 Bader Lane, Kingston, Ontario, K7L 3N6, Canada

3.1 Introduction

The discovery of N-heterocyclic carbenes (NHCs) as ligands for transition metal complexes has been likened by some to the discovery of triphenylphosphine. These interesting ligands provide a novel coordination environment for the metals to which they are bound, very distinct from the ubiquitous phosphine ligands.[1] Being very electron-rich, but considerably less sensitive to oxidation than phosphines, NHCs also provide new opportunities in catalysis.[2]

With bond strengths to transition metals estimated at approximately double those of phosphines,[3] NHC ligands have been considered to be mainly inert. Despite many early examples demonstrating that NHCs can be displaced from the metal coordination sphere, the concept of substitutionally and chemically inert NHC–metal bonds persists in the current literature. Additionally, the unique bonding mode of NHCs makes them more susceptible to certain decomposition reactions such as C–H activation and reductive elimination/migratory insertion reactions.[4] In addition to a description of the various decomposition pathways available to [(NHC)M] complexes, methods to

RSC Catalysis Series No. 6
N-Heterocyclic Carbenes: From Laboratory Curiosities to Efficient Synthetic Tools
Edited by Silvia Díez-González
© Royal Society of Chemistry 2011
Published by the Royal Society of Chemistry, www.rsc.org

enhance the stability of NHC ligands will be presented in the present Chapter. The use of some of these decomposition reactions to generate catalytically active NHC–metal complexes will also be described.

3.2 Synthetic Approaches to NHC-containing Metal Complexes

3.2.1 Early Investigations

After the initial publications of Öfele[5] and Wanzlick[6] describing different methods for the preparation of metal–carbene complexes from the corresponding imidazolium salts, Lappert carried out considerable research on the synthesis of NHC-containing metal complexes starting from electron-rich enetetramines,[7] synthesizing NHC complexes of Fe, Ru, Os, Rh, Ir, Ni, Pd, Pt and Au.[8] The major limitations of this approach are that the reactions had to be performed at relatively high temperatures, making complexes of limited stability inaccessible, and that enetetramines with bulky substituents cannot be prepared.[9] Interestingly, the isolation of free carbenes by Bertrand[10] and Arduengo[11] in the late 1980s/early 1990s, was actually assisted by the use of bulky substituents. The remarkable stability of these species spurred significant interest in this field.[12]

3.2.2 Reaction with Isolated Carbenes

In 1991, Arduengo and co-workers isolated and characterized the first 'bottleable' N-heterocyclic carbene, IAd.[11] Following this report, Arduengo described the simple coordination of these new compounds to a series of transition metals and main group elements, and Herrmann was the first to demonstrate the considerable potential of these complexes in catalysis.[13] For example, the addition of IMe to $[Rh(COD)Cl]_2$ (COD = 1,5-cyclooctadiene), broke the dimer to give complex **1**, which could react with a second equivalent of carbene to displace the chloro ligand giving cationic complex **2**, or alternatively, with carbon monoxide to displace COD and give **3** (Scheme 3.1).[14]

In a significant advance, Herrmann also demonstrated that imidazolium ions themselves could be employed with a suitably basic metal precursor. For example, reaction of IMe·HI with $[Pd(OAc)_2]$ or $[Rh(CO)_2(acac)]$ (acac = acetylacetonato) cleanly generated $[(IMe)_2PdI_2]$ and $[(IMe)RhI(CO)_2]$, respectively with expulsion of the protonated anionic ligands.[14,15] An extension of this approach involved generating alkoxy-bridged metal dimers from their corresponding chloro-bridged dimers, followed by reaction with imidazolium salts (Scheme 3.2).[16]

Cyclopentadienyl ligands (Cp) of metallocenes could also be sacrificed in the deprotonation of imidazolium salts generating [(NHC)M(Cp)] complexes and 1,3-cyclopentadiene.[17]

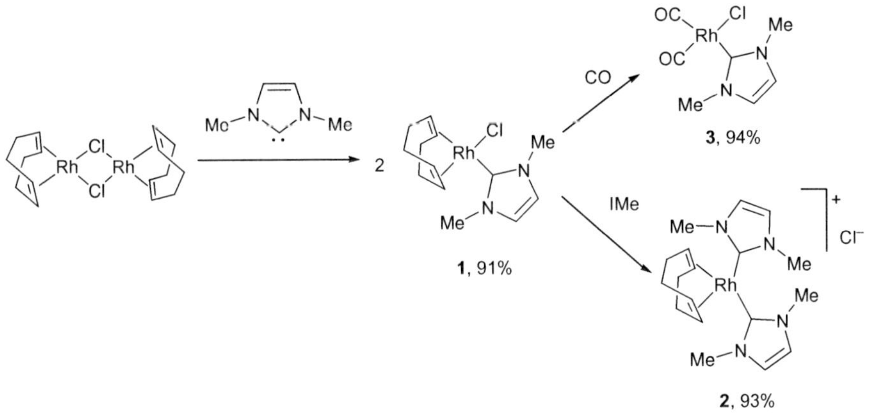

Scheme 3.1 Synthesis of [(NHC)Rh] complexes from pre-generated free NHCs.

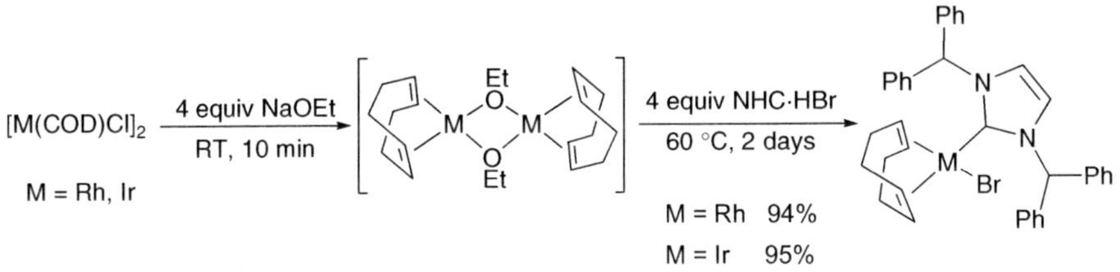

Scheme 3.2 Alternative preparation of [(NHC)M] complexes.

3.2.3 The Silver Oxide Method

The use of NHC–silver complexes as transmetalating agents has become one of the most popular methods for the generation of transition metal carbene complexes.[18] These reagents are readily prepared by reaction of Ag_2O with the corresponding azolium ion. Displaying fluxional behaviour on the NMR time scale, these reagents have proved to be valuable for exchange reactions with other metals that bind the NHC more tightly.

The first NHC–silver complex was reported in 1993, obtained by the reaction of free IMes with AgOTf.[19] Although this reaction proceeded cleanly and in good yield, it did require the isolation of a free carbene. Thus, the *in situ* preparation of the silver NHC transfer reagent by reaction of imidazolium salts with basic silver sources is more convenient.[20] Lin first found that reacting Ag_2O with *N*,*N′*-di(ethyl)benzimidazolium bromide yielded **4** (Scheme 3.3), a suitable transfer reagent for generating [(NHC)AuX] and [(NHC)PdL_n] compounds. Also, the AgBr formed could be used to regenerate **4** under basic phase-transfer conditions.[21] Remarkably, this reaction can be performed in many solvents without rigorous exclusion of air or moisture, even in water.[22]

While higher temperatures were required for the preparation of these reagents from bulkier carbenes,[20b] the generation and transmetalation of silver

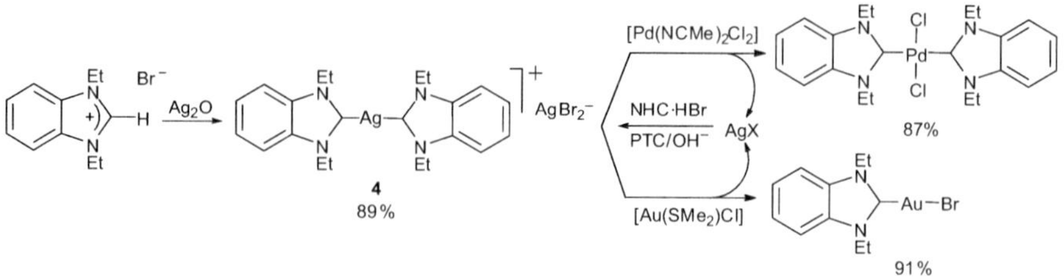

Scheme 3.3 Synthesis of [(NHC)Pd] and [(NHC)Au] complexes using the silver oxide method.

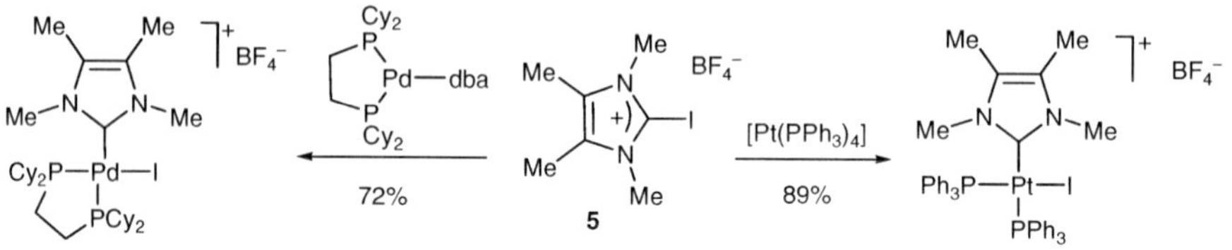

Scheme 3.4 Oxidative addition of the C2–X (X=H, I) bond of imidazolium salts.

NHC species is remarkably general. Like the formation of the [(NHC)AgX] species itself, subsequent reaction with transition metal precursors can be carried out in air without rigorous purification of solvents, with some exceptions.[23] In an interesting application, Crabtree and co-workers used sequential additions of two different silver NHC reagents in order to synthesize iridium complexes of the formula [(NHC^1)(NHC^2)Ir(COD)].[24]

3.2.4 Oxidative Addition

The direct oxidative addition of the C2–H to electron-rich late transition metals is another method for the preparation of NHC–metal complexes. Mild bases can also promote this reaction. Considering the large difference in pK_a between the imidazolium ion and the weak bases such as triethylamine or metal carbonates that are employed, it is likely that the base serves to deprotonate the oxidatively added imidazolium ion driving the reaction forwards.[25]

Early forays into this reaction were described by Cavell and Yates. Experimental studies demonstrated that oxidative addition of both C2–H and C2–I imidazolium ions to platinum was feasible.[26] The C2–I substituted imidazolium ion **5** also underwent oxidative addition to [Pd(PPh_3)$_4$]; however, an attempted reaction with the C2–H imidazolium met with failure (Scheme 3.4).[27] Palladium has been shown to oxidatively add into C2–H imidazolium ions in bidentate systems resulting in palladium complexes **6** and **7** (Scheme 3.5).[28] The oxidative addition of imidazolium ions to iridium, generating (NHC)Ir^{III}-hydrides was also reported.[25a,29]

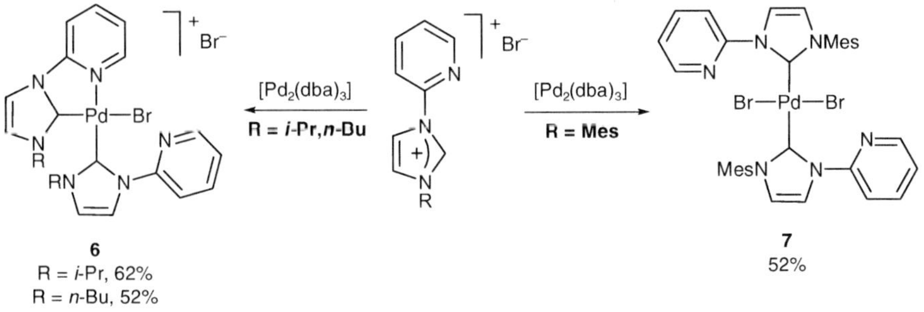

Scheme 3.5 Oxidative addition of N'-pyridyl functionalized NHCs to $[Pd_2(dba)_3]$.

Cavell also reported the oxidative addition of imidazolium salts to coordinatively unsaturated bis-IMes complexes of Pd and Ni (Equation (3.1)).[30] Reactions occurred cleanly, generating tris-NHC metal hydrides, which were remarkably stable to reductive elimination, likely due to the steric constraints of NHCs, which prevent the orbital overlap between the hydride and the carbene carbon required for reductive elimination.

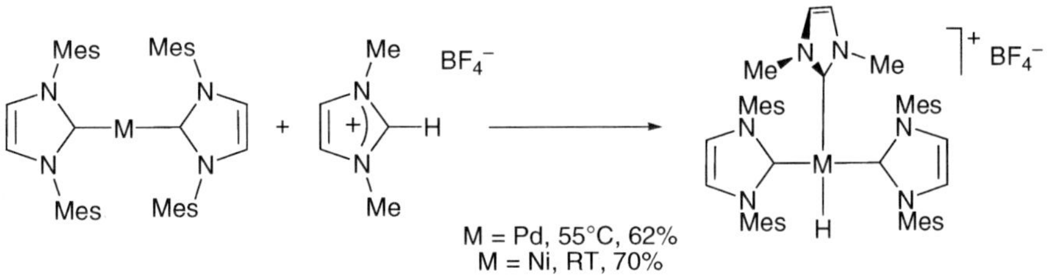

(3.1)

Also, Fürstner *et al.* demonstrated that oxidative addition of C2–Cl imidazolium ions to $[Pd(PPh_3)_4]$ generated mixed phosphine–NHC complexes.[31] Baker described the interesting C–C oxidative addition reaction of diimidazolium ion **8** to $[Pd(PPh_3)_4]$ yielding the bis-NHC cyclophane complex **9** (Equation (3.2)).[32]

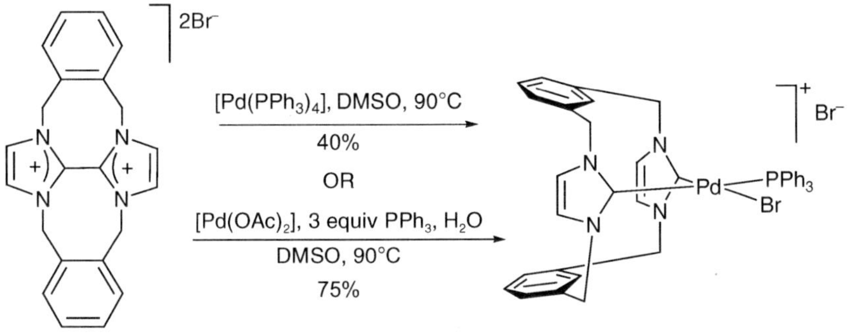

(3.2)

3.2.5 Generation of [(NHC)M] Complexes in Ionic Liquids

In recent years, ionic liquids have become extremely popular as a 'green' alternative to conventional reaction solvents.[33] Since a number of the most popular ionic liquids used are *N,N'*-dialkylated imidazolium salts, their true role in transition metal catalysed reactions, particularly in the presence of base, must be scrutinized.[34] Following the initial report of a successful Heck reaction performed in an ionic liquid by Earle and co-workers,[35] Xiao and co-workers were able to isolate a number of isomeric $[(NHC)_2PdBr_2]$ complexes from the reaction mixture (Equation (3.3)).[36] Under stoichiometric conditions, heating $[Pd(OAc)_2]$ and NaOAc in the ionic liquid, [BMIM]Br, the dimeric palladium carbene complex, $[(IBuMe)PdBr_2]_2$ could be readily obtained, which, upon further heating, generated bis-NHC–Pd complexes.

NaOAc, 100°C, 4 h

63% overall yield
cis/trans = 6:1
cis-anti/cis-syn = 1:1.7
trans-anti/trans-syn = 1.5:1

(*cis-anti*) (*cis-syn*) (*trans-syn*) (*trans-anti*)

(3.3)

Welton and co-workers reported a detailed study of the Suzuki reaction performed in imidazolium ionic liquids using palladium phosphine based catalysts,[37] and found that mixed phosphine–NHC palladium complexes of the formula $[(IBuMe)Pd(PPh_3)_2X]$ were formed. All catalytic conditions leading to formation of these complexes were successful in affecting the Suzuki reaction, and those conditions in which they could not be detected showed no conversion. This strongly suggests that [(NHC)Pd] complexes are relevant to reactions run in imidazolium-based ionic liquids.

3.2.6 Reaction with C2-functionalized Carbene Adducts

Another alternative to generate and handle the free carbene is to mask the C2-position of the imidazol-2-ylidene with various thermally labile groups, such that heating these adducts in the presence of transition metal precursors gives the desired NHC–metal complex. Enders *et al.* found that upon heating to 80 °C, a 5,5-hydromethoxy adduct released methanol producing the free 1,2,4-triazol-5-ylidene (Equation (3.4)).[38]

Ph N N OMe H N Ph Ph $\xrightarrow[\text{quant.}]{80^{\circ}\text{C, 0.1 mbar}}$ Ph N N : N Ph Ph (3.4)

Grubbs and co-workers found that heating this methanol adduct in a toluene solution of $[(PCy_3)_2(Cl)_2Ru{=}CHPh]$ displaced one equivalent of tricyclohexylphosphine generating the Enders-NHC analogue of the famous Grubbs II catalyst.[39] This methodology was extended to saturated carbenes, which can be protected as methanol, *tert*-butanol or chloroform adducts.[39,40]

Even fluorinated aromatics such as $H–C_6F_5$ can react with carbenes and serve as protecting groups to be eliminated under relatively mild conditions.[41] The resulting adducts, obtained by acid-catalyzed condensation of a fluorinated benzaldehyde and a diamine (notably without prior generation of the free carbene) are comparable in stability and reactivity to the chloroform adducts.

Crabtree and co-workers have recently reported the use of imidazole-2-carboxylates[42] as suitable precursors for the formation of a variety of transition metal NHC complexes.[43] These air and moisture stable adducts can be prepared from *N*-substituted imidazoles by reaction with dimethylcarbonate giving CO_2 adducts.[43b]

3.2.7 Templated Synthesis of [(NHC)M] Complexes

An alternative method for the synthesis of [(NHC)M] complexes is to bind ligands to a transition metal that can subsequently be converted into an NHC ligand. Using 2-azidophenyl isocyanide[44] or 2-nitrophenyl isocyanide[45] as synthons for the unstable 2-amino isocyanide, diprotic benzannulated NHC–metal complexes were obtained, and easily alkylated to give *N,N′*-disubstituted benzimidazolylidenes. This methodology was also applied to ruthenium-based catalysts (Scheme 3.6).[46]

3.3 Decomposition Reactions

Decomposition reactions of N-heterocyclic carbenes are an important aspect of NHC chemistry that needs to be considered in the design, synthesis and application of NHC complexes. In many cases, the decomposition reactions involve species on the metal that are key intermediates in the catalytic cycle, and so the relative reactivity of the desired catalytic transformations *versus* the decomposition reaction must be assessed. It is important to note that reactions occurring under stoichiometric conditions may not mirror reactivity under catalytic conditions. In addition, some decomposition reactions are reversible. The reductive elimination of an NHC and hydride ligand is an example of this, as its

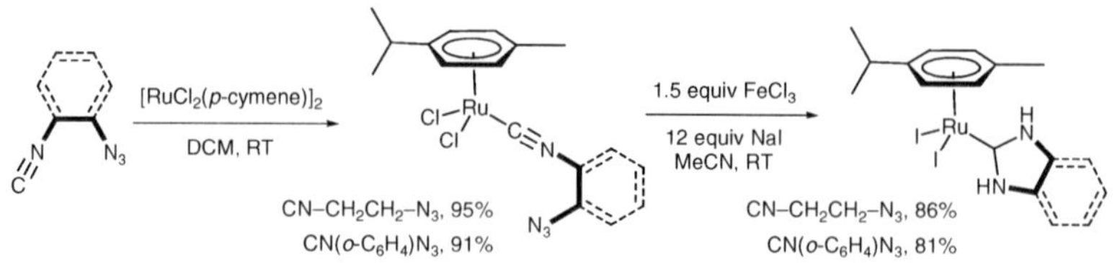

Scheme 3.6 Templated synthesis of a diprotic benzannulated NHC-ruthenium complex.

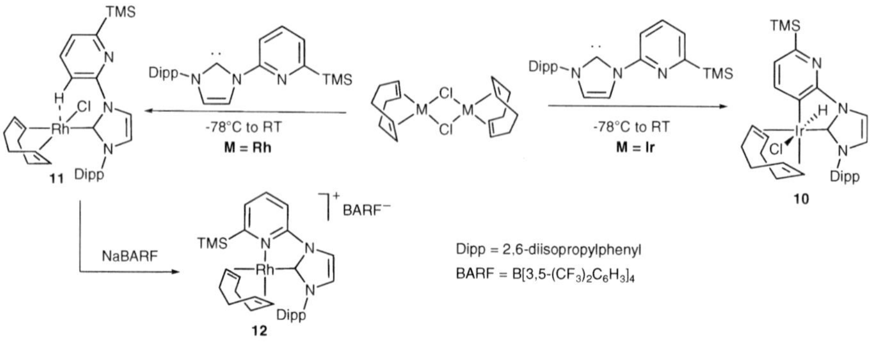

Scheme 3.7 Reactivity of Ir and Rh complexes with *N*-pyridyl functionalized NHCs.

reverse reaction—oxidative addition of an imidazolium salt—is often employed in catalyst synthesis (see Section 3.2.5). Thus, catalyst death *via* this reaction can be limited by running the reaction in the presence of additional imidazolium salt or in an ionic liquid with the same structure as the desired ligand.

3.3.1 C–H/C–C/C–N Bond Activation Reactions

The first C–H bond activation of an Ir-bound NHC was reported by Lappert in 1983. Treatment of a tetrakis-tolyl enetetramine with $[Ir(COD)Cl]_2$ in refluxing toluene resulted in a complex in which each of the three NHC substituents was orthometalated.[47] These and related complexes attracted considerable attention for their potential as Ir-based phosphorescent organic light-emitting diodes.[48]

C–H bond activation is extremely common in Ir complexes, possibly due to the high electron density on Ir, and its stability in higher oxidation states. NHCs bearing *N*-aryl substituents lacking *ortho* blocking groups are particularly susceptible.[49] For example, in an attempted synthesis of a donor-functionalized NHC–Ir complex, Danopoulos and co-workers found that NHCs bearing α-pyridyl ligands underwent C–H activation at the unprotected *ortho* position of the pyridyl ring yielding **10** upon reaction with $[Ir(COD)Cl]_2$ (Scheme 3.7). In the analogous reaction with $[Rh(COD)Cl]_2$, complex **11** featuring an agostic interaction with the same C–H bond was obtained instead.

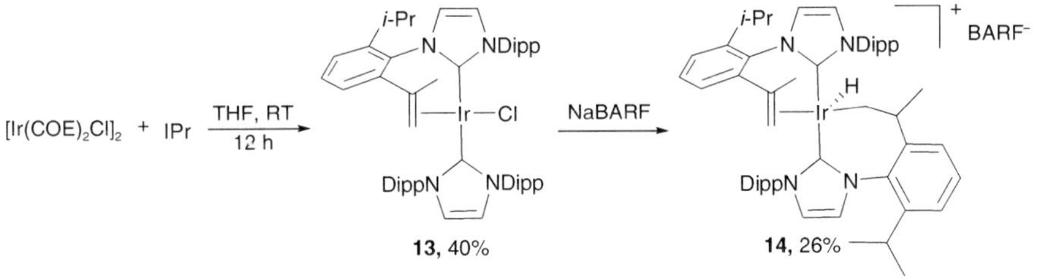

Scheme 3.8 Sequential C–H activations of the *i*-Pr group of the two IPr ligands by $[(IPr)_2IrCl]$ complexes.

Even chloride abstraction from **11** did not yield the C–H activated Rh complex, but promoted bond formation between Rh and the pyridyl nitrogen.[50]

Employing NHCs functionalized at the *ortho* positions of the *N*-aryl substituents (*i.e.* IMes or IPr) can increase stability, but in place of *N*-aryl C–H activation, the C–H bonds of the methyl or even isopropyl groups might react. Sometimes, the C–H activation is reversible and thus goes undetected, however the presence of a molecule that can accept hydrogen from the metal is often used to drive this reaction to completion.[51]

The C–H activation of IMes ligands is relatively common; however, activation of the isopropyl substituent in IPr occurs more rarely. For example, the reaction of $[Ir(COE)_2Cl]_2$ (COE = cyclooctene) with IPr at room temperature, produced complex **13** (Scheme 3.8). Upon chloride abstraction, C–H activation of a methyl hydrogen from the second IPr ligand was observed to give **14**.[52] Sigman and co-workers have observed a similar decomposition reaction with [(IPr)Ni] complexes.[53]

Although a variety of *N*-alkyl substituted NHCs were susceptible to C–H activation, as illustrated by Herrmann,[54] huge differences in reactivity were observed depending on the substituent. For example, treatment of the *N*-ethyl or *N*-isopropyl complexes **15** and **16** with base generated cyclometalated iridium hydrides **17** and **18** (Equation (3.5)).[55] Interestingly, the related *N*-propyl and *N*-butyl carbene complexes did not yield such cyclometalated complexes. Yamaguchi proposed that the accessibility of the β-hydrogens to the metal was responsible for this remarkable difference.[56]

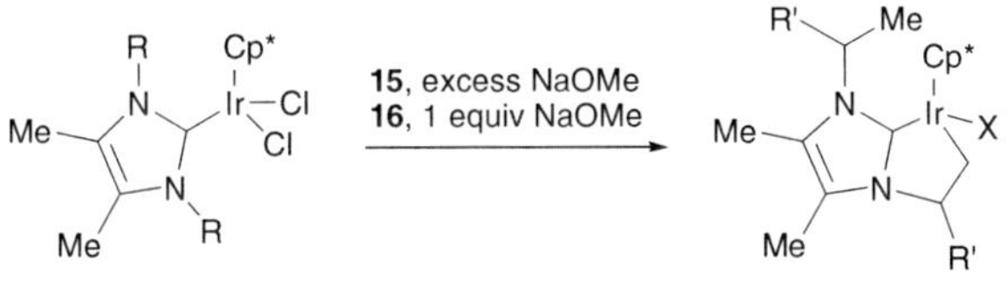

(3.5)

Also, [(NHC)Ir(Cp*)] (Cp* = 1,2,3,4,5-pentamethylcyclopentadienyl) complexes bearing *N*-benzyl substituents underwent C–H activation at the aromatic ring generating a six-membered ring.[57] When a chiral NHC was employed, the C–H activation took place with complete diastereoselectivity resulting in **19**, a highly chemoselective catalyst for the diboration of olefins such as styrene, although poor enantioselectivity was observed (Equation (3.6)).[58]

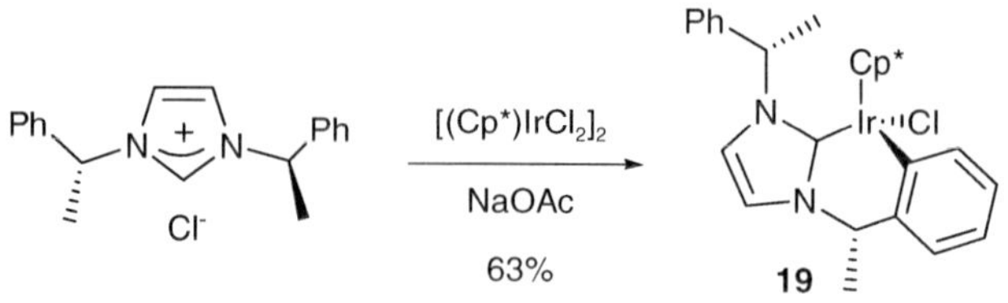

$$[(Cp^*)IrCl_2]_2,\ NaOAc,\ 63\% \rightarrow \mathbf{19} \quad (3.6)$$

N-*t*-Bu substituents are readily metalated, often in the presence of thermodynamically and kinetically favored aromatic C–H bonds. *i*-Pr substituents, on the other hand, are more difficult to metalate, which may be due to the ability of the *i*-Pr group to "escape' the sphere of the Ir, while the *t*-Bu C–H's are forced to interact with the metal, and thus succumb to metalation.[57b] This reactivity was exploited in the design of a catalyst for the deuteration of a wide range of molecules including ketones, ethers, alcohols, terminal alkenes and aromatic groups.[57a] As for Rh, the reaction of $[Rh(COE)_2Cl]_2$ with IMes resulted in *ortho* metalation of the methyl substituents, whereas with ItBu, a variety of cyclometalated products were observed depending on the solvent employed (Scheme 3.9).[59]

C–H activation of the *N*-substituents on NHC ligands was extensively documented in a wide variety of Ru complexes. The first report was from the Lappert group in 1977, who showed that treatment of $[RuCl_2(PPh_3)_3]$ with an enetetramine containing *N*-tolyl substituents led to an orthometalation of one of the *N*-tolyl groups (Scheme 3.10), and more than 20 complexes of general structure

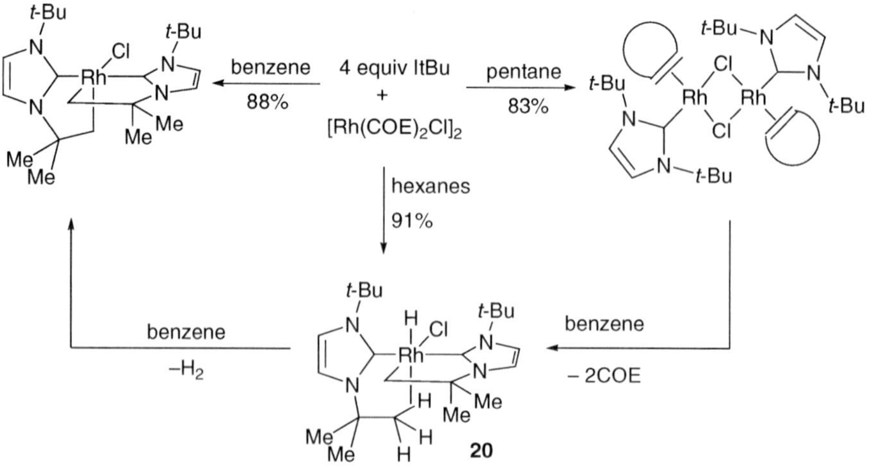

Scheme 3.9 Solvent effects on the metal/ligand stoichiometry and C–H activation.

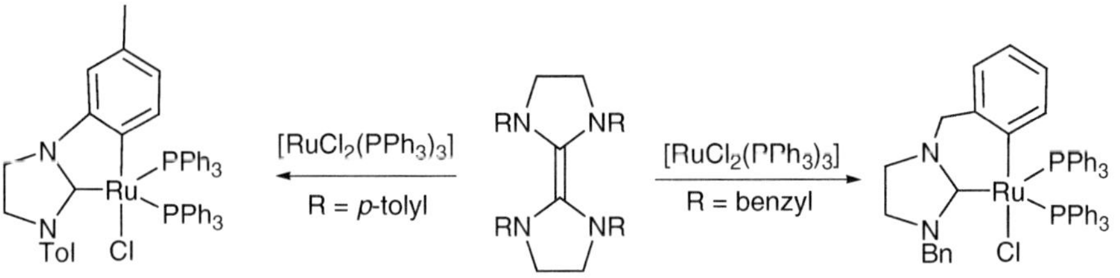

$[RuCl_2(PPh_3)_3]$, R = *p*-tolyl → **22**; **21**; $[RuCl_2(PPh_3)_3]$, R = benzyl → **23**

Scheme 3.10 C–H and C–N activation of ItBu by $[Ni(COD)_2]$.

22 were prepared.[60] Interestingly, the corresponding *N*-alkyl enetetramines reacted without *ortho* metalation illustrating the increased propensity for metalation of *N*-aryl substituents.[61] Similarly, the *N*-benzyl derivative underwent orthometalation at the sp^2 carbon, resulting in the formation of a six-membered ring chelate **23**.[62] However, related [(NHC)Ru(*p*-cymene)] complexes only underwent such activation with *N*-Ph substituents, not with *N*-Bn.[63]

The *N*-phenyl version of the Grubbs second-generation metathesis catalyst undergoes extensive decomposition *via* C–H activation under relatively mild conditions.[64] This species is even less stable than the first-generation bis-phosphine catalyst, even though NHC-derived catalysts are typically considerably more stable than bis-phosphine catalysts.[65] Although the NHC remained attached to the metal throughout this process, the alkylidene was compromised. Similarly, Blechert reported that derivatives of the Hoveyda catalyst bearing aryl ligands without substitution in the *ortho* position underwent C–H activation in air, resulting in new, unreactive alkylidenes.[66] Restricting the rotation of the *N*-aryl substituents by placement of bulky substituents on the backbone of the NHC has proven to be a promising strategy to decrease this undesired reaction.[67]

Numerous C–H activations of the NHC ligand have also been observed in ruthenium clusters, where extensive decomposition could be observed.[68] As observed with Rh and Ir complexes, mesityl substituents were more stable than phenyl substituents, but can be sensitive to C–H activation at the methyl group.[69] In 2003, Leitner reported the facile exchange of hydrogen atoms on the methyl groups of an IMes-based Ru complex with deuterated solvent. This phenomenon was used in the design of a catalyst for the deuteration of aromatic molecules.[70]

In addition to C–H activation, C–C activations and further degradation of the NHC ligand can be observed. For example, complex **24** underwent C–C cleavage and loss of methane upon extensive heating, yielding **25** in 96% yield (Equation (3.7)).[71] Related complex $[(IMes)Ru(PPu_3)_2(H)_2(Co)]$ does not undergo C–C activation, a fact explained computationally by its lower rate of phosphine dissociation of **24** relative to its bis-phosphine analogue.[72] Phosphine dissociation, when followed by H_2 elimination, would give a highly unsaturated Ru species that could more easily undergo C–C activation.

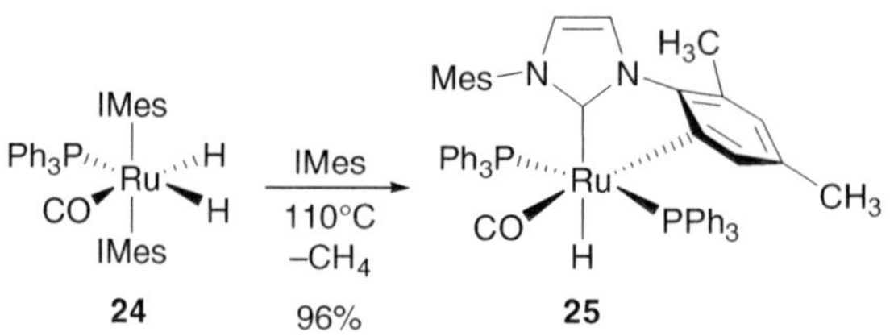

(3.7)

Activation of *N*-alkyl C–H groups was also observed; for example, in the reaction of [Ru(PPh_3)$_3$(CO)H_2] with ^{Me}IiPr in toluene at 70 °C to produce **26**. This is the first example in the Ru series studied by Whittlesey where C–H activation was observed without the use of a hydrogen acceptor (Scheme 3.11).[73] The *N,N′*-diethyl derivative of **26** also underwent cyclometalation, but like other carbene complexes studied, yielded the Ru-alkyl hydride only after reaction with a hydrogen acceptor. Changing the *N,N′*-substituents to an *n*-Pr or methyl groups completely suppressed the C–H activation reaction.

When [Ru(PPh_3)$_3$(CO)Cl(H)] was employed instead, the same C–H activation was observed, accompanied by C–N activation (Scheme 3.11). Compound **30** represents one of only a few examples of *N*-H carbenes that do not spontaneously isomerize to the corresponding *N*-bound species **31**. Instead, the isomerization required base, such as the free carbene.[74]

C–H, C–C and even C–N activation reactions were also observed in Ni complexes of NHC ligands. An extreme example was reported by Caddick and Cloke during the attempted synthesis of [(NHC)$_2$Ni] complexes where, like for **26**, C–N cleavage was preceded by C–H activation (Scheme 3.12).[75]

Although not as frequent as in Ir or Ru complexes, C–H activation of *N*-aryl substituents in [(NHC)Pd]$_n$ complexes is also known to occur and was shown to have an influence on catalytic reactions.[76] Danopoulos and co-workers showed that NHC complexes in which the *N*-aryl moiety contained only one substituent underwent metalation at the aromatic C–H bond. Depending on the reaction conditions, one or both of the NHC ligands undergo metalation.[77] Even iron complexes, which are not generally susceptible to orthometalation, were shown to undergo C–H activation of the NHC ligand.[78]

Finally, complexes of the early metals can also undergo C–H activation at the ligand. For example, Fryzuk, Hall and co-workers reported C–H activation in pendant alkyl substituents of tantalum NHC species, making this reaction remarkably general.[79]

3.3.2 Reductive Elimination

A large number of examples of reductive elimination of NHC ligands and adjacent metal alkyl, acyl or hydride ligands have appeared. One critical point that should be remembered when considering these decomposition pathways is that while they may be observed under stoichiometric conditions, they are not necessarily an issue under catalytic conditions.[4a] This is because under catalytic conditions, the rates of the productive catalytic steps can be significantly accelerated relative to the decomposition reactions.[80]

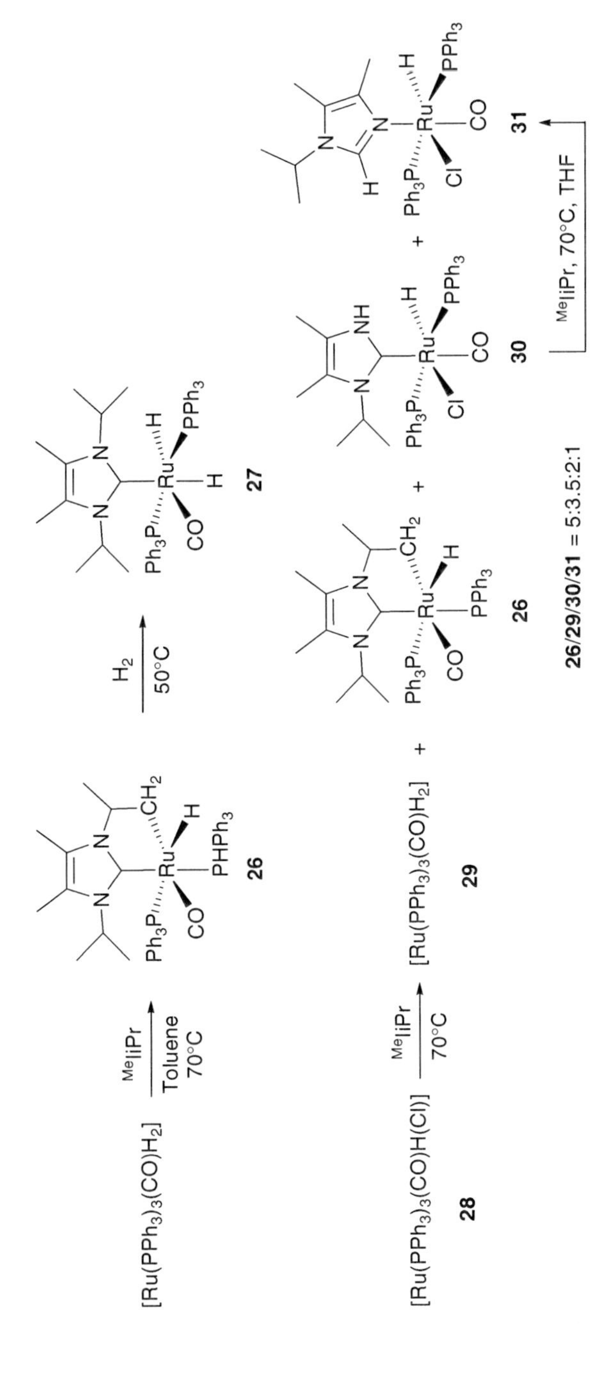

Scheme 3.11 C–H/C–N activation of ^{Me}IiPr in [(NHC)Ru] complexes.

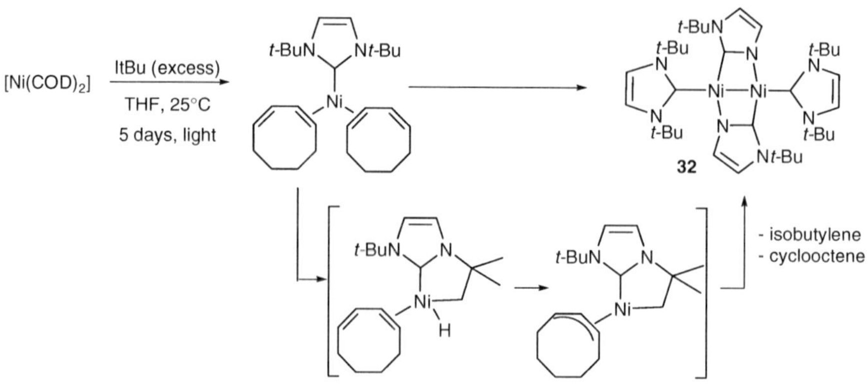

Scheme 3.12 C–H and C–N activation of ItBu by $[Ni(COD)_2]$.

When they occur under catalytically relevant conditions,[81] these reactions can lead to a loss or decrease of activity, or enantioselectivity if chiral ligands are employed. Even when small amounts of decomposition are observed, this can be problematic if the non-NHC-containing catalyst is more active than the NHC-containing species.

As early as 1998, Cavell and co-workers reported that NHC–Pd complexes with ancillary alkyl groups underwent facile reductive elimination, even at room temperature.[82] Subsequent work demonstrated that a wide variety of complexes are susceptible to this reaction, including those with ancillary aryl groups. For example, the attempted oxidative addition of iodobenzene to bis carbene complex **33** resulted in the expected product **34**, along with the C2-phenylated imidazolium ion (Equation (3.8)).[83] When p-NO_2-C_6H_4I was employed as the aryl iodide, oxidative addition complex **35** could be isolated free of the imidazolium ion in 51% yield, likely because reductive elimination was not observed at the lower temperature at which this aryl iodide underwent oxidative addition. Interestingly, complex **35** was an excellent catalyst for the Mizoroki–Heck reaction, conducting the coupling of butylacrylate and bromoacetophenone at 50 000 turnovers h^{-1}.

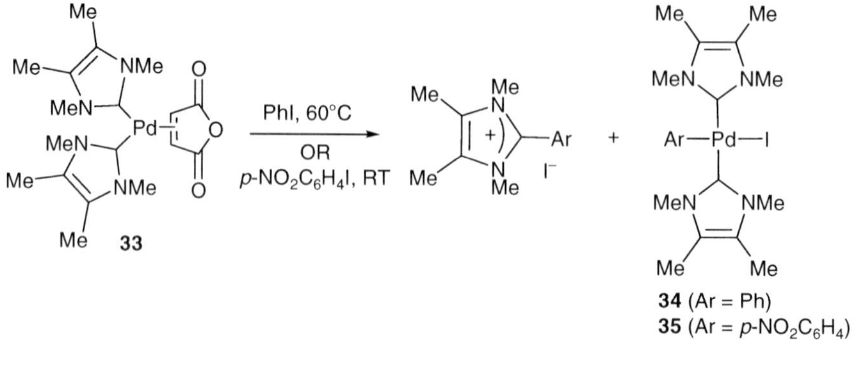

34 (Ar = Ph)
35 (Ar = p-$NO_2C_6H_4$)

(3.8)

When **35** was treated with $AgBF_4$ and butyl acrylate under stoichiometric conditions, decomposition products resulting from reductive elimination from all of the key catalytic intermediates were observed. The stability of the catalyst under the actual reaction conditions was much higher, and was attributed to faster rates of β-hydride elimination from the arylated acrylate at the elevated reaction temperatures, which removed at least two of the decomposition pathways. However, it should be noted that these decomposition pathways could be responsible for the generation of catalytically active Pd complexes containing one or zero carbene ligands in this or other systems.

Caddick and Cloke also studied the oxidative addition of aryl halides to $[(NHC)_2Pd]$ complexes.[84] If the oxidative addition of an aryl halide occurred at a low enough temperature, the immediate product $[(NHC)_2PdAr(Cl)]$ could be isolated. However, if higher temperatures were required, then reductive elimination of the NHC and aryl ligands took place to give C2-arylated imidazolium salt. Interestingly, the oxidative addition product had a *trans* relationship between the aryl and halide ligands, which led Caddick and Cloke to postulate that dissociation of one of the carbenes must have occurred since oxidative addition is expected to result in a complex with the aryl and halide ligands *cis* to one another,[84] a postulate that was later confirmed.[85]

Marshall and Grushin described the first arylation of an NHC that was prepared from a well-defined Pd complex. By using μ-hydroxo dimer **36** that already contained a Pd–aryl ligand, the high temperatures required to introduce the aryl group by oxidative addition were obviated. Complex **37** eliminated the C2-phenylated NHC upon dissolution in DCM at room temperature (Scheme 3.13).[86]

In addition to alkyl and aryl groups, allyl substituents were shown to undergo facile reductive elimination, in a process that was used for the generation of catalytically active Pd^0 species from Pd^{II} precursors (see Section 3.5.3).[87] Although they generally showed increased stability, chelating NHC ligands also underwent reductive elimination reactions.[88]

Nickel-containing carbene complexes are also known to undergo reductive elimination reactions. Depending on the nature of the carbene and the ancillary ligands, this reaction can dominate their chemistry.[83] In a study on the use of

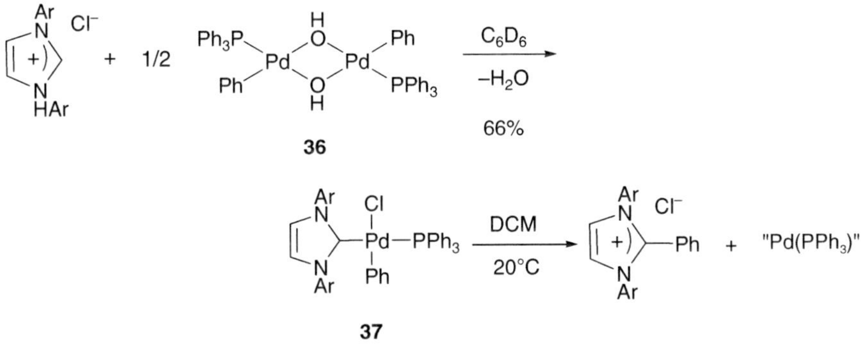

Scheme 3.13 Synthesis and decomposition of **37**.

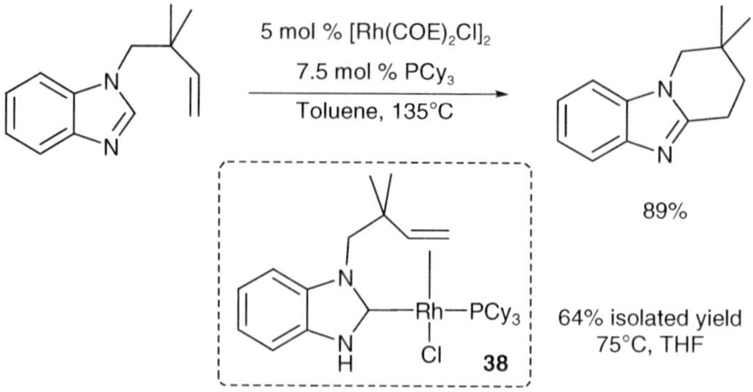

5 mol % $[Rh(COE)_2Cl]_2$
7.5 mol % PCy_3
Toluene, 135°C

89%

64% isolated yield
75°C, THF

38

Scheme 3.14 Heterocycle preparation *via* reductive elimination.

$[(NHC)_2NiX_2]$ catalysts for olefin dimerization and isomerization, Cavell showed that reductive elimination products were observed from each of the intermediates in the catalytic cycle possessing a substituent capable of undergoing such reaction.[81b]

Interestingly, one of the resulting products could be used to regenerate a catalytically active nickel hydride by oxidative addition of the C2–H of the imidazolium ion. Taking advantage of the fact that [BMIM]X is a commonly used ionic liquid, considerably higher activity was reached if the reaction was run in an ionic liquid, and this transformation could be used in the design of novel C2-alkylated ionic liquids, including unique bicyclic species.[89]

Ellman, Bergman and co-workers employed a similar C–H activation followed by addition across pendant olefins leading to a wide variety of useful heterocyclic structures.[90] The reaction proceeded *via* the intermediacy of an *N*-H based Rh N-heterocyclic carbene **38** (Scheme 3.14).[91] Although initial work focused on intramolecular additions to alkenes, intermolecular additions were also reported.[92]

In a study of NHC–Rh complexes aimed at developing an NHC-based alternative to the Monsanto acetic acid catalyst, Haynes and co-workers prepared $[(IMe)_2Rh(CO)(I)]$ and examined its reaction with MeI. Interestingly, even though several observed intermediates had alkyl and acyl groups *cis* to an NHC, reductive elimination was not observed. However, when subjected to a milder version of the industrial reaction conditions, namely 10% MeOH and 2% MeI in chlorobenzene at 120 °C under 10 atm of CO, sequential loss of the two NHCs was observed, generating the NHC-free catalyst $[Rh(CO)_2I_2]^-$, previously employed by Monsanto. This catalyst proved to be considerably more reactive than the NHC-containing species.[93]

3.3.3 Decomposition *via* Elimination of Protonated Carbene (Imidazolium Salt)

In cases where imidazolium ions are observed as decomposition products, it is not always possible to discern whether reductive elimination of a metal hydride,

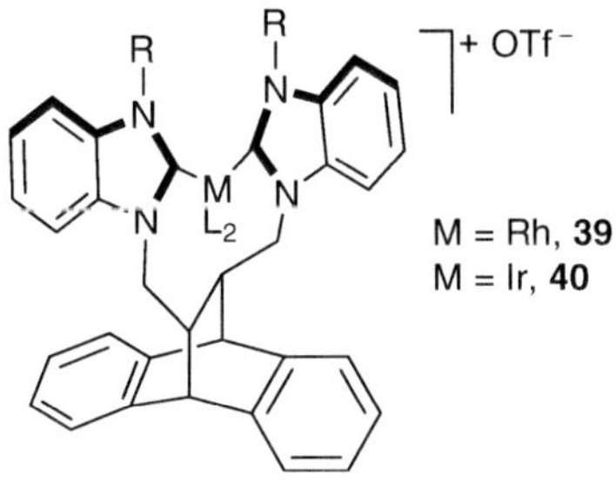

M = Rh, **39**
M = Ir, **40**

Figure 3.1 Rhodium and iridium complexes of a *C*2-symmetric chiral bidentate bis-NHC.

or reaction with a proton source was responsible, except based on reaction conditions. Mixed results were obtained in the reaction of NHC-containing complexes with acid, with high stability being obtained in some cases,[28,94] and significant decomposition observed in others.[53,93,95]

Similarly mixed results were obtained upon exposure of NHC–metal complexes to hydrogen. For example, during a study of the application of NHC–W complexes as catalysts for the hydrogenation of ketones, Bulloch and co-workers found that a significant portion of the catalyst decomposed *via* loss of $[IMes{\cdot}H]^+$. Since the reaction was run in the presence of dihydrogen, reductive elimination of hydride and NHC ligands was a reasonable explanation.[96] The dimeric compound $[(IMes)_2Co_2(CO)_6]$ was also found to decompose upon treatment with 60 bar of synthesis gas (CO/H_2) yielding $[IMesH][HCo(CO)_4]$.[97]

Rhodium complex **39** with a bidentate biscarbene ligand (Figure 3.1) was shown to decompose under hydrogenation conditions yielding Rh^0 and the corresponding protonated imidazolium species.[98] With the related $[(NHC)_2Ir]$ complex **40**, Ir^0 was also formed, despite the fact that only small amounts of decomposition product were observed. It was found that adding excess protonated NHC to the reaction led to a dramatic decrease in activity, providing evidence that the $[(NHC)_2Ir]$ species **40** displayed no catalytic activity until decomposition to give colloidal Ir takes place. This also explained why these systems gave essentially no enantioselectivity. Since highly enantioselective NHC-based Ir hydrogenation catalysts are known,[99] the inactivity of the chiral NHC–M complex observed in this study should not be extrapolated to all Rh or Ir NHC catalysts.

3.3.4 Migratory Insertion

One of the first documented stoichiometric examples of migratory insertion came from the Danopoulos and Green groups, who reacted a pincer carbene with $[(tmeda)PdMe_2]$ (tmeda = *N*,*N*,*N*′,*N*′-tetramethyl-1,2-ethylenediamine). Instead of the expected pincer complex, species **41** was observed in which one of

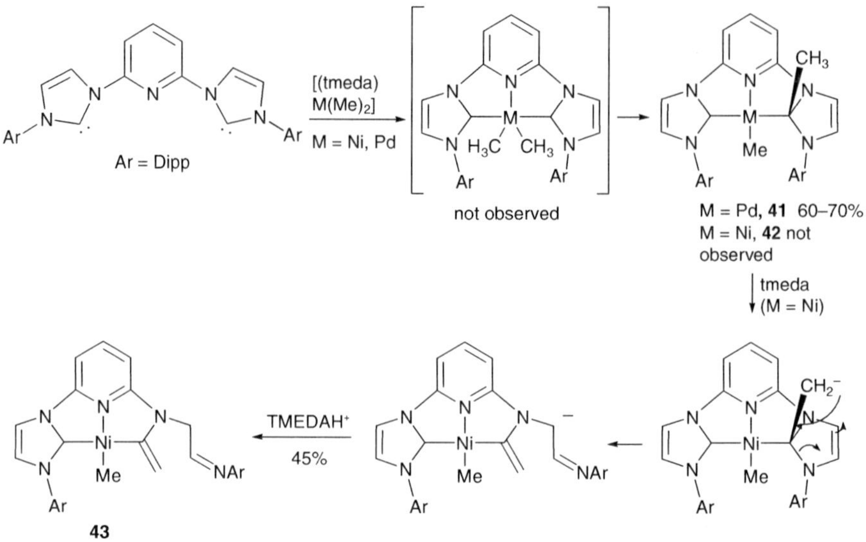

Scheme 3.15 Methyl migrations to carbenic carbon and further decomposition.

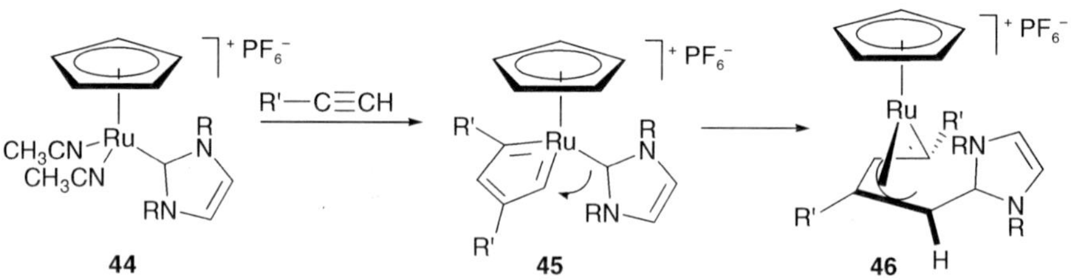

Scheme 3.16 Migratory insertion of an NHC ligand into a ruthenium alkylidene intermediate.

the methyl groups from Pd had migrated onto the NHC ligand. This process took place at room temperature in 60–70% yield (Scheme 3.15).[100]

When the same reaction was carried out with the nickel analogue, complete decomposition to compound **43**, in which one of the NHC rings had opened, took place (Scheme 3.15).[101] A similar migratory insertion of a ligand on nickel to the NHC was proposed by Hall *et al.* based on computational studies of the dehydrogenation of ammonia borane by NHC–Ni–NHC based catalysts.[102]

Also, ruthenium complex **44** reacted with alkynes yielding **45**, which underwent migratory insertion of the NHC ligand into the Ru alkylidene yielding **46** (Scheme 3.16).[103] Trzeciak and co-workers demonstrated that [(NHC)Ru(*p*-cymene)Cl_2] was able to polymerize phenyl acetylene, and that the resulting product was terminated by the imidazolium ion, likely *via* the same process.[104]

3.3.5 Displacement Behavior

Although there is no doubt that carbenes form stronger bonds to transition metals than phosphines, there are ample reports in the literature describing their displacement by other ligands. In 1978, Lappert demonstrated that when treated with CO, $[(NHC)_4RuCl_2]$ lost one carbene ligand to generate $[(NHC)_3RuCl_2(CO)]$.[47] Similarly, the same Ru complex could react with DCM at room temperature, resulting in the loss of one NHC ligand.[105]

NHC derivatives of Wilkinson's catalyst **47** succumbed to displacement of the carbene by bidentate phosphines such as 1,2-bis(diphenylphosphino)ethane (dppe), albeit under forcing conditions (Scheme 3.17).[106] A related bidentate carbene/pyridine rhodium complex also underwent quantitative ligand displacement when treated with dppe or 1,3-bis(diphenylphosphino)propane (dppp) at room temperature.[107]

Mesityl analogue **48** proved more sensitive to loss of the carbene.[105,108] While no evidence of decomposition was obtained under inert atmosphere in non-polar solvents such as toluene, when exposed to dichloromethane or dichloroethane in the presence of monodentate phosphines, complete conversion to Wilkinson's catalyst was observed. The carbene decomposition product, **49**, was isolated along with small amounts of IMes·HCl (Equation (3.9)). Although it was not clear whether this reaction took place on the metal or *via* dissociated carbene, treatment of the free carbene with DCE at 60 °C led to the same product along with IMes·HCl, providing strong evidence that free carbene was generated. Since Wilkinson's catalyst was shown to be more active in certain reactions than the carbene analog, understanding conditions under which the carbene complex is stable is critical.

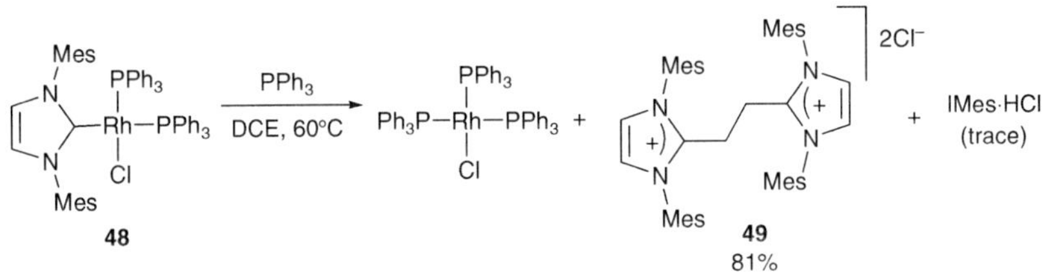

(3.9)

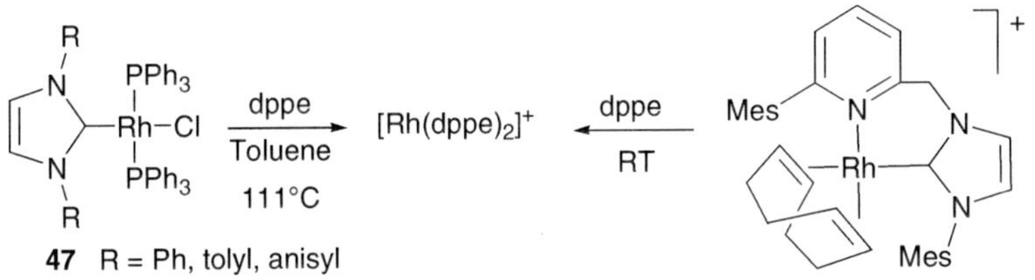

Scheme 3.17 Displacement of NHC ligands from rhodium complexes with dppe.

A carbonyl analogue of **48**, [(IMes)RhCl(CO)(PAr_3)] exhibited significantly greater stability and it could be purified by column chromatography, while **48** degraded by oxidation of one of the phosphines and binding of O_2 to the metal,[109] illustrating the importance of ancillary ligands and overall electron density at the metal.

Cobalt complexes are also prone to carbene displacement. In an attempted synthesis of [(IPr)Co(Cp)Me_2] from [(Ph_3P)Co(Cp)Me_2], Baird and co-workers found that an equilibrium was set up between free IPr and the phosphine complex.[110] Isolated [(IPr)Co(Cp)Me_2] reacted with PMe_3 resulting in complete displacement of the carbene ligand. The authors attributed this instability to steric crowding.

In 2008, Nolan reported that treating [(IPr)Pd(π-allyl)Cl] with four equivalents of PPh_3 in the presence of *i*-PrOH and a base at room temperature led to the quantitative formation of [Pd(PPh_3)$_4$].[111] This transformation was believed to begin by displacement of IPr from the coordination sphere of Pd with PPh_3, since [Pd(PPh_3)(π-allyl)Cl] was identified in the reaction of the IPr complex with PPh_3 at room temperature after 20 min. The fact that this displacement occurs under such mild conditions is truly remarkable, and implies that caution needs to be exercised when NHC–Pd complexes are employed along with phosphines. A derivative of [(IPr)Pd(π-allyl)Cl] in which the NHC has one mesityl substituent and one methylene pyridyl ligand was also shown to undergo decomposition upon treatment with base and PCy_3. In this case, the formation of Pd nanoparticles was observed.[112]

Caddick and Cloke reported similar exchange reactions as early as 2001. Bis-carbene Pd complexes were reacted with phosphines resulting in displacement of the carbene under surprisingly mild conditions. In the case of P(*o*-tolyl)$_3$, an equilibrium between the mixed phosphine carbene complex and the bis carbene complex was established at 60 °C (Equation (3.10)).[85,113]

(NHC)–Pd–(NHC) + PR'$_3$ $\xrightleftharpoons{C_6D_6}$ (NHC)–Pd–PR'$_3$ (3.10)

NHC = ItBu or SIPr

R' = *o*-tolyl, 60°C: 33%
= Cy: 100%

3.3.6 Miscellaneous Decomposition Reactions

Treatment of Grubbs' second-generation metathesis catalyst with carbon monoxide results in a Buchner rearrangement in which the alkylidene inserted into one of the aryl substituents of the NHC (Scheme 3.18).[114] It was proposed that CO, being a strong π-acceptor, decreases the electron density available for back donation and stabilization of the alkylidene ligand. The alkylidene thus becomes more electrophilic, and is attacked by the adjacent aryl ring.

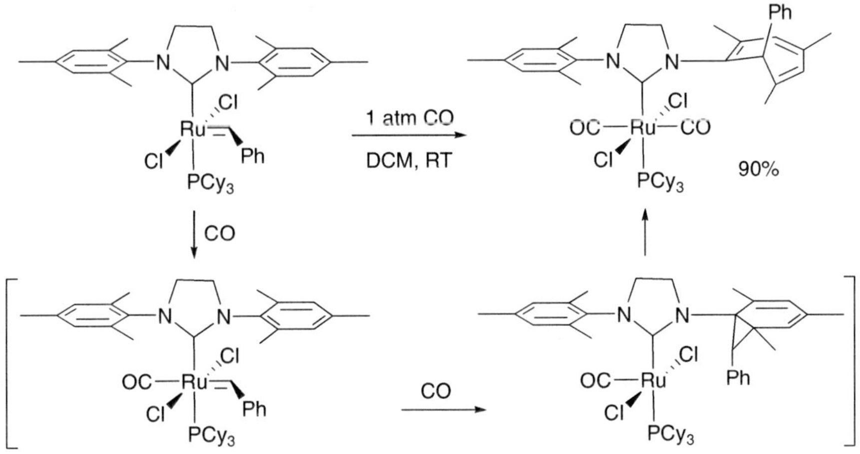

Scheme 3.18 CO-induced decomposition of Grubbs II catalyst.

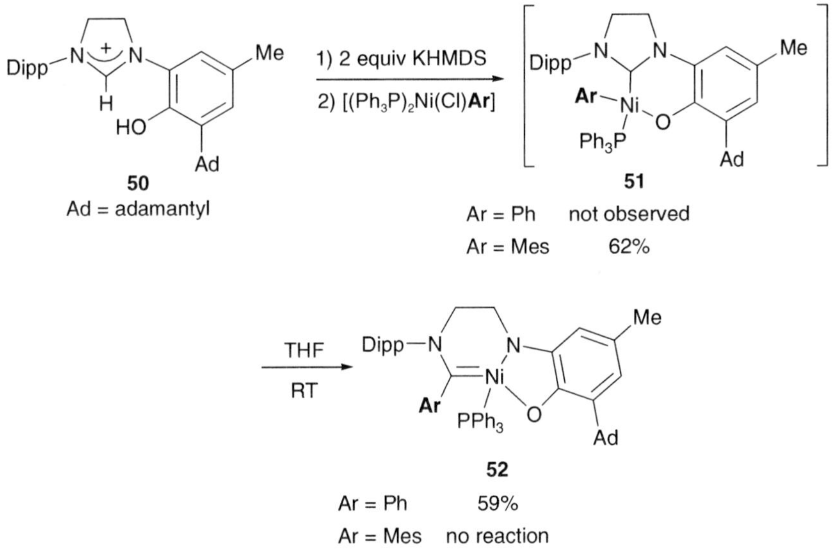

Scheme 3.19 Decomposition reaction of an [(NHC)Ni(Ar)] complex.

These postulates were supported by a DFT study carried out by Cavallo and co-workers.[115]

In 2006, Grubbs and co-workers described a high yielding decomposition product formed from the reaction of ligand **50** with potassium hexamethyldisilazane (KHMDS) and then $[(Ph_3P)_2Ni(Cl)Ph]$. Instead of the expected NHC–Ni complex **51**, species **52** was isolated, which resulted from the reaction of the Ni–Ph moiety with the carbene carbon (Scheme 3.19). The corresponding

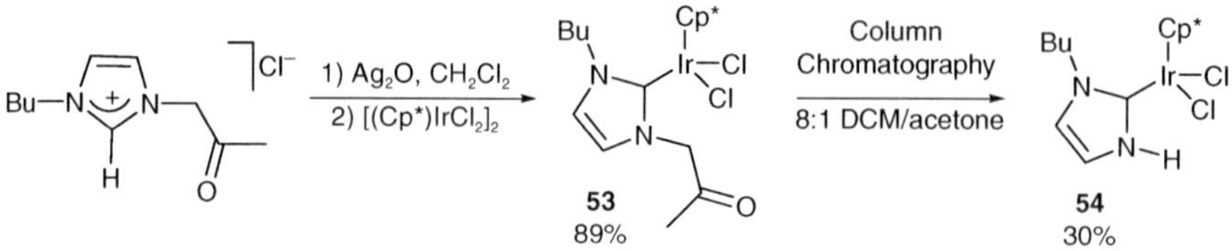

Scheme 3.20 Synthesis of an [(NHC)Ir] complex and its decomposition to an *N*-H NHC.

mesityl-substituted Ni complex, $[(Ph_3P)_2Ni(Cl)Mes]$, was resistant to this migration and gave instead the expected product **51** (Ar = Mes). It was proposed that the increased steric bulk of the mesityl group made migration more difficult, preventing decomposition.[116]

Rourke and co-workers showed that the reaction of a simple *N*-aryl imidazolium ion with $[Pd(OAc)_2]$ in DMSO led to a Pd–carbene complex, but the presence of an aniline ligand derived from the initial imidazolium salt in the coordination sphere of the Pd implied the presence of a significant decomposition process.[117]

Li and co-workers described the cleavage of a ketone substituent from NHC–Ir complex **53** resulting in an unusual *N*-H carbene complex **54** in 30% yield after purification with silica gel (Scheme 3.20). This provides an interesting method for the preparation of *N*-H N-heterocyclic carbenes, since compound **54** did not undergo tautomerization to the *N*-bound imidazole form.[118]

In 2007, Nolan and co-workers presented a reaction in which a Pt-coordinated olefin was attacked by exogenous free carbene.[119] Considering that binding of an olefin to metals such as Pt dramatically increased their electrophilicity, it is surprising this type of reaction has not been more commonly observed in the attempted preparation of carbene complexes from this class of precursors.

3.4 Stability

It is readily apparent that C–H insertion (orthometalation), reductive elimination, and displacement by ancillary ligands are some of the main pathways for decomposition of NHC ligands. In this Section, we will address ligand design elements that can be employed to minimize or eliminate such decomposition reactions.

3.4.1 Preventing Reductive Elimination

Several of the factors that appear to be important in preventing reductive elimination include the use of multidentate ligands, the employment of ancillary ligands that restrict the bite angle and the steric size of the substituents on the NHC and the electronics of these substituents.

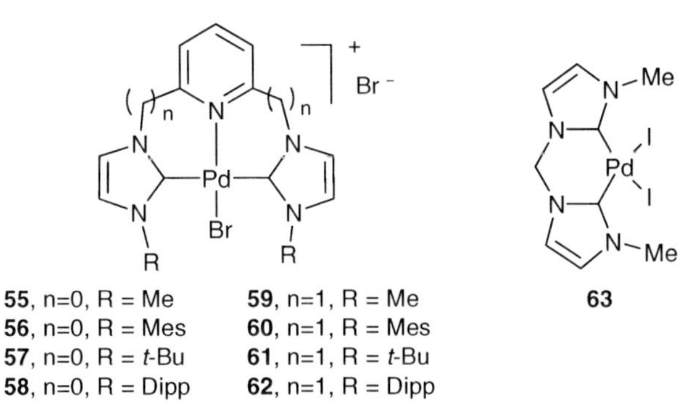

Figure 3.2 Pd complexes bearing CNC pincer or bidentate NHC ligands.

Some of the most stable NHC–Pd complexes reported to date feature a pincer type CNC or CCC ligands, where the two terminal ligands are NHCs (Figure 3.2).[120] For example, complex **63** decomposed in refluxing *N,N*-dimethylacetamide (DMA) (bp = 165 °C) depositing Pd black after 8 h, while **55** was unchanged after 24 h at this temperature. In Heck reactions catalyst **55** had no induction period and no loss of catalytic activity was found in the presence of metallic mercury.

Cavell and co-workers elucidated that, overall, the nature of the substituent on the nitrogen of the NHC is the most important factor in determining stability, with the *N*-Mes containing complexes (**56**, **60**) exhibiting the highest stability and the *N-t*-Bu systems (**57**, **61**) being the least stable.[121] This general pattern of stability could be extended to a wide number of complexes.[122]

It was found that ligands containing a methylene spacer between the NHC and the pyridine further increased the stability of the complexes,[123] since they minimized the steric interaction between substituents within the plane.

When these complexes were tested for catalytic activity, the presence or absence of the methylene spacer showed no influence on the catalytic activity. More interestingly, the least stable *N-t*-Bu complexes gave the highest activity. This is one of several examples in which a complex that decomposes under stoichiometric studies is highly active under catalytic conditions.[124]

Bidentate phosphines also had a retarding effect on reductive elimination. For example, in **64**, reductive elimination occurred rapidly at 20 °C (Figure 3.3). However, the corresponding dppp complex **65** was stable at room temperature and required heating to 65 °C for 6 h to induce decomposition. The dppe analog **66** was even more stable and very little decomposition was observed after this time.[125] These results are consistent with the well-known fact that widening the bite angle on ancillary ligands promotes reductive elimination.[126]

DFT studies have shown that as P–Pd–P bite angle increases from 80° to 130°, the activation energy for reductive elimination decreases from 26.7 to 12.7 Kcal mol^{-1}.[123] Complexes bearing ancillary ligands with bite angles less than 110° will actually prefer to undergo oxidative addition of imidazolium in

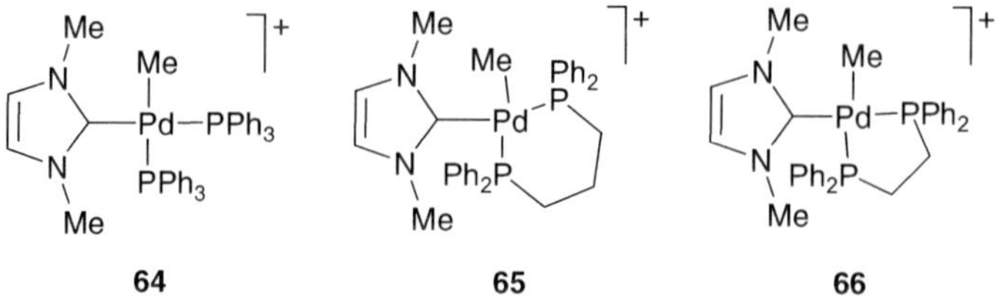

64 65 66

Figure 3.3 [(NHC)Pd] for reductive elimination.

preference to reductive elimination.[123] For substituents on phosphorous, with bulkier systems leading to more facile reductive elimination,[123] while substituents on nitrogen which remove electron density from the p-π orbital promote elimination, while those that donate into it, retard this process. Interestingly, the introduction of bulky substituents in the *ortho* positions of the aromatic group will likely reduce the interaction between the aromatic group and the carbene p-π orbital and lead to greater stabilization of the complex.

An alternative to guarding against reductive elimination of NHC, is to perform the reactions in ionic liquids. For example, NHC–Ni-based catalysts were shown to have significantly greater activity in ionic liquids, which was attributed to the reformation of NHC–Ni species by oxidative addition into the C–H bond of the imidazolium ion, preventing decomposition to metallic Ni.[81b] These types of C–H addition reactions might also explain the general stabilizing effect ionic liquids have on low-ligated metal complexes.[127]

On the contrary, Albrecht and co-workers postulated that the ease of reductive elimination of metal NHC hydrides was responsible for the limited application of NHC complexes in hydrogenation chemistry. Thus, carboxylate-functionalized catalyst **67** showed enhanced stability *versus* non-chelating complexes, with only 15% decomposition after complete conversion of styrene at 140 min (Equation (3.11)).[128] Interestingly, the complex showed greater stability in the presence of styrene, with high-pressure NMR experiments indicating that hydride transfer to the carbene ligand was 10 times slower than to styrene. Thus at a 1000:1 substrate to catalyst ratio, negligible decomposition should be expected.

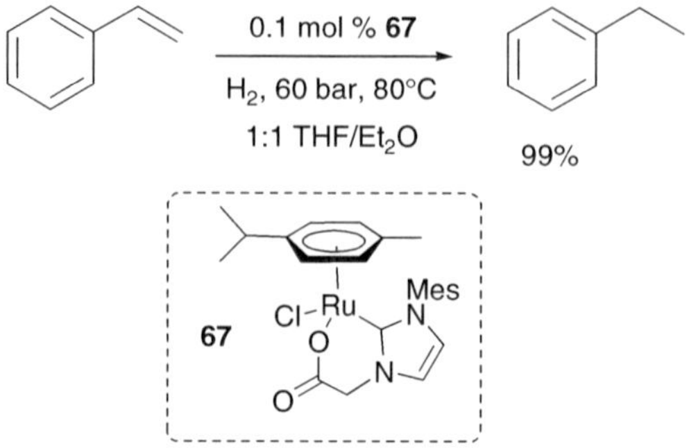

0.1 mol % **67**; H_2, 60 bar, 80°C; 1:1 THF/Et_2O → 99%

67

(3.11)

3.4.2 Inhibiting C–H Bond Activation

Iridium and ruthenium complexes of N-heterocyclic carbenes are particularly sensitive to decomposition by C–H bond activation of the substituents on nitrogen. As described in Section 3.5.2, this reaction can be used productively if reversible. However, it often leads to degradation and loss of catalytic activity.

Ru complexes bearing *N*-aryl groups without *ortho* substituents are also sensitive towards decomposition by C–H activation. This reaction can be suppressed by limiting the rotational freedom of this group through the introduction of substituents on the backbone of the carbene. However, alkylation of the *ortho* positions, as in IMes, is a better approach. Depending on the complex, C–H activation on the methyl groups in IMes might also become a competing pathway. Nevertheless, this reaction is reversible and tends to favor the starting material. Further substitution, as in IPr-type NHCs, leads to even greater stability, since the number of C–Hs adjacent to the aryl ring is minimized, as is their interaction with the metal. C–H activation of the isopropyl proton is, however, occasionally observed.

Whenever possible, metalations that produce five-membered rings can be expected over those that producing six-membered rings.[63] In the case of aliphatic activation, the substituent that appears most sensitive to metalation is the *t*-Bu group, likely because it is unable to escape the coordination sphere of the metal. Conformational issues likely also play a role in the increased stability of longer chain alkyl groups, and metalation of *N*-Et groups is often observed whereas *N*-propyl substituents are inert.[73]

The introduction of fluorine substituents instead of hydrogen may be a useful strategy for the synthesis of metalation-resistant NHCs. For example, *N*-pentafluorophenylmethylene substituted NHC complexes of Ir and Rh showed no evidence of activation of C–H, C–C or C–F bonds.[129] Grubbs and co-workers also prepared a NHC ligand designed to be resistant to orthometalation by virtue of the presence of *ortho* fluorine substituents, and the metathesis catalysts prepared from these ligands displayed increased activity when compared to traditional IMes-based catalysts.[130]

3.4.3 Preventing NHC Dissociation

Dissociation of the NHC ligand from the metal is often driven by steric strain. Thus one strategy for preventing such dissociation is to decrease their overall steric impact on the coordination sphere of the metal. The use of ancillary ligands that remove electron density from very electron rich systems may also prevent dissociation, although these two effects are sometimes difficult to separate.

For example, the Crudden group showed that $[(IMes)RhCl(PR_3)_2]$ complexes underwent dissociation of the carbene upon heating to 60–80 °C.[105,108] Treating this species with CO led to the less electron-rich and less sterically crowded complex $[(IMes)RhCl(PR_3)(CO)]$, which was stable for weeks at 80 °C. The latter complex also displayed significantly improved oxidative stability compared with the bis carbene analogue.[131]

Nickel complexes provide a valuable platform for the study of steric effects since shorter Ni–L bonds magnify these effects. When [Ni(CO)$_4$] was treated with various carbenes, the expected [(NHC)Ni(CO)$_3$] complexes were isolated for the less sterically hindered carbenes. However when carbenes featuring larger *N*-substituents, such as adamantyl or *t*-Bu, were employed, an equilibrium was established that involved dissociation of the NHC from the coordination sphere of the metal.[132]

In accord with this, the attempted preparation of [(ItBu)$_2$Ni] from ItBu and various Ni species gave different products depending on the reaction conditions, but not the expected [(ItBu)$_2$Ni] complex.[75,133] The synthesis of [(IPr)$_2$Ni] *via* reaction of [Ni(COD)$_2$] with the free IPr carbene was reported.[134] However, in their attempted synthesis, Louie *et al.* reported that the reaction between [Ni(COD)$_2$] and IPr actually resulted in the establishment of an equilibrium between this complex and free IPr with a $K_{eq} = 1$.[135]

Similar dissociative equilibria were observed in NHC–Pd complexes, particularly bis-NHC complexes (see Section 3.3.5). Herrmann and co-workers showed that under the conditions of the Suzuki–Miyaura reaction, [(NHC)$_2$Pd] complexes decomposed generating Pd black with the exception of mesityl-substituted complex **68** (Equation (3.12)).[124] Interestingly, despite this observation, complex **71** was the most active of those examined.

PhB(OH)$_2$ + Me–C$_6$H$_4$–Cl → (3 mol % [(NHC)$_2$Pd], Cs$_2$CO$_3$, dioxane, 80°C) → Me–C$_6$H$_4$–Ph (3.12)

68, R = Mes *highest stability*
69, R = *i*-Pr
70, R = Cy
71, R = *t*-Bu *highest activity*

In terms of ligand displacement reactions, exposure to ancillary bidentate ligands is more likely to lead to dissociation of the carbene than exposure to monodentate ligands, thus these species should be avoided as additives in catalytic reactions employing NHC complexes.[106,107,136]

3.5 Generation of the Active Species for NHC-containing Catalysts

Although a variety of methods can be employed to prevent decomposition reactions of NHC–metal complexes, it is also possible to use these reactions to generate catalytically active species in a controlled manner, especially in the case of reversible reactions such as C–H activation at *ortho* alkyl positions. The presence of the NHC in the coordination sphere of the metal can also change the way catalytically active species are generated, for example in the metathesis of olefins, which is where this Section begins.

3.5.1 NHC-containing Ruthenium Catalysts for Olefin Metathesis

The area of catalysis in which the application of NHCs as ancillary ligands has had the largest impact is likely ruthenium-catalyzed olefin metathesis.[137] Mechanistic studies performed on the Grubbs II catalyst [(SIMes)Cl_2(PCy_3)Ru=CHPh] revealed that the increase in catalytic activity was not due to a faster generation of the active species through dissociation of the phosphine as had been anticipated.[138] Instead, it was shown that the more active NHC-containing analogues dissociated phosphine almost two orders of magnitude *more slowly* than the bis-phosphine analogues. However, the same study showed a difference in the preference of binding phosphines *versus* olefins (k_{-1}/k_2) of greater than four orders of magnitude for bis-phosphine *versus* NHC-containing catalysts, explaining the remarkable activity of Grubbs II catalysts by their preferential binding of olefins *versus* phosphines, driving the reaction forwards.

Thus considerable effort has been put towards the design of NHC-containing ligands with faster initiation rates. This includes the *o*-isopropoxybenzylidene Ru catalysts by Hoveyda and co-workers, in which the oxygen occupies the coordination site *trans* to the NHC, negating the requirement for a second neutral donor ligand[139] and the pyridine containing compounds synthesized by Grubbs and co-workers.[140] These latter catalysts are capable of initiating metathesis processes up to three orders of magnitude faster than Hoveyda catalysts and up to six orders of magnitude faster than Grubbs II catalysts.

A recent noteworthy example is the 14-electron metathesis catalyst **72** reported by Piers, which have no ancillary ligand to dissociate.[141] As expected, this catalyst initiates extremely rapidly, displaying initial rates of ring-closing metathesis (RCM) better than the bis-pyridine substituted Grubbs catalysts. Furthermore, reaction of the catalyst with 2.2 equivalents of ethylene at $-50\,^{\circ}C$ in CD_2Cl_2 allowed the observation and characterization of the metallocyclobutane **73** formed on expulsion of the vinyl phosphonium salt (Equation (3.13)).[142] This represents the first observation of the generally accepted intermediate in the catalytic cycle.

72 $B(C_6F_5)_4^-$ — 2.2 equiv $H_2C{=}CH_2$ → **73** (> 95%) + $B(C_6F_5)_4^-$ vinyl-PCy_3^+

(3.13)

3.5.2 Other Ruthenium-catalyzed Reactions Featuring NHCs

Arisawa and Nishida showed that treating the Grubbs II catalyst with silyl enol ethers produced a catalyst capable of cycloisomerizing *N*-allyl-*o*-vinylaniline into 3-methylene-dihydroindoles.[143] Investigation into the role of the vinyloxytrimethylsilane revealed the identity of the active catalyst in this transformation, which was [(SIMes)Ru(PCy_3)(CO)(H)Cl], previously identified by Grubbs and co-workers as a decomposition product of the catalyst.[39] Under the reaction conditions, this complex was proposed to form through the intermediacy of a Fischer carbene complex generated by reaction of the pre-catalyst with the silyl enol ether.

Fogg has carried out a series of detailed studies aimed at transforming active metathesis catalysts into effective hydrogenation catalysts such that these two important reactions can be performed in tandem.[144] [Ru(H)(Cl)(CO)(PCy_3)$_2$] **74** is an effective alkene hydrogenation catalyst; however, initial attempts at replacing a PCy_3 ligand in this catalyst with IMes, gave a catalyst that had low efficiency.[145] The analogous complexes with PPh_3,[146] known to bind more weakly to metal centres than PCy_3, provided more effective hydrogenation catalysts.

Interestingly, **76a** also proved more effective at hydrogenation of alkenes than **77** in which the CO ligand was replaced by dihydrogen. The superior performance of the CO analogue was attributed to its greater stability, which maintains higher concentrations of active catalyst over the course of the reaction than **77** (Equation (3.14)).[147] A similar effect was observed by Fogg *et al.* in the development of tandem ring-opening metathesis polymerization (ROMP)-hydrogenation processes, comparing Grubbs I, II and the pyridine substituted derivative. While [(SIMes)(pyr)$_2$(Cl)$_2$Ru=CHPh] was the most active ROMP catalyst, in the subsequent hydrogenation reaction it faired poorly compared to Grubbs I and II, an observation again attributed to low catalyst stability under hydrogenation conditions, this time remedied by the addition of PCy_3.[148]

0.05 mol % [Ru], 140 psi H_2, Toluene, 80°C, 1 h

74: 33%
75: 8%
76a: 96%
76b: 97%

(3.14)

74: [Ru(H)(Cl)(CO)(PCy_3)$_2$]; **75**: [Ru(H)(Cl)(CO)(PCy_3)(IMes)]; **76a**: NHC = IMes; **76b**: NHC = SIMes ([Ru(H)(Cl)(CO)(PPh_3)(NHC)]); **77**: [Ru(H)(Cl)(H_2)(PCy_3)(IMes)]

Similar behavior was observed with [*cis*-(IMes)RhCl(P(*p*-tolyl)$_3$)$_2$], which exchanged the phosphine *trans* to IMes an order of magnitude more slowly than in the homoleptic complex [RhCl(P(*p*-tolyl)$_3$)$_3$], resulting in poorer hydrogenation efficiency with the NHC complex.[105,108] The dissociative mechanism suggested by these observations was further corroborated by catalytic results obtained using CuCl (an effective phosphine sponge) as an additive,

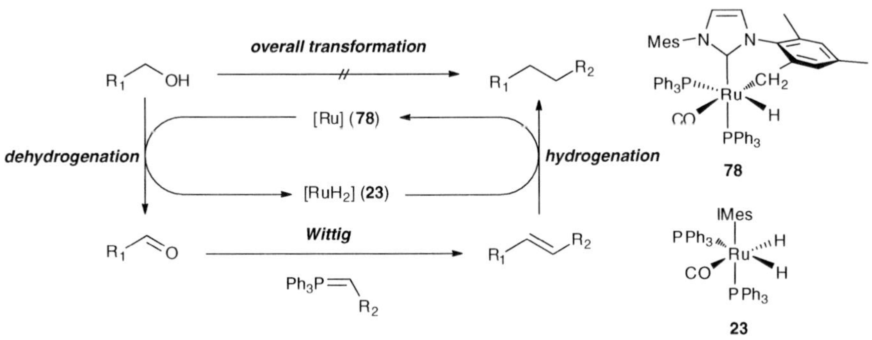

Scheme 3.21 Catalytic cycle of an indirect Wittig reaction.

which increased the activity of the IMes-modified catalyst [*cis*-(IMes)RhCl$(PPh_3)_2$] (**48**) to the point where it was more active than the all-phosphine catalyst [RhCl$(PPh_3)_3$]. Not surprisingly, analogous bis-NHC complexes were poor hydrogenation catalysts, displaying low conversion with concomitant formation of colloidal rhodium.[149]

An interesting [(NHC)Ru] system that takes advantage of an otherwise unfavorable C–H activation of the NHC ligand was [(IMes)Ru$(PPh_3)_2$(CO)$(H)_2$], described by Whittlesey and Williams.[150] This dihydride complex **23** was shown to efficiently hydrogenate alkenes in the presence of hydrogen donators such as 2-propanol or 1-phenylethanol. Interestingly, the product of the hydrogenation of a variety of alkenes was complex **78**, in which the C–H bond of an *o*-methyl group of the *N*-mesityl of the IMes ligand has undergone orthometalation. The activity of the complex is dependant on this activation, which stabilized what would otherwise be a highly unsaturated 14-electron metal centre. The unique reactivity of this complex was then exploited to affect the indirect Wittig reaction of alcohols in a 'borrowed hydrogen' manifold (Scheme 3.21). Following activation of the catalyst by vinyltrimethylsilane, dehydrogenation of a primary alcohol to an aldehyde regenerates the ruthenium dihydride. An added phosphorous ylide then undergoes a Wittig reaction with the aldehyde to generate the corresponding alkene, which serves as a hydrogen acceptor.[73]

3.5.3 Palladium Catalysis

Commonly, the first step in palladium-catalyzed cross-coupling reactions is the oxidative addition of an organohalide or triflate to a Pd^0 species. While numerous catalytic systems exist for the activation of C(sp^2)–I and C(sp^2)–Br bonds, due to their increased stability, activation of C(sp^2)–Cl bonds is a considerably greater challenge. The use of NHCs as ancillary ligands in these types of reactions has allowed the use of aryl chlorides as coupling partners under relatively mild conditions. From these substrates, experimental[85,151] and theoretical[152] studies have indicated that a dissociative pathway was favored, in which highly coordinatively unsaturated PdL complexes are the active species. Ligands favoring this type of dissociative pathway tend to be bulky, promoting dissociation of the second ligand, and preventing formation of palladium black and endowing the low coordinate Pd–L species with adequate reducing power to activate the aryl chloride. These characteristics are obviously perfectly met with NHC ligands.[153] For example, both [(IPr)Pd(OAc)$_2$] and [(NHC)Pd(allyl)X] complexes were able to affect coupling reactions of aryl chlorides at room temperature, presumably *via* [(NHC)Pd] as the active catalyst.[154]

Due to the base required in coupling reactions, the reactions are often performed with catalysts generated *in situ* from imidazolium salts. However, this approach can lead to significant discrepancies between catalytic studies depending on the reaction conditions. For example, the isolated complex [(IMes)$_2$Pd] was found to be completely inactive in the Suzuki coupling of an aryl chloride, while under the same conditions the use of [Pd$_2$(dba)$_3$] and two equivalents of IMes·HCl gave excellent yields.[155] In another study, the use of [Pd(OAc)$_2$] gave inferior results to catalysts generated with [Pd$_2$(dba)$_3$].[156] Lebel and co-workers showed that the reaction of IMes·HCl with [Pd(OAc)$_2$] in dioxane at 80 °C gave 74% yield of abnormally bonded [*trans*-(IMes)(IMes′)PdCl$_2$] complex **80**, and none of the expected product, **79**, while reaction of PdCl$_2$ or [Pd(OAc)$_2$] with IMes·HCl in the presence of Cs$_2$CO$_3$ gave **79** (Equation (3.15)).[157] Interestingly, complex **79** proved to be inactive in both standard Heck and Suzuki coupling reactions, illustrating the importance of understanding the nature of the catalytic species employed or generated in a reaction.

	79	**80**
[Pd(OAc)$_2$], dioxane, 80°C:	< 1 %	74 %
PdCl$_2$, Cs$_2$CO$_3$, dioxane, 80°C:	68 %	< 1%

(3.15)

Consistent with the high catalytic activity of PdL type systems, a multitude of *in situ* generated catalytic systems have shown repeatedly that the optimal Pd/NHC ratio is 1:1.[153,158] However, there are scattered reports in which the use of excess NHC precursor leads to better results, likely due to the ability of this excess ligand to regenerate catalytically active [(NHC)Pd] species after loss of the NHC *via* decomposition reactions, rather than any catalytic activity of more highly ligated Pd complexes. This was suggested when Fagnou showed that the addition of excess IPr·HCl in systems employing pre-catalyst [(IPr)Pd$(OAc)_2(H_2O)$] led to increases in turnover numbers for intramolecular direct arylation of aryl chlorides.[76] It was demonstrated that even though Pd/C showed no activity itself, the addition of IPr·HCl under catalytic conditions resulted in the formation of an active catalyst (Equation (3.16)).

30 mol % Pd Black
10 mol % IPr·HCl
2 equiv K_2CO_3, DMA
130°C, 24 h

25 % Conversion (3.16)

Though methods exist for the synthesis of [(NHC)Pd^0] complexes,[155,159] palladium(0) species are rarely employed as pre-catalysts due to the requirement for rigorously anaerobic and anhydrous conditions. Amongst the few examples of air-stable [(NHC)Pd^0] catalysts are those reported by Beller and co-workers featuring diene[160] or quinone[161] auxiliary ligands. Pd^{II} precursors, on the other hand, are often far more robust and can be handled with greater ease, but then of course must be activated *in situ* to produce the catalytically active [(NHC)Pd^0] species.

For this purpose, [(NHC)Pd(allyl)Cl] species can be treated with equimolar amounts of nucleophilic alkoxide bases to displace the allyl moiety and produce allyl ethers and the desired (NHC)–Pd^0 catalyst. The presence of this complex was proven by stoichiometric reaction of the Pd^{II} precursor with KO*t*-Bu in the presence of PCy_3.[162] However, as described in Section 3.3, this reaction can lead to displacement of the carbene ligand under exceedingly mild conditions, if performed in the presence of phosphine ligand. The active catalyst could also be generated in 2-propanol in the presence of a base.[111] Use of more substituted allyl moieties, such as cinnamallyl, provided a more facile activation of the catalyst allowing rapid Suzuki–Miyaura coupling of aryl chlorides at room temperature.[154b] Similar pre-catalysts that feature Cp ligands instead of allyls are readily activated at room temperature in basic isopropanol.[163]

The generation of [(NHC)Pd^0] catalysts can be achieved by activation of [(NHC)Pd$(OAc)_2$] complexes under basic conditions, which is believed to occur through coordination of a secondary alkoxide such as isopropoxide, which could then undergo β-hydride elimination to give a Pd–H that could eliminate acetic acid.[164] Alternative methods are possible under acidic conditions as described in Chapter 12.[165] Another popular method for the generation of Pd^0 catalysts *in situ* from Pd^{II} pre-catalysts is the sacrificial elimination of a carbon-based ligand.[166]

There are some examples in which NHCs became the sacrificial ligand in the generation of Pd^0 from Pd^{II} pre-catalysts. An interesting example was reported by Cavell, in which $[(NHC)Pd(allyl)(PR_3)]^+$ complexes were synthesized with the intention of being used as Pd^0–PR_3 sources. Heating of [(^{Me}IMe) Pd(allyl)(PCy_3)] induced elimination of 2-propenylimidazolium salts; however, attempts to trap the supposed Pd^0–PCy_3 product with aryl iodide as [Pd(PCy_3)(Ph)I] instead resulted in formation of 2-phenylimidazolium.[87]

3.5.4 Nickel and Platinum NHC Catalysts

Herrmann and co-workers reported the activation of aryl C–F bonds by [(NHC)Ni] systems under Kumada–Corriu-type coupling conditions.[134] Screening of catalytic conditions revealed that yields with $[(IPr)_2Ni]$ were poorer than when the catalyst was generated *in situ* from $[Ni(acac)_2]$/IPr·HBF_4. Interestingly, the electronic influence of the aryl fluorides on the reaction rate as demonstrated by Hammett σ^--values indicated different pathways for the NHC-modified catalytic system *versus* the analogous Ni-phosphine systems.

Ni-catalyzed aryl aminations involving NHC ligands are best performed at a Ni/NHC ratio of 1:1.[167] To ensure the efficient generation of Ni^0 species *in situ* from the Ni^{II} precursor, NaH/*t*-BuOH were employed in THF.[168] Additional NaO*t*-Bu generated with NaH served to generate the free carbene from the imidazolium chloride. Isolated $[(NHC)_2Ni^0]$ species demonstrated better activity in this reaction with certain substrates, leading the authors of the study to conclude that these are very close to the active species in solution.[169]

NHC complexes of platinum were shown to be efficient and selective catalysts for the hydrosilylation of internal and terminal alkynes, giving the β-(*E*)-alkene selectively.[170] Optimal results were obtained when using $[(IPr)Pt^0$(diallylether)] **81** as the pre-catalyst (Equation (3.17)). Studies suggested that the active catalyst was formed by hydrosilylation of the diallylether ligand generating the actual active species. The catalyst could be deactivated by coordination of two equivalents of the alkyne ligand prior to Si–H addition, which would promote dissociation of the IPr ligand, generating a less active and very unselective catalyst.

R^1—≡—R^2 → **81** (0.05–0.1 mol %), THF, 60°C, 30 min, $HSiR_3$ → (SiR_3, R^1, R^2) + (R_3Si, R^1, R^2)

selectivity up to 100:1

i-Pr, i-Pr, N, N, i-Pr, i-Pr, Pt, **81**

(3.17)

3.5.5 Activation by Extraction of Anionic Ligands

The use of NHCs as ancillary ligands in iridium-catalyzed Oppenauer-type oxidation of alcohols to carbonyls has led to some of the most active catalysts for this class of transformation. In 2005, Yamaguchi and co-workers reported the synthesis of a number of [(Cp*)Ir] complexes featuring NHCs as the ancillary ligands.[171] In addition to neutral complexes of the formula

[(NHC)Ir(Cp*)Cl_2], cationic analogues [(NHC)Ir(Cp*)$(MeCN)_2$][OTf]$_2$ **82** were found to be more active (Equation (3.18)). The analogous complex featuring a 2-(dimethyl)-amino functionalized Cp* ligand **83**,[55a] does not need external base.

[(NHC)Ir] **82** or **83**
K_2CO_3, acetone
40°C

(3.18)

3.6 Conclusions

The development of well-defined NHC-containing complexes has been made a significant impact on catalysis in the past 15 years. Virtually every type of reaction catalyzed by a transition metal has been attempted with NHC-based complexes, in many cases providing unique reactivity by comparison to more traditional phosphine-based catalysts. Despite this fact, there is still substantial room for development in this area, particularly in the design of chiral NHC catalysts that react with high enantioselectivity, and also in terms of understanding the stability and reactivity of the NHC itself. Significant advances have made in this last area in the last 5 years, such that a greater understanding of the limitations of these ligands is coming about. We hope that this Chapter has provided the reader with an overview of the types of decomposition reactions to which NHCs are susceptible, and the various methods that can be used to prevent, limit or harness such reactions.

References

1. F. E. Hahn and M. C. Jahnke, *Angew. Chem., Int. Ed.*, 2008, **47**, 3122–3172.
2. (a) S. Díez-González, N. Marion and S. P. Nolan, *Chem. Rev.*, 2009, **109**, 3612–3676; (b) W. A. Herrmann, *Angew. Chem., Int. Ed.*, 2002, **41**, 1290–1309.
3. J. Huang, H.-J. Schanz, E. D. Stevens and S. P. Nolan, *Organometallics*, 1999, **18**, 2370–2375.
4. (a) K. J. Cavell and D. S. McGuinness, *Coord. Chem. Rev.*, 2004, **248**, 671–681; (b) C. M. Crudden and D. P. Allen, *Coord. Chem. Rev.*, 2004, **248**, 2247–2273.
5. K. Öfele, *J. Organomet. Chem.*, 1968, **12**, P42–P43.

6. (a) H. W. Wänzlick, *Angew. Chem., Int. Ed. Engl.*, 1962, **1**, 75–80; (b) H. W. Wänzlick and H. J. Schönherr, *Angew. Chem., Int. Ed. Engl.*, 1968, **7**, 141–142.
7. D. J. Cardin, M. J. Doyle and M. F. Lappert, *J. Chem. Soc., Chem. Commun.*, 1972, 927–928.
8. (a) M. F. Lappert, *J. Organomet. Chem.*, 1975, **100**, 139–159; (b) M. F. Lappert, *J. Organomet. Chem.*, 1988, **358**, 185–213.
9. P. B. Hitchcock, M. F. Lappert and P. L. Pye, *J. Chem. Soc., Dalton Trans.*, 1977, 2160–2172.
10. A. Igau, H. Grützmacher, A. Baceiredo and G. Bertrand, *J. Am. Chem. Soc.*, 1988, **110**, 6463–6466.
11. A. J. Arduengo III, R. L. Harlow and M. Kline, *J. Am. Chem. Soc.*, 1991, **113**, 361–363.
12. (a) J. Vignolle, X. Cattoën and D. Bourissou, *Chem. Rev.*, 2009, **109**, 3333–3384; (b) A. J. Arduengo III, *Acc. Chem. Res.*, 1999, **32**, 913–921.
13. K. Öfele, W. A. Herrmann, D. Mihalios, M. Elison, E. Herdtweck, W. Scherer and J. Mink, *J. Organomet. Chem.*, 1993, **459**, 177–184.
14. W. A. Herrmann, M. Elison, J. Fischer, C. Köcher and G. R. J. Artus, *Chem.–Eur. J.*, 1996, **2**, 772–780.
15. W. A. Herrmann, M. Elison, J. Fischer, C. Köcher and G. R. J. Artus, *Angew. Chem., Int. Ed. Engl.*, 1995, **34**, 2371–2374.
16. C. Köcher and W. A. Herrmann, *J. Organomet. Chem.*, 1997, **532**, 261–265.
17. (a) M. H. Voges, C. Romming and M. Tilset, *Organometallics*, 1999, **18**, 529–533; (b) C. D. Abernethy, H. Alan, Cowley and R. A. Jones, *J. Organomet. Chem.*, 2000, **596**, 3–5.
18. I. J. B. Lin and C. S. Vasam, *Coord. Chem. Rev.*, 2007, **251**, 642–670.
19. A. J. Arduengo III, H. V. Rasika Dias, J. C. Calabrese and F. Davidson, *Organometallics*, 1993, **12**, 3405–3409.
20. (a) O. Guerret, S. Solé, H. Gornitzka, M. Teichert, G. Trinquier and G. Bertrand, *J. Am. Chem. Soc.*, 1997, **119**, 6668–6669; (b) A. A. D. Tulloch, A. A. Danopoulos, S. Winston, S. Kleinhenz and G. Eastham, *J. Chem. Soc., Dalton Trans.*, 2000, 4499–4506.
21. H. M. J. Wang and I. J. B. Lin, *Organometallics*, 1998, **17**, 972–975.
22. (a) J. C. Garrison, R. S. Simons, C. A. Tessier and W. J. Youngs, *J. Organomet. Chem.*, 2003, **673**, 1–4; (b) A. Kascatan-Nebioglu, M. J. Panzner, J. C. Garrison, C. A. Tessier and W. J. Youngs, *Organometallics*, 2004, **23**, 1928–1931; (c) C. A. Quezada, J. C. Garrison, M. J. Panzner, C. A. Tessier and W. J. Youngs, *Organometallics*, 2004, **23**, 4846–4848.
23. X. Hu, I. Castro-Rodriguez, K. Olsen and K. Meyer, *Organometallics*, 2004, **23**, 755–764.
24. L. N. Appelhans, C. D. Incarvito and R. H. Crabtree, *J. Organomet. Chem.*, 2008, **693**, 2761–2766.
25. (a) M. Viciano, E. Mas-Marzá, M. Poyatos, M. Sanaú, R. H. Crabtree and E. Peris, *Angew. Chem., Int. Ed.*, 2005, **44**, 444–447; (b) M. Raynal,

C. S. J. Cazin, C. Vallée, H. Olivier-Bourbigou and P. Braunstein, *Organometallics*, 2009, **28**, 2460–2470.
26. D. S. McGuinness, K. J. Cavell and B. F. Yates, *Chem. Commun.*, 2001, 355–356.
27. D. S. McGuinness, K. J. Cavell, B. F. Yates, B. W. Skelton and A. H. White, *J. Am. Chem. Soc.*, 2001, **123**, 8317–8328.
28. S. Gründemann, M. Albrecht, A. Kovacevic, J. W. Faller and R. H. Crabtree, *J. Chem. Soc., Dalton Trans.*, 2002, 2163–2167.
29. E. Mas-Marzá, M. Sanaú and E. Peris, *Inorg. Chem.*, 2005, **44**, 9961–9967.
30. N. D. Clement, K. J. Cavell, C. Jones and C. J. Elsevier, *Angew. Chem., Int. Ed.*, 2004, **43**, 1277–1279.
31. (a) A. Fürstner, G. Seidel, D. Kremzow and C. W. Lehmann, *Organometallics*, 2003, **22**, 907–909; (b) D. Kremzow, G. Seidel, C. W. Lehmann and A. Fürstner, *Chem.–Eur. J.*, 2005, **11**, 1833–1853.
32. M. V. Baker, D. H. Brown, V. J. Hesler, B. W. Skelton and A. H. White, *Organometallics*, 2006, **26**, 250–252.
33. (a) P. Wasserscheid and T. Welton, *Ionic Liquids in Synthesis*, Wiley-VCH, Weinheim, 2007; (b) T. Welton, *Chem. Rev.*, 1999, **99**, 2071–2084.
34. S. Chowdhury, R. S. Mohan and J. L. Scott, *Curr. Org. Synth.*, 2007, **4**, 381–389.
35. A. J. Carmichael, M. J. Earle, J. D. Holbrey, P. B. McCormac and K. R. Seddon, *Org. Lett.*, 1999, **1**, 997–1000.
36. L. Xu, W. Chen and J. Xiao, *Organometallics*, 2000, **19**, 1123–1127.
37. F. McLachlan, C. J. Mathews, P. J. Smith and T. Welton, *Organometallics*, 2003, **22**, 5350–5357.
38. D. Enders, K. Breuer, G. Raabe, J. Runsink, J. H. Teles, J.-P. Melder, K. Ebel and S. Brode, *Angew. Chem., Int. Ed. Engl.*, 1995, **34**, 1021–1023.
39. T. M. Trnka, J. P. Morgan, M. S. Sanford, T. E. Wilhelm, M. Scholl, T.-L. Choi, S. Ding, M. W. Day and R. H. Grubbs, *J. Am. Chem. Soc.*, 2003, **125**, 2546–2558.
40. A. J. Arduengo III, J. C. Calabrese, F. Davidson, H. V. Rasika Dias, J. R. Goerlich, R. Krafczyk, W. J. Marshall, M. Tamm and R. Schmutzler, *Helv. Chim. Acta*, 1999, **82**, 2348–2364.
41. G. W. Nyce, S. Csihony, R. M. Waymouth and J. L. Hedrick, *Chem.–Eur. J.*, 2004, **10**, 4073–4079.
42. H. A. Duong, T. N. Tekavec, A. M. Arif and J. Louie, *Chem. Commun.*, 2004, 112–113.
43. (a) A. M. Voutchkova, L. N. Appelhans, A. R. Chianese and R. H. Crabtree, *J. Am. Chem. Soc.*, 2005, **127**, 17624–17625; (b) A. M. Voutchkova, M. Feliz, E. Clot, O. Eisenstein and R. H. Crabtree, *J. Am. Chem. Soc.*, 2007, **129**, 12834–12846.
44. (a) F. E. Hahn, V. Langenhahn, N. Meier, T. Lügger and W. P. Fehlhammer, *Chem.–Eur. J.*, 2003, **9**, 704–712; (b) F. E. Hahn, V. Langenhahn, T. Lügger, T. Pape and D. Le Van, *Angew. Chem., Int. Ed.*, 2005, **44**, 3759–3763.

45. F. E. Hahn, C. G. Plumed, M. Münder and T. Lügger, *Chem.–Eur. J.*, 2004, **10**, 6285–6293.
46. O. Kaufhold, A. Flores-Figueroa, T. Pape and F. E. Hahn, *Organometallics*, 2009, **28**, 896–901.
47. P. B. Hitchcock, M. F. Lappert and P. Terreros, *J. Organomet. Chem.*, 1982, **239**, C26–C30.
48. (a) T. Sajoto, P. I. Djurovich, A. Tamayo, M. Yousufuddin, R. Bau, M. E. Thompson, R. J. Holmes and S. R. Forrest, *Inorg. Chem.*, 2005, **44**, 7992–8003; (b) C.-F. Chang, Y.-M. Cheng, Y. Chi, Y.-C. Chiu, C.-C. Lin, G.-H. Lee, P.-T. Chou, C.-C. Chen, C.-H. Chang and C.-C. Wu, *Angew. Chem., Int. Ed.*, 2008, **47**, 4542–4545; (c) C. H. Chien, S. Fujita, S. Yamoto, T. Hara, T. Yamagata, M. Watanabe and K. Mashima, *Dalton Trans.*, 2008, 916–923.
49. (a) A. A. Danopoulos, D. Pugh and J. A. Wright, *Angew. Chem., Int. Ed.*, 2008, **47**, 9765–9767; (b) M. Raynal, R. Pattacini, C. S. J. Cazin, C. Vallée, H. Olivier-Bourbigou and P. Braunstein, *Organometallics*, 2009, **28**, 4028–4047.
50. A. A. Danopoulos, S. Winston and M. B. Hursthouse, *J. Chem. Soc., Dalton Trans.*, 2002, 3090–3091.
51. O. Torres, M. Martín and E. Sola, *Organometallics*, 2009, **28**, 863–870.
52. C. Y. Tang, W. Smith, D. Vidovic, A. L. Thompson, A. B. Chaplin and S. Aldridge, *Organometallics*, 2009, **28**, 3059–3066.
53. B. R. Dible, M. S. Sigman and A. M. Arif, *Inorg. Chem.*, 2005, **44**, 3774–3776.
54. M. Prinz, M. Grosche, E. Herdtweck and W. A. Herrmann, *Organometallics*, 2000, **19**, 1692–1694.
55. (a) F. Hanasaka, K.-I. Fujita and R. Yamaguchi, *Organometallics*, 2006, **25**, 4643–4647; (b) Y. Tanabe, F. Hanasaka, K.-I. Fujita and R. Yamaguchi, *Organometallics*, 2007, **26**, 4618–4626.
56. F. Hanasaka, Y. Tanabe, K.-I. Fujita and R. Yamaguchi, *Organometallics*, 2006, **25**, 826–831.
57. (a) R. Corberán, M. Sanaú and E. Peris, *J. Am. Chem. Soc.*, 2006, **128**, 3974–3979; (b) R. Corberán, M. Sanaú and E. Peris, *Organometallics*, 2006, **25**, 4002–4008.
58. R. Corberán, V. Lillo, J. A. Mata, E. Fernandez and E. Peris, *Organometallics*, 2007, **26**, 4350–4353.
59. N. M. Scott, R. Dorta, E. D. Stevens, A. Correa, L. Cavallo and S. P. Nolan, *J. Am. Chem. Soc.*, 2005, **127**, 3516–3526.
60. (a) P. B. Hitchcock, M. F. Lappert and P. L. Pye, *J. Chem. Soc., Chem. Commun.*, 1977, 196–198; (b) P. B. Hitchcock, M. F. Lappert, P. L. Pye and S. Thomas, *J. Chem. Soc., Dalton Trans.*, 1979, 1929–1942.
61. P. B. Hitchcock, M. F. Lappert and P. L. Pye, *J. Chem. Soc., Dalton Trans.*, 1978, 826–836.
62. M. A. Owen, P. L. Pye, B. Piggott and M. V. Capparelli, *J. Organomet. Chem.*, 1992, **434**, 351–362.
63. C. Y. Zhang, Y. Zhao, B. Li, H. B. Song, S. S. Xu and B. Q. Wang, *Dalton Trans.*, 2009, 5182–5189.

64. J. M. Berlin, K. Campbell, T. Ritter, T. W. Funk, A. Chlenov and R. H. Grubbs, *Org. Lett.*, 2007, **9**, 1339–1342.
65. (a) S. H. Hong, A. Chlenov, M. W. Day and R. H. Grubbs, *Angew. Chem., Int. Ed.*, 2007, **46**, 5148–5151; (b) J. Mathew, N. Koga and C. H. Suresh, *Organometallics*, 2008, **27**, 4666–4670.
66. K. Vehlow, S. Gessler and S. Blechert, *Angew. Chem., Int. Ed.*, 2007, **46**, 8082–8085.
67. (a) C. K. Chung and R. H. Grubbs, *Org. Lett.*, 2008, **10**, 2693–2696; (b) F. Grisi, A. Mariconda, C. Costabile, V. Bertolasi and P. Longo, *Organometallics*, 2009, **28**, 4988–4995; (c) K. M. Kuhn, J.-B. Bourg, C. K. Chung, S. C. Virgil and R. H. Grubbs, *J. Am. Chem. Soc.*, 2009, **131**, 5313–5320.
68. (a) C. E. Cooke, M. C. Jennings, M. J. Katz, R. K. Pomeroy and J. A. C. Clyburne, *Organometallics*, 2008, **27**, 5777–5799; (b) J. A. Cabeza, I. del Río, J. M. Fernández-Colinas, E. Pérez-Carreño, M. G. Sánchez-Vega and D. Vázquez-García, *Organometallics*, 2009, **28**, 1832–1837.
69. (a) K. Abdur-Rashid, T. Fedorkiw, A. J. Lough and R. H. Morris, *Organometallics*, 2003, **23**, 86–94; (b) D. Burtscher, B. Perner, K. Mereiter and C. Slugovc, *J. Organomet. Chem.*, 2006, **691**, 5423–5430; (c) E. M. Leitao, S. R. Dubberley, W. E. Piers, Q. Wu and R. McDonald, *Chem.–Eur. J.*, 2008, **14**, 11565–11572; (d) J. H. Lee, K. S. Yoo, C. P. Park, J. M. Olsen, S. Sakaguchi, G. K. S. Prakash, T. Mathew and K. W. Jung, *Adv. Synth. Catal.*, 2009, **351**, 563–568.
70. D. Giunta, M. Hölscher, C. W. Lehmann, R. Mynott, C. Wirtz and W. Leitner, *Adv. Synth. Catal.*, 2003, **345**, 1139–1145.
71. R. F. R. Jazzar, S. A. Macgregor, M. F. Mahon, S. P. Richards and M. K. Whittlesey, *J. Am. Chem. Soc.*, 2002, **124**, 4944–4945.
72. R. A. Diggle, S. A. Macgregor and M. K. Whittlesey, *Organometallics*, 2008, **27**, 617–625.
73. S. Burling, B. M. Paine, D. Nama, V. S. Brown, M. F. Mahon, T. J. Prior, P. S. Pregosin, M. K. Whittlesey and J. M. J. Williams, *J. Am. Chem. Soc.*, 2007, **129**, 1987–1995.
74. S. Burling, M. F. Mahon, R. E. Powell, M. K. Whittlesey and J. M. J. Williams, *J. Am. Chem. Soc.*, 2006, **128**, 13702–13703.
75. S. Caddick, F. G. N. Cloke, P. B. Hitchcock and A. K. de K. Lewis, *Angew. Chem., Int. Ed.*, 2004, **43**, 5824–5827.
76. L.-C. Campeau, P. Thansandote and K. Fagnou, *Org. Lett.*, 2005, **7**, 1857–1860.
77. N. Stylianides, A. A. Danopoulos, D. Pugh, F. Hancock and A. Zanotti-Gerosa, *Organometallics*, 2007, **26**, 5627–5635.
78. A. A. Danopoulos, D. Pugh, H. Smith and J. Saßmannshausen, *Chem.–Eur. J.*, 2009, **15**, 5491–5502.
79. L. P. Spencer, C. Beddie, M. B. Hall and M. D. Fryzuk, *J. Am. Chem. Soc.*, 2006, **128**, 12531–12543.
80. K. Cavell, *Dalton Trans.*, 2008, 6676–6685.
81. (a) D. S. McGuinness, N. Saendig, B. F. Yates and K. J. Cavell, *J. Am. Chem. Soc.*, 2001, **123**, 4029–4040; (b) D. S. McGuinness, W. Mueller, P.

Wasserscheid, K. J. Cavell, B. W. Skelton, A. H. White and U. Englert, *Organometallics*, 2001, **21**, 175–181.
82. D. S. McGuinness, M. J. Green, K. J. Cavell, B. W. Skelton and A. H. White, *J. Organomet. Chem.*, 1998, **565**, 165–178.
83. D. S. McGuinness, K. J. Cavell, B. W. Skelton and A. H. White, *Organometallics*, 1999, **18**, 1596–1605.
84. S. Caddick, F. G. N. Cloke, P. B. Hitchcock, J. Leonard, A. K. de K. Lewis, D. McKerrecher and L. R. Titcomb, *Organometallics*, 2002, **21**, 4318–4319.
85. A. K. de K. Lewis, S. Caddick, F. G. N. Cloke, N. C. Billingham, P. B. Hitchcock and J. Leonard, *J. Am. Chem. Soc.*, 2003, **125**, 10066–10073.
86. W. J. Marshall and V. V. Grushin, *Organometallics*, 2003, **22**, 1591–1593.
87. (a) A. T. Normand, A. Stasch, L.-L. Ooi and K. J. Cavell, *Organometallics*, 2008, **27**, 6507–6520; (b) A. T. Normand, M. S. Nechaev and K. J. Cavell, *Chem.–Eur. J.*, 2009, **15**, 7063–7073.
88. (a) R. E. Douthwaite, M. L. H. Green, P. J. Silcock and P. T. Gomes, *Organometallics*, 2001, **20**, 2611–2615; (b) D. J. Nielsen, K. J. Cavell, B. W. Skelton and A. H. White, *Inorg. Chim. Acta*, 2002, **327**, 116–125; (c) T. Steinke, B. K. Shaw, H. Jong, B. O. Patrick, M. D. Fryzuk and J. C. Green, *J. Am. Chem. Soc.*, 2009, **131**, 10461–10466.
89. (a) N. D. Clement and K. J. Cavell, *Angew. Chem., Int. Ed.*, 2004, **43**, 3845–3847; (b) A. T. Normand, K. J. Hawkes, N. D. Clement, K. J. Cavell and B. F. Yates, *Organometallics*, 2007, **26**, 5352–5363.
90. (a) K. L. Tan, R. G. Bergman and J. A. Ellman, *J. Am. Chem. Soc.*, 2001, **123**, 2685–2686; (b) R. K. Thalji, K. A. Ahrendt, R. G. Bergman and J. A. Ellman, *J. Am. Chem. Soc.*, 2001, **123**, 9692–9693.
91. (a) K. L. Tan, R. G. Bergman and J. A. Ellman, *J. Am. Chem. Soc.*, 2002, **124**, 3202–3203; (b) K. J. Hawkes, K. J. Cavell and B. F. Yates, *Organometallics*, 2008, **27**, 4758–4771.
92. J. C. Lewis, R. G. Bergman and J. A. Ellman, *Acc. Chem. Res.*, 2008, **41**, 1013–1025.
93. H. C. Martin, N. H. James, J. Aitken, J. A. Gaunt, H. Adams and A. Haynes, *Organometallics*, 2003, **22**, 4451–4458.
94. M. Muehlhofer, T. Strassner and W. A. Herrmann, *Angew. Chem., Int. Ed.*, 2002, **41**, 1745–1747.
95. (a) E. D. Blue, T. B. Gunnoe, J. L. Petersen and P. D. Boyle, *J. Organomet. Chem.*, 2006, **691**, 5988–5993; (b) A. K. D. Lewis, S. Caddick, O. Esposito, F. G. N. Cloke and P. B. Hitchcock, *Dalton Trans.*, 2009, 7094–7098; (c) O. Saker, M. F. Mahon, J. E. Warren and M. K. Whittlesey, *Organometallics*, 2009, **28**, 1976–1979.
96. F. Wu, V. K. Dioumaev, D. J. Szalda, J. Hanson and R. M. Bullock, *Organometallics*, 2007, **26**, 5079–5090.
97. H. van Rensburg, R. P. Tooze, D. F. Foster and A. M. Z. Slawin, *Inorg. Chem.*, 2004, **43**, 2468–2470.

98. M. S. Jeletic, M. T. Jan, I. Ghiviriga, K. A. Abboud and A. S. Veige, *Dalton Trans.*, 2009, 2764–2776.
99. M. C. Perry, X. Cui, M. T. Powell, D.-R. Hou, J. H. Reibenspies and K. Burgess, *J. Am. Chem. Soc.*, 2002, **125**, 113–123.
100. A. A. Danopoulos, N. Tsoureas, J. C. Green and M. B. Hursthouse, *Chem. Commun.*, 2003, 756–757.
101. D. Pugh, A. Boyle and A. A. Danopoulos, *Dalton Trans.*, 2008, 1087–1094.
102. (a) X. Yang and M. B. Hall, *J. Am. Chem. Soc.*, 2008, **130**, 1798–1799; (b) X. Yang and M. B. Hall, *J. Organomet. Chem.*, 2009, **694**, 2831–2838.
103. (a) E. Becker, V. Stingl, G. Dazinger, M. Puchberger, K. Mereiter and K. Kirchner, *J. Am. Chem. Soc.*, 2006, **128**, 6572–6573; (b) E. Becker, V. Stingl, G. Dazinger, K. Mereiter and K. Kirchner, *Organometallics*, 2007, **26**, 1531–1535.
104. P. Csabai, F. Jó, A. M. Trzeciak and J. J. Ziólkowski, *J. Organomet. Chem.*, 2006, **691**, 3371–3376.
105. D. P. Allen, C. M. Crudden, L. A. Calhoun and R. Wang, *J. Organomet. Chem.*, 2004, **689**, 3203–3209.
106. M. J. Doyle, M. F. Lappert, P. L. Pye and P. Terreros, *J. Chem. Soc., Dalton Trans.*, 1984, 2355–2364.
107. C.-Y. Wang, Y.-H. Liu, S.-M. Peng and S.-T. Liu, *J. Organomet. Chem.*, 2006, **691**, 4012–4020.
108. D. P. Allen, C. M. Crudden, L. A. Calhoun, R. Wang and A. Decken, *J. Organomet. Chem.*, 2005, **690**, 5736–5746.
109. J. M. Praetorius, D. P. Allen, R. Wang, J. D. Webb, F. Grein, P. Kennepohl and C. M. Crudden, *J. Am. Chem. Soc.*, 2008, **130**, 3724–3725.
110. R. W. Simms, M. J. Drewitt and M. C. Baird, *Organometallics*, 2002, **21**, 2958–2963.
111. S. Fantasia and S. P. Nolan, *Chem.–Eur. J.*, 2008, **14**, 6987–6993.
112. C.-Y. Wang, Y.-H. Liu, S.-M. Peng, J.-T. Chen and S.-T. Liu, *J. Organomet. Chem.*, 2007, **692**, 3976–3983.
113. L. R. Titcomb, S. Caddick, F. G. N. Cloke, D. J. Wilson and D. McKerrecher, *Chem. Commun.*, 2001, 1388–1389.
114. (a) B. R. Galan, M. Gembicky, P. M. Dominiak, J. B. Keister and S. T. Diver, *J. Am. Chem. Soc.*, 2005, **127**, 15702–15703; (b) B. R. Galan, M. Pitak, M. Gembicky, J. B. Keister and S. T. Diver, *J. Am. Chem. Soc.*, 2009, **131**, 6822–6832.
115. A. Poater, F. Ragone, A. Correa and L. Cavallo, *J. Am. Chem. Soc.*, 2009, **131**, 9000–9006.
116. A. W. Waltman, T. Ritter and R. H. Grubbs, *Organometallics*, 2006, **25**, 4238–4239.
117. K. Randell, M. J. Stanford, G. J. Clarkson and J. P. Rourke, *J. Organomet. Chem.*, 2006, **691**, 3411–3415.
118. X. Wang, H. Chen and X. Li, *Organometallics*, 2007, **26**, 4684–4687.
119. S. Fantasia, H. Jacobsen, L. Cavallo and S. P. Nolan, *Organometallics*, 2007, **26**, 3286–3288.

120. E. Peris, J. A. Loch, J. Mata and R. H. Crabtree, *Chem. Commun.*, 2001, 201–202.
121. D. J. Nielsen, K. J. Cavell, B. W. Skelton and A. H. White, *Inorg. Chim. Acta*, 2006, **359**, 1855–1869.
122. H. Ren, X. Zhao, S. Xu, H. Song and B. Wang, *J. Organomet. Chem.*, 2006, **691**, 4109–4113.
123. D. C. Graham, K. J. Cavell and B. F. Yates, *Dalton Trans.*, 2005, 1093–1100.
124. V. P. W. Böhm, C. W. K. Gstöttmayr, T. Weskamp and W. A. Herrmann, *J. Organomet. Chem.*, 2000, **595**, 186–190.
125. A. M. Magill, B. F. Yates, K. J. Cavell, B. W. Skelton and A. H. White, *Dalton Trans.*, 2007, 3398–3406.
126. (a) P. C. J. Kamer, P. W. N. M. van Leeuwen and J. N. H. Reek, *Acc. Chem. Res.*, 2001, **34**, 895–904; (b) P. N. M. van Leeuwen, P. C. J. Kamer, J. N. H. Reek and P. Dierkes, *Chem. Rev.*, 2000, **100**, 2741–2770.
127. U. Hintermair, T. Gutel, A. M. Z. Slawin, D. J. Cole-Hamilton, C. C. Santini and Y. Chauvin, *J. Organomet. Chem.*, 2008, **693**, 2407–2414.
128. C. Gandolfi, M. Heckenroth, A. Neels, G. Laurenczy and M. Albrecht, *Organometallics*, 2009, **28**, 5112–5121.
129. S. Burling, M. F. Mahon, S. P. Reade and M. K. Whittlesey, *Organometallics*, 2006, **25**, 3761–3767.
130. T. Ritter, M. W. Day and R. H. Grubbs, *J. Am. Chem. Soc.*, 2006, **128**, 11768–11769.
131. A. C. Chen, D. P. Allen, C. M. Crudden, R. Y. Wang and A. Decken, *Can. J. Chem.–Rev. Can. Chim.*, 2005, **83**, 943–957.
132. (a) R. Dorta, E. D. Stevens, C. D. Hoff and S. P. Nolan, *J. Am. Chem. Soc.*, 2003, **125**, 10490–10491; (b) N. M. Scott, H. Clavier, P. Mahjoor, E. D. Stevens and S. P. Nolan, *Organometallics*, 2008, **27**, 3181–3186.
133. B. R. Dible and M. S. Sigman, *Inorg. Chem.*, 2006, **45**, 8430–8441.
134. V. P. W. Böhm, C. W. K. Gstöttmayr, T. Weskamp and W. A. Herrmann, *Angew. Chem., Int. Ed.*, 2001, **40**, 3387–3389.
135. J. Louie, J. E. Gibby, M. V. Farnworth and T. N. Tekavec, *J. Am. Chem. Soc.*, 2002, **124**, 15188–15189.
136. R. E. Douthwaite, M. L. H. Green, P. J. Silcock and P. T. Gomes, *J. Chem. Soc., Dalton Trans.*, 2002, 1386–1390.
137. R. H. Grubbs, Ed., *Handbook of Metathesis*, Wiley-VCH, Weinheim, 2003.
138. (a) M. S. Sanford, J. A. Love and R. H. Grubbs, *J. Am. Chem. Soc.*, 2001, **123**, 6543–6554; (b) M. S. Sanford, M. Ulman and R. H. Grubbs, *J. Am. Chem. Soc.*, 2001, **123**, 749–750; (c) J. A. Love, M. S. Sanford, M. W. Day and R. H. Grubbs, *J. Am. Chem. Soc.*, 2003, **125**, 10103–10109.
139. (a) J. S. Kingsbury, J. P. A. Harrity, P. J. Bonitatebus and A. H. Hoveyda, *J. Am. Chem. Soc.*, 1999, **121**, 791–799; (b) S. B. Garber, J. S. Kingsbury, B. L. Gray and A. H. Hoveyda, *J. Am. Chem. Soc.*, 2000, **122**, 8168–8179.
140. (a) M. S. Sanford, J. A. Love and R. H. Grubbs, *Organometallics*, 2001, **20**, 5314–5318; (b) J. A. Love, J. P. Morgan, T. M. Trnka and R. H. Grubbs, *Angew. Chem., Int. Ed.*, 2002, **41**, 4035–4037.

141. P. E. Romero, W. E. Piers and R. McDonald, *Angew. Chem., Int. Ed.*, 2004, **43**, 6161–6165.
142. P. E. Romero and W. E. Piers, *J. Am. Chem. Soc.*, 2005, **127**, 5032–5033.
143. M. Arisawa, Y. Terada, K. Takahashi, M. Nakagawa and A. Nishida, *J. Org. Chem.*, 2006, **71**, 4255–4261.
144. D. E. Fogg and E. N. dos Santos, *Coord. Chem. Rev.*, 2004, **248**, 2365–2379.
145. H. M. Lee, D. C. Smith, Z. He, E. D. Stevens, C. S. Yi and S. P. Nolan, *Organometallics*, 2001, **20**, 794–797.
146. U. L. Dharmasena, H. M. Foucault, E. N. dos Santos, D. E. Fogg and S. P. Nolan, *Organometallics*, 2005, **24**, 1056–1058.
147. N. J. Beach, J. M. Blacquiere, S. D. Drouin and D. E. Fogg, *Organometallics*, 2009, **28**, 441–447.
148. K. D. Camm, N. Martinez Castro, Y. Liu, P. Czechura, J. L. Snelgrove and D. E. Fogg, *J. Am. Chem. Soc.*, 2007, **129**, 4168–4169.
149. X.-Y. Yu, H. Sun, B. O. Patrick and B. R. James, *Eur. J. Inorg. Chem.*, 2009, **2009**, 1752–1758.
150. (a) M. G. Edwards, R. F. R. Jazzar, B. M. Paine, D. J. Shermer, M. K. Whittlesey, J. M. J. Williams and D. D. Edney, *Chem. Commun.*, 2004, 90–91; (b) S. Burling, M. K. Whittlesey and J. M. J. Williams, *Adv. Synth. Catal.*, 2005, **347**, 591–594.
151. F. Barrios-Landeros and J. F. Hartwig, *J. Am. Chem. Soc.*, 2005, **127**, 6944–6945.
152. (a) J. C. Green, B. J. Herbert and R. Lonsdale, *J. Organomet. Chem.*, 2005, **690**, 6054–6067; (b) K. C. Lam, T. B. Marder and Z. Lin, *Organometallics*, 2006, **26**, 758–760.
153. E. A. B. Kantchev, C. J. O'Brien and M. G. Organ, *Angew. Chem., Int. Ed.*, 2007, **46**, 2768–2813.
154. (a) R. Singh, M. S. Viciu, N. Kramareva, O. Navarro and S. P. Nolan, *Org. Lett.*, 2005, **7**, 1829–1832; (b) N. Marion, O. Navarro, J. Mei, E. D. Stevens, N. M. Scott and S. P. Nolan, *J. Am. Chem. Soc.*, 2006, **128**, 4101–4111.
155. V. P. W. Böhm, C. W. K. Gstöttmayr, T. Weskamp and W. A. Herrmann, *J. Organomet. Chem.*, 2000, **595**, 186–190.
156. G. A. Grasa, M. S. Viciu, J. Huang, C. Zhang, M. L. Trudell and S. P. Nolan, *Organometallics*, 2002, **21**, 2866–2873.
157. H. Lebel, M. K. Janes, A. B. Charette and S. P. Nolan, *J. Am. Chem. Soc.*, 2004, **126**, 5046–5047.
158. (a) U. Christmann and R. Vilar, *Angew. Chem., Int. Ed*, 2005, **44**, 366–374; (b) N. Marion and S. P. Nolan, *Acc. Chem. Res.*, 2008, **41**, 1440–1449.
159. (a) S. Caddick, F. G. N. Cloke, G. K. B. Clentsmith, P. B. Hitchcock, D. McKerrecher, L. R. Titcomb and M. R. V. Williams, *J. Organomet. Chem.*, 2001, **617–618**, 635–639; (b) C. W. K. Gstöttmayr, V. P. W. Böhm, E. Herdtweck, M. Grosche and W. A. Herrmann, *Angew. Chem., Int. Ed.*, 2002, **41**, 1363–1365; (c) R. Jackstell, M. G. Andreu, A. Frisch,

K. Selvakumar, A. Zapf, H. Klein, A. Spannenberg, D. Röttger, O. Briel, R. Karch and M. Beller, *Angew. Chem., Int. Ed.*, 2002, **41**, 986–989; (d) A. C. Frisch, A. Zapf, O. Briel, B. Kayser, N. Shaikh and M. Beller, *J. Mol. Catal. A: Chem.*, 2004, **214**, 231–239; (e) K. Arentsen, S. Caddick, F. G. N. Cloke, A. P. Herring and P. B. Hitchcock, *Tetrahedron Lett.*, 2004, **45**, 3511–3515.

160. R. Jackstell, S. Harkal, H. Jiao, A. Spannenberg, C. Borgmann, D. Röttger, F. Nierlich, M. Elliot, S. Niven, K. Cavell, O. Navarro, M. S. Viciu, S. P. Nolan and M. Beller, *Chem.–Eur. J.*, 2004, **10**, 3891–3900.
161. K. Selvakumar, A. Zapf, A. Spannenberg and M. Beller, *Chem.–Eur. J.*, 2002, **8**, 3901–3906.
162. M. S. Viciu, R. F. Germaneau, O. Navarro-Fernandez, E. D. Stevens and S. P. Nolan, *Organometallics*, 2002, **21**, 5470–5472.
163. Z. Jin, S.-X. Guo, X.-P. Gu, L.-L. Qiu, H.-B. Song and J.-X. Fang, *Adv. Synth. Catal.*, 2009, **351**, 1575–1585.
164. (a) M. S. Viciu, E. D. Stevens, J. L. Petersen and S. P. Nolan, *Organometallics*, 2004, **23**, 3752–3755; (b) M. J. Schultz, S. S. Hamilton, D. R. Jensen and M. S. Sigman, *J. Org. Chem.*, 2004, **70**, 3343–3352.
165. J. A. Mueller, C. P. Goller and M. S. Sigman, *J. Am. Chem. Soc.*, 2004, **126**, 9724–9734.
166. R. B. Bedford, C. S. J. Cazin, S. J. Coles, T. Gelbrich, P. N. Horton, M. B. Hursthouse and M. E. Light, *Organometallics*, 2003, **22**, 987–999.
167. C. Desmarets, R. Schneider and Y. Fort, *J. Org. Chem.*, 2002, **67**, 3029–3036.
168. J. J. Brunet, D. Besozzi, A. Courtois and P. Caubere, *J. Am. Chem. Soc.*, 2002, **104**, 7130–7135.
169. K. Matsubara, S. Miyazaki, Y. Koga, Y. Nibu, T. Hashimura and T. Matsumoto, *Organometallics*, 2008, **27**, 6020–6024.
170. (a) G. De Bo, G. Berthon-Gelloz, B. Tinant and I. E. Markó, *Organometallics*, 2006, **25**, 1881–1890; (b) G. Berthon-Gelloz, J.-M. Schumers, G. De Bo and I. E. Markó, *J. Org. Chem.*, 2008, **73**, 4190–4197.
171. (a) F. Hanasaka, K.-I. Fujita and R. Yamaguchi, *Organometallics*, 2004, **23**, 1490–1492; (b) F. Hanasaka, K.-I. Fujita and R. Yamaguchi, *Organometallics*, 2005, **24**, 3422–3433.

CHAPTER 4

Biologically Active N-Heterocyclic Carbene–Metal Complexes

MICHAEL C. DEBLOCK,[a] MATTHEW J. PANZNER,[a] CLAIRE A. TESSIER,[a] CAROLYN L. CANNON[b] AND WILEY J. YOUNGS[a,*]

[a] Center for Silver Therapeutics Research, Department of Chemistry and Integrated Bioscience Program, Buchtel College of Arts and Sciences, University of Akron, Akron, Ohio 44325-3601, USA; [b] School of Medicine, Pediatrics, University of Texas Southwestern Medical Center, Dallas, Texas 75390, USA

4.1 Introduction

In 1968 the first reported syntheses of N-heterocyclic carbenes (NHCs) were published by Öfele and Wanzlick.[1] When Arduengo isolated the first free NHC in 1991,[2] this breakthrough significantly increased interest in this area of research. Since then the study of organometallic complexes of N-heterocyclic carbenes has continued to grow and the number of their applications is increasing.[3] NHCs can bind to both soft and hard metals, increasing their diversity and versatility.[4] A new area of research for metal NHC complexes is their use as broad-spectrum antimicrobials and anti-tumour agents.[5] Although most of these results are preliminary, metal NHC species have certainly shown the ability to treat a wide range of biological afflictions from multi-drug resistant pathogens gathered from cystic fibrosis patients to *in vivo* treatment of ovarian cancer.

RSC Catalysis Series No. 6
N-Heterocyclic Carbenes: From Laboratory Curiosities to Efficient Synthetic Tools
Edited by Silvia Díez-González
© Royal Society of Chemistry 2011
Published by the Royal Society of Chemistry, www.rsc.org

4.2 N-Heterocyclic Carbene–Silver Complexes

Many papers showed that NHC–silver(I) could be effective broad-spectrum antimicrobials. NHC–Ag^I compounds were used in *in vivo* studies to treat multidrug-resistant pulmonary infections found in cystic fibrosis patients,[6] and *in vitro* and *in vivo* studies on several strains of cancerous cells.[7]

4.2.1 Medicinal Uses of Silver Complexes

Silver nitrate was first used as an antimicrobial agent in the seventeeth century, a practice that was continued for open wounds and skin ulcers through the twentieth century.[8] The use of silver compounds began to wane with the discovery of organic antibiotics such as penicillin. However, in the mid 1900s organisms such as *Pseudomonas aeruginosa*, *Proteus mirabilis* and *Proteus morgani* began to show increased resistance to penicillin and sulfa drugs. In the 1960s, Fox introduced a response to these highly resistant organisms by developing silver sulfadiazine, a combination of a sulfa antibiotic and silver(I). Marketed as Silvadene®, this effective antimicrobial cream is still widely used in burns wards today.[9] Despite the wide use of silver compounds and the isolation of resistant strains of *Salmonella* and *Escherichia coli*, the amount of resistant organisms has not significantly increased.[10]

4.2.2 Antimicrobial Properties of NHC–Silver(I) Complexes

In 2004, Youngs reported NHC–Ag^I complexes that possessed antimicrobial properties against *E. coli*, *Staph. aureus* and *P. aeruginosa*.[11] These pincer compounds **1** and **2** (Scheme 4.1) had better minimum inhibitory concentrations (MIC) in lysogeny broth (LB) than $AgNO_3$. Soon after, a NHC cyclophane *gem*-diol salt was reacted with Ag_2O to form the corresponding silver complex **3** (Scheme 4.2).[12] This NHC–Ag^I compound was combined with Techophilic, a medical-grade polymer. Electrospun fibres were then manipulated to produce a fibrous mat and tested for their antimicrobial properties. *E. coli*, *Staph. aureus* and *P. aeruginosa* were compared to $AgNO_3$ after a 48 h incubation period. The silver Techophilic mat demonstrated the lowest MIC values for all three pathogens tested.

The next biologically active NHC–Ag^I complex developed, **4**, was derived from a modified caffeine complex combined with silver acetate (Scheme 4.3).[13] Caffeine, a xanthine derivative, was chosen due to its low toxicity, cost, and high availability. The *in vitro* activity of this silver acetate complex was evaluated against a wide variety of pathogens recovered from cystic fibrosis patients. Complex **4** was very effective against all tested strains of *P. aeruginosa*, *B. cepacia*, *B. multivorans*, *B. cenocepacia*, *B. stabilis*, *B. vietnamiensis*, *B. dolosa*, *B. ambifaria*, *B. anthina* and *B. pyrocinia*. MIC values were between 1 and 10 $\mu g\,mL^{-1}$, and resistance was observed for one strain of *E. coli* with pMG101 plasmid, a plasmid known to cause silver resistance.[10b] This complex

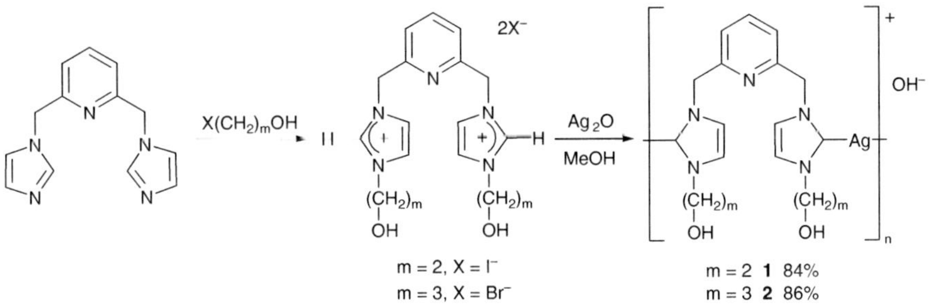

Scheme 4.1 Synthesis of **1** and **2**.

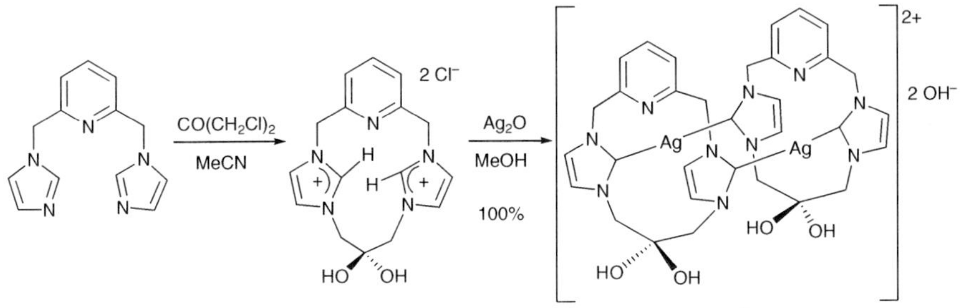

Scheme 4.2 Synthesis of **3**.

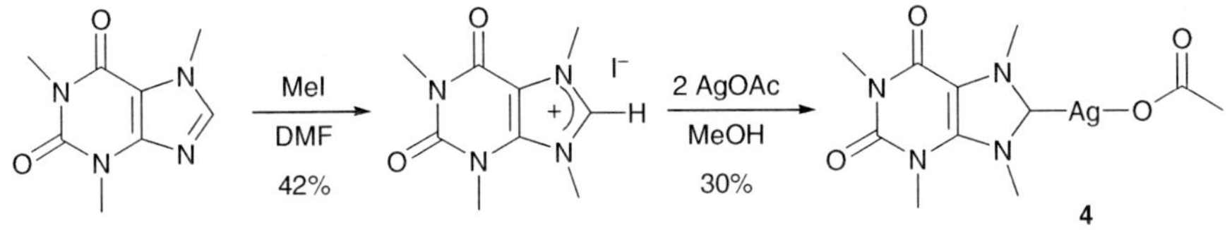

Scheme 4.3 Synthesis of **4**.

was also effective against fungal strains of *Aspergillus niger* and *Saccharomyces cerevisiae* with MIC values below 13 $\mu g\,mL^{-1}$.

Compound **4** was also found efficient against a completely antibiotic resistant strain of *Burkholderia*, specifically *Burkholderia dolosa*, which is resistant to clinical antibiotics in concentrations up to 5 $mg\,mL^{-1}$. However, *B. dolosa* was found to be susceptible to **4** at significantly lower concentrations. Transmission electronic microscopy (TEM) photos of *B. dolosa* obtained after a 1 h treatment of 5 $\mu g\,mL^{-1}$ of **4** showed the powerful antimicrobial properties of the NHC–Ag^{I}. Figure 4.1 shows the morphology of the cell before and after treatment, when the cell was annihilated.

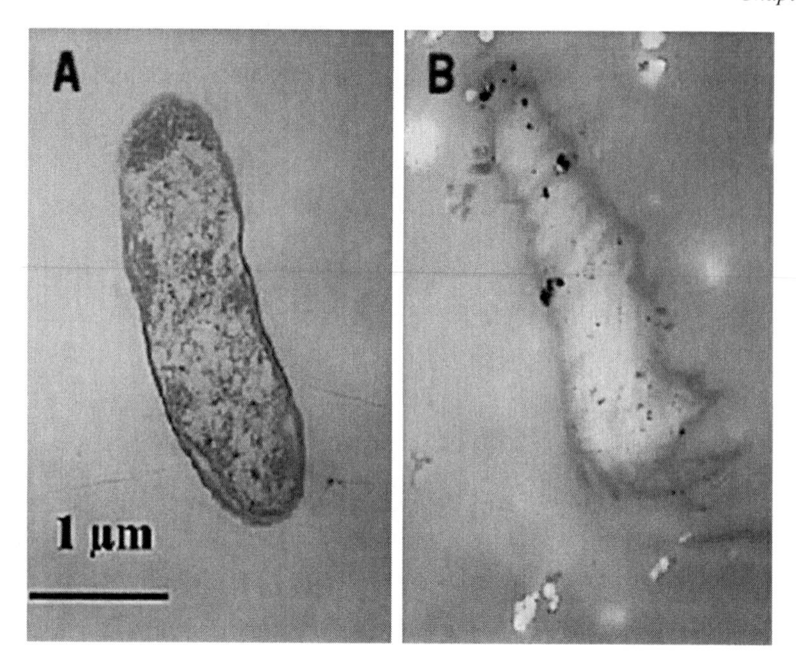

Figure 4.1 TEM of *B. dolosa* strain AU4459 of (A) normal cell and (B) treated cell with **4**. Reproduced, with permission, from Kascatan-Nebioglu *et al.*[13]

The observation that the electron-withdrawing groups of xanthene compounds stabilized NHC–Ag^I bonds led the Youngs group to investigate imidazole-based carbenes with different electron-withdrawing groups on the 4 and 5 positions, **5** and **6** (Figure 4.2).[14] Compound **5**, based on 4,5-dichloroimidazole, was stable for at least 17 weeks in D_2O as determined by ^1H NMR, compared to 3 days for xanthine derivatives. Stability in aqueous solutions is of importance for medicinal applications as the silver complex needs to be stable enough to be delivered to the area of interest without degrading.

The Ghosh group reported a NHC–Ag^I **7** synthesized from 1-benzyl-3-*tert*-butylimidazolium chloride (Scheme 4.4).[15] This ligand precursor showed no antimicrobial properties against both *E. coli* and *B. subtilis*, while the NHC–Ag^I compound showed inhibition of *B. subtilis* but not *E. coli*.

4.2.2.1 Treatment of Multidrug-resistant Infections in Cystic Fibrosis Models

In recent years multidrug-resistant bacteria have become more prevalent. Infections caused by these pathogens are exceptionally difficult to cure and can be life threatening amongst susceptible populations, including people with compromised immune systems and disorders such as cystic fibrosis. Such infections require novel therapeutics and NHC–metal complexes were shown to

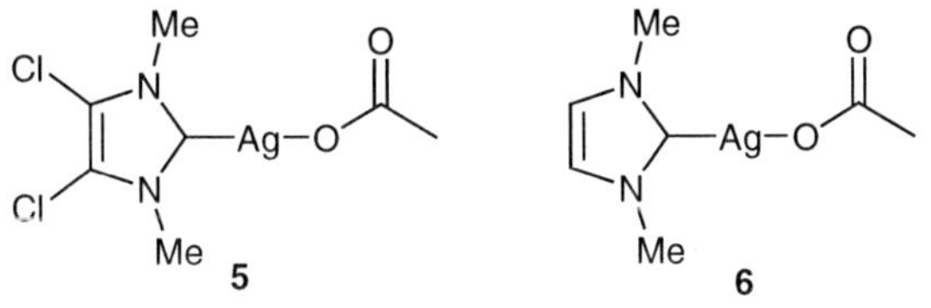

Figure 4.2 Structures of **5** and **6**.

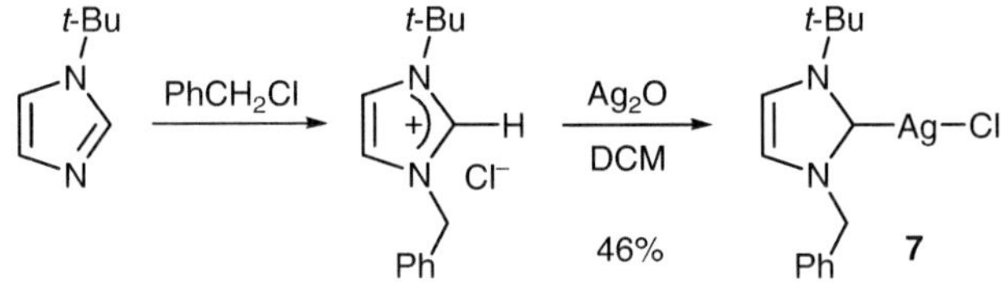

Scheme 4.4 Synthesis of 7.

be a potential effective treatment.[16] NHC–Ag^{I} complexes **4** and **8** successfully treated *P. aeruginosa* infections in mice (Figure 4.3). *In vitro* testing with **8** showed the complex to be very effective against a wide variety of cystic fibrosis (CF) relevant pathogens at concentrations of 2 μg mL^{-1} or less. In this study strains of *P. aeruginosa*, *A. xyloxoxidans*, *S. maltophilia*, *S. aureus*, *B. multivorans*, *B. cenocepacia*, *B. dolosa*, *B. thailandensis*, *B. gladioli*, and *E. coli* were tested, using *E. coli* (J53 + pGM101 plasmid) that exhibited silver resistance as a control.

Three groups of six mice were each given a lethal dose of PA M57-15, a clinically isolated strain of *P. aeruginosa*. Then each mouse was treated once every 12 h for a total of 48 h. Each group was treated with either water, **4** or **8** and evaluated after 72 h (Figure 4.3).[16] These *in vivo* studies showed a very large survival advantage for mice treated with either **4** or **8**. These compounds along with many other NHC–Ag^{I} could become a life-saving treatment for people with multidrug-resistant infections.

4.2.3 Anti-tumour Activities of NHC–Silver(I) Complexes

The first report on the anti-cancer properties of NHC–Ag^{I} complexes was made by the Youngs group and involved the use of 4,5-dichloroimidazole derived NHC–Ag^{I}. Three compounds, **5**, **9** and **10**, were tested against three cancer lines, OVCAR-3 (ovarian), MB157 (breast), and HeLa (cervical) (Figure 4.4).[7] Cell viability was determined by using the MTT assay (MTT = 3-(4,5-dimethylthiazol-2-yl)-2,5-diphenyltetrazolium bromide). IC_{50} values for **5**, **9** and **10** on OVCAR-3 were observed to be less than 35 μM and less than 20 μM on MB157. HeLa cells on the other hand were not affected in notable

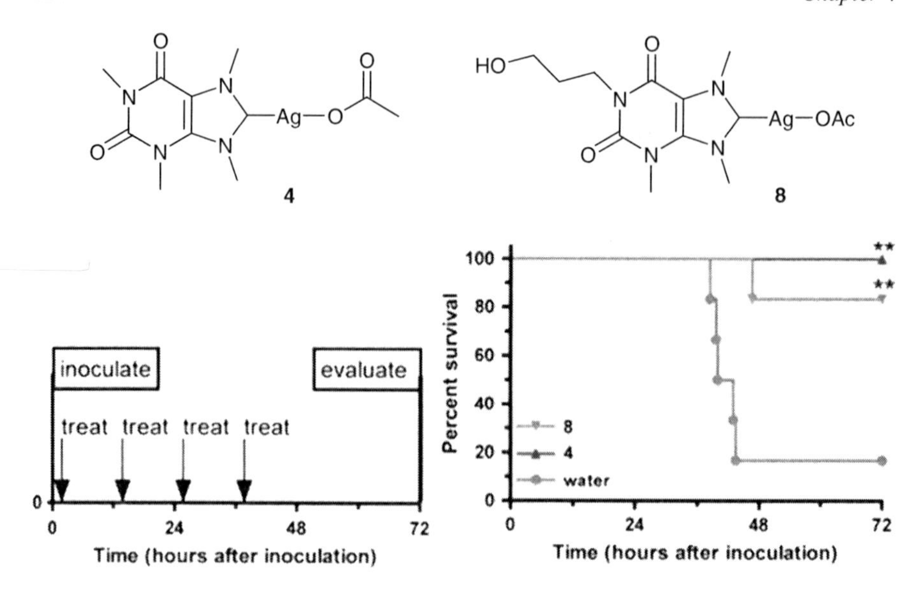

Figure 4.3 Treatment and survival advantage of mice after treatment with **4** and **8**.

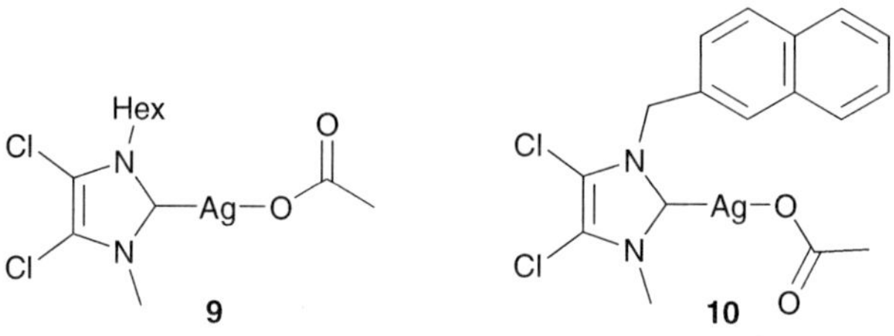

Figure 4.4 Structures of **9** and **10**.

concentrations, demonstrating some selectivity for the action of silver on cells. The NHC precursors of these compounds showed no effective change in cell viability.[7] Reported preliminary *in vivo* tests showed that subcutaneous tumours treated with **5** at the site of the tumour became necrotic, probing the ability of NHC–Ag^{I} to effectively treat actively growing tumours (Figure 4.5).

4.3 N-Heterocyclic Carbene–Gold Complexes

4.3.1 Medicinal Uses of Gold

In the late 1890s, Koch discovered that gold cyanide was effective against *Mycobacterium tuberculosis*, more commonly known as TB.[17] Although gold was used to treat pulmonary tuberculosis, it was not a very effective therapy and was discontinued. Years after the discovery that gold exhibited

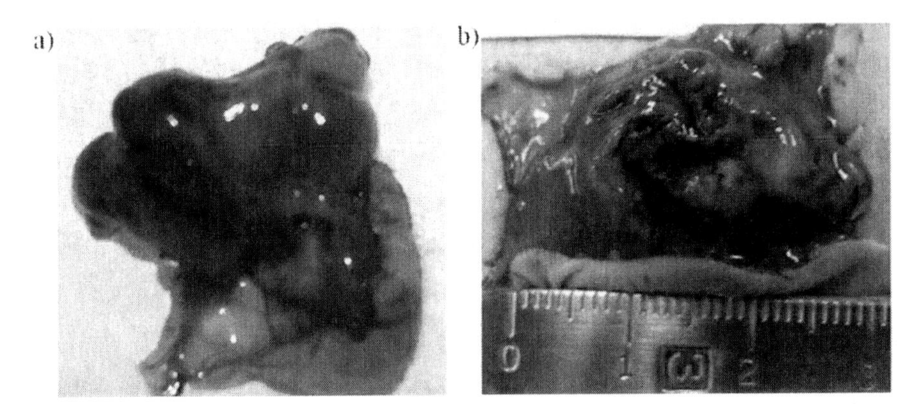

Figure 4.5 Solid OVCAR-3 tumours. (a) Healthy tumour before SubQ injection of **5**; (b) Necrotic tumour after three SubQ injections of 333 mg kg^{-1} dose of **5** over a 10 day period.

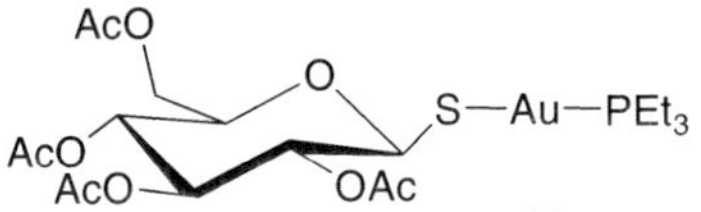

AcO
O
S—Au—PEt_3
AcO
AcO
OAc
11

Figure 4.6 Structure of Auranofin **11**.

bacteriostatic activity, gold(I) thiolates first, and Auranofin **11** after, were used to treat rheumatoid arthritis (Figure 4.6).[18] Treatments with gold compounds are currently being explored with promising research for the treatment of malaria, cancer and HIV.[18]

4.3.2 Antimicrobial Properties of NHC–Gold(I) Complexes

In 2004, Çetinkaya reported the antimicrobial activity of a series of NHC–Au^I complexes **12**–**17** (Figure 4.7).[19] The *in vitro* antimicrobial activity of these complexes was determined for a variety of Gram-positive and Gram-negative bacteria, using ampicillin and flucytosine as standards. These gold complexes showed effective antimicrobial activity against both bacteria types, but little activity against yeast and fungi. Ghosh and co-workers also reported the antimicrobial activity of an NHC–Au^I complex **18**, an analogue of the silver compound **7** (Scheme 4.5).[15] Similarly to the silver compound, the gold derivative **18** showed no activity against *E. coli*, but did show antimicrobial activity against *B. subtilis*. Morphology of the *B. subtilis* was examined after treatment with the NHC–Au^I and an elongation of the cells was observed. Ghosh suggested that the cause of this elongation was the blocking of the cytokinesis step of cell devision.

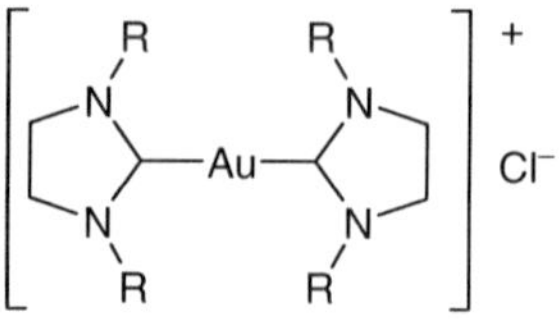

12 R = Mes
13 R = $CH_2C_6H_4OMe$-*p*
14 R = $CH_2C_6H_4(NMe_2)$-*p*
15 R = $CH_2C_6H_2Me_3$-2,4,6
16 R = Cyp
17 R = Et

Figure 4.7 Synthesis of **12**–**17**.

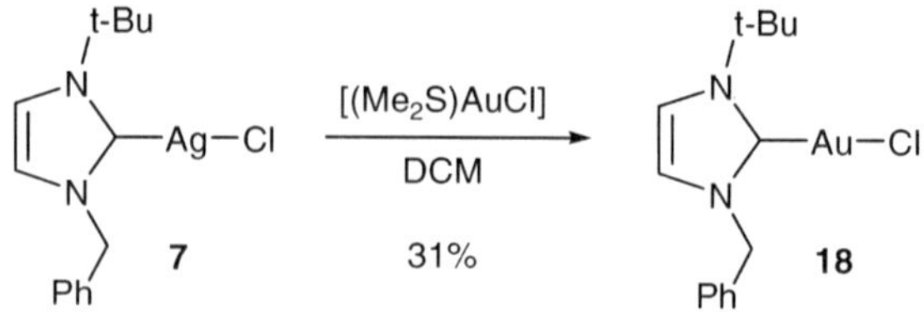

$[(Me_2S)AuCl]$, DCM, 31%: **7** → **18**

Scheme 4.5 Synthesis of **18**.

4.3.3 Anti-tumour Properties of NHC–Gold(I) and NHC–Gold(III) Complexes

Major contributions to NHC–Au^I anti-cancer research were produced by Berners-Price and co-workers.[5c] They evaluated dinuclear NHC–Au^I complexes **19**–**25** and mono-nuclear NHC–Au^I complexes **26**–**30** (Figure 4.8) for their mitochondrial permeability.[20] These compounds displayed different lipophilicity and were evaluated for their ability to induce mitochondrial membrane permeabilisation (MMP). Due to this unique property, gold complexes have the potential to be exceptional anti-tumour agents because when a cell undergoes apoptosis it does not harm surrounding cells.[21] Cancerous cells have elevated plasma and mitochondrial membrane potentials relative to normal cells. By targeting MMP it might be possible to overcome two major problems, the occurrence of drug-resistant cells and the selectivity between cancerous and non-cancerous cells.[22] The mono-nuclear NHC–Au^I compounds showed a direct correlation of anti-mitochondrial activity based on their lipophilicity. The dinuclear NHC–Au^I compounds also showed effective anti-mitochondrial activity.

Berners-Price also described NHC–Au^I compounds that targeted the mitochondria and exhibited thioredoxin reductase (TrxR) inhibition properties.[23] Compounds **27** and **31**–**33** showed toxicity to breast cancer cell lines MDA-MB-231 and MDA-MB-468, but did not show toxicity on normal human mammary epithelia cells (Figure 4.8). Compound **28**, an intermediate lipophilic complex, showed the best selectivity and cytotoxicity potential. This compound, while being inactive on glutathione reductase, was observed to

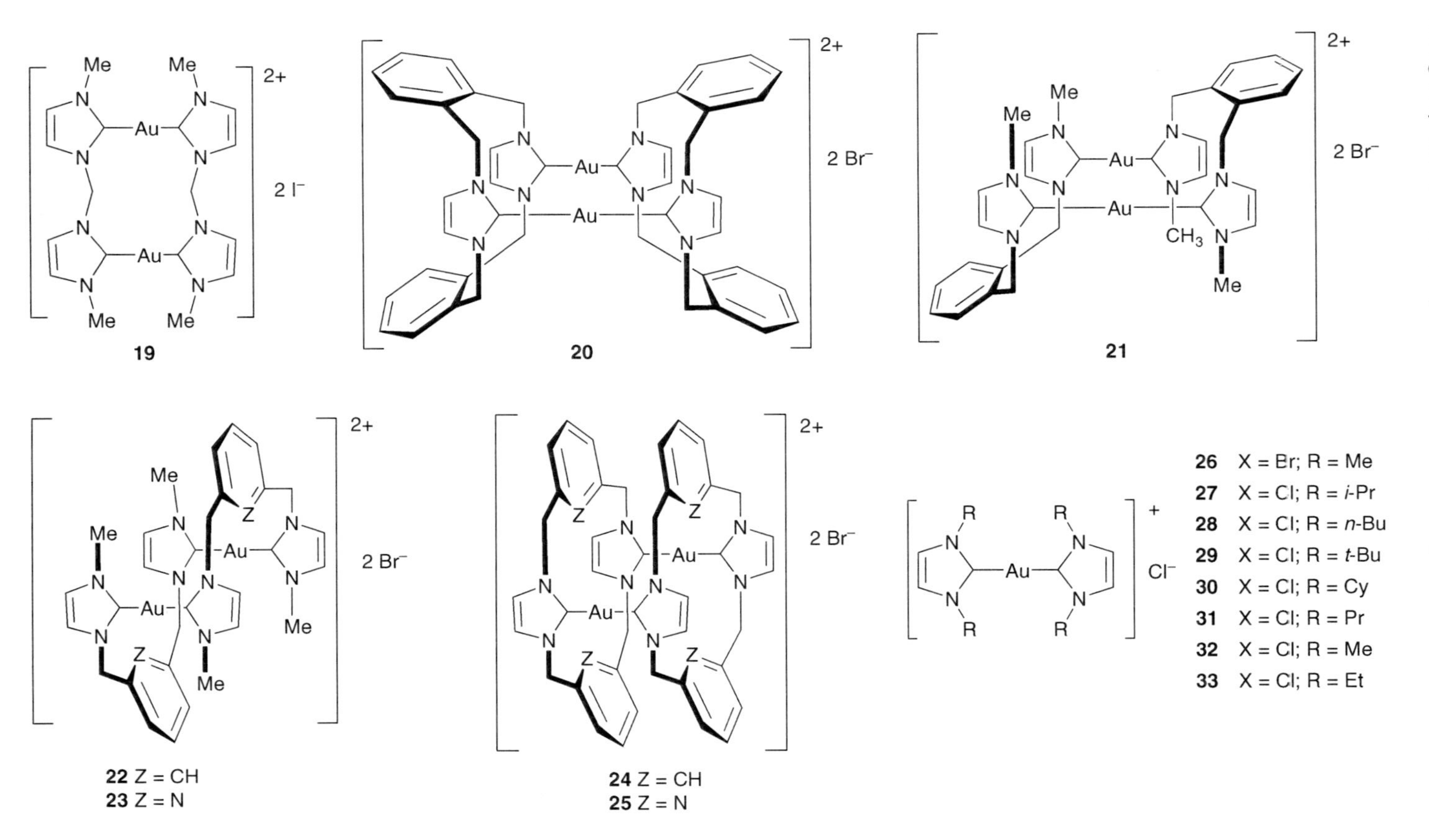

Figure 4.8 Complexes **19**–**33**.

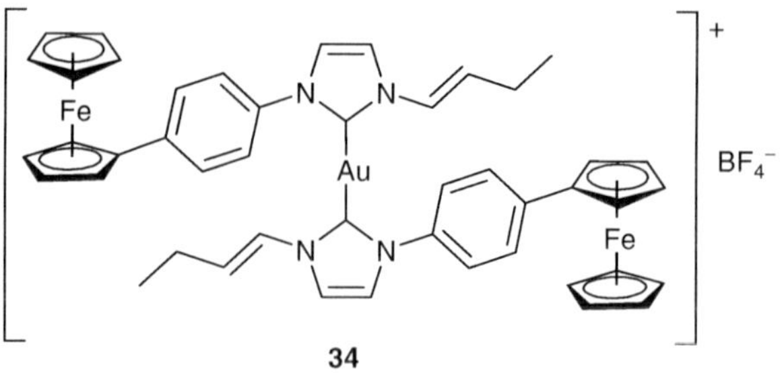

Figure 4.9 Complex **34**.

selectively target mitochondrial selenoproteins, such as TrxR. When TrxR was inhibited, apoptosis occurred.

Raubenheimer reported the anti-cancer activity of compound **34** (Figure 4.9).[24] This complex was tested on human cervical epithelioid carcinoma cell line HeLa, human colon adenocarcinoma cell line CoLo 320 DM, leukaemia cell line Jurkat, and breast cancer cell line MCF-7. IC_{50} values determined by the MTT assay showed human lymphocytes were less sensitive than the cancer cells tested, indicating some selectivity for the cancer cells.

Metzler-Nolte reported a variety NHC–Au^{I} compounds **35**–**39** and NHC–Au^{III} compounds **40**–**42** (Figure 4.10).[25] All compounds were tested on HeLa, HT-29 and HepG2 cancer cell lines, and showed some anti-cancer activity, with IC_{50} values between 3 and 100 μM. The highest resistance was observed for HT-29 cancer cells, while the HeLa cells were the most sensitive (interestingly, this is in contrast to NHC–Ag^{I} compounds in Section 4.2.3). Also, when comparing compounds **38** and **39** with **37**, no increase of activity could be observed, even if compound **38** showed IC_{50} values similar to **37**, and compound **39** had a lower activity.

4.4 Other Medically Relevant NHC–Metal Complexes

4.4.1 Antimicrobial Properties of NHC–Ruthenium(II) Complexes

There are three main properties that make ruthenium compounds suited for medicinal applications: the rate of ligand exchange, the range of accessible oxidation states, and the ability to mimic iron binding to biological molecules.[26] Antimicrobial activity of NHC–Ru^{II} complexes was first investigated by Çetinkaya.[27] A variety of ruthenium(II) complexes based on **43** and **44** were evaluated for their antimicrobial and antifungal activities (Figure 4.11).[27,28] Different NHC–Ru^{II} compounds were tested against *Ec. faecalis, S. aureus, E. coli* and *P. aeruginosa*, and against *C. albicans* and *C. tropicalis* to observe their antifungal activities. Their overall antibacterial activity was limited, and none of the compounds tested showed antibacterial activity equivalent to the

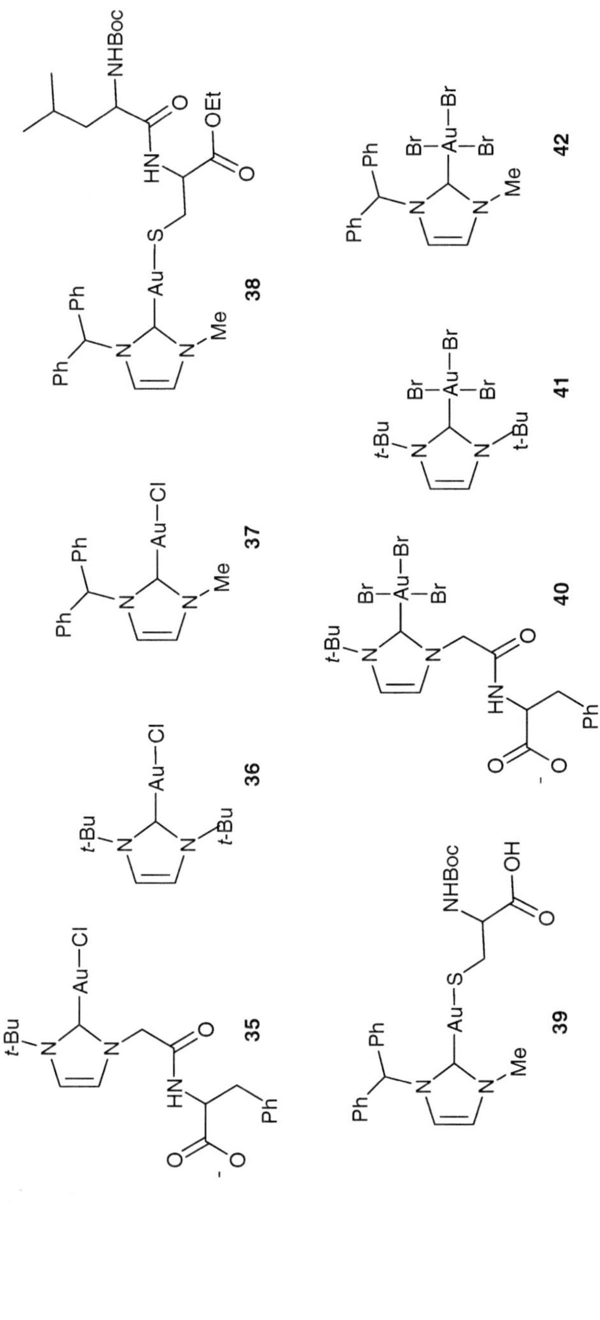

Figure 4.10 Complexes **35**–**42**.

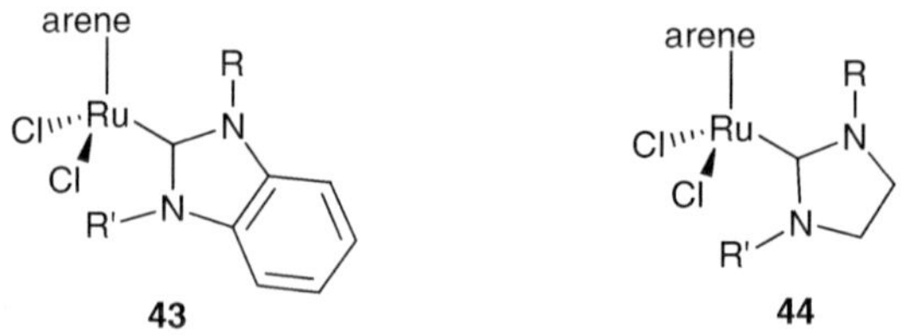

Figure 4.11 Compounds **43** and **44**.

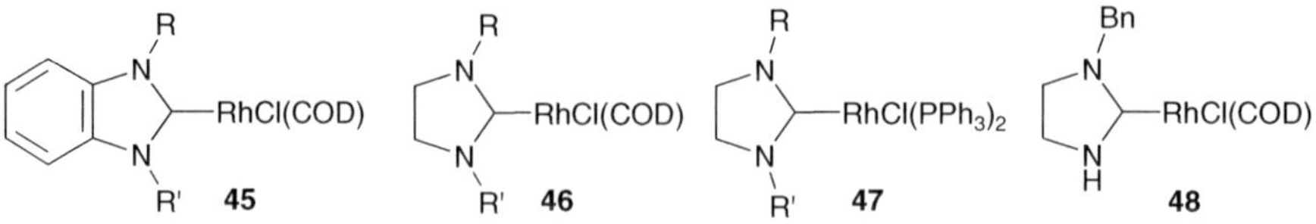

Figure 4.12 Compounds **45**–**48**.

control compound ampicillin. The best activity was observed against Gram-positive *Ec. faecalis* and *S. aureus* with one compound as low as 6.25 μg mL^{-1}. All NHC–Ru^{II} compounds tested had minimal effect on Gram-negative bacteria *E. coli* and *P. aeruginosa*. Some of them also showed minimal effects on fungal strains of *C. albicans* and *C. tropicalis*.

4.4.2 Antimicrobial Properties of NHC–Rhodium(I) Complexes

Çetinkaya investigated the antimicrobial activity of NHC–Rh^{I} complexes **45**–**48** against *Ec. faecalis, S. aureus, E. coli* and *P. aeruginosa* (Figure 4.12).[27] These NHC–Rh^{I} complexes showed greater antimicrobial activity than the ruthenium complexes **43** and **44**. Two compounds showed effectiveness of 5 μg mL^{-1} against Gram-positive *Ec. faecalis* and *S. aureus*. Similarly to the ruthenium compounds, these rhodium derivatives showed minimal effects against Gram-negative bacteria *E. coli* and *P. aeruginosa*.

4.4.3 Anti-tumour Activity of NHC–Palladium(II) Complexes

In 2007, Panda, Ghosh and co-workers studied the anti-cancer activity of two palladium compounds, **49** and **50** (Figure 4.13).[15] IC$_{50}$ values for **50** were calculated for three different cancerous cell lines: 4 (HeLa), 0.8 (HCT 116) and 1 μM (MCF-7). Detailed mechanistic studies conducted on the most potent complex **50** suggested that its mode of action involved arresting the cells at the G2 phase, thereby preventing the cells from entering the mitotic phase. The blocked cells then underwent cell death by a P53-dependent pathway.

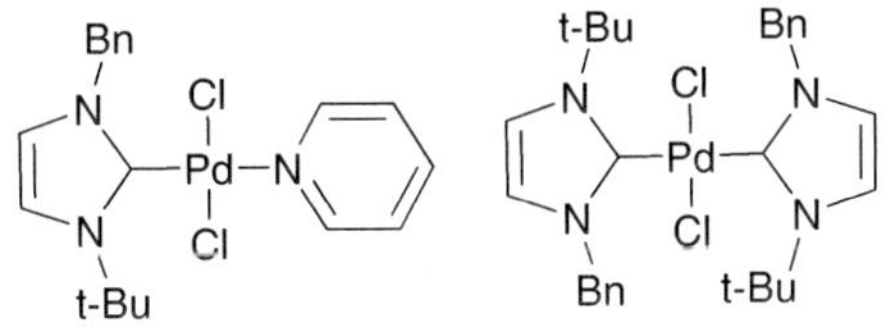

Figure 4.13 Compounds **49** and **50**.

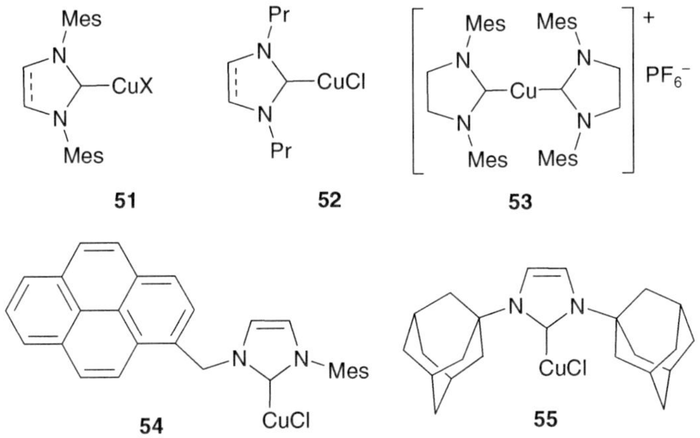

Figure 4.14 Compounds **51**–**55**. SEQ CHAPTER

4.4.4 Anti-tumour Activity of NHC–Copper(I) Complexes

Teyssot *et al.* reported the anti-cancer properties of NHC–Cu^I compounds **51**–**55** (Figure 4.14).[5b] These were tested against MCF-7 breast cancer cells and exhibited substantially higher cytotoxicities than cisplatin. Compound **51** (NHC = SIMes, X = Cl) was also evaluated on four other cancer cell lines: HL60, KB, MCF-7R and LNCaP, exhibiting IC_{50} values up to a 150 times lower than cisplatin.[29] The authors also studied the cell-cycle and apoptotic levels of **51** (NHC = SIMes, X = Cl) in MCF-7 to discover that **51** acted like a chemical nuclease, preventing cell cycle progression in the G1 phase and in turn inducing apoptosis.

4.5 Conclusion

A variety of N-heterocyclic metal carbenes, although mostly known for their use in catalysis, are currently being investigated for new and exciting roles in medicine. Many of these NHC–metal complexes have shown the ability to treat multi-drug resistant infections and cancers for which there are currently no effective therapies. Although this is only a relatively small research area in

NHC chemistry, the results look promising. Further efforts should therefore lead to the development on several effective drugs.

References

1. (a) K. Öfele, *J. Organomet. Chem.*, 1968, **12**, P42–P43; (b) W.-H. Wanzlick and J.-H. Schönberr, *Angew. Chem., Int. Ed. Engl.*, 1968, **7**, 141–142.
2. A. J. Arduengo III, R. L. Harlow and M. Kline, *J. Am. Chem. Soc.*, 1991, **113**, 361–363.
3. (a) J. C. Garrison and W. J. Youngs, *Chem. Rev.*, 2005, **105**, 3978–4008; (b) Special Issue Devoted to Carbenes, *Chem. Rev.*, 2009, **109**, 3209–3884.
4. X. Hu, I. Castro-Rodriguez, K. Olsen and K. Meyer, *Organometallics*, 2004, **23**, 755–764.
5. (a) K. M. Hindi, M. J. Panzner, C. A. Tessier, C. L. Cannon and W. J. Youngs, *Chem. Rev.*, 2009, **109**, 3859–3884; (b) M.-L. Teyssot, A.-S. Jarrousse, M. Manin, A. Chevry, S. Roche, F. Norre, C. Beaudoin, L. Morel, D. Boyer, R. Mahiou and A. Gautier, *Dalton Trans.*, 2009, 6894–6902; (c) P. J. Barnard and S. J. Berners-Price, *Coord. Chem. Rev.*, 2007, **251**, 1889–1902.
6. C. L. Cannon, L. A. Hogue, R. K. Vajravelu, G. H. Capps, A. Ibricevic, K. M. Hindi, A. Kascatan-Nebioglu, M. J. Walter, S. L. Brody and W. J. Youngs, *Antimicrob. Agents Chemother.*, 2009, **53**, 3285–3293.
7. D. A. Medvetz, K. M. Hindi, M. J. Panzner, A. J. Ditto, Y. H. Yun and W. J. Youngs, *Met.-Based Drugs*, 2008, 384010.
8. (a) H. J. Klasen, *Burns*, 2000, **26**, 117–130; (b) S. Silver, L. T. Phung and G. Silver, *J. Ind. Microbiol. Biotechnol.*, 2006, **33**, 627–634.
9. H. J. Klasen, *Burns*, 2000, **26**, 131–138.
10. (a) G. L. Mchugh, R. C. Moellering, C. C. Hopkins and M. N. Swartz, *Lancet*, 1975, **305**, 235–240; (b) A. Gupta, K. Matsui, J.-F. Lo and S. Silver, *Nat. Med.*, 1999, **5**, 183–188; (c) S. Silver, *FEMS Microbiol. Rev.*, 2003, **27**, 1–353.
11. A. Melaiye, R. S. Simons, A. Milsted, F. Pingitore, C. Wesdemiotis, C. A. Tessier and W. J. Youngs, *J. Med. Chem.*, 2004, **47**, 973–977.
12. A. Melaiye, Z. Sun, K. Hindi, A. Milsted, D. Ely, D. H. Reneker, C. A. Tessier and W. J. Youngs, *J. Am. Chem. Soc.*, 2005, **127**, 2285–2291.
13. A. Kascatan-Nebioglu, A. Melaiye, K. Hindi, S. Durmus, M. J. Panzner, L. A. Hogue, R. J. Mallett, C. E. Hovis, M. Coughenour, S. D. Crosby, A. Milsted, D. L. Ely, C. A. Tessier, C. L. Cannon and W. J. Youngs, *J. Med. Chem.*, 2006, **49**, 6811–6818.
14. K. Hindi, T. J. Siciliano, S. Durmus, M. J. Panzner, D. A. Medvetz, D. V. Reddy, L. A. Hogue, C. E. Hovis, J. K. Hilliard, R. J. Mallet, C. A. Tessier, C. L. Cannon and W. J. Youngs, *J. Med. Chem.*, 2008, **51**, 1577–1583.

15. S. Ray, R. Mohan, J. K. Singh, M. K. Samantaray, M. M. Shaikh, D. Panda and P. Ghosh, *J. Am. Chem. Soc.*, 2007, **129**, 15042–15053.
16. M. J. Panzner, K. M. Hindi, B. D. Wright, J. B. Taylor, D. S. Han, W. J. Youngs and C. L. Cannon, *Dalton Trans.*, 2009, 7308–7313.
17. S. P. Fricker, *Gold Bull.*, 1996, **29**, 53–60.
18. C. F. Shaw III, *Chem. Rev.*, 1999, **99**, 2589–2600.
19. Ý. Özdemir, A. Denizci, H. T. Öztürk and B. Çetinkaya, *Appl. Organomet. Chem.*, 2004, **18**, 318–322.
20. M. V. Baker, P. J. Barnard, S. J. Berners-Price, S. K. Brayshaw, J. L. Hickey, B. W. Skelton and A. H. White, *Dalton Trans.*, 2006, 3708–3715.
21. P. Bernardi, L. Scorrano, R. Colonna, V. Petronilli and F. Di Lisa, *Eur. J. Biochem.*, 1999, **264**, 687–701.
22. L. B. Chen, *Ann. Rev. Cell Biol.*, 1988, **4**, 155–181.
23. J. L. Hickey, R. A. Ruhayel, P. J. Barnard, M. V. Baker, S. J. Berners-Price and A. Filipovska, *J. Am. Chem. Soc.*, 2008, **130**, 12570–12571.
24. U. E. I. Horvath, G. Bentivoglio, M. Hummel, H. Schottenberger, K. Wurst, M. J. Nell, C. E. J. van Rensburg, S. Cronje and H. G. Raubenheimer, *New J. Chem.*, 2008, **32**, 533–539.
25. J. Lemke, A. Pinto, P. Niehoff, V. Vasylyeva and N. Metzler-Nolte, *Dalton Trans.*, 2009, 7063–7070.
26. C. S. Allardyce and P. J. Dyson, *Platinum Metals Rev.*, 2001, **45**, 62–69.
27. B. Çetinkaya, E. Çetinkaya H. Küçükbay and R. Durmaz, *Arzneim.-Forsch.*, 1996, **46**, 821–823.
28. B. Çetinkaya, Ý. Özdemir, B. Binbaşioğlu, R. Durmaz and S. Günal, *Arzneim.-Forsch.*, 1999, **49**, 538–540.
29. M.-L. Teyssot, A.-S. Jarrousse, A. Chevry, A. De Haze, C. Beaudoin, M. Manin, S. P. Nolan, S. Díez-González, L. Morel and A. Gautier, *Chem.–Eur. J.*, 2009, **15**, 314–318.

CHAPTER 5

Non-classical N-Heterocyclic Carbene Complexes

ANNEKE KRÜGER AND MARTIN ALBRECHT*

University College Dublin, School of Chemistry and Chemical Biology, Belfield, Dublin 4, Ireland

5.1 Introduction

The transition metal chemistry of N-heterocyclic carbenes (NHCs) is dominated by imidazol-2-ylidenes and their derivatives including oxazoles, thiazoles, 1,2,4-triazoles, and pyrimidines, as well as related saturated systems.[1] These carbenes have found widespread application as strongly donating ligands in transition metal-catalysed reactions and also as organocatalysts in their own right.[2] They have become highly popular because the free carbenes are stabilised by two adjacent heteroatoms, which renders the carbene often sufficiently stable to be manipulated and isolated under ambient conditions.[3]

In contrast to these classical types of NHCs, isomers with less extensive heteroatom stabilisation, termed here 'non-classical carbenes', initially received much less attention.[4] Non-classical carbenes can contain just one, or even no heteroatom in positions α to the carbene carbon atom. Due to the lower number of heteroatoms, π stabilising effects are substantially diminished when compared to classical carbenes, which in turn reduces the stability of the free carbene. On the other hand, the electron-withdrawing inductive influence of the heteroatoms is significantly decreased. As a consequence, these non-classical carbenes are surmised to be stronger donors than their classical analogues.[5] Lately, they have gained interest especially as ligands in transition metal complexes because of their distinct electronic properties, which open up new synthetic and catalytic opportunities.

RSC Catalysis Series No. 6
N-Heterocyclic Carbenes: From Laboratory Curiosities to Efficient Synthetic Tools
Edited by Silvia Díez-González
© Royal Society of Chemistry 2011
Published by the Royal Society of Chemistry, www.rsc.org

This Chapter focuses specifically on the catalytic applications of complexes comprising non-classical carbene ligands, including work published until late 2009. General and synthetic aspects have been reviewed comprehensively elsewhere,[4] and only a brief summary is given here. We use the term 'non-classical' as a general term for all NHC-type carbenes that are not stabilised by two adjacent heteroatoms.[6] Hence, non-classical carbene ligands are for example derived from imidazolium, thiazolium and oxazolium, 1,2,3-triazolium, pyrazolium, isothiazolium, pyridinium and quinolinium salts, and they also include cyclic alkyl(amino)carbenes (CAACs) and amino(ylide)carbenes (AYCs). Moreover, this definition also encompasses acyclic systems with reduced heteroatom stabilisation. Despite not belonging to the family of heterocyclic ligands, acyclic carbenes with reduced heteroatom stabilisation are chemically strongly related to the cyclic analogues and they have therefore been included in this Chapter. The class of acyclic carbenes is of significant broadness and includes, amongst others,[7] alkyl(amino)-, amino(aryl)- and amino(-silyl)carbenes.[8] Different kinds of non-classical carbene complexes, including representative acyclic systems with reduced heteroatom stabilisation, are summarised in Figure 5.1.

Various terminologies have been used for specific subclasses of non-classical carbenes: 'normal' carbenes include all carbenes that can be represented by a neutral canonical resonance form and 'abnormal' carbenes those for which a valence bond representation requires the introduction of formal charges on some nuclei (for example, C4-bound imidazolylidenes, Scheme 5.1).[4] Carbenes with no heteroatom in the position α to the carbene carbon atom are denoted 'remote' carbenes. Hence, remote carbenes can be either normal or abnormal (**E** or **I** respectively, Figure 5.1).

The nature of the metal–carbene bond and the degree of $M–C_{carbene}$ π interaction vary considerably within non-classical carbene complexes. For example, in abnormal imidazol-4-ylidenes, the question arises as to whether contributions from mesoionic limiting resonance forms prevail, including a metal-bound vinylic M–C=C fragment and an intramolecularly stabilising cationic amidinium NCN unit.[4] Recent crystallographic and computational analysis of an abnormal free carbene did not indicate substantial differences when compared to classical NHCs,[9] thus supporting a similar bonding situation.[10] Moreover evaluation of geometrical features from an X-ray structure and calculations on non-classical carbene rhodium(I) complexes comprising acyclic aminocarbene ligands revealed structural features that did not deviate significantly from Fischer-type carbene complexes. These studies illustrate the strong relationship of non-classical carbenes both to normal imidazolium-derived and to Fischer-type carbene ligands.[11] It should be noted that structural analyses based on NHC $M–C_{carbene}$ bond lengths generally provide little insight on the bond order, as the M–C bond is relatively insensitive to the carbene bonding mode. Spectroscopic analyses using ^{13}C NMR techniques are also often inconclusive, since chemical shifts appear to be much more sensitive to substituent effects than to the carbene bonding mode. Recently, an approach using an isopropyl benzimidazol-2-ylidene (BIMY) ancillary ligand

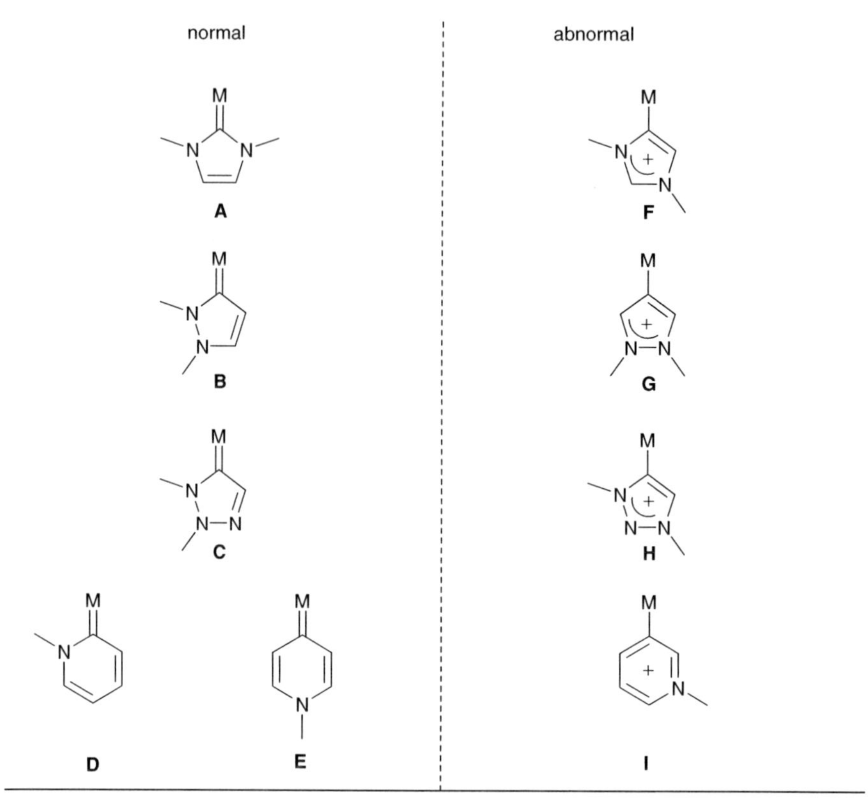

Figure 5.1 *Metal complexes comprising the classical imidazol-2-ylidene ligand (**A**) and representative non-classical carbene ligands (**B**–**N**), including normal carbenes (**B**–**E**), abnormal carbenes (**F**–**I**), remote carbenes (**E**, **G**, **I**), cyclic alkyl(amino)carbenes (**J**), acyclic carbenes (**K**, **L**, **M**) and amino(ylide)-carbenes (**N**). Substituted nitrogen centres may be replaced by oxygen or sulfur. The M=C bond representation—while strongly over-emphasizing the differences in the nature of the metal–carbon bond in these non-classical carbene complexes—was used to accentuate normal and abnormal bonding.

as ^{13}C NMR reporter group in complexes [(BIMY)PdBr$_2$(carbene)] was suggested as an alternative technique for measuring the donor properties of carbenes.[12] Preliminary data indicated that the probe may be at least as sensitive as

*The graphical representation of the carbene complexes in this Chapter has been adapted for the sake of consistency throughout this book, yet it does not reflect the authors' preference.

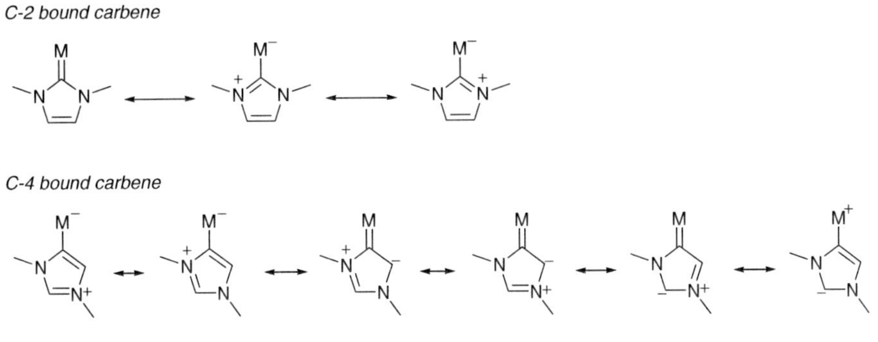

C-2 bound carbene

C-4 bound carbene

Scheme 5.1 Limiting resonance forms of C2- and C4-bound imidazolylidenes.

commonly used IR spectroscopy of carbonyl complexes, though further data is required to make this method broadly applicable.

Despite these formal ambiguities, non-classical variations of NHCs constitute an attractive class of ligands with great opportunities for catalysis and synthesis in general. This Chapter aims to overview the most recent advances in using non-classical NHC ligands in transition metal-mediated catalysis and to illustrate the attractive prospects that emerge from these achievements.

5.2 Synthesis of Non-classical Carbene Complexes

A variety of methods have been used for the formation of classical NHC complexes, and most of these methods are also applicable to the synthesis of non-classical NHC metal systems. Most frequently used methods for metalating cyclic non-classical carbene precursors include direct metalation *via* C–H and C–X bond activation,[13] as well as C–H bond oxidative addition.[14] Less common are transmetalation protocols, firstly because the relevant metal carbene precursors, especially the Ag–carbene salts, have only limited stability.[15] In addition, argentation suffers from a low regioselectivity due to the weak acidity of the proton to be removed. For example, exocyclic C–H bond activation was observed in C2-methylated imidazolium salts, a reaction pathway that was successfully prevented when using C2-arylated imidazolium salts.[16]

Another key route towards metal carbene complexes involves the generation of a free carbene, either *in situ* or isolated, and subsequent metal coordination. While the stability of classical carbenes makes the free carbene route the method of choice for complex synthesis, the reduced heteroatom stabilisation in non-classical carbenes generally precludes the formation of free carbenes.[17] For example, theoretical investigations using energy decomposition analyses showed that imidazol-4-ylidene is about 20 kcal mol^{-1} less stable than its classical counterpart, imidazol-2-ylidene.[18]

While free carbenes with low heteroatom stabilisation such as **1** (Figure 5.2) were pioneered by Bertrand and co-workers as early as in the late 1980s,[19] non-classical carbenes remained elusive until recently. The isolation of the free alkyl(amino)carbene **2** demonstrated that a single electron-active substituent

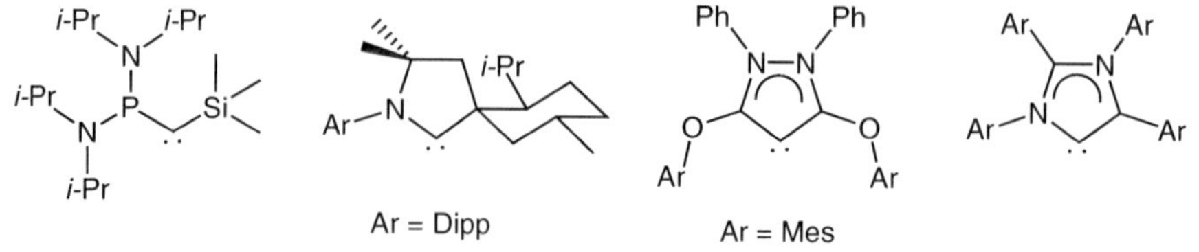

Figure 5.2 Stable free carbenes with reduced heteroatom stabilisation (Dipp = 2,6-diisopropylphenyl; Mes = mesityl).

was in fact sufficient for stabilising a carbene.[20] Building on this success and on the remote effect of oxygen atoms as π stabilisers, the first abnormal pyrazolylidene **3**, which may also be formulated with an allene resonance structure, was crystallised.[21] Currently, a range of persistent carbenes are known that may find use as non-classical carbene ligands.[6] Further achievements include the stabilisation of an all-carbon four-membered ring allene,[22] and, more relevant in the context of this Chapter, carbene **4** as the first abnormal imidazol-4-ylidene stable at room temperature,[9] indicating that the free carbene route may be more feasible than initially thought for metalating imidazolium salts at the abnormal position. The presence of aryl substituents on the imidazolium precursor presumably constituted a critical parameter for ensuring sufficient stability of the free carbene.

Only a few cases are known where non-classical carbene bonding is competitive with classical carbene complex formation. When using imidazolium salts in which the normal position was sterically encumbered, in combination with iridium or osmium polyhydride precursors, abnormal carbenes were preferentially formed.[23] Theoretical calculations further attributed a pivotal role to the anion in the imidazolium salt precursor,[23a] since C–H bond activation in the formation of abnormal carbene complexes was computed to proceed *via* oxidative addition, favoured by non-coordinating anions. In contrast, normal carbene complexation occurred through a heterolytic C–H bond cleavage process (deprotonation), promoted by small coordinating anions that support the stabilisation of the dissociating proton. Preferential abnormal carbene formation was also accomplished with $[IrCl(COD)]_2$ (COD = 1,5-cyclooctadiene),[24] a precursor that shows a high affinity to olefins.

Interestingly, abnormal imidazol-4-ylidene complexes were also obtained *via* C–H activation starting from free imidazol-2-ylidene. Little mechanistic information is available so far, but in most cases sterics were reasoned to be responsible for the preferred abnormal metalation.[25]

Although steric factors and counter-ion effects can, to some degree, promote abnormal carbene formation, in most cases protection of the C2-position by substitution with an alkyl or aryl group is necessary to ensure the selective formation of the desired product. It must be noted, however, that in metalation reactions using iridium or silver precursors, exocyclic $C(sp^3)$–H bond activation of C2-methylated imidazolium cations were observed, thus yielding metal–alkyl complexes.[15] Therefore, protection of the C2 position with an aryl group is generally more reliable.

Metal-mediated activation of the C–H bond in non-classical carbene precursors is often promoted by the presence of acetate as an additive (Scheme 5.2).[26,27] The underlying principle was thought to be strongly related to electrophilic cyclometalations. Thus, the availability of two Lewis basic sites in close proximity—one for metal coordination and the second for proton abstraction—prearranged the reactants, thus favouring M–C bond formation either *via* agostic interactions or *via* a six-membered transition state involving hydrogen bonding. Generally, C–H bond activation for the synthesis of non-classical carbene complexes requires slightly more forcing conditions than the metalation of classical carbene precursors, which is in agreement with the different C–H bond strengths in the starting materials.

Oxidative addition provided an alternative method towards the synthesis of non-classical carbene complexes (Scheme 5.3),[28] and also avoided protection protocols in order to specifically inhibit normal carbene formation. This methodology was developed for the metalation of non-classical imidazolylidene, pyrazolylidene, and pyridylidene ligand precursors as well as for CAACs and allowed for installing various metals including molybdenum and the metals of the group 10.[13]

Moreover, the carbene may be constructed directly in the metal coordination sphere, much like Fischer carbene complex synthesis. For example, the cycloaddition of alkynes to amino-functionalised Fischer carbene chromium

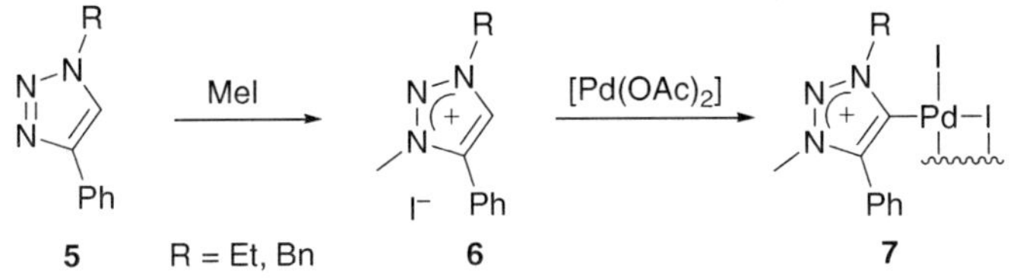

R
MeI
[Pd(OAc)$_2$]
Ph
I$^-$
Pd
5 R = Et, Bn 6 7

Scheme 5.2 Metalation of triazolium salts using an acetate-containing metal precursor.

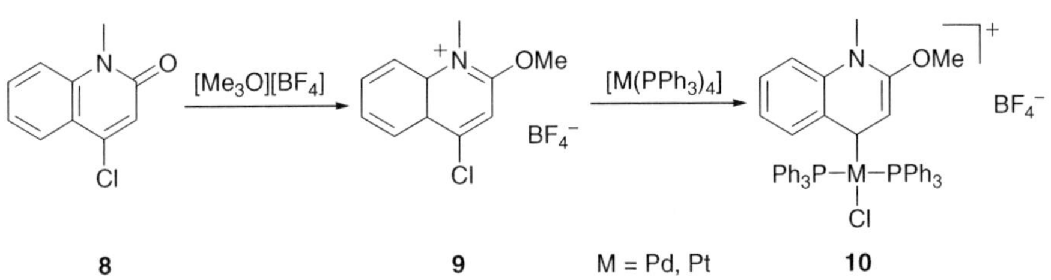

[Me$_3$O][BF$_4$]
OMe
[M(PPh$_3$)$_4$]
BF$_4^-$
Cl
Ph$_3$P–M–PPh$_3$
8 9 M = Pd, Pt 10

Scheme 5.3 Preparation of remote carbene complexes by oxidative addition.

complexes furnished non-classical pyridylidene complexes, which were demonstrated to be useful precursors for transmetalation to rhodium(I) and gold(I).[29]

5.3 Reactivity and Stability

An elegant study by Cavell and co-workers pointed towards abnormal imidazolylidenes being more prone to reductive elimination than normal imidazolylidenes.[14] In the presence of an alkene ligand the mixed bis-carbene complex **11** underwent reductive elimination of the 4-imidazolylidene exclusively, thus yielding platinum(0) diolefin complex **12** and imidazolium salt **13** (Equation (5.1)). No reductive elimination of the normally bound imidazol-2-ylidene ligand was observed under these conditions.

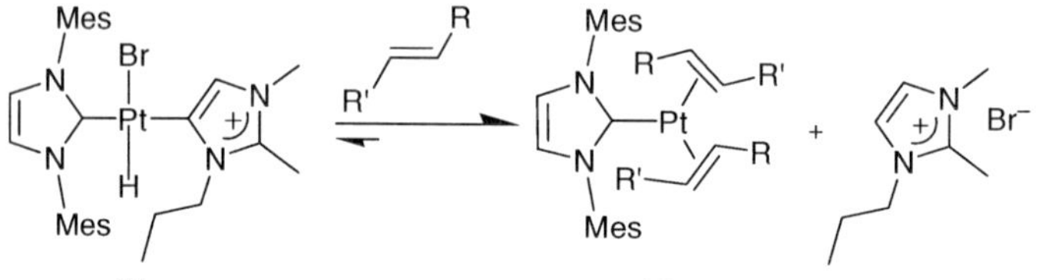

(5.1)

Similar trends were observed in a complementary study using the palladium complexes **14** and **15** comprising sterically identical normal and abnormal bis(imidazolylidene) ligands (Scheme 5.4).[30] In the presence of chlorine, complex **14** was stable and did not react, whereas the abnormal carbene complex decomposed to $[PdCl_4]^{2-}$ and a doubly chlorinated bisimidazolium dication **16**. This outcome was explained by oxidative Cl_2 addition to complex **15**, followed by reductive $C_{carbene}$–Cl bond formation. Obviously, this process was unfavourable with normally bound imidazolylidenes. It is worth noting that an analogue of complex **15** with no alkyl substituents at the C5 and C5′ positions induced reductive $C_{carbene}$–$C_{carbene}$ bond formation. The higher propensity of abnormal carbenes to be reductively cleaved was rationalised by the enhanced electron donor properties of the non-classical carbenes, which made them more susceptible towards elimination processes. Evidently, steric factors could be ruled out in these systems.

The increased donor properties of non-classical carbenes relative to their classical analogues were demonstrated both theoretically and experimentally. Complexes **14** and **15** were analysed by X-ray photoelectron spectroscopy.[30] Both the palladium 3d and 3p electron binding energies in the abnormal complex **15** were lower by 0.5 eV than in the normal complex **14**, which reflected the stronger donor capabilities of the abnormal carbene ligand. Furthermore, X-ray diffraction and infrared spectroscopic studies were used to demonstrate the larger *trans* influence of abnormal carbenes when compared to their normal analogues.[16,31]

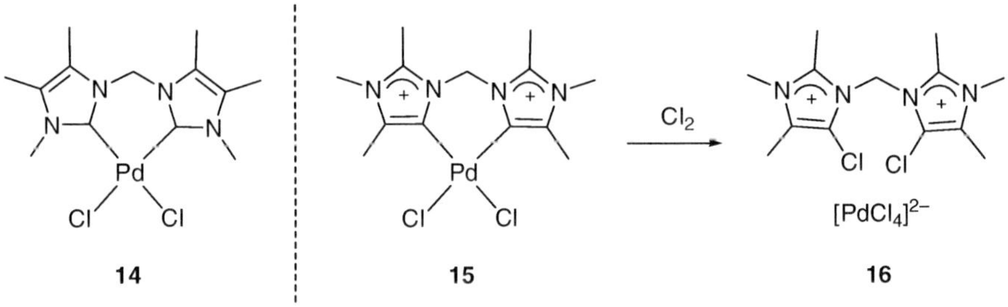

Scheme 5.4 Reactivity of normal and abnormal carbene palladium complexes towards chlorine.

The increased donor ability of abnormally bound imidazolylidenes increased the nucleophilicity of the metal centre. Abnormal NHC–palladium complexes were thus shown reactive towards Lewis acids. When the abnormal NHC complex **15** was treated with $AgBF_4$, the adduct **18** was formed while normal carbene complexes underwent the expected halide abstraction to form **17** (Scheme 5.5). Crystallographic analysis revealed short Ag···Pd distances of 2.8701 Å, suggesting a strong metal–metal interaction. Theoretical calculations indicated that the palladium centre acted as a Lewis base in this adduct, despite its formal dipositive charge. No such adduct formation was observed with analogous normal NHC–palladium complexes.

Adduct **18** may be considered as model intermediate for reactions of the palladium complex with other electrophiles. For example, strong Brønsted acids like HCl, which reacted with the abnormal complex **15** but not with its normal counterpart, induced $C_{carbene}$–Pd bond cleavage to yield the imidazolium salt **20**. However, the rates of this acidolysis were too fast to allow the detection of an intermediate similar to **18**. Notably, the coordination of two abnormal carbenes inverts the electronic properties of the palladium centre. While palladium(II) is generally considered to be electrophilic, it is obviously nucleophilic in complex **15** and related species. This fact offers a plethora of new synthetic possibilities and unprecedented reactivity schemes, which cannot be accessed by using classical carbene or phosphine ligands.

5.4 Application in Catalysis

Since non-classical carbene ligands increase the electron density at the metal centre, catalytic processes that are rate-limited by an oxidative addition step should benefit particularly from this class of ligands. For example, enhanced catalytic activity may be expected for the transition metal-mediated activation of dihydrogen and aryl–halogen bonds. Taking into account also that it is more difficult to find catalytically silent palladium than catalytically active species, it is not surprising that the largest body of catalysis using non-classical carbene metal complexes encompasses palladium-mediated cross-coupling reactions. In

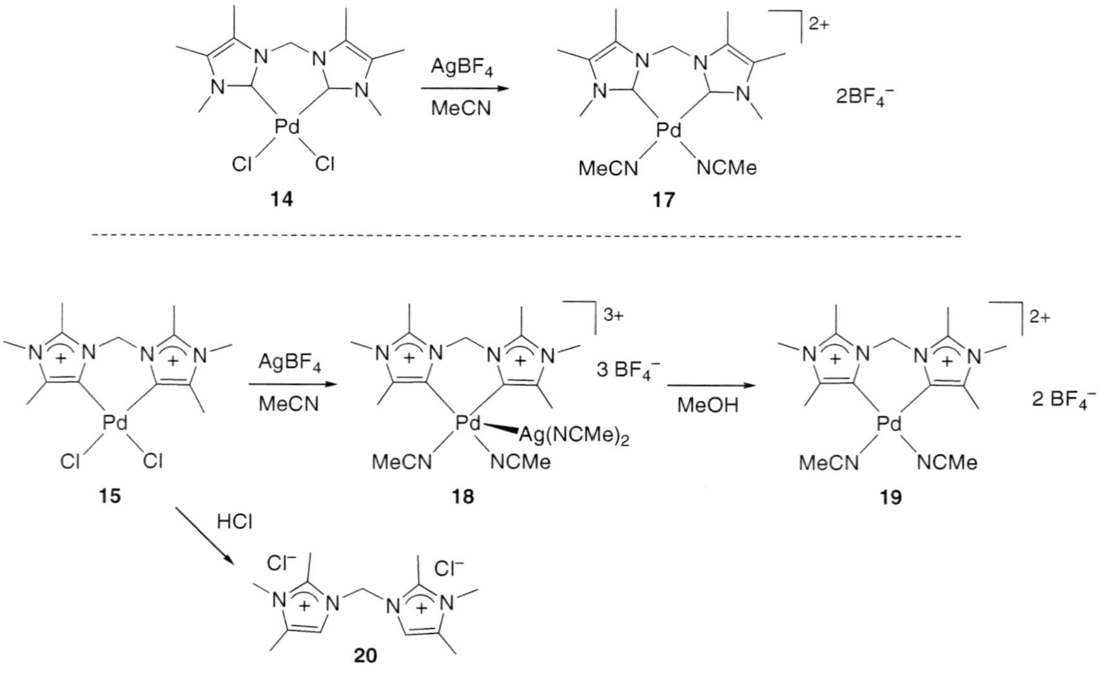

Scheme 5.5 Reactivity of normal and abnormal carbene palladium complexes towards acids.

parallel, a number of other attractive catalytic applications have been disclosed, including direct and transfer hydrogenation as well as alkylation reactions. A most recent study further suggested that non-classical NHC ligands may find use in rhodium-catalysed C–H bond activation processes.

5.4.1 Cross-coupling Reactions

5.4.1.1 Mizoroki–Heck and Suzuki–Miyaura Cross-couplings

Several studies reported on the catalytic activity of non-classical carbene palladium complexes in Mizoroki–Heck and in Suzuki–Miyaura coupling reactions (Scheme 5.6). Here, we grouped them into three categories, including (i) comparative studies between non-classical and classical carbene complex catalysts, (ii) accounts comparing different types of non-classical carbene complexes, and (iii) investigations to elucidate mechanistic details.

In a number of cases palladium-catalysed cross-coupling reactions have served as tool to demonstrate the beneficial effect of non-classical carbene ligands in catalysis when compared to classical imidazol-2-ylidenes. For example, the catalytic activity of the normal/abnormal dicarbene complex **21** was much higher than that of its normal/normal analogue **22**, both in the Heck and Suzuki couplings (Figure 5.3).[32] It is interesting to note that *in situ* formation of the catalyst from the dimesityl imidazolium salt and $[Pd(OAc)_2]$ in the presence of a base ensued similar catalytic activity than that of **21**. This result suggested the formation of complex **21** (at least in minor quantities) even though stoichiometric reactions under catalytic conditions afforded the catalytically less active complex **22**.

A complex closely related to **21** (R = Dipp) performed considerably worse than both classical imidazol-2-ylidene complexes and mixtures $[Pd(OAc)_2]$/imidazolium salt, when used in intramolecular arylation (Equation (5.2)).[33] In this case, the abnormal binding mode had apparently a detrimental effect on the catalytic activity.

Cl
1 mol % **21** (R = Dipp)
2 equiv K_2CO_3, DMA, 130°C
O
(5.2)

The catalytic activity of the pyrazolin-2-ylidene complex **23a** (Figure 5.4) in the Heck coupling of aryl bromides with styrene was twice that of its normal imidazol-2-ylidene analogue.[13b] In addition, the pyrazolin-2-ylidene complex required a shorter induction period than its imidazol-2-ylidene counterpart. The catalytic activity of the catalyst was slightly increased by increasing the steric bulk around the metal centre, through modification of the *ortho*-positioned nitrogen substituent from a methyl to a phenyl group (**23b**).

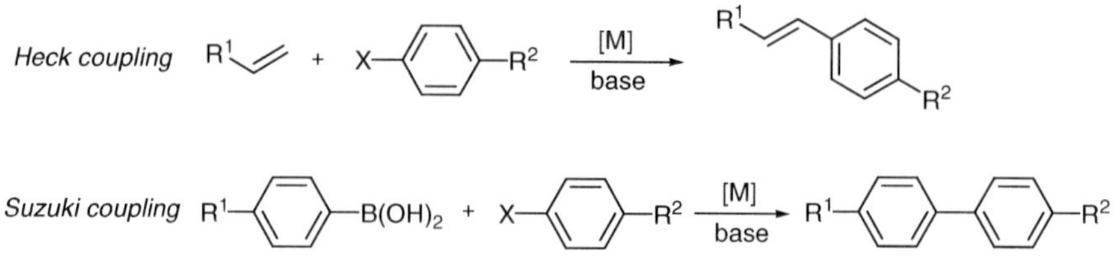

Heck coupling R1 + X–C6H4–R2 → [M], base

Suzuki coupling R1–C6H4–B(OH)2 + X–C6H4–R2 → [M], base → R1–C6H4–C6H4–R2

Scheme 5.6 Mizoroki–Heck and Suzuki–Miyaura coupling reactions.

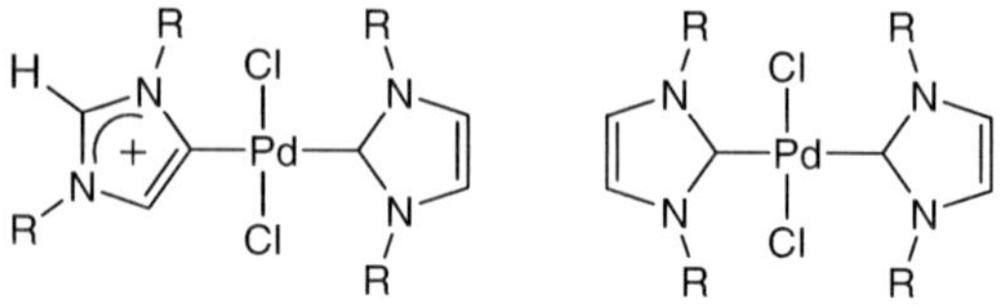

R = Mes	21	22
Heck coupling (Ar-Br)	77%	<5%
Suzuki coupling (Ar-Cl)	44%	<5%

Figure 5.3 Catalysts used in Mizoroki–Heck (2 mol% [Pd], 2 equiv $CsCO_3$, *N,N*-dimethylacetamide (DMA), 120°C) and Suzuki–Miyaura (2 mol% catalyst loading, 2 equiv $CsCO_3$, dioxane, 80°C) coupling reactions.

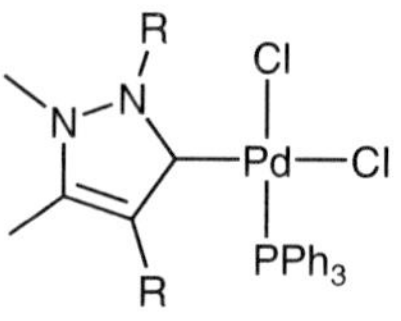

23 a R = Me
b R = Ph

4-MeCO-ArBr	**a**	98%
	b	98%
ArBr	**a**	86%
	b	94%
4-MeO-ArBr	**a**	59%
	b	69%

Figure 5.4 Catalysts used in Mizoroki–Heck coupling reactions (1 mol% [Pd], 1.5 equiv NaOAc, DMA, 130°C).

The performance of complex **10** bearing a quinolinylidene as a remote carbene ligand (Figure 5.5) was compared to that of simple imidazol-2-ylidene and phosphine-containing palladium catalysts in the Heck coupling of activated and non-activated aryl bromides with butyl acrylate.[34] The remote carbene complex showed much higher activity in these reactions as well as in the Suzuki coupling of a deactivated aryl bromide with phenylboronic acid. Since the complexes tested were structurally very different, purely electronic comparisons

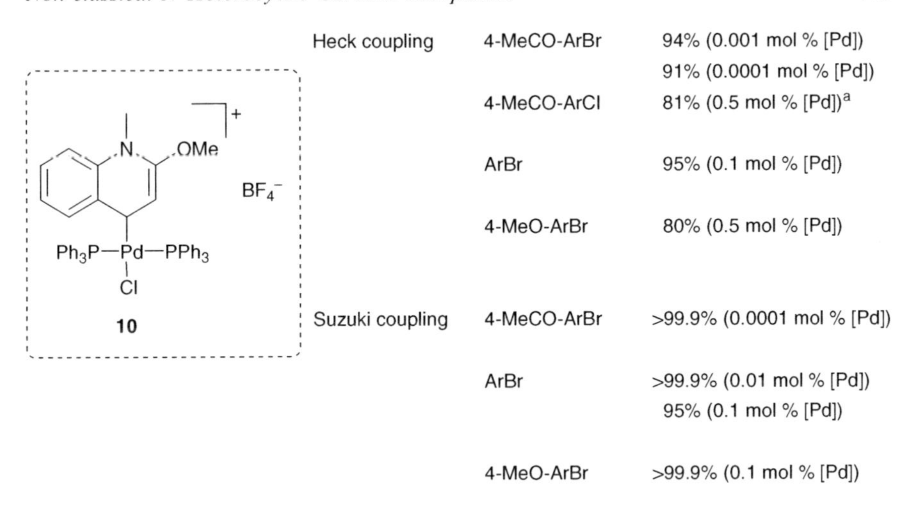

Figure 5.5 Catalyst used in Mizoroki–Heck (1.5 equiv NaOAc, DMA, 145°C) and Suzuki–Miyaura (1.5 equiv K_2CO_3, xylene, 130°C) coupling reactions.[a] 0.2 equiv $[R_4N]Br$, 150°C.

were difficult. Nevertheless, the non-classical carbene ligand appeared to have a beneficial effect on the catalytic performance.

In some cases, the catalytic performance was improved upon abstraction of metal-bound halides. Such improvement is illustrated by the enhanced activity achieved upon halide abstraction in complex **24** to yield the cationic complex **25** (Figure 5.6).[35]

Acyclic aminocarbenes have not been used widely as ligands for transition metal catalysts. The amino(aryl)carbene complexes **26** and **27** are rare examples that showed moderate activity in the Suzuki coupling of aryl iodides and bromides with phenylboronic acid (Figure 5.7).[36]

Even though the different reaction conditions precluded comparison with the non-classical carbene complexes, these examples demonstrated the potential of non-classical carbenes as spectator ligands. Further improvements may be accomplished by adapting some of the principles developed for classical carbene complexes, such as introduction of sterically demanding substituents in the *ortho* position, perhaps paired with coordinatively labile ligands *trans* to the non-classical carbene.[37]

Raubenheimer and co-workers compared the catalytic activity of structurally related 2-, 3- and 4-pyridylidene complexes in Suzuki coupling involving activated aryl chlorides as substrates.[38] Complex **30**, a normal remote carbene complex, gave the highest yields (Figure 5.8), followed by complex **29**, an abnormal remote carbene complex, while the normal carbene complex **28** showed the lowest performance. Even though the differences were relatively small, these studies suggested that the reduction of the inductive effect exerted by the nitrogen atom has some influence on the activity of the catalyst.

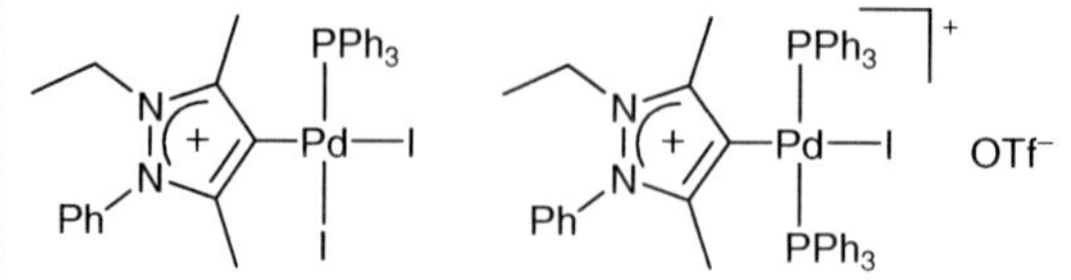

	24	25
4-HCO-ArBr[a]	80%	93%
4-MeCO-ArBr[a]	80%	95%
4-MeO-ArBr[b]	82%	96%
4-HCO-ArCl[b]	20%	37%

Figure 5.6 Catalysts used in Suzuki–Miyaura coupling reactions ($PhB(OH)_2$, 1 mol% [Pd], 1.5 equiv K_2CO_3, H_2O). [a]RT. [b]1.5 equiv $[(n\text{-}Bu)_4N]Br$, 80°C.

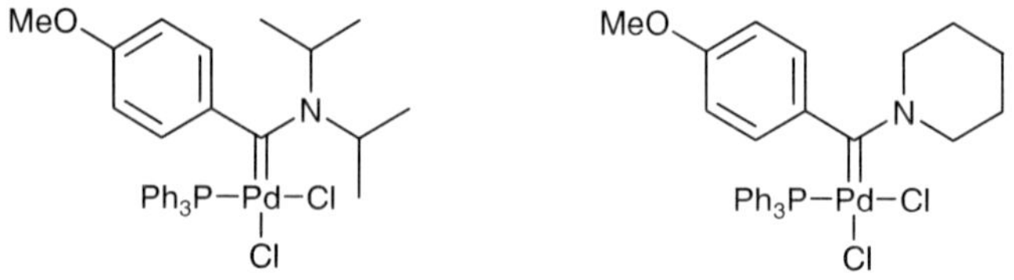

	26		27
4-MeCO-ArI	68%	4-MeCO-ArBr	55%

Figure 5.7 Catalysts used in Suzuki–Miyaura coupling reactions (1 mol% [Pd], 1.5 equiv K_2CO_3, THF, 66°C).

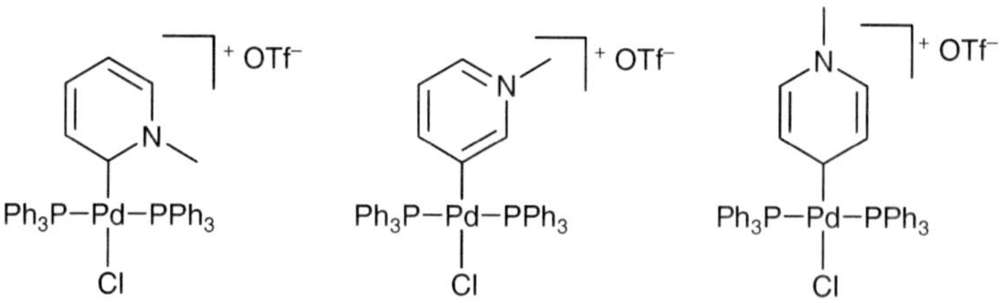

	28	29	30
4-MeCO-ArCl	86%	91%	99%

Figure 5.8 Catalysts used in Suzuki–Miyaura coupling reactions (0.1 mol% [Pd], 2.2 equiv K_2CO_3, DMA, 130°C).

In a similar study, the catalytic activity of the classical carbene complex **31** and its abnormal thiazolylidene analogues **32** and **33**, obtained by oxidative addition, were tested in the Suzuki coupling of activated aryl bromides (Figure 5.9).[38] At 70 °C, the catalytic activity decreased in the order **31** > **32** > **33**. The abnormal thiazolylidene complexes were thus less active than their normal counterpart. It is worth noting that the steric impact of the ligand in the normal carbene complex **31** and in the abnormal system **32** was quite similar, differing only by the absence and presence, respectively, of a hydrogen atom in the *ortho* position. Further tuning of the nitrogen substituents might give more insight into the effects of the abnormal binding on the catalytic activity. In particular, complex **33** was much less affected by steric modifications than **31** and **32**.

Mechanistic insights on non-classical carbene palladium-catalysed reactions were obtained from kinetic measurements. For example, time-dependent analysis of the conversions obtained with the quinolylidene complex **10** and with the pyridylidene complexes **28**–**30** indicated no induction period,[34,38] thus pointing to a homogeneous action mode. In agreement with this assumption, no palladium black formation was observed in any of the catalytic runs. In most other cases, however, no mechanistic analyses were reported and both homogeneous and heterogeneous working modes of the catalysts need to be considered. This ambiguity is exemplified by the chelating pyridylidene palladium complex **34** (Figure 5.10), which showed catalytic activity in Suzuki-type coupling comparable to that of *in situ* prepared systems of [Pd(OAc)$_2$] in pyridinium salt ionic liquids.[39] This similarity evoked the hypothesis that the catalytically active species is the same in both processes. Whether this was a palladium aggregate stabilised by pyridinium salts or a pyridylidene–palladium complex obtained from C–H bond activation in the ionic liquid remains to be established. Poisoning tests using mercury(0) did not affect the catalytic activity

	31	32	33
4-MeCO-ArCl	83%	72%	65%

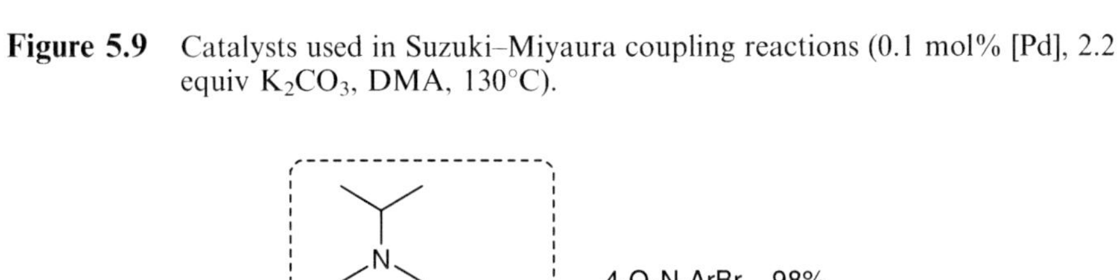

Figure 5.9 Catalysts used in Suzuki–Miyaura coupling reactions (0.1 mol% [Pd], 2.2 equiv K_2CO_3, DMA, 130°C).

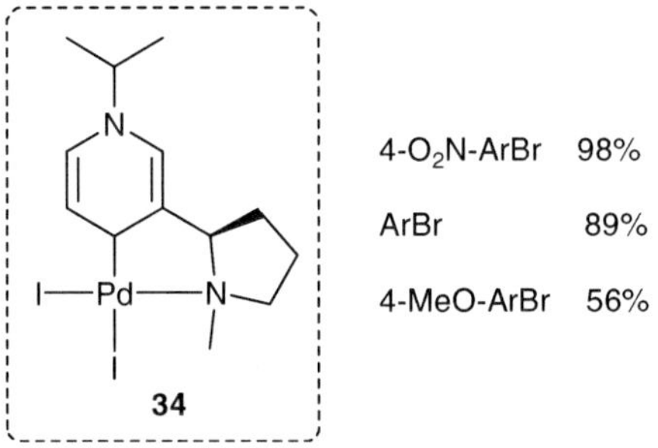

34	
4-O$_2$N-ArBr	98%
ArBr	89%
4-MeO-ArBr	56%

Figure 5.10 Catalyst for Suzuki–Miyaura coupling reactions (PhB(OH)$_2$, 5 mol% [Pd], 2 equiv K_2CO_3, DMSO, 100°C).

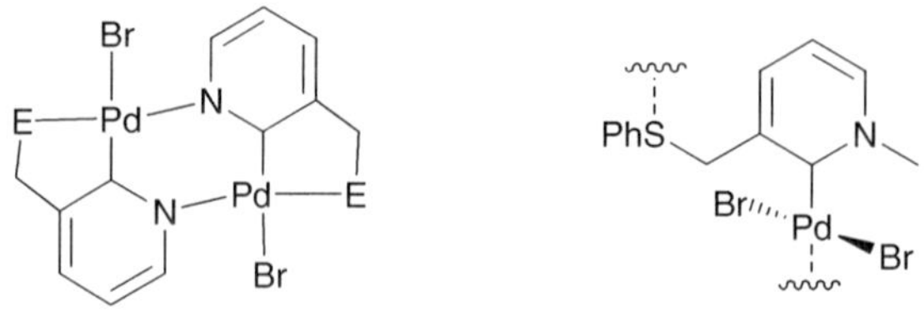

35 a E = NMe_2
b E = SMe
c E = SPh

36

4-MeCO-ArBr ~96% ~96%

Figure 5.11 Catalysts used in Mizoroki–Heck coupling reactions (styrene, 0.2 mol% [Pd], 1.1 equiv K_2CO_3, DMA, 140°C).

of complex **34**, suggesting a homogeneous process that involving a well-defined, molecular active site.[40] It should be noted, however, that the mercury test has not been validated for C–C bond forming reactions, and a systematic study may be required to make this test also reliable for cross-coupling catalysis.

When using the pyridylidene complex **36** or the dimeric compounds **35** in Heck coupling (Figure 5.11), sigmoidal kinetics were measured.[41] No effect was noted when changing the donor group E of the carbene chelate, or the catalyst precursor from the polymeric **36** to dimeric **35**, containing a palladium centre as highly labile substituent at the nitrogen. These findings and the required high reaction temperatures suggested a heterogeneous process, reinforced in this case by the mercury test, since conversions abruptly stop upon addition of mercury(0) to the catalytic runs.

The abnormal or remote binding of the carbene ligand did not always have as pronounced an effect on the catalytic activity of the catalyst as in the examples mentioned above. Imidazol-4-ylidene chelate **37** (Figure 5.12) was tested in the Heck coupling of aryl bromides with styrene[13c] to show a comparable activity to that of classical imidazol-2-ylidene complexes.[42] The forcing conditions required (140 °C) to obtain catalytic conversion and the absence of conversion with less activated substrates raised the question of whether the catalysis is in fact homogeneous or heterogeneous, implying initial decomposition of the carbene complexes. Since conversions drop significantly in the presence of excess mercury(0), again a heterogeneous action was indicated. In line with this hypothesis, complex **21** (R = Dipp) featuring a similar abnormal ligand produced mixtures of palladium black and imidazolium salt during catalytic arylation experiments.[33]

An indirect proof of ligand dissociation was provided by the catalytic runs carried out with the normal 2-pyridylidene palladium complex **38** (Figure 5.12).[43] Unspectacular Heck coupling results were obtained since **38** did not display better activity than $[Pd_2(dba)_3]$ (dba = dibenzylideneacetone). Again, formation of colloidal palladium might need to be taken into account, making the carbene ligand inefficient for tailoring selectivity in product formation.

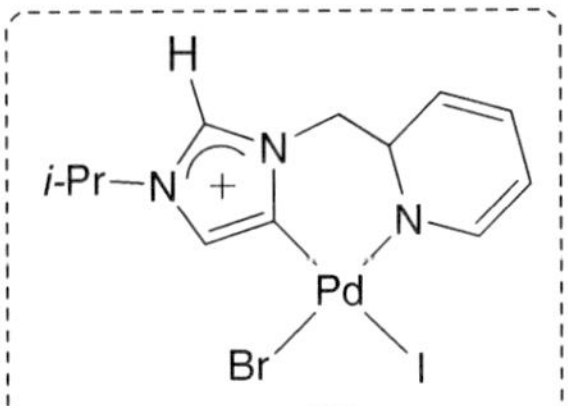

37

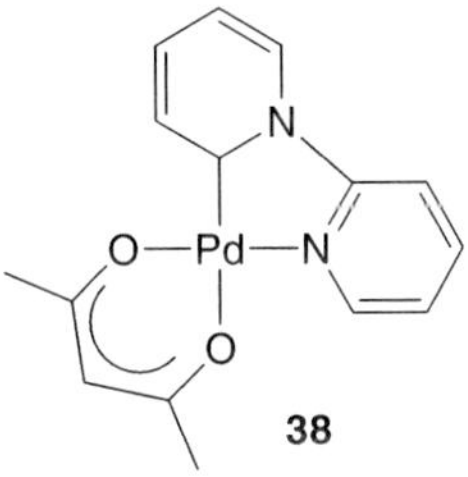

38

4-O_2N-ArBr 94% (TON = 188)

4-HCO-ArBr 95% (TON = 190)

Figure 5.12 Catalysts used in Mizoroki–Heck coupling reactions (0.5 mol % [Pd], 1.1 equiv NaOAc, DMA, 140 °C)

5.4.1.2 Kumada–Corriu Cross-coupling

The nickel(II) complexes **39**–**42** containing non-classical 2- and 4-quinolinylidene ligands were active in Kumada–Corriu cross-coupling of aryl chlorides and Grignard reagents (Scheme 5.7).[44] Their catalytic activity was compared to that of 2- and 4-pyridylidene, 2-imidazolylidene, a phosphane-containing complex and the most active catalyst system known to date, consisting of an *in situ* formed catalyst from IMes·HBF_4 and $[Ni(acac)_2]$ (acac = acetylacetonato). Remarkably, the quinolinylidene complex **41c** compared well with the benchmark system and at higher catalyst loadings, even deactivated aryl chlorides could be converted. Time monitoring of the conversion with various quinolinylidene catalysts did not reveal any induction period, which is in line with fast formation of the catalytically active species and with a homogeneous mode of action. Complex **41c** displayed the highest catalytic activity and best product selectivity, but other than that no obvious correlation was observed between the catalytic performance and the carbene ligand (*e.g.* remote *vs.* non-remote). The pyridinylidene- and quinolinylidene systems all gave better conversions than the simple imidazolylidene and phosphine systems. While these results may suggest that the stronger donor properties of non-classical carbene ligands and hence the higher electron density at the nickel centre may enhance the catalytic performance, it should be noted that the steric properties of these systems are quite different. In particular, the absence and presence of shielding *ortho* substituents in the carbene ligand may affect the catalytic activity substantially (cf. palladium-mediated cross-coupling in the previous section).

5.4.1.3 Hydroamination and Buchwald–Hartwig Amination

The CAAC gold complex **43a** catalysed the amination of terminal and internal alkynes (Scheme 5.8).[45] When using ammonia or a primary amine as substrate, imines were the major product. For ammonia fixation, the reaction was also successfully carried out using the preformed ammonia complex **43b** as

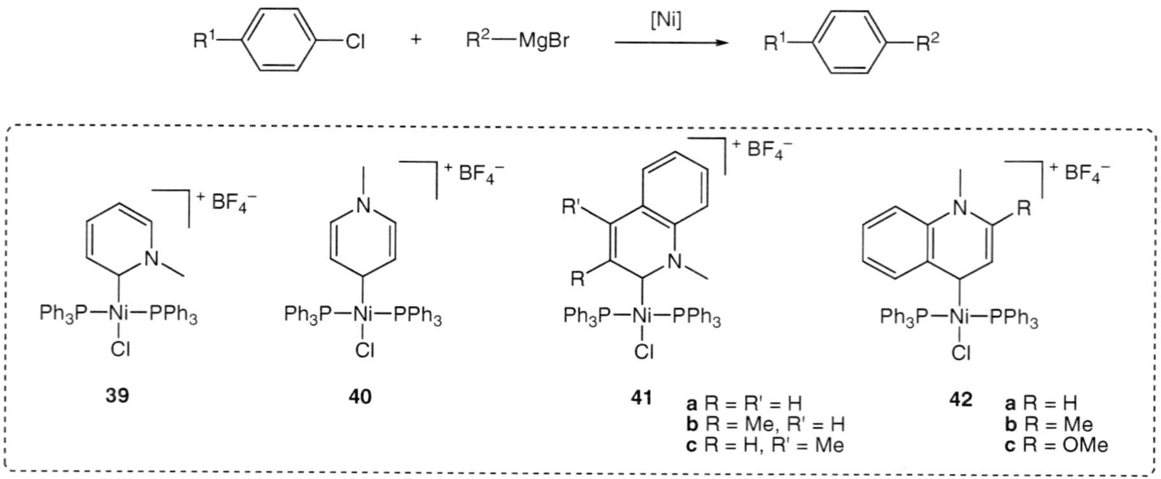

$R^1-C_6H_4-Cl + R^2-MgBr \xrightarrow{[Ni]} R^1-C_6H_4-R^2$

$Ph_3P-Ni(Cl)-PPh_3$ $^+$ BF_4^-

39

40

41 **a** R = R' = H; **b** R = Me, R' = H; **c** R = H, R' = Me

42 **a** R = H; **b** R = Me; **c** R = OMe

	39	40	41	42
o-Tol-MgBr, ArCl	59%	68%	**a** 65% **b** 78% **c** 92%	**a** 81% **b** 83% **c** 42%
Ph-MgBr, 4-Me-ArCl			**c** 51% (1 mol % [Ni]) 79% (3 mol % [Ni])	
Ph-MgBr, 4-MeO-ArCl			**c** 70% (3 mol % [Ni])	

Scheme 5.7 Kumada–Corriu coupling reactions (1 mol% [Ni], THF, RT).

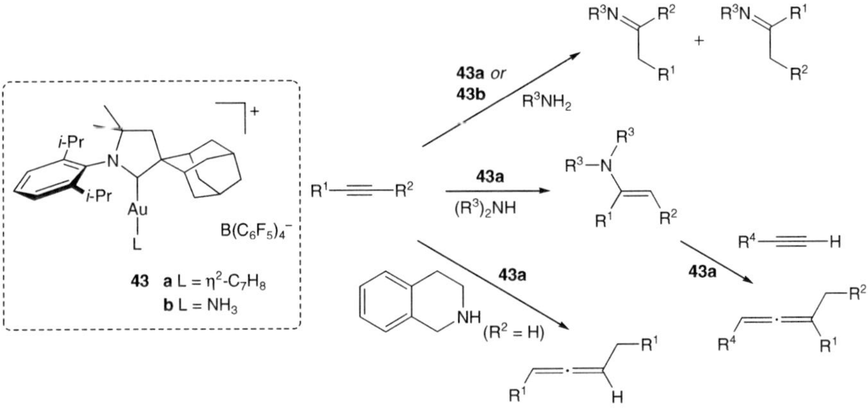

Scheme 5.8 Hydroamination reactions catalysed by **43a** and **b** (5 mol% [Au], C_6D_6, >70 °C). R^1, R^4 = alkyl, aryl; R^2, R^3 = H, alkyl, aryl.

catalyst.[45b] At elevated temperatures (> 130 °C), two acetylenes were coupled to ammonia, which provided an interesting synthetic approach to substituted pyrroles starting from diynes. Secondary amines underwent similar hydroamination of acetylenes and afford enamines in good yields.[45c] Remarkably, the gold carbene complex **43a** also catalysed the coupling of enamines to terminal acetylenes, producing substituted allenes.[45a] Based on these results, Bertrand and co-workers developed a one-pot procedure for the formation of allenes from terminal alkynes by using tetrahydroisoquinoline as sacrificial secondary amine.[45c] This process yielded 1,3-disubstituted allenes as final products. Unlike these species, terminal, *i.e.* 1,1-disubstituted allenes were efficiently hydroaminated in the presence of primary or secondary amines and catalytic amounts of complex **43a** (Equation (5.3)).[45d]

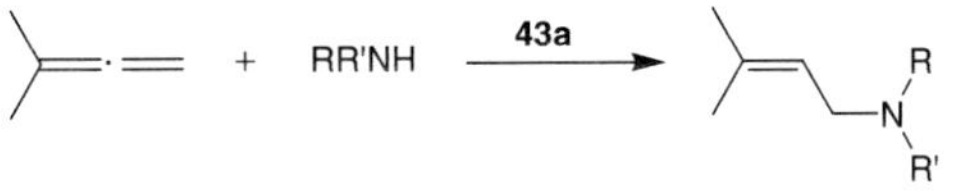

(5.3)

These catalytic C–N bond formation processes were postulated to involve a gold(III) intermediate, which was supposed to be readily accessible owing to the strong donor properties of the non-classical carbene ligand. For more applications of non-classical carbene-bearing gold complexes, please refer to Chapter 11.

Finally, the acyclic amino(aryl)carbene palladium complex **26** (see Figure 5.7) gave high conversions in the coupling of aryl bromides with morpholine at ambient temperature. For pyridyl chloride substrates, higher temperatures (70 °C) were required in order to achieve an appreciable 73% yield.[36]

5.4.1.4 Ketone α-Arylation

The palladium complexes **44**–**46** containing CAAC ligands as well as the acyclic amino(aryl)carbene complex **47** were investigated in the room temperature α-arylation of propiophenone with sterically hindered, non-activated aryl chlorides (Scheme 5.9).[46] Complex **46** showed superior activity, which was rationalised by the steric bulk of the *ortho*-positioned cyclohexyl moiety. The substituted cyclohexyl group is flexible and bulky, thus destabilising *cis*-coordinating ligands and, in contrast to the unsubstituted cyclohexyl fragment in **45**, resistant towards engaging in stabilising agostic interactions with the palladium centre. This situation pronounced the coordinative unsaturation at the metal centre, thus rendering it particularly reactive. Complex **46** favourably competed with the most active phosphane-based systems known to date.

Despite being less activating than the CAAC complexes, the amino(aryl)carbene ligand in **47** is of considerable interest in its own right (Scheme 5.9).[46] Due to its strong topological relationship with Buchwald's biaryl phosphine, this carbene ligand may become a useful ligand for a wide range of metal-mediated coupling reactions.

5.4.2 Hydrogenation and Hydrosilylation Reactions

5.4.2.1 Direct Hydrogenation

The abnormal imidazol-4-ylidene complex **49** was found to catalyse the hydrogenation of cyclooctene as a model substrate at ambient temperature and atmospheric H_2 pressure (Scheme 5.10).[47] Its normal analogue **48** showed a far lower catalytic activity, which was explained in terms of higher electron density at palladium in **49** due to stronger σ-donation of the abnormal carbene. Consequently, activation of dihydrogen *via* an oxidative addition mechanism was expected to be favoured. Notably, while the dicationic bissolvento complex **49** displayed catalytic activity, its neutral analogue **15** was inactive, perhaps because of the lower tendency of **15** to coordinate a substrate molecule. Consistent with this model, strongly coordinating solvents such as MeCN, DMF, and DMSO had a detrimental effect on catalyst performance, and polar solvents with weak coordination ability such as alcohols were found to work best.

Preliminary mechanistic investigations suggested that the catalytic activity was associated with the formation of particles in the 100–500 nm range. It is thus conceivable that the formation of the catalytically active species occurred through a mechanism strongly related to the oxidative addition/reductive elimination cycle described for the reaction of **15** with Cl_2 (see Scheme 5.4),[30] thus affording an imidazolium salt and a palladium polyhydride species, which may aggregate to the hydrogenation-active species. In such a model, the abnormal carbene ligand would have the role of a mediator, since only in abnormal carbene complexes reductive elimination, and hence catalyst activation, takes place while normal carbene palladium complexes may be less

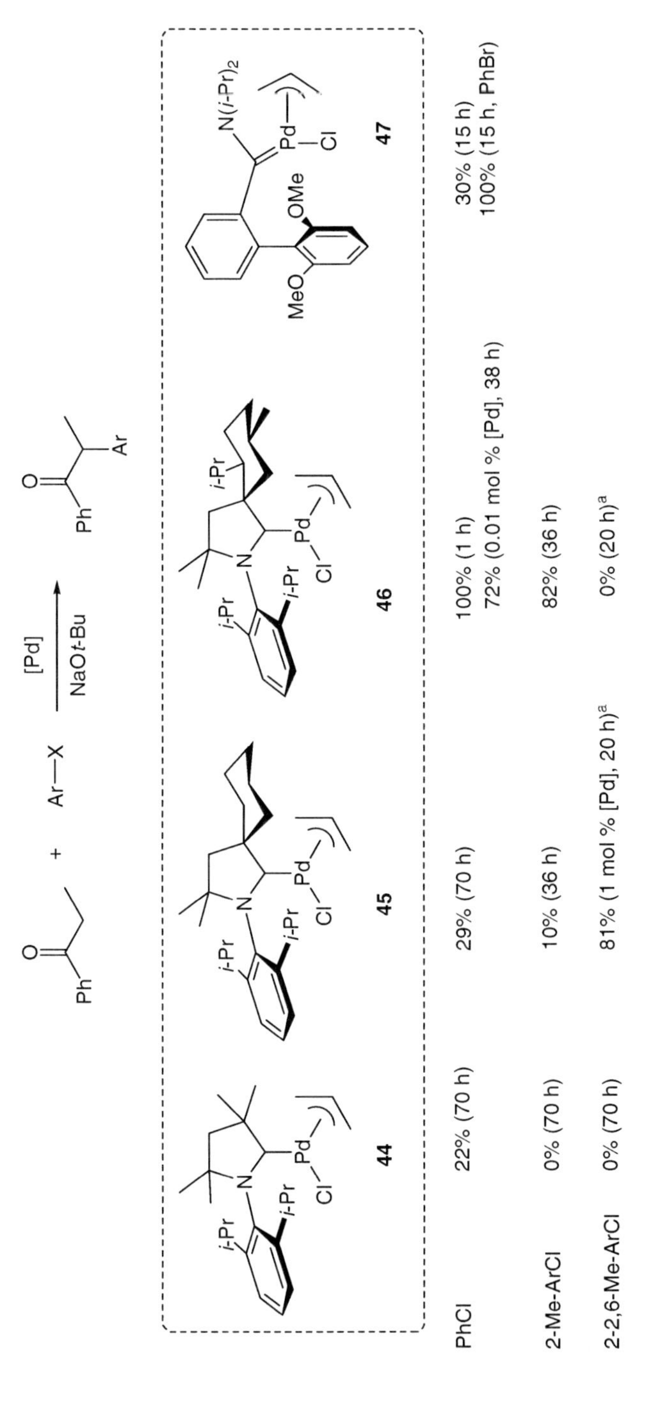

Scheme 5.9 Catalysts for α-arylation reactions (**44**–**46**: 0.5 mol% [Pd], 1.1 equiv NaO*t*-Bu, THF, RT; **47**: 2 mol% [Pd], 1.1 equiv KO*t*-Bu, THF, RT). [a]50 °C.

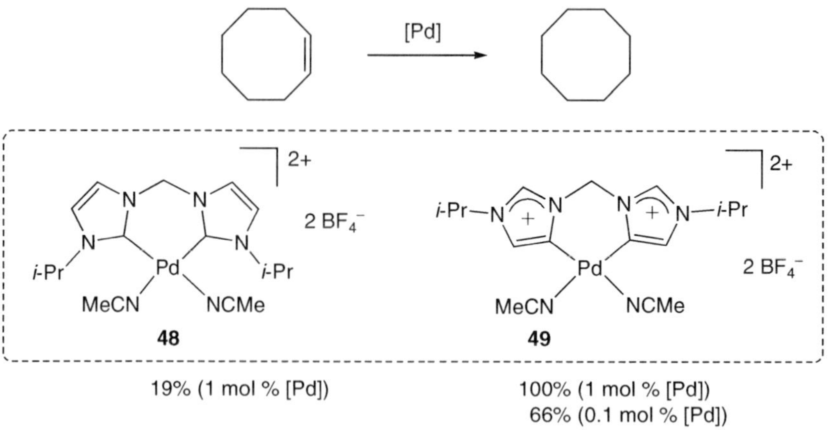

[Pd]

48

49

19% (1 mol % [Pd])

100% (1 mol % [Pd])
66% (0.1 mol % [Pd])

Scheme 5.10 Hydrogenation reactions catalysed by **48** and **49** (1 bar H_2, EtOH, RT).

active as a consequence of their pronounced stability towards reductive elimination reactions.

5.4.2.2 Transfer Hydrogenation

Biscarbene complexes of rhodium(III) **50**–**52**, prepared by NaOAc-assisted double C–H bond activation with $RhCl_3$, showed catalytic activity in the transfer hydrogenation of ketones (Scheme 5.11).[48] When using *i*-PrOH as hydrogen source, conversions were good, while their normal counterparts were essentially inactive. Not unexpectedly, the substitution pattern on the heterocyclic nitrogen atoms did not influence the catalytic activity significantly, and electronic differences appeared to be more relevant. For example, exchanging the iodide ligands for chlorides as in **51b** increased the catalytic activity by a factor of three, reflected by the TOF_{50} ~300 h^{-1} for **51b** as opposed to TOF_{50} ~100 h^{-1} for **51a**. A more electron-rich metal centre was believed to accelerate the product release step in the catalytic cycle, which may also account for the significant increase in catalytic activity when switching from normal carbene ligands to stronger σ-donating abnormal carbenes.

Due to the microscopic reversibility of transfer hydrogenation, this reaction was particularly suitable for cascade processes. For example, β-alkylation of secondary alcohols with primary alcohols was demonstrated (Scheme 5.12), including the dehydrogenation of the alcohols followed by aldol condensation, which was catalysed by the base typically used in transfer hydrogenations.[49] Subsequent re-hydrogenation of the aldol condensation product provided the β-alkylated alcohol. The normal and abnormal imidazolylidene and the non-classical, normal pyrazol-2-ylidene ruthenium complexes **53**–**56** were investigated as catalysts in such β-alkylations (Scheme 5.12).[50] In line with the

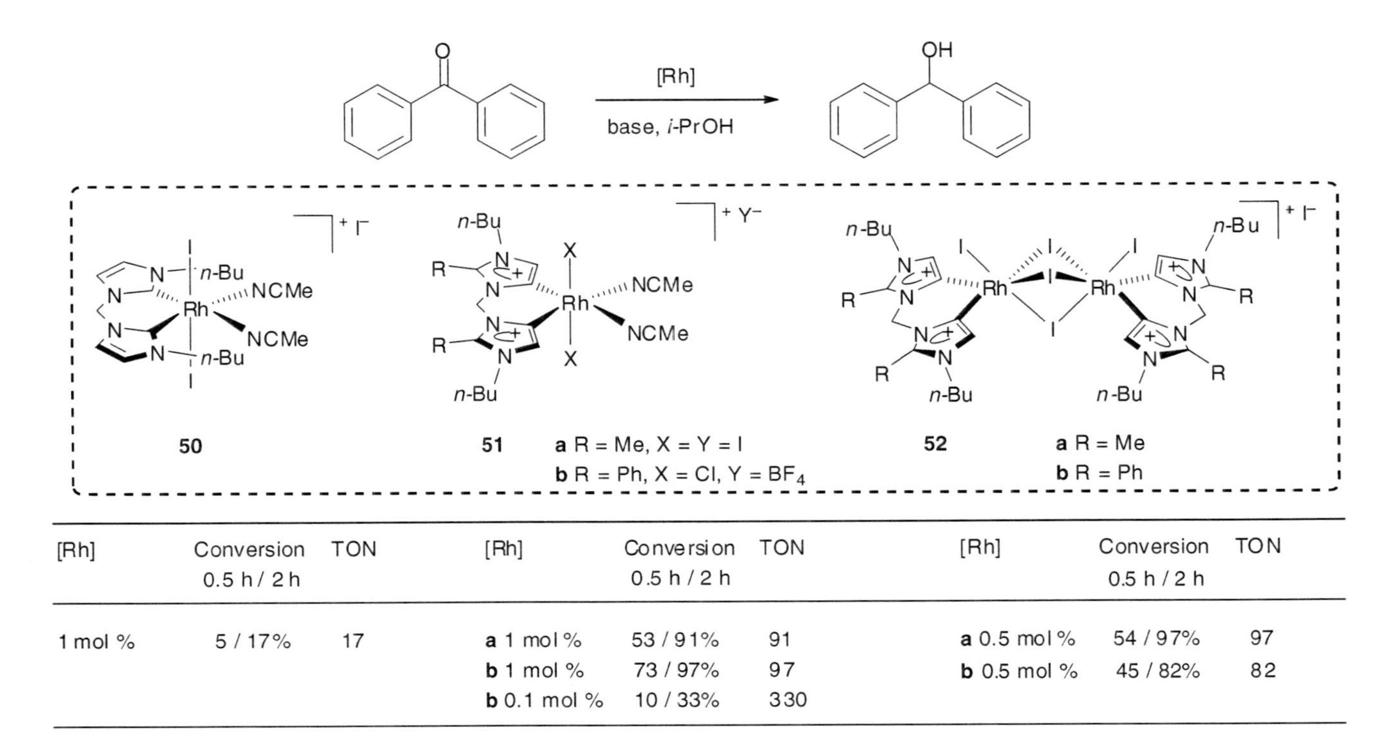

[Rh]	Conversion 0.5 h / 2 h	TON	[Rh]	Conversion 0.5 h / 2 h	TON	[Rh]	Conversion 0.5 h / 2 h	TON
1 mol %	5 / 17%	17	**a** 1 mol %	53 / 91%	91	**a** 0.5 mol %	54 / 97%	97
			b 1 mol %	73 / 97%	97	**b** 0.5 mol %	45 / 82%	82
			b 0.1 mol %	10 / 33%	330			

Scheme 5.11 Transfer hydrogenation reactions catalysed by **50–52** (0.1 equiv KOH, 82 °C).

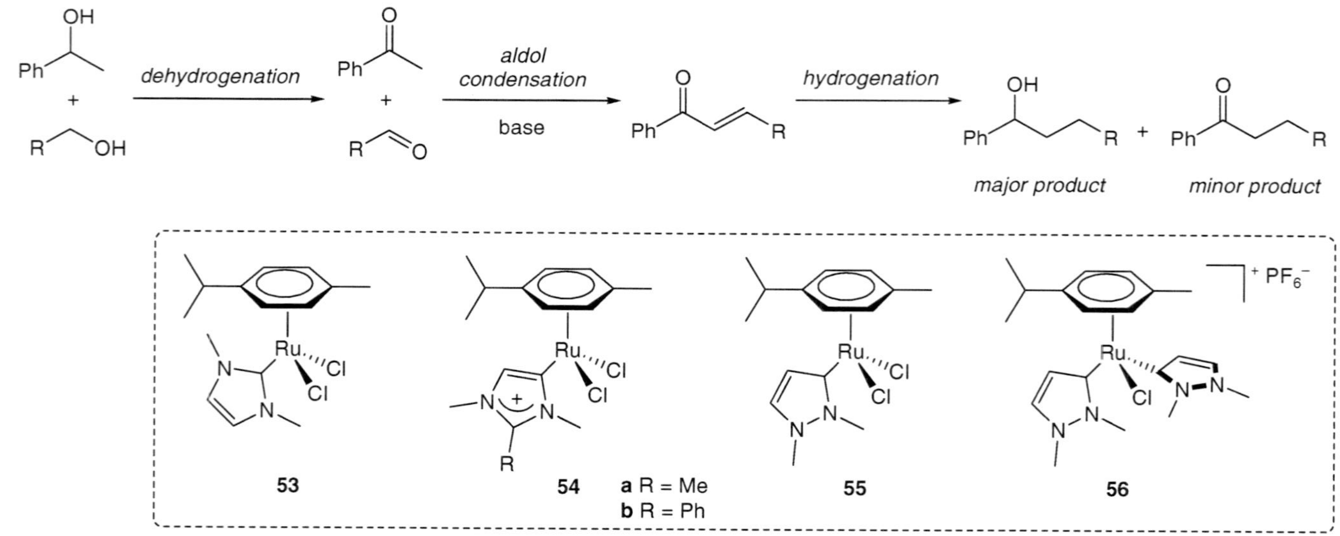

dehydrogenation → aldol condensation, base → hydrogenation → major product + minor product

		53	54 a	54 b	55	56
R = Pr	Yield	60% (22 h)	>95% (22 h)	86% (22 h)	>95% (13 h)	95% (10 h)
	alcohol:ketone	78:22	90:10	91:9	90:10	90:10
R = 3-Cl-(C_6H_4)	Yield	95% (24 h)	94% (24 h)	86% (24 h)	>95% (10 h)	>95% (8 h)
	alcohol:ketone	92:8	81:19	100:0	100:0	85:15
R = Ph	Yield	57% (24 h)	>95% (24 h)	>95% (8 h)	>95% (8h)	>95% (8h)
	alcohol:ketone	100:0	88:12	93:7	97:3	88:12

54: **a** R = Me; **b** R = Ph. 56: PF_6^- salt.

Scheme 5.12 Catalysts used in β-alkylations (1 mol % [Ru], toluene, 110 °C).

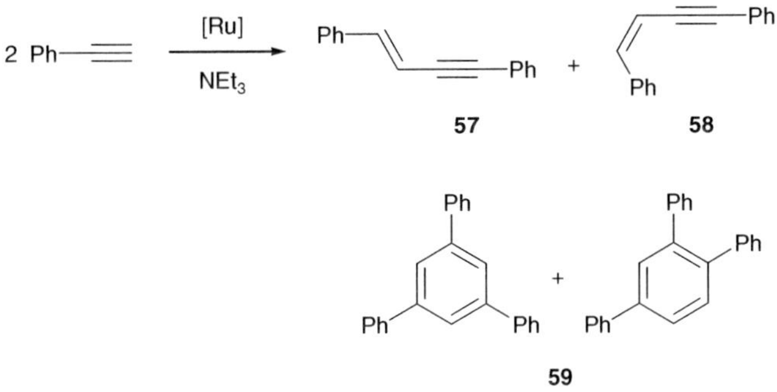

[Ru]	57:58:59	*Yield*
53	*33:9:58*	*97%*
54a	*14:5:81*	*88%*
54b	*26:10:64*	*79%*
55	*21:8:63*	*95%*
56	*28:9:63*	*81%*

Scheme 5.13 Dimerisation of phenylacetylene (5 mol% [Ru], 0.25 equiv NEt_3, MeCN, 70 °C).

previously mentioned results, the non-classical carbene complexes performed better than the classical imidazolylidene complex **53**. The pyrazol-2-ylidene complexes **55**, and especially biscarbene complex **56**, showed the highest catalytic activity, while abnormal imidazolylidene complex **54** afforded comparable conversions but required longer reaction times.

These ruthenium complexes were also investigated as catalysts for the dimerisation of phenylacetylene (Scheme 5.13).[50] The trimerisation of acetylene is typically a competitive process and when compared to $[RuCl_2(p\text{-cymene})]_2$, catalysts **53**–**56** afforded higher dimerisation yields. However, the results obtained did not allow a clear distinction between the effects of the different types and bonding modes of the carbene ligands.

5.4.2.3 *Hydrosilylation*

The rhodium(I) isoquinolinylidene complexes **60a**–**c** were used as catalysts in the hydrosilylation of acetophenone (Scheme 5.14).[51] The carbene ligands were modified by introducing an electron-withdrawing nitro substituent or electron-donating methoxy group at the conjugated benzene ring. Displacing the COD ligand in these complexes by CO allowed for analysis of the donor ability of the ligands by infrared spectroscopy. Not surprisingly, the methoxy-substituted ligand was the most basic in this series, and the nitro-substituted ligand slightly less basic than the unsubstituted ligand. These results indicated that fine-tuning of the electronic properties of the carbene ligands *via* aryl substitution was efficient.

The catalytic activity of the slightly more electron-rich complex **60a** was higher than that of **60b**. Further increase of the electron density on the metal centre promoted, however, undesired reactions as catalysis using complex **60c** produced a complex mixture within minutes.

The performance of normal and abnormal carbene complexes as *in situ* formed hydrosilylation catalysts was investigated by using $[Pt(nbe)_3]$ (nbe = norbornene) and the imidazolium salts **61** and **62** containing a

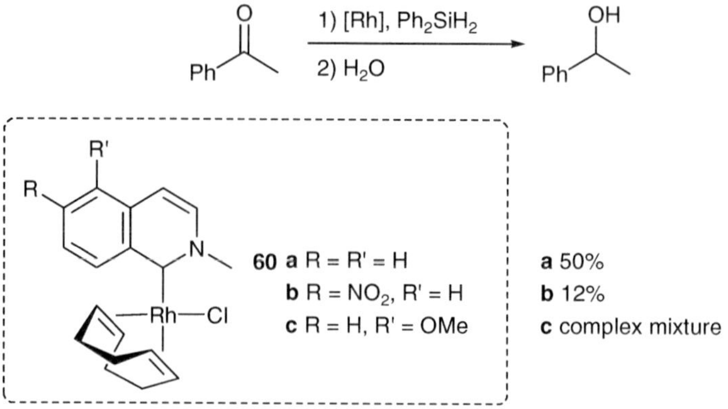
O
1) [Rh], Ph2SiH2
2) H2O
OH
Ph
Ph
R'
R
N
Rh
Cl
60 a R = R' = H
b R = NO2, R' = H
c R = H, R' = OMe
a 50%
b 12%
c complex mixture

Scheme 5.14 Hydrosilylation of acetophenone catalysed by **60** (0.5 mol% [Rh], MeOH, RT).

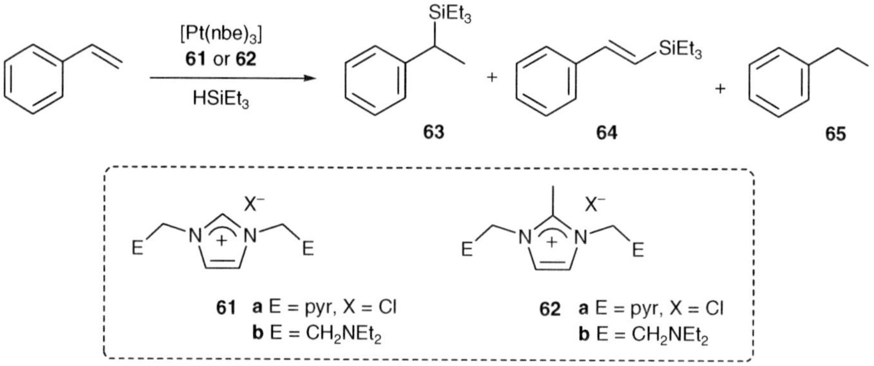
[Pt(nbe)3]
61 or 62
HSiEt3
SiEt3
SiEt3
+
+
63
64
65
X⁻
X⁻
E
N
+
N
E
61 a E = pyr, X = Cl
b E = CH2NEt2
62 a E = pyr, X = Cl
b E = CH2NEt2

Scheme 5.15 Hydrosilylation reactions catalysed by *in situ* generated Pt catalysts (toluene, 120 °C).

unprotected and a methyl-protected C2 position, respectively (Scheme 5.15).[52] Styrene was converted in good yields, while ketones appeared to be only moderately appropriate substrates. Interestingly, the different imidazolium salts and hence presumably the different carbene bonding mode had a pronounced influence on the product selectivity. The normal carbene precursor **61** induced the predominant formation of the branched hydrosilylation product **63**. In contrast, the abnormal carbene precursor **62** afforded the substitution product **64** resulting from dehydrogenative hydrosilylation. The different selectivity in C–Si bond formation and the propensity of the $[Pt(nbe)_3]$/abnormal carbene precursor towards dehydrogenation pointed to fundamentally different modes of action and hold great promise for further catalytic transformation of olefins with non-classical carbene metal complexes.

5.4.3 Olefin Metathesis

CAACs were explored as ligands in ruthenium-catalysed olefin metathesis.[53,54] The carbene ruthenium complexes **66**–**68** were prepared *via* free carbenes and are air- and moisture-stable. Complexes **66** and **67** afforded di- and tri-substituted olefins in high yields (Scheme 5.16). Analysis of the steric situation around the ruthenium centre and, in particular, the Ru–$C_{carbene}$ bond, shorter in the CAAC complexes than in second-generation metathesis catalysts with classical carbene ligands, suggested that the steric bulk at the carbene ligand may reduce the reaction rates. Slower olefin conversion may originate from a limited flexibility in the four-membered metalacycle intermediate, especially when using substituted olefin substrates. Hence, sterically less demanding carbenes such as in **68** increased the catalytic activity to a level that compared favourably with the activity of commercially available second- and third-generation Grubbs' catalysts. Yet, tetra-substituted olefins were not accessible with this class of carbene ligands.

The kinetic selectivity of the CAAC-based catalysts was investigated by probing the *E*/*Z* diastereoselectivity in the cross-metathesis of *cis*-1,4-diaceto-2-butene with allylbenzene (Equation (5.4)).[55] Compared to the commercially available Grubbs' catalysts, **66**–**68** afforded lower *E*/*Z* ratios (3:1 at 70%

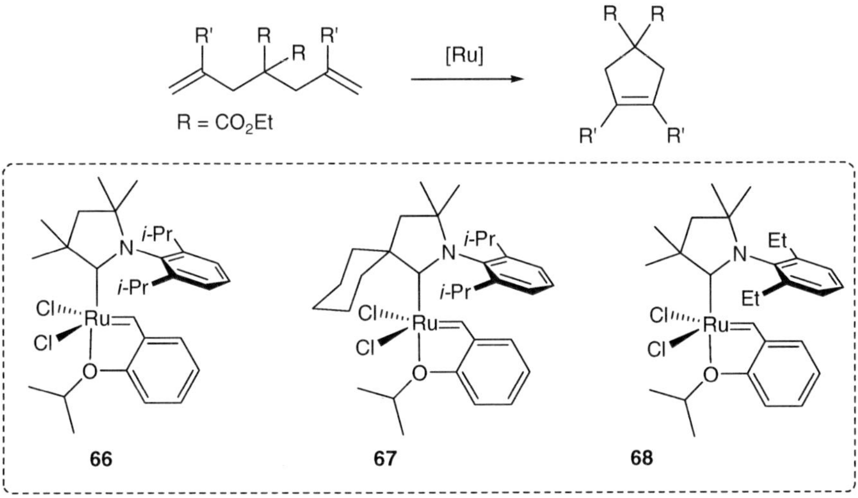

R' R R R'
[Ru]
R R
R = CO_2Et
R' R'

i-Pr N *i*-Pr Cl Ru Cl O **66**
i-Pr N *i*-Pr Cl Ru Cl O **67**
Et N Et Cl Ru Cl O **68**

	R' = H	Yield	R' = Me	Yield
66	3.3 h, 60°C	97%	20 h, 60°C	95%
67	10 h, 60°C	95%	48 h, 60°C	96%
68	15 min, 30°C	95%	1 h, 30°C	95%

Scheme 5.16 Ring-closing metathesis reactions catalysed by **66**–**68** (R′ = H: 5 mol% [Ru], C_6D_6; R′ = Me: 1 mol% [Ru], C_6D_6).

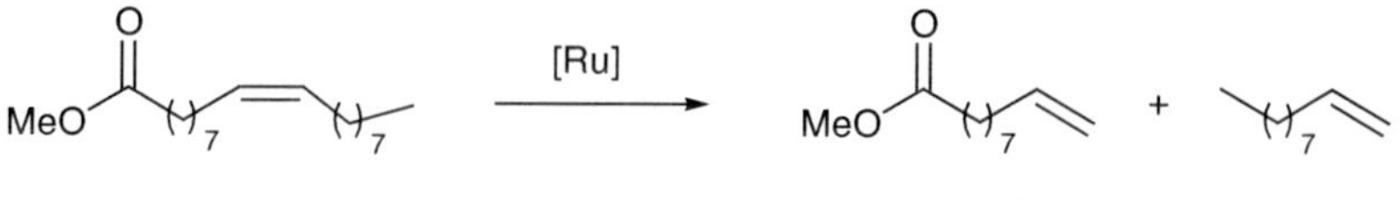

Catalyst	Conversion	Selectivity
66	61% (22 h)	92%
67	46% (6 h)	94%
68	73% (5 h)	73%

Scheme 5.17 Ethenolysis of methyl oleate catalysed by **66**–**68** (0.01 mol% [Ru], 10 bar ethylene, 40 °C).

conversion, ~2:1 at <60% conversion). The activity of **68** was significantly higher than that of the more sterically hindered catalysts **66** and **67**.

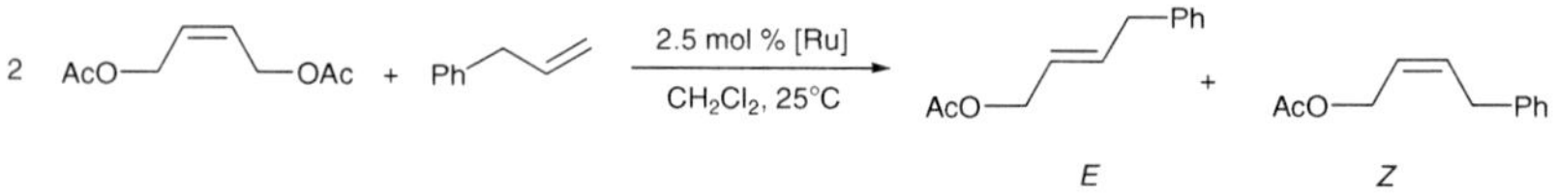

2.5 mol % [Ru]
CH_2Cl_2, 25°C

(5.4)

Catalysts **66**–**68** also displayed high selectivities for the formation of terminal olefins in the ethenolysis of methyl oleate (Scheme 5.17). Notably, at low catalyst loadings (10 ppm) **68** achieved the highest TONs (35 000) reported to date.

5.4.4 Bond Activation

The activation of strong bonds such as H–H and, in particular, of C–H bonds has thus far been achieved only in stoichiometric reactions and catalytic processes need yet to be developed. In this context, it is interesting to note that the intramolecular activation of an inactivated alkyl C–H bond was achieved under remarkably mild conditions by a rhodium centre coordinated to abnormal imidazolylidene ligands (Equation (5.5)).[56] Complex **70** comprised a unique C,C,C-tridentate facially coordinating dicarbene ligand, which was expected to be strongly donating because of the presence of two abnormal carbene and one anionic alkyl ligand. Studies using the analogous imidazolium salt **69** (R′ = H) predisposed to normal carbene bonding indicate that the C–H bond activation leading to complex **70** was not a mere consequence of a constrained ligand geometry but a direct result of the electron density at the rhodium centre due to different carbene bonding modes. Even if the reaction was not catalytic, reversible C_{alkyl}–H bond cleavage was suggested based on isotope labelling studies. Such behaviour should be useful for future catalytic activation of C_{alkyl}–H bonds.

1) $RhCl_3$, NaOAc
2) KI

R = *i*-Pr, *n*-Bu
R' = Me, H

69 70

(5.5)

5.5 Conclusions

The achievements described in this Chapter provide unambiguous evidence that non-classical carbenes can have a remarkable influence on the catalytic activity and the selectivity of transition metal complexes. This influence may in most cases be attributed to the increased σ-donor capacity of these ligands. The performance of the catalyst can be further improved by tailoring the steric bulk of the non-classical carbene ligand as well as by adjusting its donor capacity by incorporating electron-withdrawing and electron-donating groups on the ligand. Nevertheless, the nature of the catalytically active species should be carefully assigned since the determination of the mechanism by poisoning experiments and kinetics is not always straightforward, and especially in cross-coupling and hydrogenation catalysis, some results have been contradictory.

The unique donor properties and versatile synthetic accessibility of non-classical carbenes make this class of ligands highly attractive in transition metal chemistry in general and in metal-mediated catalysis in particular. Despite the short time that has elapsed since the discovery of this class of ligands, numerous catalytic applications have been disclosed. Specifically, various remarkable reactivity patterns have been identified, including the activation of typically unreactive C–H bonds. Based on the potential to further develop, modify and optimise non-classical carbenes, and with the perspective of generating a range of bottleable non-classical free carbenes, it is highly likely that this class of ligands will have a major impact on catalysis.

References and Notes

1. (a) F. E. Hahn and M. C. Jahnke, *Angew. Chem., Int. Ed.*, 2008, **47**, 3122–3172; (b) S. P. Nolan (ed.), *N-Heterocyclic Carbenes in Synthesis*, Wiley-VCH, Weinheim, 2006.

2. (a) D. Enders, O. Niemeier and A. Henseler, *Chem. Rev.*, 2007, **107**, 5606–5655; (b) N. E. Kamber, W. Jeong, R. M. Waymouth, R. C. Pratt, B. G. G. Lohmeijer and J. L. Hedrick, *Chem. Rev.*, 2007, **107**, 5813–5840.
3. A. J. Arduengo III, H. V. R. Dias, R. L. Harlow and M. Kline, *J. Am. Chem. Soc.*, 1992, **114**, 5530–5534.
4. O. Schuster, L. Yang, H. G. Raubenheimer and M. Albrecht, *Chem. Rev.*, 2009, **109**, 3445–3478.
5. M. Albrecht, *Chem. Commun.*, 2008, 3601–3610.
6. This definition of 'non-classical carbenes' is in agreement with previous proposals (cf. references 4 and 5). Worth noting, it also includes most Fischer-type carbene ligands, which are often acyclic and stabilised only by one heteroatom, typically an oxygen, sometimes a sulfur or a nitrogen atom. The catalytic activity of Fischer carbenes is quite limited because of the synthetic constraints connected to Fischer carbene preparation, and the use of catalytically less active molybdenum, tungsten, or chromium. While it might be appropriate to use more specific terms (*e.g.* normal, abnormal, remote, or cyclic, acyclic, see reference 4), we have used here for pragmatic reasons the more inclusive definition of 'non-classical' carbenes, as it allows for discussion of the chemically strongly related acyclic carbene metal complexes. Despite the intrinsic relationship between Fischer-type carbene complexes and the non-classical carbene complexes discussed here, different styles of formulae representation have emerged, including a single bond between the metal and the carbene-type carbon in NHC-type systems, as opposed to double bond representation in Fischer-type and in most acyclic carbene complexes. These distinct representations undoubtedly overemphasise the (only minor) bonding mode differences.
7. (a) C. A. Dyker, V. Lavallo, B. Donnadieu and G. Bertrand, *Angew. Chem., Int. Ed.*, 2008, **47**, 3206–3209; (b) A. Fürstner, M. Alcarazo, R. Goddard and C. W. Lehmann, *Angew. Chem., Int. Ed.*, 2008, **47**, 3210–3214.
8. J. Vignolle, X. Cattoën and D. Bourissou, *Chem. Rev.*, 2009, **109**, 3333–3384.
9. E. Aldeco-Perez, A. J. Rosenthal, B. Donnadieu, P. Parameswaran, G. Frenking and G. Bertrand, *Science*, 2009, **326**, 556–559.
10. M. Albrecht, *Science*, 2009, **326**, 532–533.
11. (a) X. Cattoën, H. Gornitzka, D. Bourissou and G. Bertrand, *J. Am. Chem. Soc.*, 2004, **126**, 1342–1343; (b) K. Miqueu, E. Despagnet-Ayoub, P. W. Dyer, D. Bourissou and G. Bertrand, *Chem.–Eur. J.*, 2003, **9**, 5858–5864.
12. H. V. Huynh, Y. Han, R. Jothibasu and J. A. Yang, *Organometallics*, 2009, **28**, 5395–5404.
13. (a) P. J. Fraser, W. R. Roper, F. G. A. Stone and A. Gordon, *J. Chem. Soc., Dalton Trans.*, 1974, 760–764; (b) J. Schütz, E. Herdtweck and W. A. Herrmann, *Organometallics*, 2004, **23**, 6084–6086; (c) E. Kluser, A. Neels and M. Albrecht, *Chem. Commun.*, 2006, 4495–4497; (d) R. Jazzar, J.-B. Bourg, R. D. Dewhurst, B. Donnadieu and G. Bertrand, *J. Org. Chem.*,

2007, **72**, 3492–3499; (e) H. G. Raubenheimer and S. Cronje, *Dalton Trans.*, 2008, 1265–1272.
14. D. Bacciu, K. J. Cavell, I. A. Fallis and L. Ooi, *Angew. Chem., Int. Ed.*, 2005, **44**, 5282–5284.
15. (a) A. R. Chianese, B. M. Zeglis and R. H. Crabtree, *Chem. Commun.*, 2004, 2176–2177; (b) M. Viciano, M. Feliz, R. Corberán, J. A. Mata, E. Clot and E. Peris, *Organometallics*, 2007, **26**, 5304–5314.
16. (a) A. R. Chianese, A. Kovacevic, B. M. Zeglis, J. W. Faller and R. H. Crabtree, *Organometallics*, 2004, **23**, 2461–2468; (b) G. Song, X. Li, Z. Song, J. Zhao and H. Zhang, *Chem.–Eur. J.*, 2009, **15**, 5535–5544.
17. (a) D. J. Lavorato, J. K. Terlouw, G. A. McGibbon, T. K. Dargel, W. Koch and H. Schwartz, *Int. J. Mass Spectrom.*, 1998, **179/180**, 7–14; (b) O. Holloczki and L. Nyulaszi, *J. Org. Chem.*, 2008, **73**, 4794–4799.
18. (a) G. Sini, O. Eisenstein and R. H. Crabtree, *Inorg. Chem.*, 2002, **41**, 602–604; (b) R. Tonner, G. Heydenrych and G. Frenking, *Chem. Asian J.*, 2007, **2**, 1555–1567.
19. (a) A. Igau, H. Grützmacher, A. Baceiredo and G. Bertrand, *J. Am. Chem. Soc.*, 1988, **110**, 6463–6466; (b) S. Sole, H. Gornitzka, W. W. Schoeller, D. Bourissou and G. Bertrand, *Science*, 2001, **292**, 1901–1903.
20. (a) V. Lavallo, Y. Canac, C. Präsing, B. Donnadieu and G. Bertrand, *Angew. Chem., Int. Ed.*, 2005, **44**, 5705–5709; (b) V. Lavallo, Y. Canac, A. DeHope, B. Donnadieu and G. Bertrand, *Angew. Chem., Int. Ed.*, 2005, **44**, 7236–7239.
21. V. Lavallo, C. A. Dyker, B. Donnadieu and G. Bertrand, *Angew. Chem., Int. Ed.*, 2008, **47**, 5411–5414.
22. M. Melaimi, P. Parameswaran, B. Donnadieu, G. Frenking and G. Bertrand, *Angew. Chem., Int. Ed.*, 2009, **48**, 4792–4795.
23. (a) L. N. Appelhans, D. Zuccaccia, A. Kovacevic, A. R. Chianese, J. R. Miecznikowski, A. Macchioni, E. Clot, O. Eisenstein and R. H. Crabtree, *J. Am. Chem. Soc.*, 2005, **127**, 16299–16311; (b) M. Baya, B. Eguillor, M. A. Esteruelas, M. Olivan and E. Onate, *Organometallics*, 2007, **26**, 6556–6563; (c) B. Eguillor, M. A. Esteruelas, M. Olivan and M. Puerta, *Organometallics*, 2008, **27**, 445–450.
24. G. Song, X. Wang, Y. Li and X. Li, *Organometallics*, 2008, **27**, 1187–1192.
25. (a) C. E. Ellul, M. F. Mahon, O. Saker and M. K. Whittlesey, *Angew. Chem., Int. Ed.*, 2007, **46**, 6343–6345; (b) A. A. Danopoulos, N. Tsoureas, J. A. Wright and M. E. Light, *Organometallics*, 2004, **23**, 166–168; (c) X. Hu, I. Castro-Rodriquez and K. Meyer, *Organometallics*, 2003, **22**, 3016–3018.
26. M. Heckenroth, E. Kluser, A. Neels and M. Albrecht, *Dalton Trans.*, 2008, 6242–6249.
27. P. Mathew, A. Neels and M. Albrecht, *J. Am. Chem. Soc.*, 2008, **130**, 13534–13535.
28. W. H. Meyer, M. Deetlefs, M. Pohlmann, R. Scholz, M. W. Esterhuysen, G. R. Julius and H. G. Raubenheimer, *Dalton Trans.*, 2004, 413–420.

29. C. E. Strasser, E. Stander-Grobler, O. Schuster, S. Cronje and H. G. Raubenheimer, *Eur. J. Inorg. Chem.*, 2009, 1905–1912.
30. M. Heckenroth, A. Neels, M. G. Garnier, P. Aebi, A. W. Ehlers and M. Albrecht, *Chem.–Eur. J.*, 2009, **15**, 9375–9386.
31. G. Song, Y. Zhang and X. Li, *Organometallics*, 2008, **27**, 1936–1943.
32. H. Lebel, M. K. Janes, A. B. Charette and S. P. Nolan, *J. Am. Chem. Soc.*, 2004, **126**, 5046–5047.
33. L.-C. Campeau, P. Thansandote and K. Fagnou, *Org. Lett.*, 2005, **7**, 1857–1860.
34. (a) S. K. Schneider, P. Roembke, G. R. Julius, C. Loschen, H. G. Raubenheimer, G. Frenking and W. A. Herrmann, *Eur. J. Inorg. Chem.*, 2005, 2973–2977; (b) S. K. Scheneider, P. Roembke, G. R. Julius, H. G. Raubenheimer and W. A. Herrmann, *Adv. Synth. Catal.*, 2006, **348**, 1862–1873.
35. (a) Y. Han, H. V. Huynh and G. K. Tan, *Organometallics*, 2007, **26**, 6581–6585; (b) Y. Han, L. J. Lee and H. V. Huynh, *Organometallics*, 2009, **28**, 2278–2786.
36. D. Kremzow, G. Seidel, C. W. Lehmann and A. Fürstner, *Chem.–Eur. J.*, 2005, **11**, 1833–1853.
37. E. A. B. Kantchev, C. J. O'Brien and M. G. Organ, *Angew. Chem., Int. Ed.*, 2007, **46**, 2768–2813.
38. E. Stander, PhD thesis, Stellenbosch University, South Africa, 2008.
39. M. Albrecht and H. Stoeckli-Evans, *Chem. Commun.*, 2005, 4705–4707.
40. D. R. Anton and R. H. Crabtree, *Organometallics*, 1983, **2**, 855–859.
41. A. Poulain, A. Neels and M. Albrecht, *Eur. J. Inorg. Chem.*, 2009, **13**, 1871–1881.
42. J. A. Loch, M. Albrecht, E. Peris, J. Mata, J. W. Faller and R. H. Crabtree, *Organometallics*, 2002, **21**, 700–706.
43. N. A. Piro, J. S. Owen and J. E. Bercaw, *Polyhedron*, 2004, **23**, 2797–2804.
44. S. K. Schneider, C. F. Rentzsch, A. Krüger, H. G. Raubenheimer and W. A. Herrmann, *J. Mol. Catal. A: Chem.*, 2007, **265**, 50–58.
45. (a) V. Lavallo, G. D. Frey, S. Kousar, B. Donnadieu and G. Bertrand, *Proc. Natl. Acad. Sci. U. S. A.*, 2007, **104**, 13569–13573; (b) V. Lavallo, G. D. Frey, B. Donnadieu, M. Soleilhavoup and G. Bertrand, *Angew. Chem., Int. Ed*, 2008, **47**, 5224–5228; (c) X. Zeng, G. D. Frey, S. Kousar and G. Bertrand, *Chem.–Eur. J.*, 2009, **15**, 3056–3060; (d) X. Zeng, M. Soleilhavoup and G. Bertrand, *Org. Lett.*, 2009, **11**, 3166–3169.
46. J. Vignolle, H. Gornitzka, B. Donnadieu, D. Bourissou and G. Bertrand, *Angew. Chem., Int. Ed.*, 2008, **47**, 2271–2274.
47. M. Heckenroth, E. Kluser, A. Neels and M. Albrecht, *Angew. Chem., Int. Ed.*, 2007, **46**, 6293–6296.
48. L. Yang, A. Krüger, A. Neels and M. Albrecht, *Organometallics*, 2008, **27**, 3161–3171.
49. (a) C. S. Cho, B. T. Kim, T.-J. Kim and S. C. Shim, *J. Org. Chem.*, 2001, **66**, 9020–9022; (b) K. Fujita, C. Asai, T. Yamaguchi, F. Hanasaka and R. Yamaguchi, *Org. Lett.*, 2005, **7**, 4017–4019.

50. A. Prades, M. Viciano, M. Sanaú and E. Peris, *Organometallics*, 2008, **27**, 4254–4259.
51. S. Gómez-Bujedo, M. Alcarazo, C. Pichon, E. Álvarez, R. Fernández and J. M. Lassaletta, *Chem. Commun.*, 2007, 1180–1182.
52. D. Bacciu, PhD thesis, Cardiff University, United Kingdom, 2007.
53. V. Lavallo, Y. Canac, A. DeHope, B. Donnadieu and G. Bertrand, *Angew. Chem., Int. Ed.*, 2005, **44**, 7236–7239.
54. D. R. Anderson, V. Lavallo, D. J. O'Leary, G. Bertrand and R. H. Grubbs, *Angew. Chem., Int. Ed.*, 2007, **46**, 7262–7265.
55. D. R. Anderson, T. Ung, G. Mkrtumyan, G. Bertrand, R. H. Grubbs and Y. Schrodi, *Organometallics*, 2008, **27**, 563–566.
56. A. Krüger, A. Neels and M. Albrecht, *Chem. Commun.*, 2010, 315–317.

CHAPTER 6

Early Transition and Rare Earth Metal Complexes with N-Heterocyclic Carbenes

LARS-ARNE SCHAPER,* EVANGELINE TOSH* AND WOLFGANG A. HERRMANN*

Technische Universität München, Lichtenbergstr. 4, D-85747 Garching bei München, Germany

6.1 Introduction

N-Heterocyclic carbenes (NHCs) have gained outstanding importance in organometallic chemistry and homogenous catalysis.[1] In 1968, Öfele and Wanzlick reported highly unusual compounds obtained from imidazolium salts providing the basis for NHC chemistry.[2,3] Lappert's extensive research considerably broadened the organometallic fundamentals in this area,[4] and Bertrand first demonstrated that stable free carbenes were accessible, with the isolation of phosphinocarbenes from α-diazophosphines in 1988.[5] Arduengo isolated an electronically stabilized nucleophilic N-heterocyclic carbene in 1991 and showed that bulky substituents were not implicitly required to stabilize such species.[6,7] Soon, the significance of NHC complexes as a new structural motif in homogenous catalysis was perceived.[8] Nowadays, NHC complexes with virtually any metal are known providing a plethora of new complexes and a new ground for organometallic chemistry.[9]

In comparison to late transition metals (LTMs), reports of early transition metal (ETM) complexes with phosphane ligands are scarce. On the contrary, amides are more common ligands for ETMs than for LTMs. This can be explained by the electronic demands of the hard Lewis acidic ETMs, which

RSC Catalysis Series No. 6
N-Heterocyclic Carbenes: From Laboratory Curiosities to Efficient Synthetic Tools
Edited by Silvia Díez-González
© Royal Society of Chemistry 2011
Published by the Royal Society of Chemistry, www.rsc.org

prefer hard basic ligands. Hence the soft base nature of the phosphane donor ligands provides a mismatch of donor and acceptor, responsible for lower stability of the resulting complexes.[10] This could also explain the scarcity of NHC–ETM complexes, but yet NHCs have no necessary requirement for backbonding (*vide infra*) and their strongly nucleophilic singlet lone pair makes them suitable for a wider range of metals.[11]

As aforementioned, Öfele explored this area already in 1968 with pentacarbonyl- and tetra carbonyl–chromium(0) NHC ligated compounds.[2] After Arduengo's groundbreaking report, various NHC–ETM compounds were reported.[11,12] Even so, to date they are relatively seldom and only a few applications of NHC–ETM in catalysis are known. McGuinness *et al.* investigated the reactivity of ETMs in ethene polymerization and oligomerization reactions and obtained very high activities for Ti, V and Cr complexes; such studies may arouse more interest in this challenging research area.[13]

On the other hand, two presumptions are responsible for the late development of organometallic chemistry with lanthanoids and actinoids. One is that 'rare-earth metals' implies sparsity and high expenses as impediments for the use of these metals. However, the element concentrations in the continental crust (Figure 6.1) show that these elements are certainly very seldom compared to iron (abundance: 43 200 ppm) yet the rarest, uranium and thulium, are far more common than the precious metals, *e.g.* silver, platinum, or frequently used transition metals such as palladium, rhodium or iridium.[14]

The second assumption is that lanthanoids were deemed to generate uninteresting chemistry because they lack the orbital interaction and backbonding capability of the transition metals as the 4f orbitals of the lanthanoids have a

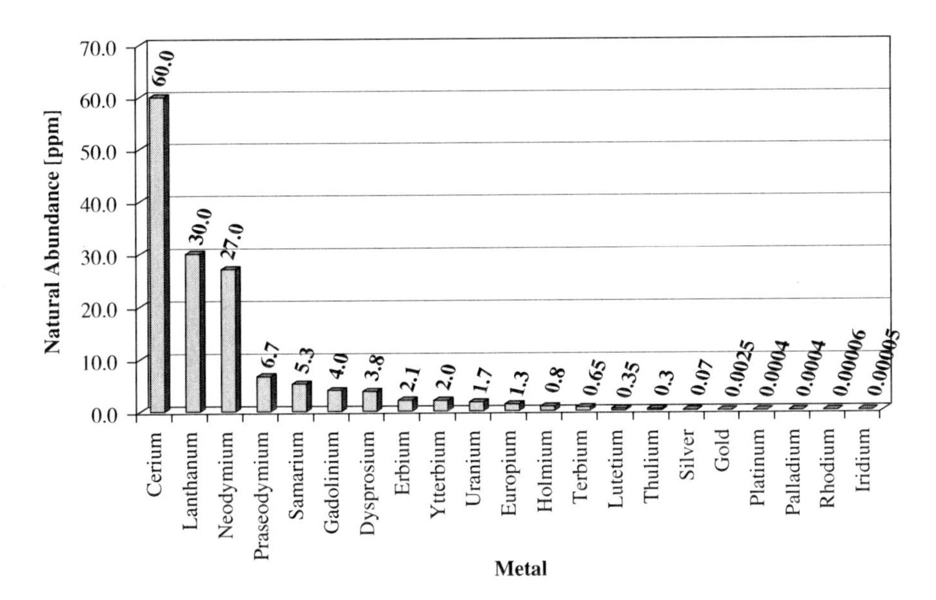

Figure 6.1 Natural abundances of selected metals in the continental crust.

limited radial extension compared to the d orbitals of the transition metals. Based on Wilkinson's early observations,[15] Evans enunciated a few simple rules for lanthanoid reactivity:[16]

- The reactivity of these elements is not strongly dependent on the $4f^n$ configuration.
- The ligand geometries mostly derive from optimized electrostatic interactions.
- The reactivity can be strongly affected by steric factors, with sterically saturated complexes being the most stable.

Because NHC ligands may act as Lewis base donors, it was not surprising they could also coordinate metals incapable of π-backbonding such as the alkaline earth metals Be or Mg.[17] Today there are many examples known of N-heterocyclic carbene ligands tethered to early transition metals, lanthanoid and actinoid metals.[18]

This Chapter provides a general overview of typical syntheses, depicts the bonding situation between the metal and the carbene ligand, and describes the applications of these metal complexes.

6.2 Structural Survey and Typical Syntheses

6.2.1 Complexes with Monodentate NHC Ligands

The first early transition and f-block metal NHC compounds were simple monodentate NHC adducts. These complexes derived from corresponding tetrahydrofuran precursors offering one coordination site.

6.2.1.1 *Carbene Complexes via Ligand Substitution*

One of the most common methods to date for the preparation of early transition and rare earth metal NHC complexes is the substitution of weakly coordinating donor molecules by a free NHC (isolated or *in situ* generated). Accordingly, ETM complexes were prepared by Öfele, Herrmann *et al.* in 1994. Encouraged by the successful substitution of CO in the hexacarbonyl compounds with stable IMe giving the mono-adducts $[(IMe)M(CO)_5]$ (M = Cr, Mo, W), a series of adducts were prepared by simple substitution of the corresponding solvent substituted metal halides at room temperature (Scheme 6.1).[12a,b] Due to its high nucleophilicity, IMe displaced *O*- and *N*-donor ligands, normally used to prevent further nucleation, and enhanced their stability towards hydrogenolysis.

Concomitantly, Schumann *et al.* reported the first organolanthanoid–carbene adducts, prepared by treatment of $[Yb(C_5Me_4Et)_2(THF)]$ with ^{Me}IMe or ^{Me}IiPr in THF (Equation (6.1)).[19] This class of complexes was quickly expanded to include the analogous Sm^{II} compounds $[(NHC)Sm(C_5Me_4Et)_2]$.[20]

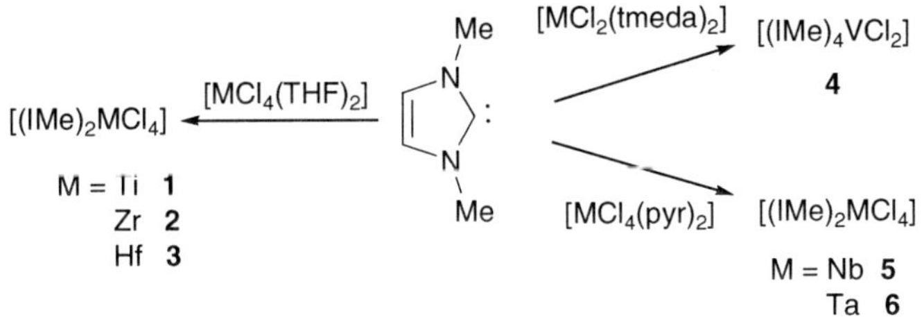

Scheme 6.1 Preparation of complexes **1–6** (tmeda = *N*,*N*,*N'*,*N'*-tetramethylethylenediamine).

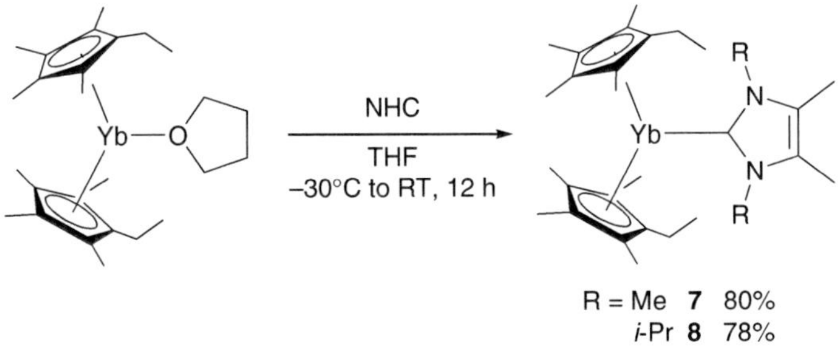

(6.1)

Arduengo also reported some lanthanoid carbene complexes *via* the reaction of ^{Me}IMe with [Sm(Cp*)$_2$(THF)], yielding [(^{Me}IMe)Sm(Cp*)$_2$] **9**, which could react with another carbene molecule to form the bis-NHC adduct [(^{Me}IMe)$_2$ Sm(Cp*)$_2$] **10**.[21]

The first example of an actinoid–carbon bond was also obtained by replacement of the THF ligand with IMes in the uranium(VI) precursor [UO$_2$Cl$_2$(THF)$_3$] giving [(IMes)$_2$UO$_2$Cl$_2$] (**11**) (Equation (6.2)).[22]

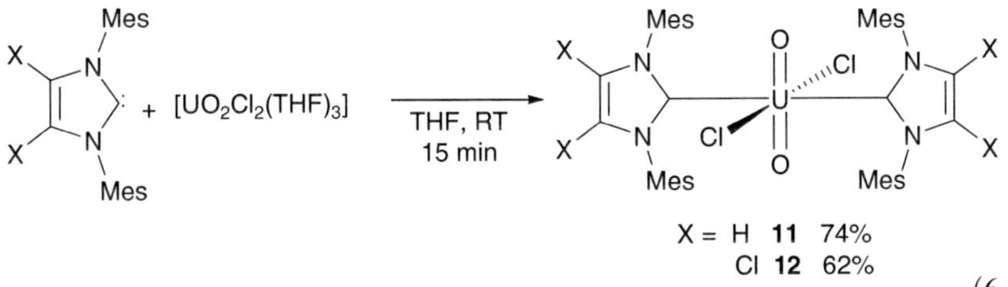

(6.2)

6.2.1.2 Carbene Complexes via Coordination of a Free NHC

Simple ligand association is commonly used especially for metal complexes with low electron counts where stability benefits from additional electron density. With the first NHC–samarium complexes, Arduengo reported that coordination of ^{Me}IMe to a europium(III) complex with a tris-(2,2,6,

6-tetramethylheptane-3,5-dionato)-ligand (thd) yielded the corresponding adduct [(^{Me}IMe)Eu(thd)$_3$] **13** (Equation (6.3)).[21] The carbenic carbon and one oxygen atom formed a linear axis and the remaining five oxygen atoms spanned around the europium centre. Despite smaller atomic radius, the yttrium(III) analogue [(^{Me}IMe)Y(thd)$_3$] **14** could be formed under the same conditions.

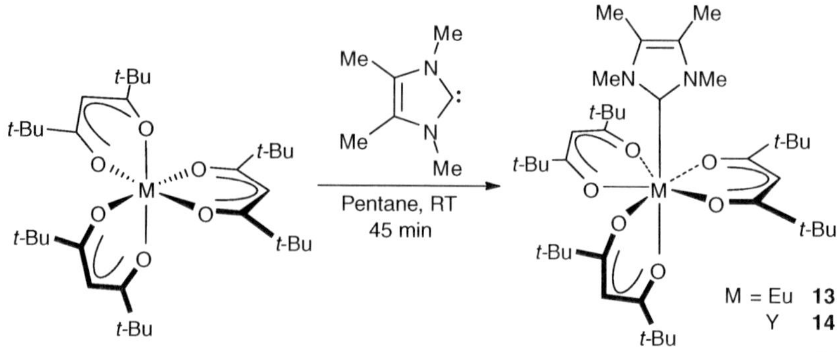

Pentane, RT
45 min

M = Eu **13**
Y **14**

(6.3)

Simple addition of a NHC to yield an early transition metal carbene complex in a high oxidation state was applied to synthesize [(IMes)VCl$_3$(O)] **15** (Figure 6.2). The NHC ligand conferred a high stability to the complex, and while other trichloro-oxo-vanadium(V) adducts (*e.g.* [VCl$_3$(O)(MeCN)]])[23] were easily hydrolyzed in air, compound **15** was air-stable over months even in a dichloromethane solution.[24] A major contribution to this stability may arise from the highly unusual Cl–C$_{carbene}$ interactions determined from the crystal structure (*vide infra*).

Nikiforov *et al.* reported the coordination of ^{Me}IMe and ^{Me}IiPr to TiF$_4$ yielding [(^{Me}IMe)$_2$TiF$_4$] and [(^{Me}IiPr)$_2$TiF$_4$] respectively, whereas Kuhn *et al.* reported that from TiCl$_4$ only the mono-adducts were formed.[25] Hahn *et al.* synthesized a related complex [(NHC)TiCl$_4$] with a benzimidazol-2-ylidene ligand,[26] whereas by addition of IiPr to the related Hf precursor HfCl$_4$, Niehues *et al.* synthesized the mono substituted compound [(IiPr)HfCl$_4$].[27]

6.2.1.3 Complexes via Ligand Dissociation

Before the first free stable carbene could be isolated, the *in situ* preparation of carbenes was implicitly required for synthesis of transition NHC–metal

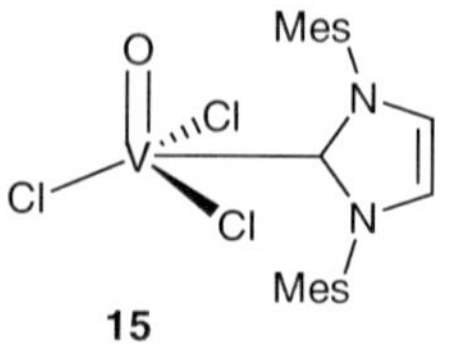

15

Figure 6.2 [(IMes)VCl$_3$(O)] complex.

compounds. Yet, Abernethy was the first in applying this strategy to ETM chemistry by employing imidazolium salts in the synthesis (Equation (6.4)).[28] However, in 2008, Lorber and Vendier isolated a number of side products, while examining this reaction. In addition, they synthesized **16** in higher yields by simple addition of the free IMes to $[Ti(NMe_2)Cl_2]$,[29] and similar behaviour was observed for the V analogue.[30]

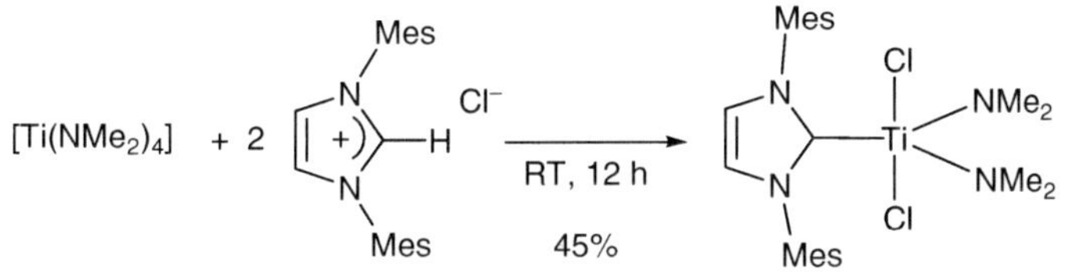

$[Ti(NMe_2)_4]$ + 2 [IMes·H]$^+$ Cl^- → (RT, 12 h, 45%) **16**

(6.4)

A similar reaction was reported by Zhang *et al.* (Equation (6.5)),[31] and $[V(=NR)(CH_2SiMe_3)_3]$ (R = adamantyl or 2,6-dimethylphenyl) precursors were treated with free IPr giving the corresponding alkyl-alkylidene complexes **17** and **18** *via* α-H-elimination from two adjacent alkyl ligands.

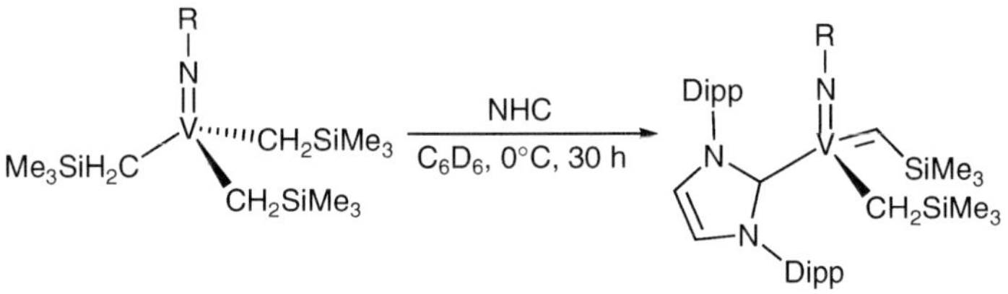

$[V(=NR)(CH_2SiMe_3)_3]$ → (NHC, C_6D_6, 0°C, 30 h)

R = adamantyl **17**
2,6-(Me)$_2$Ph **18**

(6.5)

6.2.2 Complexes with Anionic Multidentate NHC Ligands

One approach to achieve more stable metal–ligand bonds is to employ multidentate ligands. Multiple bound ligands are particularly desirable for catalytic applications because they can liberate coordination sites on the metal without complete dissociation. The common degradation of catalysts over time is very often result of dissociation or side reactions of the supporting ligand. The stability of early transition and rare-earth metal complexes clearly benefited from the combination of an anionic donor and a NHC in one ligand generating electropositive hard metal centres with strong and short carbene–metal bonds.[11] Concomitantly, the applied anionic anchored bi- or tridentate ligands enhanced complex stability through formation of five- or six-membered metalaheterocycles.

6.2.2.1 Complexes via Salt Metathesis

NHCs with an alkoxy-anchor coordinated to alkali metals serve as excellent ligand transfer reagents.[11] The reagent synthesized by Arnold *et al.* also showed an extraordinary stability and was successfully used in the preparation of [(NHC)Ti(O*i*-Pr)$_3$] **19** through a salt metathesis reaction (Equation (6.6)).[32] This transfer reagent was also employed for synthesizing a related [(NHC)$_3$UI] complex.[33]

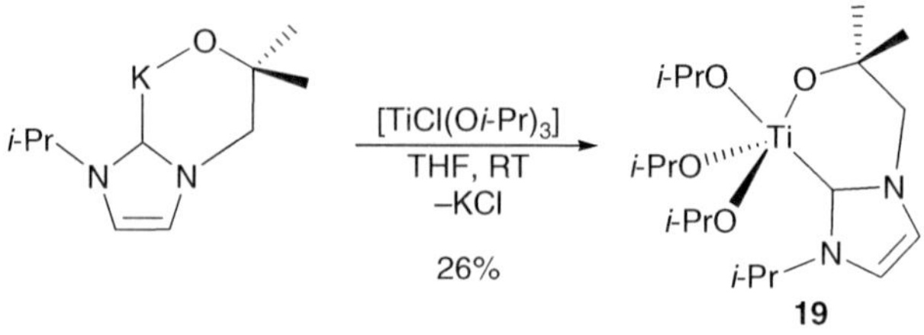

(6.6)

Aryloxo-NHC-containing complexes could also be produced from amido precursors where the ligands also acted as internal base as shown by the Shen group. Hence, a hydroxyaryl-imidazolium ligand reacted with [LiY{N(*i*-Pr$_2$)}$_4$] and BuLi at – 78 °C to give the NHC–yttrium complex [(NHC)$_3$Y] **21** (Scheme 6.2).[34] From [LiYb{N(*i*-Pr$_2$)}$_4$], a bis-substituted ytterbium compound [(NHC)$_2$Yb {N(*i*-Pr$_2$)}] **22** was prepared.[35] Of note, all attempts to prepare the monosubstituted complex were unsuccessful.

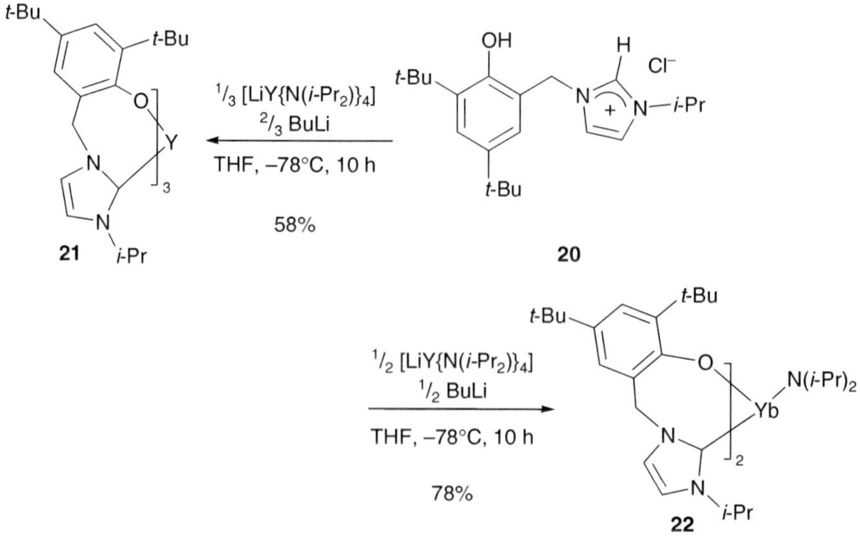

Scheme 6.2 Preparation of complexes **21** and **22**.

Other examples of metal complexes prepared by salt metathesis are **23** and **24**. The disodium salt was unstable at room temperature, but it could be *in situ* generated and transferred to $[MCl_4(THF)_2]$ (M = Ti, Zr) (Equation (6.7)).[36]

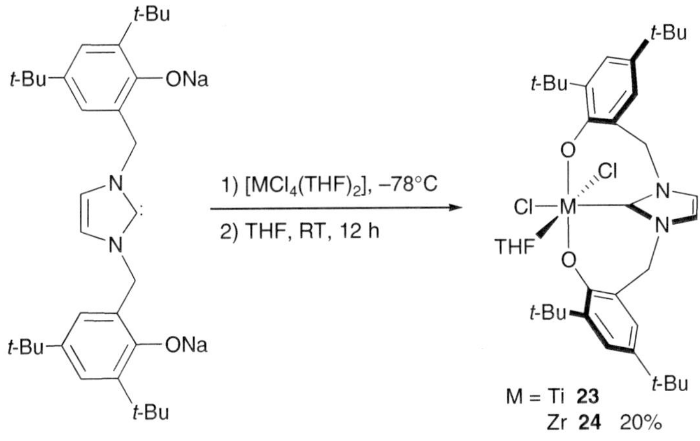

1) $[MCl_4(THF)_2]$, –78°C
2) THF, RT, 12 h

M = Ti **23**
Zr **24** 20%

(6.7)

Alternatively, an amido-lithium-NHC dimer reported by Arnold and co-workers could be attached to different metals such as uranium or titanium, to form derivatives **25**[37] and **26**–**28**[38] respectively (Scheme 6.3).

Danopoulos described the preparation of indenyl- and fluorenyl-functionalized NHC–ETM complexes **29** and **30** through a salt-elimination reaction (Figure 6.3).[39] By a different synthetic method Wang *et al.* reported closely related complexes $[(NHC)M(CH_2SiMe_3)_2]$ **31**–**33** bearing Y, Lu or Sc, which represented the first examples of functionalized NHC covalently bonding to rare earth metal alkyl complexes.[40]

6.2.2.2 *Complexes via Substitution*

Fryzuk and co-workers prepared a bis-amido NHC ligand (Scheme 6.4).[41] Addition of $[Zr(NMe_2)_4]$ afforded zirconium complex $[(NHC)Zr(NMe_2)_2]$ **34**

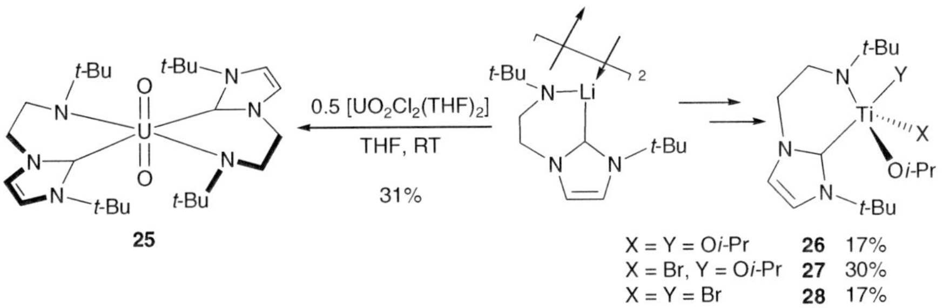

0.5 $[UO_2Cl_2(THF)_2]$
THF, RT
31%

X = Y = O*i*-Pr **26** 17%
X = Br, Y = O*i*-Pr **27** 30%
X = Y = Br **28** 17%

Scheme 6.3 Preparation of complexes **25**–**28**.

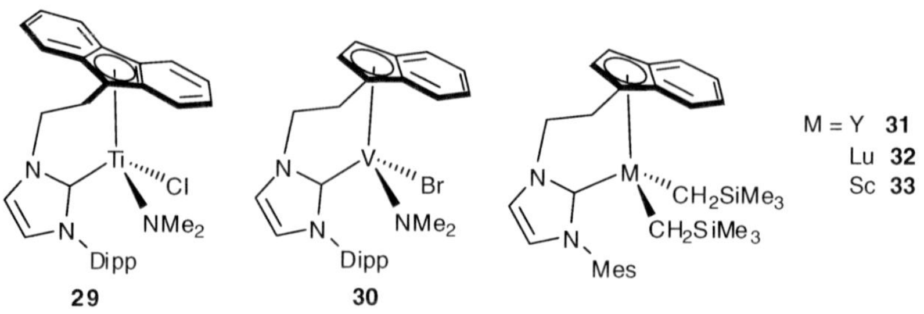

Figure 6.3 Indenyl- and fluorenyl-bridged NHC early transition and rare earth metal complexes (Dipp = 2,6-diisopropylphenyl).

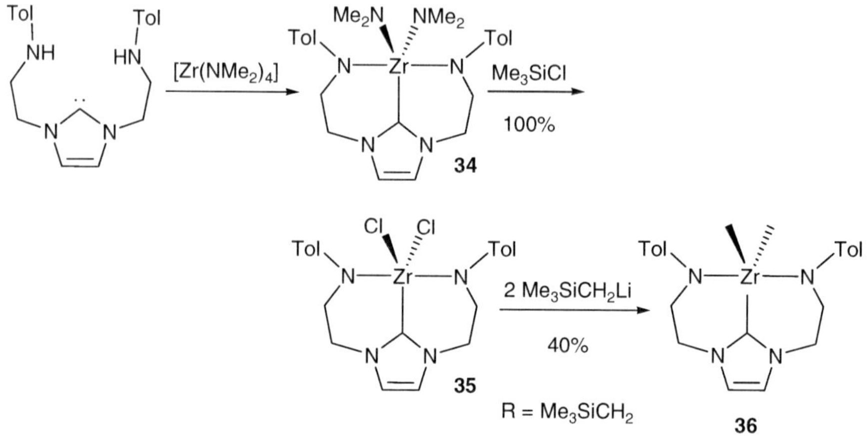

Scheme 6.4 Ligand substitution for tridentate-amine NHC complexes.

without any side products. Its reaction with Me_3SiCl provided dichloride **35** in good yields, which was reacted with Me_3SiCH_2Li to yield the dialkyl derivative **36**. The same complex could be synthesized in one step by replacing the tetrakis (dimethylamido)zirconium by $[Zr(CH_2SiMe_3)_4]$. Also, a series of Zr and Hf compounds was readily accessible for the mesityl-substituted NHC ligands. Of note, while the Zr compounds decomposed within 24 h, the Hf complexes displayed excellent thermal stability.[42] In later reports by Fryzuk, the corresponding Ta compounds were also shown to be accessible *via* this synthetic route.[43]

In 2003, Danopoulos *et al.* successfully synthesized a CNC pincer ligand and used it to prepare a Pd complex.[44] This ligand was later applied to the synthesis of ETM and actinoid complexes. Substitution of tmeda with the CNC pincer yielded $[(NHC)VCl_2(THF)]$ **37** (Equation (6.8)). This route was alternatively applied to the preparation of titanium, chromium, niobium and uranium complexes.[45]

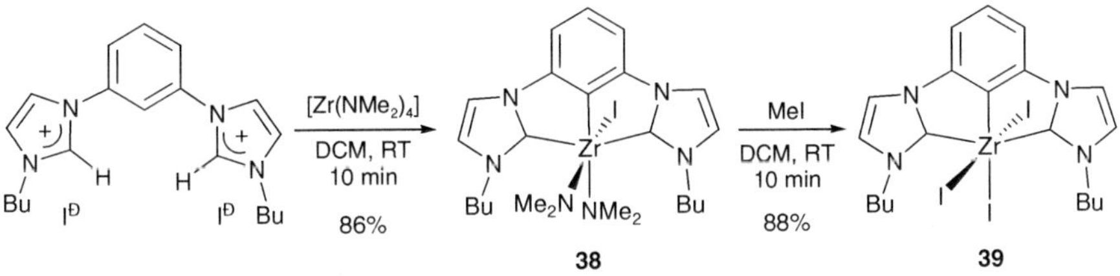

Scheme 6.5 Tridentate CCC pincer complex through deprotonation.

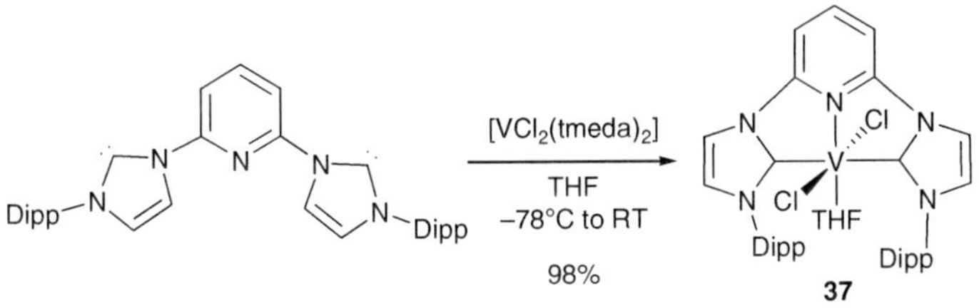

(6.8)

6.2.2.3 Complexes via Deprotonation

Another N-heterocyclic carbene pincer ligand was introduced to ETM chemistry by Hollis and co-workers.[46] This tridentate ligand had two NHCs connected *via* a phenyl bridge, and it could easily be deprotonated by the amido ligands in $[Zr(NMe_2)_4]$ acting as internal base to form the tridentate Zr complex **38**. Then, treatment with methyl iodide yielded the expected triiodide $[(NHC)ZrI_3]$ **39** at room temperature (Scheme 6.5).

6.2.3 Bimetallic Complexes with N-Heterocyclic Carbenes

Early transition and f-block metal complexes with bonds to another metal are scarce.[47] Roughly around the same time of Arduengo's discovery of the stable carbene,[6] Herrmann *et al.* reported the synthesis of a stable germanium(II) analogue of N-heterocyclic carbenes.[48] Shortly afterwards, Denk and Lappert independently reported the synthesis of stable silylenes and later Jones disclosed an anionic gallium heterocycle, valence isoeletronic to NHCs, with strong parallels.[49] Based on these reports, the Arnold group reported in 2007 the first synthesis of a heterobimetallic rare earth metal NHC complex (Equation (6.9)).[50] A salt elimination reaction between the gallium heterocycle and the neodymium dimer **40** yielded **41** at room temperature. This compound was stable in the solid state as well as in solution for several months and decomposed only after heating to 100 °C for 16 h.

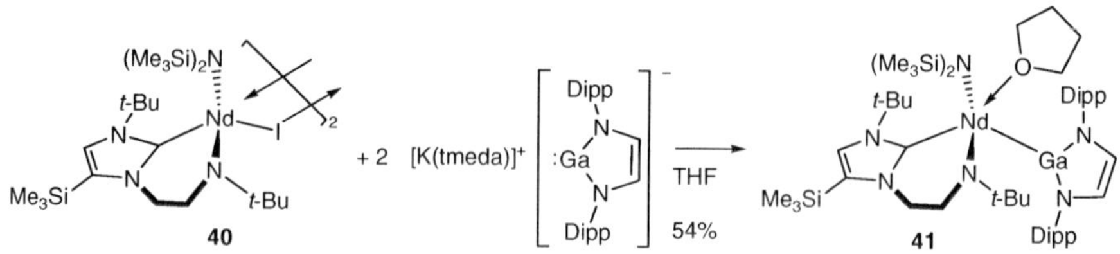

(6.9)

Gade *et al.* synthesized a series of compounds of general formula [MeC{CH_2N(TMS)}$_3$Ti–M(CO)$_2$(Cp)] and [PhC{CH_2N(TMS)}$_3$Ti–M(CO)$_2$(Cp)] (M=Fe, Ru).[51] The precursor K[Fe(CO)$_2$(Cp)] salt was successfully applied by Arnold *et al.* to [(NHC)Nd{N(TMS)$_2$}(μ-I)] **40** in a salt metathesis reaction to yield the first NHC complex with a lanthanoid–transition metal bond **42** (Equation (6.10)).[52]

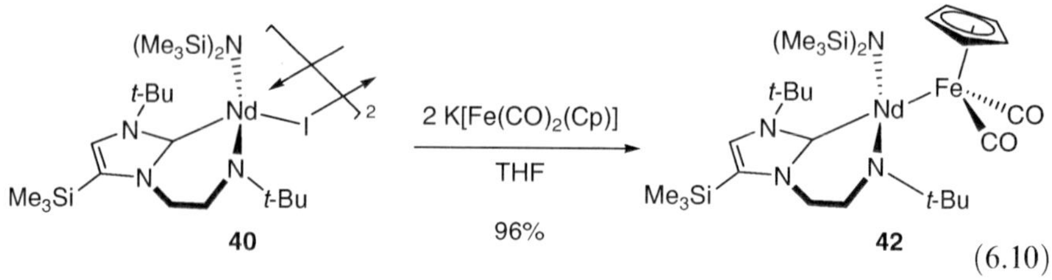

(6.10)

6.3 Structure and Bonding

6.3.1 General Trends

Most ETM complexes with N-heterocyclic carbene ligands studied to date showed no great differences compared to NHC–LTM complexes regarding the metal–carbene bond length or ^{13}C NMR carbene resonance. The typical M–$C_{carbene}$ bond length is between 2.1 Å and 2.4 Å and the corresponding chemical shifts in ^{13}C NMR are found in the range 170 to 220 ppm, where the highest downfield shifts were observed for saturated NHCs as in [(NHC)Y{N(TMS)$_2$}$_2$] **43** ($\delta(C_{carbene}) = 216$ ppm).[53] Extraordinary long metal–carbene bonds were observed in the X-ray analysis of **43** (2.599 Å) or in the tris-alkoxy-bound NHC complex **44** (2.588 Å) (Figure 6.4).

NHC–metal bonds in lanthanoid complexes were expected to be longer than in analogous transition metal complexes according to their greater ionic radii. Typical bond lengths are between 2.4 Å and 2.7 Å, with a general tendency of shortening bond lengths from cerium to lutetium caused by lanthanoid contraction. Only few ^{13}C NMR values for the $C_{carbene}$ have been reported so far, probably due to these compounds' paramagnetism, but the reported values were observed around 200 ppm. An example illustrating the paramagnetic influence in these metals is the europium compound [(IMe)Eu(thd)$_3$] **13**, which

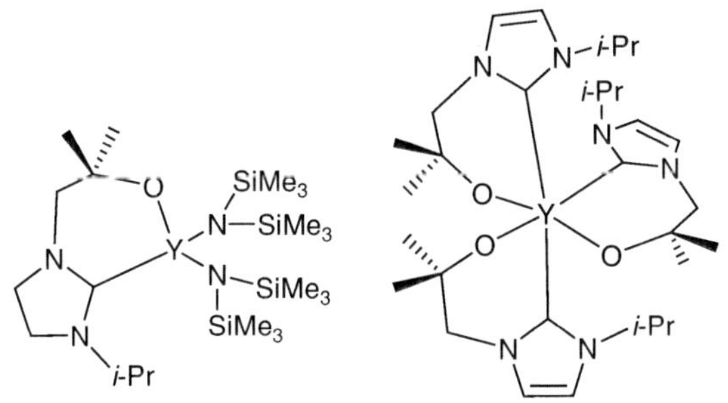

Figure 6.4 Early transition metal complexes with long metal–carbene bonds.

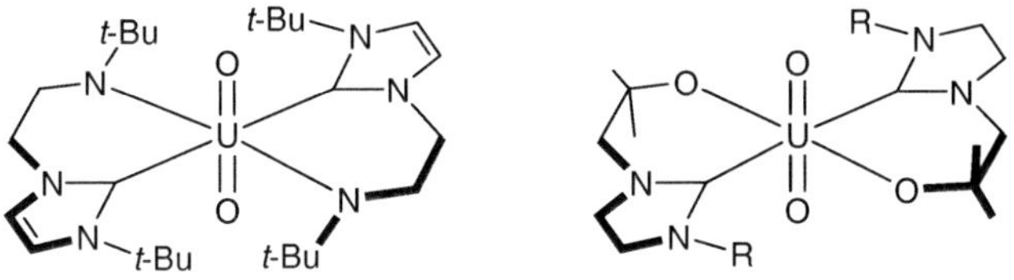

Figure 6.5 Uranium–NHC compounds characterized by NMR (Mes = mesityl).

displayed a paramagnetically shifted resonance of $\delta(C_{carbene}) = 46.5$ ppm in ^{13}C NMR. The longest carbon–metal bond length was observed for the samarocene bis-NHC complex $[(IMe)_2Sm(Cp^*)_2]$ **10**, 2.845(7) Å, which may result from steric crowding.[21]

In the only isolated example of actinoid NHC-ligated compounds, namely uranium complexes, the NHC–U bond lengths were in the range of lanthanoid compounds. The few reported ^{13}C NMR spectra exhibited extremely downfield-shifted carbene resonances. Direct comparison to NHC–lanthanoid complexes was not straightforward, due to the different oxidation states, but, for instance, compound $[(NHC)_2UO_2]$ **25** showed a $C_{carbene}$ resonance at 262.8 ppm and the saturated examples **45** and **46** at 281.6 ppm and 283.6 ppm (Figure 6.5).[37,53a]

6.3.2 Bonding

N-Heterocyclic carbenes in early transition and rare earth metals are generally regarded as pure σ-donor ligands forming stronger metal–ligand bonds than phosphane ligands by extremely polar electrostatic interactions. Nonetheless, the bonding situation in metal NHC complexes is still under discussion,[54] and detailed calculations by Green *et al.* suggested that for electron rich metals the π-interaction between metal and NHC contributed 25–30% to the total orbital interaction.[55] Such interactions should not be possible for the electron-deficient early transition and f-block metals, in particular for those in high oxidation states. Still, many single crystal diffraction studies exhibit unusual close carbene–metal or carbene–ligand distances, which may be explained by π-interactions. Computational studies were conducted to clarify the nature of the carbene–metal bond in these compounds, and selected examples will be discussed in the following section.

6.3.2.1 *Searching for Backbonding in NHC–ETM Complexes*

An early analysis of the bonding situation between ETM and carbenes was conducted by Niehus *et al.*[56] Cationic pseudotetrahedral titanocene **47** showed an in-plane orientation of the ligand compared to the cyclopentadienyl rings in the crystal structure as well as in solution, which was confirmed by the signals corresponding to two diastereotopic isopropyl groups in NMR (Figure 6.6). The results from DFT calculations compared the observed in-plane orientation to a perpendicular orientation of the ligand, but only small energetic differences between the two ligand arrangements could be detected using H– instead of *i*-Pr–. Employing the more sterically demanding isopropyl group expectedly constrained the NHC rotation and led to a clear preference for the in-plane orientation ($\Delta E = 19.7\,\text{kcal mol}^{-1}$). Strong steric repulsions between the isopropyl groups and the cyclopentadienyl rings led to a clear increase in the bond length from 2.319 Å for the in-plane orientation to 2.569 Å in the perpendicular orientation. Mulliken population analysis showed a smaller positive charge on the carbon atom (+0.12 e *vs.* +0.20 e) and a larger one on the metal (+1.00 e *vs.* +0.93 e) for the perpendicular arrangement. At the same time the nitrogen N-centres exhibited no mentionable change, confirming the assumed lack of π-bonding contributions.

The crystal structure of [(IMes)VCl_3(O)] **15** exhibited particularly short distances between the $C_{carbene}$ and the *cis*-ligands: the C(1)–Cl(1) and C(1)–Cl(3) distances were 2.849(2) Å and 2.887(2) Å, respectively, significantly below the sum of the van der Waals radii for carbon and chlorine (3.49 Å).[24] The angles between C(1), V(1) and the adjacent Cl atoms (81.04(6)° and 82.20(6)°) indicated the existence of chloride–carbon interactions (Figure 6.7). Such interactions were not expected as the vacant carbon p-orbital should be energetically

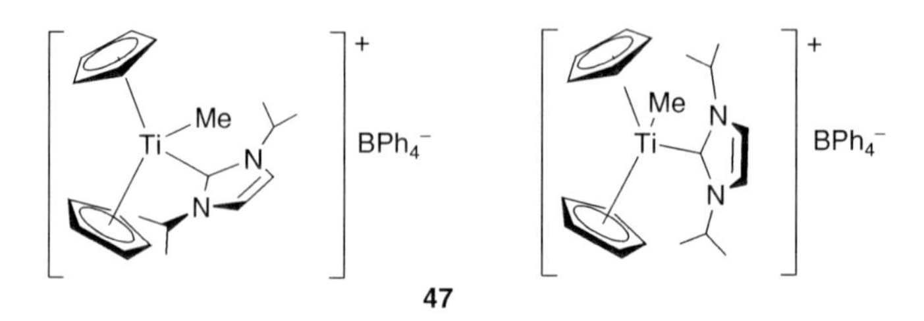

Figure 6.6 Compared in-plane (left) and perpendicular (right) ligand arrangement.

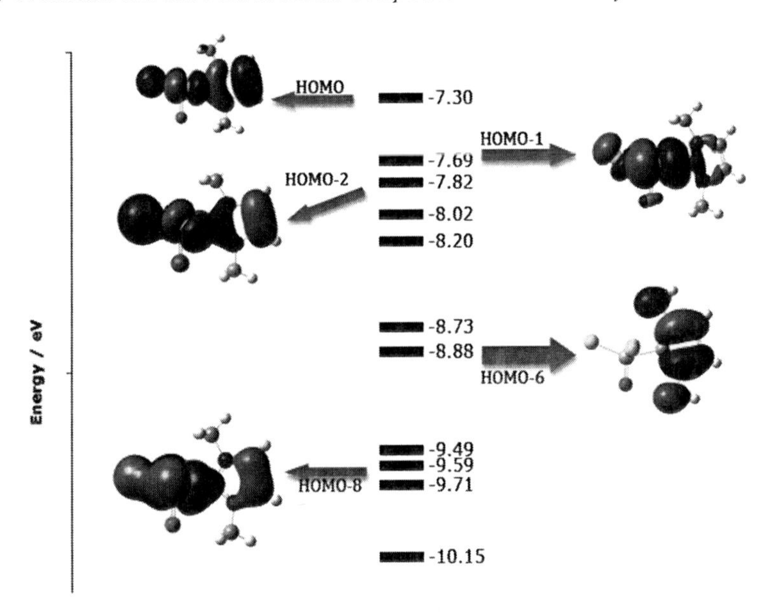

Figure 6.7 Highest occupied molecular orbitals of **15**.

increased by strong N > C π-donation. DFT calculations with a simplified model compound (methyl substituents instead of mesityl) showed similar bond distances and angles, demonstrating the observed short Cl–$C_{carbene}$ distances independent from the steric demand of the NHC substituents or crystal-packing forces. The molecular orbital (MO) comprising the Cl···carbene π-interaction (HOMO-8) is 2.29 eV below the highest occupied molecular orbital (HOMO), expressing the energetic preference of this interaction (Figure 6.7).[57]

The computed molecular orbitals showed bonding overlap between the chlorine lone pairs and the formally vacant p-orbital, which involved significant vanadium d-orbital contributions. This bonding interaction could be regarded as a form of backbonding based on electron density from the ligand lone pairs (Figure 6.8).[24]

Similar chloride–carbene interactions were found in the single crystal structure of [(IMes)$TiCl_2(NMe_2)_2$] **16**, which exhibited a distorted trigonal–bipyramidal geometry with both chlorides in the axial positions. The Cl(1)–Ti(1)–Cl(2) angle in this structure was 163.93(6)°, noticeably bent away from an ideal linear ligand coordination. The Cl···$C_{carbene}$ distances were 3.120(5) Å and 3.103(5) Å, greater than in [(IMes)VCl_3(O)] **15** as a result of the greater ionic radius of the metal centre. Yet again, these distances were well below the sum of van der Waals radii for chlorine and carbon (3.65 Å) and DFT calculations suggested chloride–carbene interactions of similar type as in **15**.[28] This dichloro–Ti–NHC series with $C_{carbene}$···Cl interactions was further expanded by Lorber and Vendier with [(IMes)Ti(=NR)Cl_2($NHMe_2$)] **48**–**50** (Figure 6.9),[29] as well as with various Ti, Zr and Hf complexes.[30]

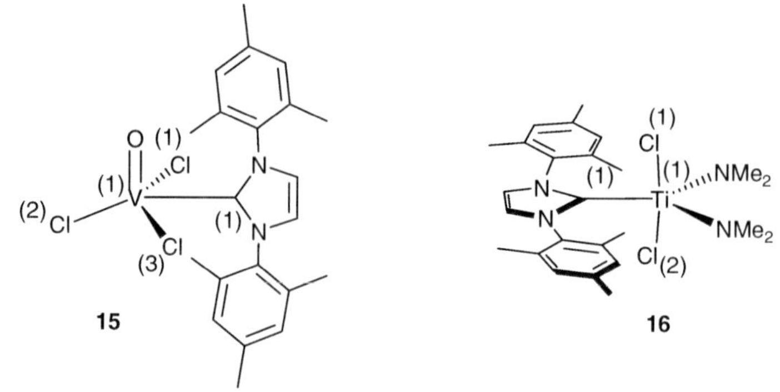

Figure 6.8 Compounds exhibiting strong chloride···NHC interactions.

In the distorted trigonal–bipyramidal crystal structure of [(NHC)TiBr(O*i*-Pr)$_2$] **27** Mungur *et al.* observed an unusual short distance between $C_{carbene}$ and the *cis*-ligands. An interaction was proposed because of the observed bending of one O*i*-Pr and one Br ligand towards the NHC with accordingly short $C_{carbene}$–Br (3.118 Å) and $C_{carbene}$–$O_{cis\text{-alkoxide}}$ (2.765 Å) distances. However, DFT calculations clarified the origin of these interactions to derive from steric interactions, presumably between the lone pairs on the Br and the alkoxide donors.[38,58]

These results showed that π-backbonding contribution by *cis*-ligands was not a general feature of NHC–ETM complexes. Nevertheless, the presence of such interactions in some complexes led to exciting implications: considering these intramolecular interactions between chloride and carbene as a nucleophilic attack by the chloride ligands to the $C_{carbene}$ would indicate that the carbene is an electrophile. This would also support the view that NHC ligands on ETM in high oxidation states should be considered as Fischer-type carbenes.[28,59]

The first experimental analysis of early transition metal backbonding was conducted by Arnold *et al.* when they compared two tris-alkoxy-NHC substituted complexes, which differed only in the metal centre, titanium or yttrium.[53b] As mentioned above, the yttrium compound [(NHC)$_3$Y] **44** showed unusual long metal–carbene bonds (averaging 2.588 Å), whereas the three titanium–carbene distances measured 2.28 Å in [(NHC)$_3$Ti] **51**. Correcting these distances by the ionic radii led to slightly shorter titanium–NHC distance (1.461 Å *vs.* 1.548 Å) (Figure 6.9). Based on these observations, it was tempting to presume backbonding contribution from the titanium(III)-d^1-metal accounting for the shorter metal–carbene bond. However, closer inspection of the DFT data showed no significant backbonding contribution. A comparison of the calculated total electron density of the two complexes suggested, that the observed M–$C_{carbene}$ bond contraction was derived from the smaller and therefore more polarizing size of the titanium centre (Figure 6.10).[60]

The investigations conducted to date could not evidence direct metal-to-carbene π-interactions for early transition metal NHC ligated complexes, but this question is still under debate.

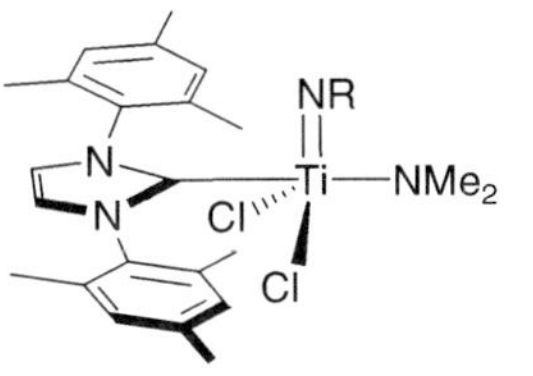

R = Dipp **48**
2,6-Cl_2-C_6H_3 **49**
t-Bu **50**

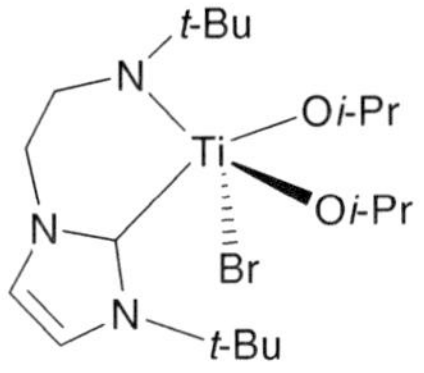

27

Figure 6.9 Complexes with short ligand–carbene distances.

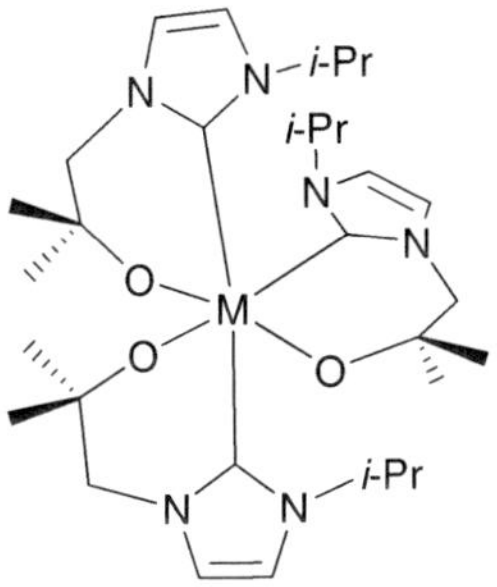

[M]	NHC–M distance (average)	Corrected distance (average)
M = Y, **44**	2.588 Å	1.548 Å
M = Ti, **51**	2.279 Å	1.461 Å

Figure 6.10 Compared yttrium and titanium complexes.

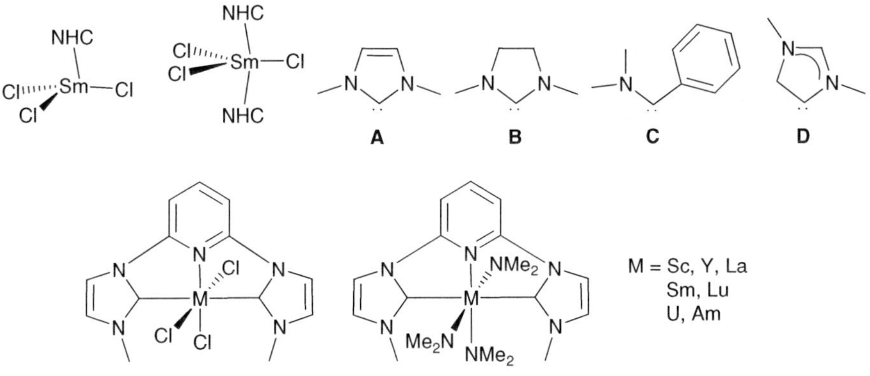

M = Sc, Y, La
Sm, Lu
U, Am

Figure 6.11 Model compounds for examination of the bonding mode.

6.3.2.2 Searching for Backbonding in NHC–Rare Earth Metal Complexes

Maron and Bourissou conducted a detailed computational investigation of the mono- and bis-adducts of $SmCl_3$ and the model carbene ligands **A**–**D** (Figure 6.11).[61] Optimized geometries of the mono-adducts deviated slightly from planarity with Cl–Sm–Cl angles ranging from 350.3° to 355.1°, indicating a distorted tetrahedral geometry. The bis-adducts of **A**, **B** and **D** displayed trigonal–bipyramidal geometry with perpendicular arrangement of the two carbene ligands for **A** and **B** while the abnormal bound carbenes **D** showed a co-planar alignment. The bis-adduct of less symmetric carbene **C** led to a significant distorted geometry with a $C_{carbene}$–Sm–$C_{carbene}$ angle of 145.2°. The bond lengths did not vary significantly (2.51 Å–2.60 Å for the mono-species and 2.63 Å–2.70 Å for the bis-adduct) and were comparable to other known Sm compounds. The calculated orbitals showed that the carbene–Sm bond was a result of carbene-to-metal σ-donation with marginal chlorine-to-carbene back-donation interactions, whereas direct metal-to-carbene π-donation was not observed.[24]

The same authors also analyzed the bonding properties of the complexes deriving from tridentate ligand 2,6-bis(methylimidazol-2-ylidyl)pyridine and Group 3, lanthanoid and actinoid trichlorides, triamides or the bare M^{III} cations (Figure 6.11). These octahedral compounds displayed stronger calculated coordination energies for more electron-deficient metal fragments. As no $C_{carbene}$ > M π-donation and no metal-to-carbene back-donations were calculated, this observation clarified that mainly ionic bonding occurred in these complexes. Only ligand interactions were found for the tris-amido species of La, Sm and Am.[62]

By addition of ^{Me}IMe to the low-valent uranium(III) precursors [U(^{Ad}ArO)$_3$(tacn)] (Ad = adamantyl, tacn = 1,4,7-triazacyclononane) or [U(N(SiMe$_3$)$_2$)$_3$], the first low valent NHC–uranium species [(^{Me}IMe)U(^{Ad}ArO)$_3$] **52** and [(^{Me}IMe)U(N(SiMe$_3$)$_2$)$_3$] **53** were synthesized (Figure 6.12).[63,64] Although the highly reactive uranium(III) was heavily shielded in the first starting compound by the tacn and adamantyl substituents, these were still flexible enough to open up and allow the NHC coordination. Both paramagnetic compounds had M–$C_{carbene}$ bond lengths within the usual dimensions, longer in **52** (2.789(14) Å) than in **53** (2.672(5) Å). The nature of the latter uranium–carbene interaction was examined by computational analysis. Inspection of the Single Occupied Molecular Orbital (SOMO) of the second most energetic electron in the system clearly indicated

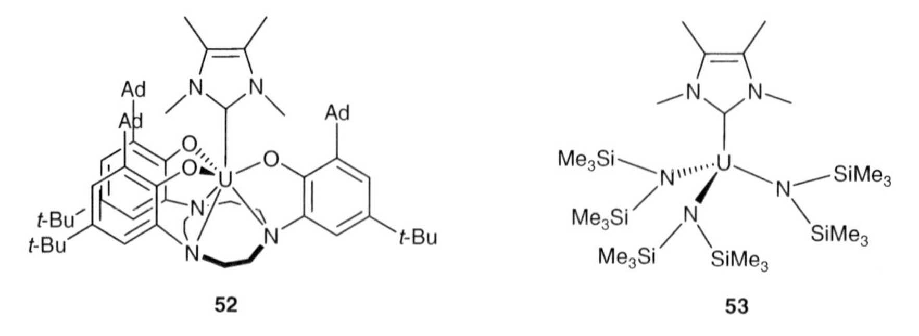

Figure 6.12 Low valent NHC–U complexes.

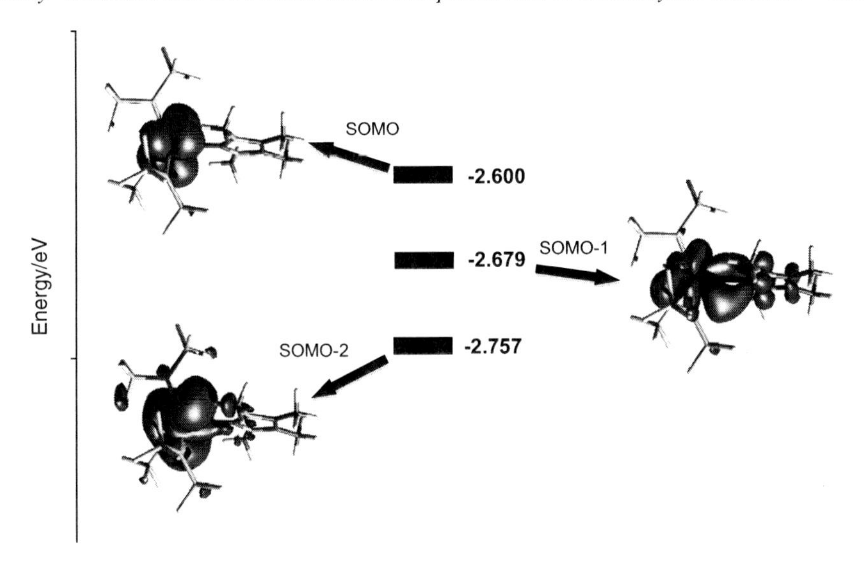

Figure 6.13 MO-diagram of the three highest SOMOs of **53**.

NHC coordination *via* π-bonding interactions between f-type uranium orbitals and π-type orbitals of the carbene ligand (Figure 6.13).[63]

The observed short $C_{carbene}\cdots Cl$ distances in $[(IPr)_2UCl_4]$ were theoretically examined by Liddle, and the results suggested that these short contacts resulted from aryl···chloride steric repulsions rather than from carbene chloride interactions.[65]

Summarizing, metal-to-carbene π-interactions are generally not observed for early transition and rare earth NHC complexes, the only example so far being uranium compound **53**. This type of backbonding, however, seems restricted to highly electron rich rare earth metal centres such as the uranium(III) ion.

6.3.3 Distorted Geometries

The out-of-plane angle (also referred to as pitch) is defined as the angle between $M–C_{carbene}$ and an imaginary C_{2v}-axis through the carbene bisecting the imidazol-2-ylidene plane. The in-plane angle measures the lateral tilt of the ligand (Figure 6.14).

In a crystallographic study of complexes with slightly twisted NHC ligands, Arnold *et al.* elucidated trends for the corresponding distortion angles.[66] Complex $[(^{Me}IMe)_2Sm(Cp^*)_2]$ **10** showed extremely elongated bonds (M–C = 2.837(7) Å and 2.845(7) Å), which could result from steric crowding around the metal centre. Europium complex **13** had an out-of-plane angle of 2.5° and in-plane distortion angle of 9.7°.[21] Herrmann *et al.* reported the bis-NHC adduct **54**, which possessed an out-of-plane angle of 9.3° and an in-plane bending of 0.9°. This geometric distortion may be caused by agostic yttrium-η^2(Si–H)

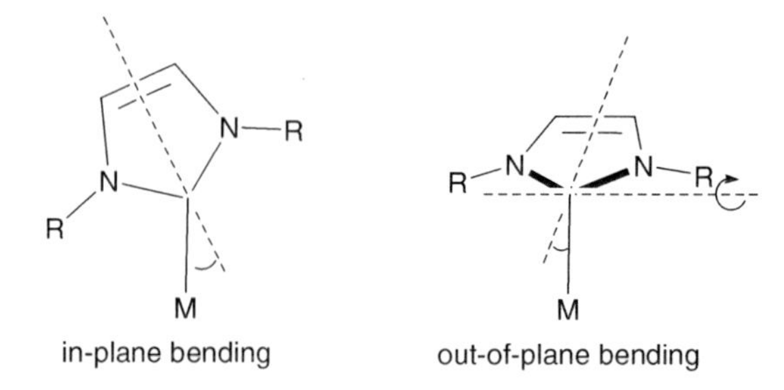

Figure 6.14 Angles used to describe the ligands geometric distortion.

interactions. The corresponding mono-adduct **55** displayed similar out-of-plane bending with an angle of 8°.[67] Clear steric influences were shown by Mungur *et al.* when they synthesized dioxo-uranium complexes **25** and **56** with similar amido-bound-NHC ligands.[37] The steric demand of the *tert*-butyl groups caused an important geometric distortion in **25**, whereas the less bulky mesityl groups produced little distortion (Table 6.1). The $C_{carbene}$–U bond distances, however, were both 2.64 Å, implying very similar bonding interactions. This assumption was supported by IR spectroscopy and both asymmetric uranyl stretch vibration values were almost identical (**25**: $\nu_{IR} = 929\,cm^{-1}$, **56**: $\nu_{IR} = 933\,cm^{-1}$).

These results implied that N-heterocyclic carbenes bound *via* polarized electrostatic interactions to the electropositive metals and that in these compounds bond strengths did not depend on the bending angle imposed by steric factors.[37,66]

6.4 Reactivity

Although early transition and rare earth metal complexes with N-heterocyclic ligands are no longer laboratory curiosities, studies on their reactivity are still extremely rare.

The strength of the metal–amide bond in yttrium complex **57** (Scheme 6.6) allowed Arnold *et al.* to monitor the reactivity of the NHC–metal bond *via* $^{1}J_{YC}$ coupling in ^{13}C NMR.[68] Triphenylphosphane or trimethylamine *N*-oxide did not displace the carbene, showing again that the NHC moiety has stronger donating properties than most other donors. Only tmeda and triphenylphosphane oxide successfully dissociated the carbene to form the corresponding metal complexes with pendant NHCs.

The reduction of compound **57** or its samarium analogue **58** by potassium naphtalenide afforded the bimetallic dimers **59** and **60** with K-bound abnormal NHCs, which also bound to yttrium or samarium *via* the regioselectively deprotonated backbone (Scheme 6.7).[69] Reaction of **58** with potassium-intercalated graphite (KC_8) in dimethoxyethane (DME) yielded the methoxy-bridged dimer **61** as a product of DME cleavage. Quenching the yttrium dimer

Table 6.1 Observed bending angles of structures exhibiting distorted geometries.

Complex	*Out-of-plane bending* / *In-plane bending*	*Complex*	*Out-of-plane bending* / *In-plane bending*
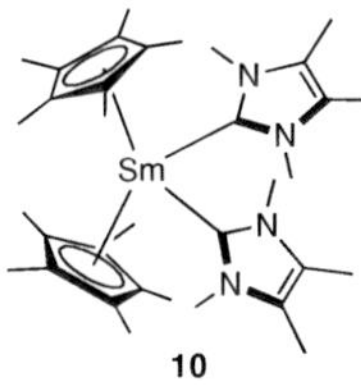 **10**	21.4° / –	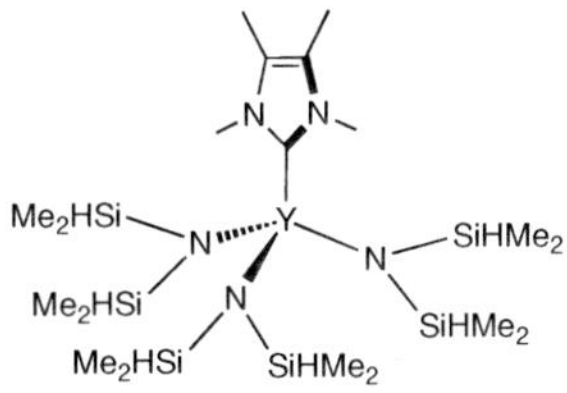 **55**	8° / –
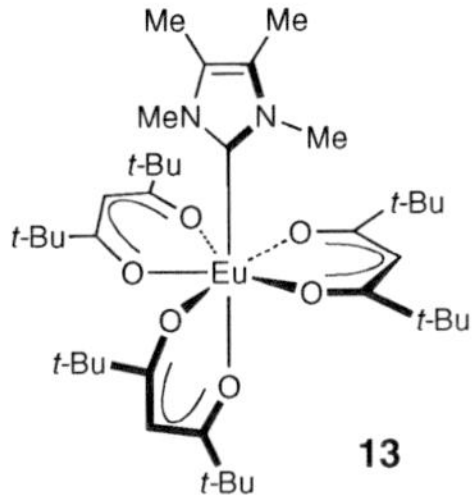 **13**	2.5° / 9.7°	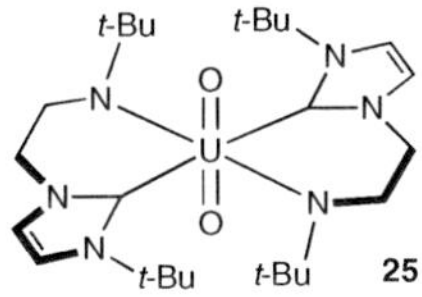 **25**	23° / 17°
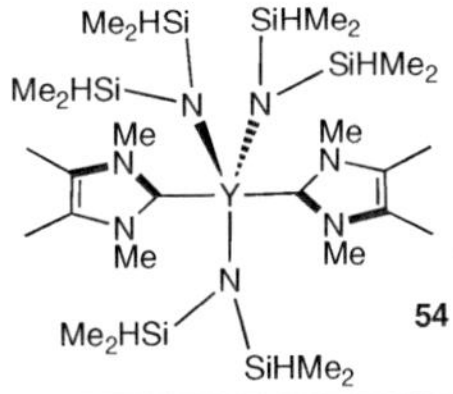 **54**	9.3° / 0.9°	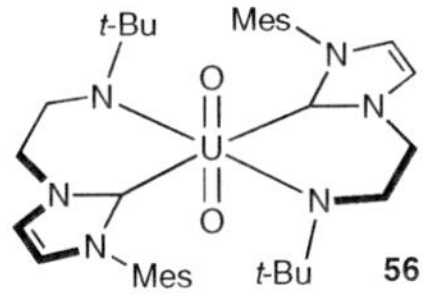 **56**	9° / 3°

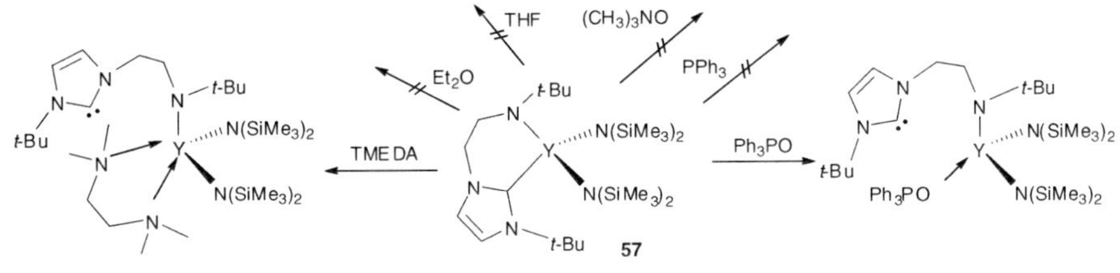

Scheme 6.6 Labilization reactions of yttrium amido-NHC compound **57**.

59 with electrophilic Me_3SiCl in THF afforded monomeric **62** *via* backbone silylation and simultaneous KCl elimination.

Another example of regioselective C4 silylation was observed with the neodymium analogue to **57** and **58**, compound **63**. Addition of Me_3SiI to **63** yielded the backbone-functionalized dimer **40** (Equation (6.11)), which showed interesting reactivity with other metals (see Section 6.2.3.).[52,70] For this regioselective silylation, a reaction mechanism starting with nucleophilic

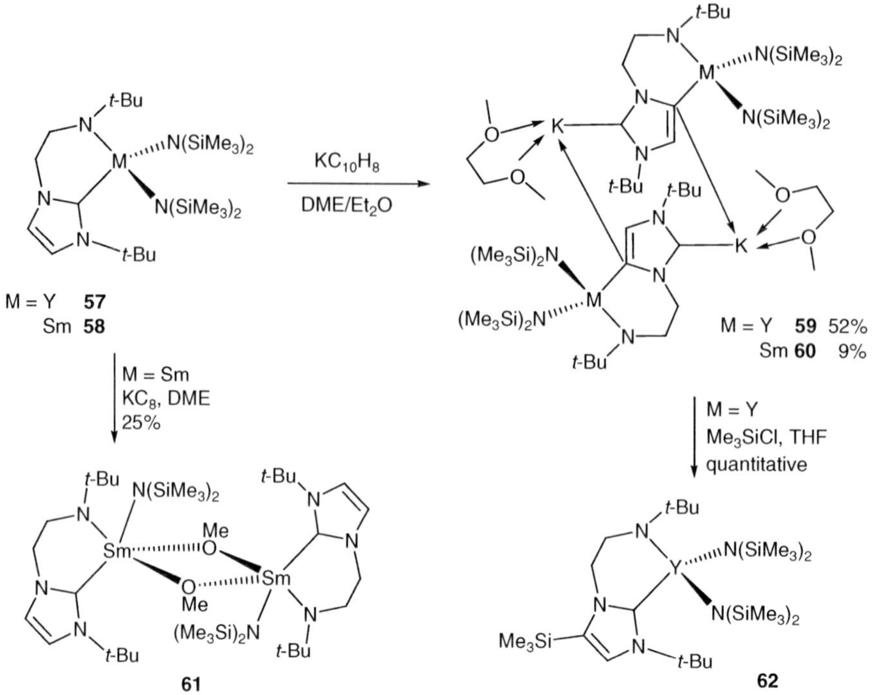

Scheme 6.7 Dimerization reactions *via* regioselective backbone deprotonation.

substitution of one silylamide was proposed. This silylamide would immediately deprotonate the NHC to generate a C4 carbanion, which then would be quenched by the trimethylsilyl cation.

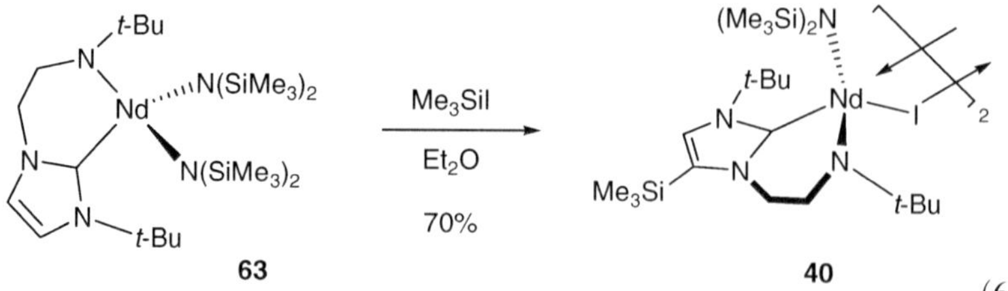

(6.11)

Maron and Bourissou observed in their calculations with different metal complexes bearing a tridentate 2,6-bis(methylimidazol-2-ylidyl)pyridine ligand an intrinsic preference of the ligand for La over U and for Sm over Am (predicted for bare ions). Such differentiation was also calculated for the tris-amido fragments, where lanthanum was preferred over uranium.[62] In contrast to the expected reactivity, Mehdoui *et al.* were able to synthesize two analogous cerium(III) **64** and uranium(III) **65** compounds [(^{Me}IMe)M(Cp*)$_2$I] by addition of ^{Me}IMe to an equimolar solution of [M(Cp*)$_2$I] (Equation (6.12)).[71]

The uranium(III) adduct **65** was preferentially formed at room temperature (molar ration 80:20) as well as at – 60 °C (molar ratio 90:10).

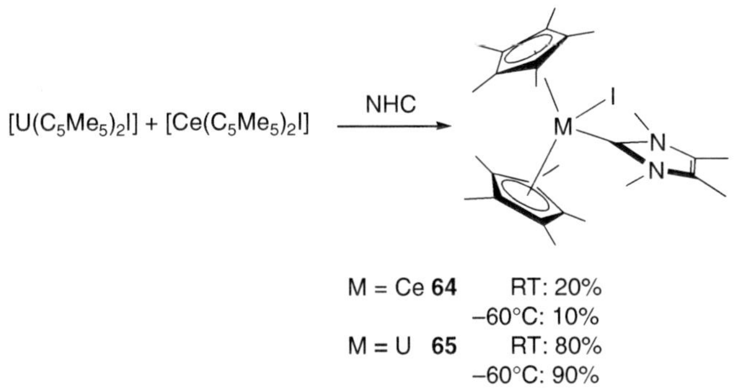

$[U(C_5Me_5)_2I] + [Ce(C_5Me_5)_2I] \xrightarrow{NHC}$

M = Ce **64** RT: 20%
–60°C: 10%
M = U **65** RT: 80%
–60°C: 90%

(6.12)

The same behaviour was observed for the addition of $C_3Me_4N_2$ to the triscyclopentadienyl compounds $[U(C_5H_4t\text{-}Bu)_3]$ and $[Ce(C_5H_4t\text{-}Bu)_3]$ (Equation (6.13)). Comparison of the four crystal structures showed 0.03 Å shorter NHC–U bonds, although the ionic radius of uranium(III) cation exceeds cerium(III) cation by 0.02 Å.[60a] These deviations together with the results from the competition experiments indicated stronger, more covalent interactions between the electron-rich actinoid and the higher affinity of the π-accepting NHC for the 5f ion over the 4f ion.

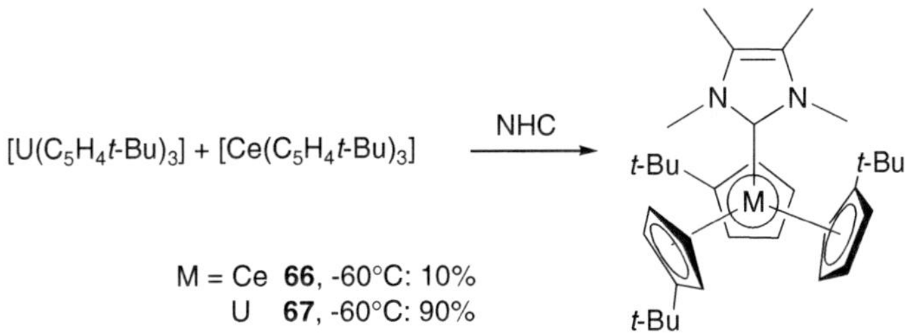

$[U(C_5H_4t\text{-}Bu)_3] + [Ce(C_5H_4t\text{-}Bu)_3] \xrightarrow{NHC}$

M = Ce **66**, -60°C: 10%
U **67**, -60°C: 90%

(6.13)

6.5 Catalytic Applications

Early transition and rare earth metal NHC complexes have primarily been applied in polymerization reactions. The ligand is often tethered in a bidentate fashion and catalysis initiated by the initial ligand dissociation of the NHC carbene moiety. Although a relatively limited amount of research has been carried out in this area, these preliminary investigations have established the utility of early transition and rare earth metal NHC complexes as catalysts in oligomerization and polymerization reactions. For an overview of alkene oligomerization and polymerization reactions catalyzed by metal NHC catalysts including those of mid to late transition metals readers are directed to a recent review by McGuinness.[72]

Table 6.2 Complexes for the polymerization of ethene (modified methylaluminoxane = MMAO, dry methyl-aluminoxane = DMAO, co-polymeric isobutyl methylaluminoxane = coMAO, triisobutylaluminium = TIBAL).

Complex	*Co-catalyst [equiv]*	*Reported activity*	*Ref.*
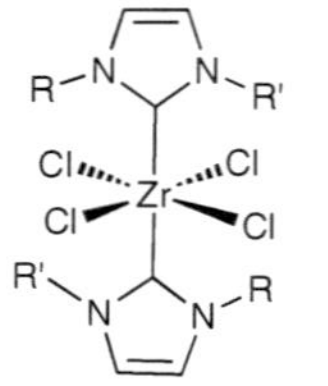 R–N, N–R', Cl, Cl, Zr, Cl, Cl, R', N, N–R R = R' = *i*-Pr R = Me, R' = Mes	MAO	7–75 kg mol^{-1} bar^{-1} h^{-1}	74
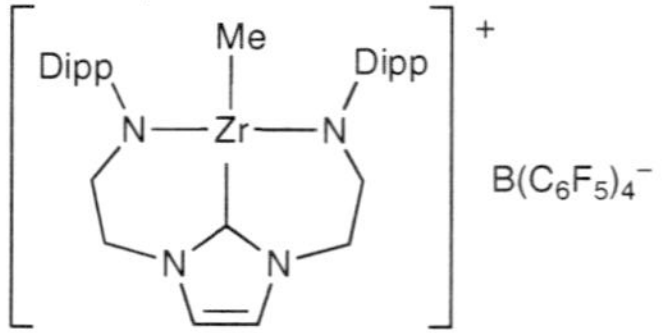 [Me, Dipp, Dipp, N–Zr–N, N, N]$^+$ $B(C_6F_5)_4^-$	—	125 kg mol^{-1} bar^{-1} h^{-1}	42
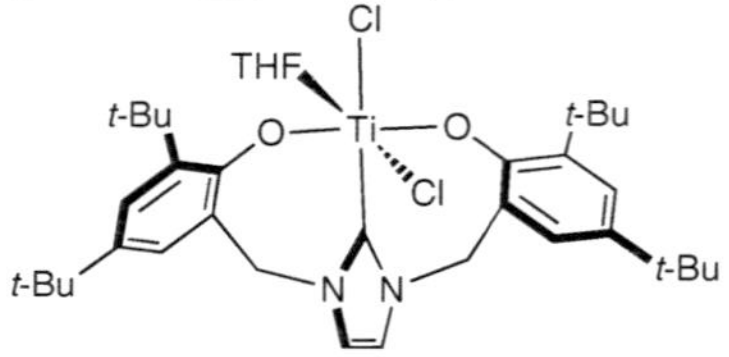 Cl, THF, *t*-Bu, O–Ti–O, *t*-Bu, Cl, *t*-Bu, N, N, *t*-Bu	MMAO MAO	290 kg mol^{-1} bar^{-1} h^{-1} 100 kg mol^{-1} bar^{-1} h^{-1}	75

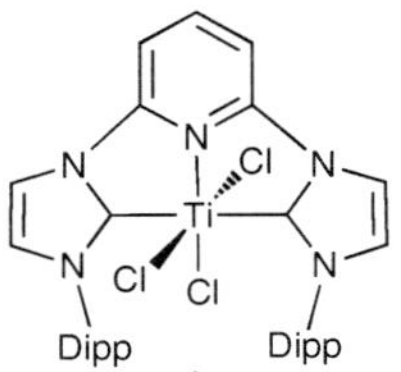

	MAO[500]	791 g product (mmol of M)$^{-1}$ bar^{-1} h^{-1}, TOF: 28 195 h^{-1}	13*b*
	MAO [1000]	652 g product (mmol of M)$^{-1}$ bar^{-1} h^{-1}, TOF: 23 240 h^{-1}	
	DMAO[500]	286 g product (mmol of M)$^{-1}$ bar^{-1} h^{-1}, TOF: 10 195 h^{-1}	
	coMAO[500]	694 g product (mmol of M)$^{-1}$ bar^{-1} h^{-1}, TOF: 24 740 h^{-1}	
	MAO[500]	1280 g product (mmol of M)$^{-1}$ bar^{-1} h^{-1} ,TOF: 45 625 h^{-1}	13*b*
	MAO[1000]	907 g product (mmol of M)$^{-1}$ bar^{-1} h^{-1} , TOF: 32 330 h^{-1}	
	coMAO[500]	1446 g product (mmol of M)$^{-1}$ bar^{-1} h^{-1} , TOF: 51 545 h^{-1}	
	[Et_2AlCl][500]	278 g product (mmol of M)$^{-1}$ bar^{-1} h^{-1}, TOF: 9 910 h^{-1}	

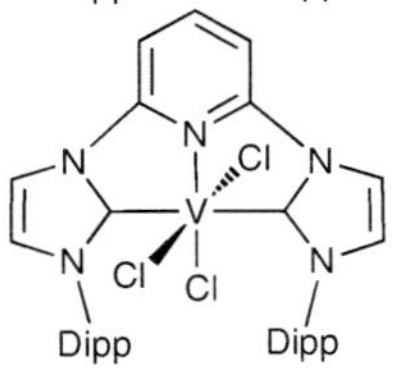

6.5.1 Polymerization of Ethene

To date, the polymerization of ethene is the most studied catalytic application in this context. Due to NHC dissociation, the overwhelming majority of the systems studied have a chelating bidentate NHC moiety. Active complexes and activities reported are shown in Table 6.2. However, comparison of these systems with an optimized standard system with methylaluminoxane (MAO) as co-catalyst for homogenous Ziegler–Natta polymerization proved the depicted compound activities as low: Kaminsky and Steiger reported for $[Zr(Cp)_2Cl_2]$/MAO an activity of 40 000 kg(PE) g(Zr)$^{-1}$ h^{-1} at 70 °C under an ethene pressure of 8 bar (456 000 kg(PE) mol^{-1} bar^{-1} h^{-1}).[73]

6.5.2 Polymerization of Isoprene

Early transition and rare earth metal NHC complexes tested in the polymerization of isoprene are shown in Figure 6.15.

Even though $[(^{Me}IiPr)Sm(C_5H_4t\text{-}Bu)_2(i\text{-}Pr)]$ **68** was used as a stereospecific catalyst for the polymerization of isoprene, it was proposed that the role of the carbene ligand was simply to dissociate from the complex to open a free coordination site.[74] Therefore, further optimization of these systems through NHC tuning is expected to meet very limited success.

For this transformation, the use of bidentate NHCs was more promising. Shown in Figure 6.15, complexes **69** and **70** of Sc, Y, Ho and Lu were screened and all of them, except the Sc complexes, were found active upon activation with $[Ph_3C][B(C_6F_5)_4]/Al(t\text{-}Bu)_3$. The fluorenyl complexes were more active and selective than the indenyl analogues. Comparing the catalytic activity in a metal series revealed that larger atomic radii corresponded to greater catalytic activity (Ho > Y > Lu ≫ Sc). This behaviour was observed for both series.[40,75] Closely related $[(NHC)'Lu(CH_2SiMe_3)_2]$, in which the fluorenyl was *t*-Bu-substituted, illustrated a rare example utilizing a rare earth metal for the co-polymerization of ethene/norbornene.[76]

Other polymerization reactions were reported with these families of complexes, and $[(IiPr)Sm(Cp^*)_2]$ and $[(IiPr)Sm(C_5H_4t\text{-}Bu)(C_3H_5)]$ were proposed to experience loss of an NHC ligand to initiate the polymerization of methylacrylate.[77] $[(NHC)Zr(CH_3)]$, active for the polymerization of ethene, was also found slightly active for the polymerization of 1-hexene upon treatment with $[Ph_3C][B(C_6F_5)_4]$. A yield of 4% atactic poly-1-hexene was observed after 1 h with 0.2 mol% catalyst loading in chlorobenzene at room temperature. Longer reaction times did not result in higher yields, which was attributed to a short-lived active species in solution.[42]

6.5.3 Ring-opening Polymerizations

Early transition and rare earth metal NHC complexes were also employed for the polymerization of D- and L-lactide. These systems are particularly noteworthy as they exploit the utility of the NHC to function not only as a

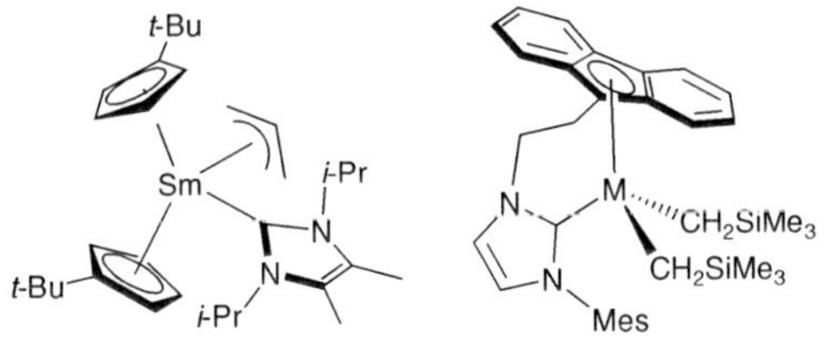

Figure 6.15 Catalysts for isoprene polymerization.

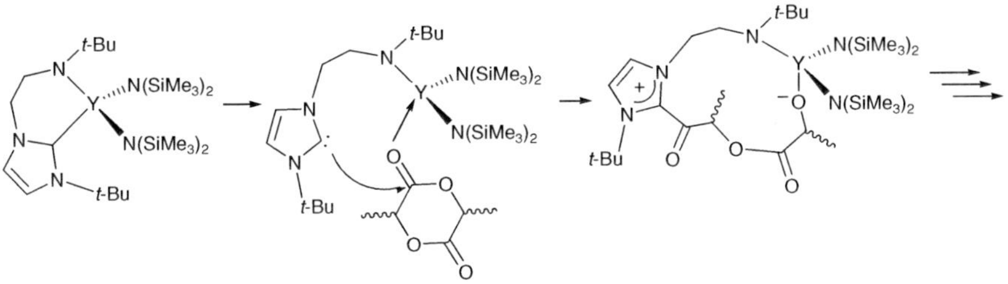

Scheme 6.8 Proposed mechanism for ring-opening polymerization.

dissociating ligand but also to play a role in bifunctional catalysis for the ring-opening polymerization producing a polymer with a low polydispersity and high heterotacticity. Both Ti and Y systems were catalytically active with the yttrium one reacting more quickly and yielding higher molecular weight polymers. The yttrium centre was proposed to act as the Lewis acid and the free NHC as the Lewis base (Scheme 6.8).[32b] It should be mentioned that also free NHCs are known organocatalysts for this transformation.[78] Again, due to the ligand dissociation, further optimization of this system is limited.

6.6 Conclusions and Outlook

Early transition metal NHC ligated complexes were among the first ones reported for these ligands as well as several rare earth metal complexes. To date, pioneering investigations have paved the way to a wealth of complexes from nearly any metal and equipped scientists with a great repertory of approaches to future compounds. Although investigations concerning bonding and compound reactivity were carried out, the knowledge is still limited here. Further studies on catalytic applications remain to be conducted as well, since to date they are limited to polymerization reactions. In particular, small molecule activation could provide further insights not only into catalytic activity but also reactivity of these under-explored complexes. Also metal complexes bearing labile carbene ligands fixed by an anionic anchor are particularly promising as much as heterobimetalic complexes, which could lead to the development of bifunctional catalysts.

References

1. (a) F. E. Hahn and M. C. Jahnke, *Angew. Chem., Int. Ed.*, 2008, **47**, 3122–3172; (b) E. Peris and R. H. Crabtree, *Coord. Chem. Rev.*, 2004, **248**, 2239–2246; (c) W. A. Herrmann, *Angew. Chem., Int. Ed.*, 2002, **41**, 1290–1309; (d) D. Bourissou, O. Guerret, F. P. Gabbaï and G. Bertrand, *Chem. Rev.*, 2000, **100**, 39–92.
2. (a) K. Öfele, *J. Organomet. Chem.*, 1968, **12**, 42–43; (b) K. Öfele and M. Herberhold, *Angew. Chem., Int. Ed. Engl.*, 1970, **9**, 739–740.
3. H.-W. Wanzlick and H.-J. Schonherr, *Angew. Chem., Int. Ed. Engl.*, 1968, **7**, 141–142.
4. (a) J. Cardin, B. Çetinkaya and M. F. Lappert, *Chem. Rev.*, 1972, **72**, 545–574; (b) M. F. Lappert, *J. Organomet. Chem.*, 1988, **358**, 185–214.
5. A. Igau, H. Grutzmacher, A. Baceiredo and G. Bertrand, *J. Am. Chem. Soc.*, 1988, **110**, 6463–6466.
6. A. J. Arduengo III, M. Kline, J. C. Calabrese and F. Davidson, *J. Am. Chem. Soc.*, 1991, **113**, 9704–9705.
7. A. J. Arduengo III, H. V. R. Dias, R. L. Harlow and M. Kline, *J. Am. Chem. Soc.*, 1992, **114**, 5530–5534.
8. (a) S. Díez-González, N. Marion and S. P. Nolan, *Chem. Rev.*, 2009, **109**, 3612–3676; (b) M. Poyatos, J. A. Mata and E. Peris, *Chem. Rev.*, 2009, **109**, 3677–3707; (c) W. A. Herrmann, M. Elison, J. Fischer, C. Köcher and G. R. J. Artus, *Angew. Chem., Int. Ed. Engl.*, 1995, **34**, 2371–2374.
9. M. Regitz, *Angew. Chem., Int. Ed. Engl.*, 1996, **35**, 725–728.
10. (a) M. D. Fryzuk, T. S. Haddad, D. J. Berg and S. J. Ret, *Pure Appl. Chem.*, 1991, **63**, 845–850; (b) L. H. Gade, *Chem. Commun.*, 2000, 173–181.
11. S. T. Liddle, I. S. Edworthy and P. L. Arnold, *Chem. Soc. Rev.*, 2007, **36**, 1732–1744.
12. (a) K. Öfele, W. A. Herrmann, D. Mihalios, M. Elison, E. Herdtweck, W. Scherer and J. Mink, *J. Organomet. Chem.*, 1993, **459**, 177–184; (b) W. A. Herrmann, K. Öfele, M. Elison, F. E. Kuehn and P. W. Roesky, *J. Organomet. Chem.*, 1994, **480**, C7–C9; (c) F. E. Hahn, L. Wittenbecher, D. L. Van and R. Fröhlich, *Angew. Chem., Int. Ed.*, 2000, **39**, 541–544.
13. (a) D. S. McGuinness, V. C. Gibson, D. F. Wass and J. W. Steed, *J. Am. Chem. Soc.*, 2003, **125**, 12716–12717; (b) D. S. McGuinness, V. C. Gibson and J. W. Stee, *Organometallics*, 2004, **23**, 6288–6292; (c) D. S. McGuinness, J. A. Suttil, M. G. Gardiner and N. W. Davi, *Organometallics*, 2008, **27**, 4238–4247.

14. K. H. Wedepohl, *Geochim. Cosmochim. Acta*, 1995, **59**, 1217–1232.
15. G. Wilkinson and J. M. Birmingham, *J. Am. Chem. Soc.*, 1954, **76**, 6210.
16. W. J. Evans, *Inorg. Chem.*, 2007, **46**, 3435–3449.
17. (a) A. J. Arduengo III, H. V. R. Dias, F. Davidson and R. L. Harlow, *J. Organomet. Chem.*, 1993, **462**, 13–18; (b) W. A. Herrmann, O. Runte and G. Artus, *J. Organomet. Chem.*, 1995, **501**, C1–C4; (c) W. A. Herrmann and C. Köcher, *Angew. Chem., Int. Ed. Engl.*, 1997, **36**, 2162–2187.
18. P. L. Arnold and I. J. Casely, *Chem. Rev.*, 2009, **109**, 3599–3611.
19. H. Schumann, M. Glanz, J. Winterfeld, H. Hemling, N. Kuhn and T. Kratz, *Angew. Chem., Int. Ed. Engl.*, 1994, **33**, 1733–1734.
20. H. Schumann, M. Glanz, J. Winterfeld, H. Hemling, N. Kuhn and T. Kratz, *Chem. Ber.*, 1994, **127**, 2369–2372.
21. A. J. Arduengo III, M. Tamm, S. J. McLain, J. C. Calabrese, F. Davidson and W. J. Marshall, *J. Am. Chem. Soc.*, 1994, **116**, 7927–7928.
22. W. J. Oldham Jr., S. M. Oldham, B. L. Scott, K. D. Abney, W. H. Smith and D. A. Costa, *Chem. Commun.*, 2001, 1348–1349.
23. J.-C. Daran, Y. Jeannin, G. Constant and R. Morancho, *Acta Crystallogr. Sect. B*, 1975, **31**, 1833–1837.
24. C. D. Abernethy, G. M. Codd, M. D. Spicer and M. K. Taylor, *J. Am. Chem. Soc.*, 2003, **125**, 1128–1129.
25. (a) N. Kuhn, T. Kratz, D. Bläser and R. Boese, *Inorg. Chim. Acta*, 1995, **238**, 179–181; (b) G. B. Nikiforov, H. W. Roesky, P. G. Jones, J. Magull, A. Ringe and R. B. Oswald, *Inorg. Chem.*, 2008, **47**, 2171–2179.
26. F. E. Hahn, T. V. Fehren and R. Fröhlich, *Z. Naturforsch.*, 2004, **59b**, 348–350.
27. M. Niehues, G. Kehr, R. Fröhlich and G. Erker, *Z. Naturforsch.*, 2003, **58b**, 1005–1008.
28. P. Shukla, J. A. Johnson, D. Vidovic, A. H. Cowley and C. D. Abernethy, *Chem. Commun.*, 2004, 360–361.
29. C. Lorber and V. Vendier, *Organometallics*, 2008, **27**, 2774–2783.
30. C. Lorber and L. Vendier, *Dalton Trans.*, 2009, 6972–6984.
31. W. Zhang and K. Nomura, *Organometallics*, 2008, **27**, 6400–6402.
32. (a) P. L. Arnold, M. Rodden, K. M. Davis, A. C. Scarisbrick, A. J. Blake and C. Wilson, *Chem. Commun.*, 2004, 1612–1613; (b) D. Patel, S. T. Liddle, S. A. Mungur, M. Rodden, A. J. Blake and P. L. Arnold, *Chem. Commun.*, 2006, 1124–1126.
33. P. L. Arnold, A. J. Blake and C. Wilson, *Chem.–Eur. J.*, 2005, **11**, 6095–6099.
34. Z. G. Wang, H. M. Sun, H. S. Yao, Q. Shen and Y. Zhang, *Organometallics*, 2006, **25**, 4436–4438.
35. Z. G. Wang, H. M. Sun, H. S. Yao, Y. M. Yao and Y. Z. Q. Shen, *J. Organomet. Chem.*, 2006, **691**, 3383–3390.
36. (a) H. Aihara, T. Matsuo and H. Kawaguchi, *Chem. Commun.*, 2003, 2204–2205; (b) D. Zhang, H. Aihara, T. Watanabe, T. Matsuo and H. Kawaguchi, *J. Organomet. Chem.*, 2007, **692**, 234–242.

37. S. A. Mungur, S. T. Liddle, C. Wilson, M. J. Sarsfield and P. L. Arnold, *Chem. Commun.*, 2004, 2738–2739.
38. S. A. Mungur, A. J. Blake, C. Wilson, J. McMaster and P. L. Arnold, *Organometallics*, 2006, **25**, 1861–1867.
39. S. P. Downing and A. A. Danopoulos, *Organometallics*, 2006, **25**, 1337–1340.
40. B. Wang, D. Wang, D. Cui, W. Gao, T. Tang, X. Chen and X. Jing, *Organometallics*, 2007, **26**, 3167–3172.
41. L. P. Spencer, S. Winston and M. D. Fryzuk, *Organometallics*, 2004, **23**, 3372–3374.
42. L. P. Spencer and M. D. Fryzuk, *J. Organomet. Chem.*, 2005, **690**, 5788–5803.
43. L. P. Spencer, C. Beddie, M. B. Hall and M. D. Fryzuk, *J. Am. Chem. Soc.*, 2006, **128**, 12531–12543.
44. A. A. Danopoulos, A. A. D. Tulloch, S. Winston and G. Eastham, *Dalton Trans.*, 2003, 1009–1015.
45. D. Pugh, J. A. Wright, S. Freeman and A. A. Danopoulos, *Dalton Trans.*, 2006, 775–782.
46. R. J. Rubio, G. T. S. Andavan, E. B. Bauer, T. K. Hollis, J. Cho, F. S. Tham and B. J. Donnadieu, *J. Organomet. Chem.*, 2005, **690**, 5353–5364.
47. S. T. Liddle and D. P. Mills, *Dalton Trans.*, 2009, 5592–5605.
48. W. A. Herrmann, M. Denk, J. Behm, W. Scherer, F. R. Klingan, H. Bock, B. Solouki and M. Wagner, *Angew. Chem., Int. Ed. Engl.*, 1992, **31**, 1485–1488.
49. (a) M. Denk, R. Lennon, R. Hayashi, R. West, A. V. Belyakov, H. P. Verne, A. Haaland, M. Wagner and N. Metzler, *J. Am. Chem. Soc.*, 1994, **116**, 2691–2692; (b) B. Gerhus, M. F. Lappert, J. Heinicke, R. Boese and D. Bläser, *J. Chem. Soc., Chem. Commun.*, 1995, 1931–1932; (c) R. J. Baker, C. Jones and J. A. Platts, *J. Am. Chem. Soc.*, 2003, **125**, 10534–10535; (d) R. J. Baker and C. Jones, *Coord. Chem. Rev.*, 2005, **249**, 1857–1869.
50. P. L. Arnold, S. T. Liddle, J. McMaster, C. Jones and D. P. Mills, *J. Am. Chem. Soc.*, 2007, **129**, 5360–5361.
51. S. Friedrich, H. Memmler, L. H. Gade, W.-S. Li, I. J. Scowen, M. Mcpartlin and C. E. Housecroft, *Inorg. Chem.*, 1996, **35**, 2433–2441.
52. P. L. Arnold, J. McMaster and S. T. Liddle, *Chem. Commun.*, 2009, 818–820.
53. (a) P. L. Arnold, I. J. Casely, Z. R. Turner and C. D. Carmichael, *Chem.–Eur. J.*, 2008, **14**, 10415–10422; (b) P. L. Arnold, S. Zlatogorsky, N. A. Jones, C. D. Carmichael, S. T. Liddle, A. J. Blake and C. Wilson, *Inorg. Chem.*, 2008, **47**, 9042–9049.
54. S. Díez-González and S. P. Nolan, *Coord. Chem. Rev.*, 2007, **251**, 874–883.
55. J. C. Green and B. J. Herbert, *Dalton Trans.*, 2005, 1214–1220.
56. M. Niehues, G. Erker, G. Kehr, P. Schwab, R. Fröhlich, O. Blacque and H. Berke, *Organometallics*, 2009, **21**, 2905–2911.

57. Based on (http://pubs.acs.org/doi/suppl/10.1021/ja0276321/suppl_file/ja0276321-2_s1.pdf), *J. Am. Chem. Soc.*, 2003, **125**, 1128–1129 (ref. 24). The MOs were calculated with B1123LYP/1126-1131 + G** for all elements including V. The energies of the MOs and their shapes were obtained by Mulliken population analysis.
58. K. B. Wiberg, *Tetrahedron*, 1968, **24**, 1083–1096.
59. T. E. Taylor and M. B. Hall, *J. Am. Chem. Soc.*, 1984, **106**, 1576–1584.
60. (a) R. D. Shannon, *Acta Crystallogr. Sect. A*, 1976, **32**, 751–767; (b) N. A. Jones, S. T. Liddle, C. Wilson and P. L. Arnold, *Organometallics*, 2007, **26**, 755–757.
61. L. Maron and D. Bourissou, *Organometallics*, 2007, **26**, 1100–1103.
62. L. Maron and D. Bourissou, *Organometallics*, 2009, **28**, 3686–3690.
63. H. Nakai, X. Hu, L. N. Zakharov, A. L. Rheingold and K. Meyer, *Inorg. Chem.*, 2004, **43**, 855–857.
64. J. L. Stewart and R. A. Andersen, *Polyhedron*, 1998, **17**, 953–958.
65. B. M. Gardner, J. McMaster and S. T. Liddle, *Dalton Trans.*, 2009, 6924–6926.
66. P. L. Arnold and S. T. Liddle, *Chem. Commun.*, 2006, 3959–3971.
67. W. A. Herrmann, F. C. Munck, G. R. J. Artus, O. Runte and R. Anwander, *Organometallics*, 1997, **16**, 682–688.
68. P. L. Arnold, S. A. Mungur, A. J. Blake and C. Wilson, *Angew. Chem., Int. Ed.*, 2003, **42**, 5981–5984.
69. P. L. Arnold and S. T. Liddle, *Organometallics*, 2006, **25**, 1485–1491.
70. P. L. Arnold and S. T. Liddle, *Chem. Commun.*, 2005, 5638–5640.
71. T. Mehdoui, J.-C. Berthet, P. Thuery and M. Ephritikhine, *Chem. Commun.*, 2005, 2860–2862.
72. D. McGuinness, *Dalton Trans.*, 2009, 6915–6923.
73. W. Kaminsky, *J. Polym. Sci. Part A: Polym. Chem.*, 2004, **42**, 3911–3921.
74. D. Baudry-Barbier, N. Andre, A. Dormond, C. Pardes, P. Richard, M. Visseaux and C. J. Zhu, *Eur. J. Inorg. Chem*, 1998, 1721–1727.
75. B. Wang, D. Cui and K. Lv, *Macromolecules*, 2008, **41**, 1983–1988.
76. B. Wang, T. Tang, Y. Lia and D. Cui, *Dalton Trans.*, 2009, 8963–8969.
77. M. Glanz, S. Dechert, H. Schumann, D. Wolff and J. Springer, *Z. Anorg. Allg. Chem.*, 2000, **626**, 2467–2477.
78. (a) N. E. Kamber, W. Jeong, R. M. Waymouth, R. C. Pratt, B. G. G. Lohmeijer and J. L. Hedrick, *Chem. Rev.*, 2007, **107**, 5813–5840; (b) S. Díez-González, S. P. Nolan and N. Marion, *Angew. Chem., Int. Ed.*, 2007, **46**, 2988–3000.

CHAPTER 7

NHC–Iron, Ruthenium and Osmium Complexes in Catalysis

LIONEL DELAUDE* AND ALBERT DEMONCEAU

Laboratory of Macromolecular Chemistry and Organic Catalysis, Institut de Chimie (B6a), Université de Liège, Sart-Tilman, 4000 Liège, Belgium

7.1 Introduction

In this Chapter, the catalytic applications of organometallic species—either pre-formed or generated *in situ*—based on Group 8 transition metals and N-heterocyclic carbene (NHC) ligands are surveyed. Thus far, the coordination chemistry of iron with NHCs has not been extensively investigated, and only a few reports on the use of NHC–Fe complexes in catalysis are available. In view of the great potentials of these systems in organic synthesis, we strove to list them all. Contrastingly, the chemistry of NHC–Ru complexes has reached an unprecedented level of maturity, thanks to the relentless research efforts thrown into the development of second- and third-generation olefin metathesis catalysts. Because this is an intensively reviewed area, only a timely update and the state of the art will be provided regarding the ruthenium-catalysed metathesis reactions. Other Ru-promoted carbon skeletal transformations such as cyclopropanation, allylation, or cycloisomerisation will be reviewed in more detail. Oxidation and reduction processes with NHC-containing ruthenium complexes will not be described here as they are discussed separately in Chapters 12 and 13, respectively. Finally, the catalytic activity of NHC-based osmium complexes in olefin metathesis and hydrogen transfer reactions will be examined.

RSC Catalysis Series No. 6
N-Heterocyclic Carbenes: From Laboratory Curiosities to Efficient Synthetic Tools
Edited by Silvia Díez-González
© Royal Society of Chemistry 2011
Published by the Royal Society of Chemistry, www.rsc.org

7.2 NHC–Iron-catalysed Reactions

Because iron is a cheap, abundant, non-toxic, and environmentally benign metal, it has attracted a lot of attention in catalysis over the last few years.[1] In particular, iron-based catalytic systems showed great promise for the formation of carbon–carbon bonds *via* cross-coupling reactions.[2] In most cases, however, simple inorganic salts or coordination compounds of Fe^{II} and Fe^{III} are employed as catalyst precursors. Recourse to NHC ligands to fine-tune the catalytic properties of iron active centres is still scarce in organic synthesis and was documented for only a handful of reactions. Other important transformations mediated by NHC–Fe species were found in the related fields of biocatalysis and organometallic synthesis, and will be discussed first.

7.2.1 Organometallic and Electrochemical Reactions

The use of NHC–iron species in a catalytic organometallic transformation leading to the formation of metal–metal bonds was brought to light in 2009 by Lavallo and Grubbs.[3] When the non-carbonyl Fe^0 complex $[Fe(cot)_2]$ (cot = cyclooctatetraene) was treated with SIMes, a stoichiometric reaction ensued, and a high yield of crystalline compound **1** was obtained (Equation (7.1)). Use of bulkier SIPr resulted in a different reaction path, which afforded the tri-iron cluster $[Fe_3(cot)_3]$ (**2**) in a catalytic fashion (Equation (7.2)).

2 [Fe(cot)$_2$] —(SIMes: Mes–N N–Mes, 79%)→ 0.5 **1** [Fe^0 / Fe^I dimer]

(7.1)

3 [Fe(cot)$_2$] —(SIPr: Dipp–N N–Dipp, 95%)→ **2**

Dipp = 2,6-diisopropylphenyl

(7.2)

A possible mechanism was proposed for these two transformations (Scheme 7.1). In both cases, the initial step would be the substitution of a single cot ligand from the iron starting material by the SIMes or SIPr carbene. The highly reactive [(NHC)Fe(cot)] intermediate would then react with a second equivalent of $[Fe(cot)_2]$ and undergo an internal electron transfer leading to the formation of bimetallic species **3** featuring a Fe–Fe bond and a ligand-centred

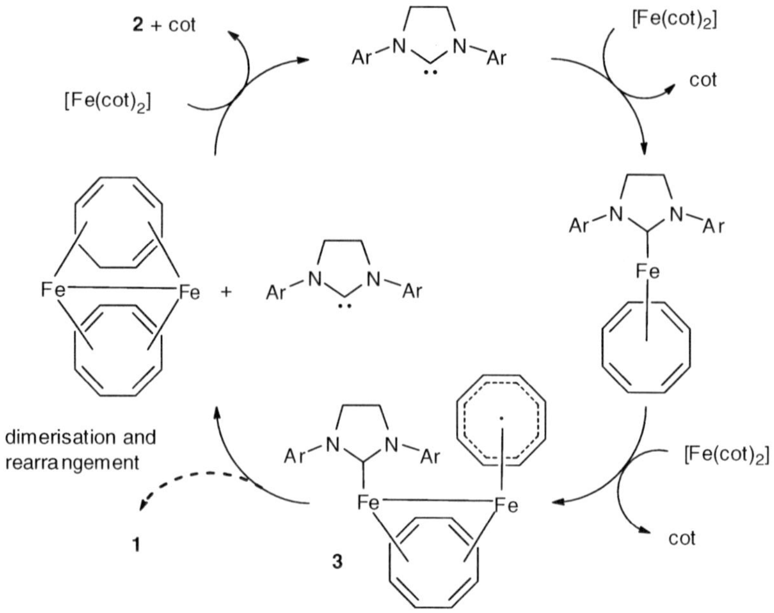

Scheme 7.1 Possible mechanism for the formation of complexes **1** and **2** from $[Fe(cot)_2]$.

radical. At this point, divergent routes may be envisaged depending on the nature of the NHC substituents. Stoichiometric formation of the tetrametallic product **1** bearing the SIMes ligand might involve a radical coupling between two bimetallic intermediates **3** followed by rearrangement. With the more sterically demanding SIPr ligand, the dimerisation process would be blocked and the unstable **3** would extrude its NHC moiety to form the $[Fe_2(cot)_2]$ complex. Further reaction with a third equivalent of $[Fe(cot)_2]$ would afford the tri-iron cluster **2** and complete the catalytic cycle.

In the field of biocatalysis, NHC–Fe species were designed to model the active site of [FeFe]-hydrogenases, an important class of metalloenzymes that promote the reversible reduction of protons into dihydrogen.[4,5] For instance, homobimetallic complex **4** catalysed the reduction of protons at –2.07 V *versus* the ferrocenium/ferrocene redox couple (Fc^+/Fc^0).[5] The proposed mechanism would start with complex **4** accepting two electrons to form dianion $\mathbf{4}^{2-}$ (Scheme 7.2). Next, the tricarbonyliron moiety in this electron-rich dianion would be protonated to afford an intermediate hydride species $(\mathbf{4}\text{–H})^-$. The iron atom bearing the IMes ligand of this monoanion then might be protonated before an intramolecular electron transfer would regenerate catalyst **4** with evolution of dihydrogen. With the prospect that hydrogen may become a major energy vector as the century progresses, recourse to natural and artificial hydrogenases might provide cheap electrocatalysts for H_2 production.[6]

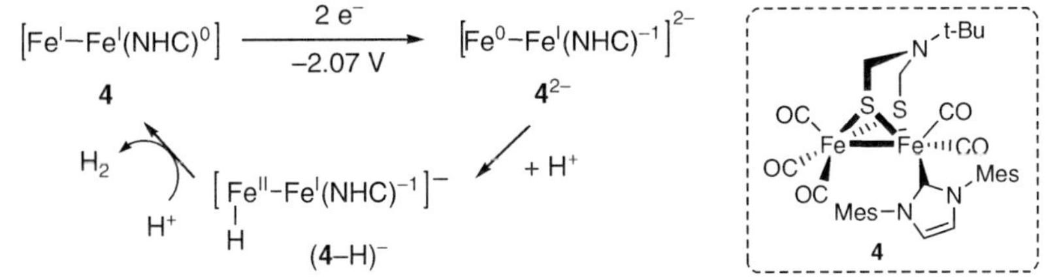

Scheme 7.2 Possible mechanism for the electrocatalytic reduction of protons by **4**.

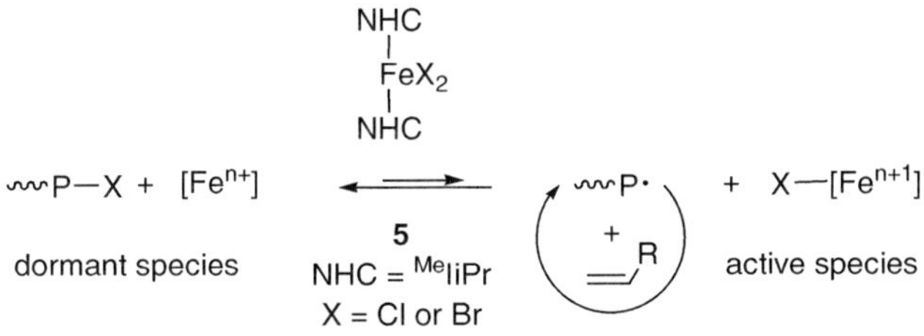

Scheme 7.3 ATRP of vinyl monomers with NHC–Fe complexes **5**.

7.2.2 Polymerisation Reactions

In 2000, Louie and Grubbs reported that iron(II) halides possessing two strongly donating ^{Me}IiPr ligands (**5**) were highly efficient catalysts for the atom transfer radical polymerisation (ATRP) of styrene and methyl methacrylate (Scheme 7.3).[7] Indeed, the reaction rates recorded with complexes **5** either pre-formed or generated *in situ* rivalled those obtained previously with copper-based systems, while ensuring very narrow molecular weight distributions. The high Lewis basicity of the imidazolylidene ligands may lower the redox potential of the iron(II) centres (thereby easing the halide abstraction from dormant polymer chains), while stabilising the iron(III) species and therefore enhancing the rapid exchange of halides between dormant and active polymer chain ends.

In 2004, Gibson and co-workers synthesized a small range of Fe^{II} and Fe^{III} complexes bearing a bis(NHC)-pyridine tridentate ligand, **6** and **7** (Figure 7.1), but they were inactive in ethylene oligomerisation and polymerisation.[8] In 2006, two NHC–phenoxide chelates (**8**) were prepared by Shen *et al.*[9] One of them was an active initiator for the ring-opening polymerisation of ε-caprolactone, albeit with poor control over the chain length and molecular weight distribution.

7.2.3 Cyclisation Reactions

In 2005, Hilt and co-workers disclosed the facile intermolecular ring expansion of styrene oxide with several dienes, acrylates, enynes or styrenes under iron catalysis to afford polysubstituted tetrahydrofuran derivatives in a highly chemo- and regioselective fashion.[10] The authors used $FeCl_2$ mixed with a

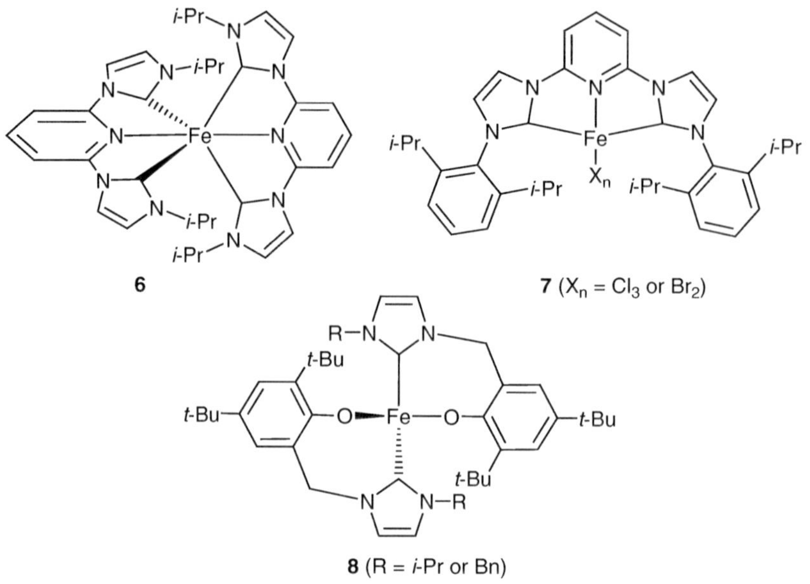

6

7 (X_n = Cl_3 or Br_2)

8 (R = *i*-Pr or Bn)

Figure 7.1 Chelating NHC–Fe complexes tested as catalysts for polymerisation reactions.

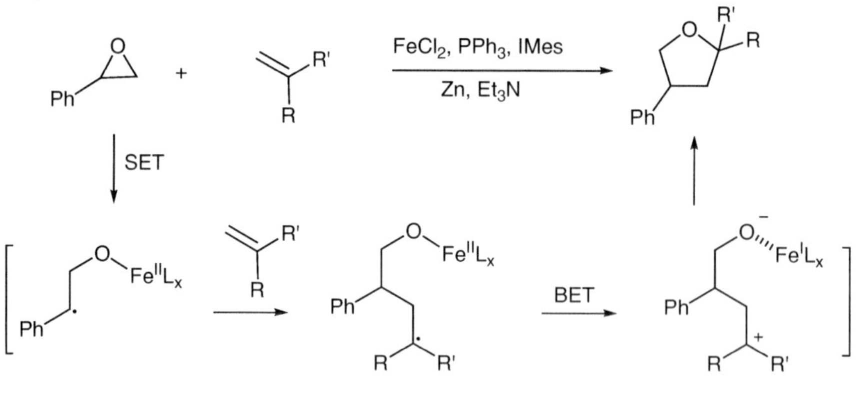

$FeCl_2$, PPh_3, IMes

Zn, Et_3N

SET

BET

Scheme 7.4 Iron-catalysed ring expansion of styrene oxide.

phosphine and a NHC in the presence of a reducing agent such as zinc powder (Scheme 7.4). Triethylamine was also added to prevent the undesired dimerisation of the epoxide under the reaction conditions. The authors proposed a radical mechanism initiated by a single electron transfer (SET) from an iron(I) species, generated *in situ via* Zn reduction of a Fe^{II} complex, to the epoxide. Subsequent addition to the alkene and back electron transfer (BET) regenerating the Fe^{I} catalyst would then afford a zwitterionic intermediate prone to cyclisation.

Recently, Okamoto *et al.* reported that NHC–Fe species derived from $FeCl_2$ or $FeCl_3$ by *in situ* reduction with zinc powder in the presence of IMes or IPr were efficient catalysts for the intramolecular cyclotrimerisation of triynes into annulated benzenes (Equation (7.3)).[11] Although no well-defined complex could be isolated from this mixture, its activity was preserved for a few days when kept under inert atmosphere.

$FeCl_3$, IPr
Zn, THF

R, R' = H, $SiMe_3$, Aryl, Alkyl, CH_2OH, CH_2OBn
X, Y = O, $C(CO_2Et)_2$

(7.3)

7.2.4 C–C Bond Forming Reactions

Despite the growing interest in NHCs for Pd- and Ni-promoted cross-coupling reactions (see Chapters 9 and 10, respectively), there are only two reports making use of these ligands in iron-catalysed cross-coupling.[12] In 2006, Bedford and co-workers showed that NHCs were efficient supporting ligands for the Kumada–Corriu cross-coupling of primary and secondary alkyl halides with 4-tolylmagnesium bromide.[13] In addition to a well-defined complex **7** ($X_n = Br_2$, see Figure 7.1), simple mixtures of iron(III) chloride with two equivalents of an imidazolin-2-ylidene species generated *in situ* from the corresponding azolium salt (SItBu, SICy, SIMes) or pentafluorobenzene adduct (SIMes and SIPr) led to highly efficient catalytic systems that matched or even out-performed known systems. The authors proposed a highly simplified radical mechanism with an active iron species in the oxidation state *n* first reacting with the alkyl halide *via* SET to form an alkyl radical possibly associated with the iron centre. Next, transmetalation with the Grignard reagent would give an iron–aryl complex, which might be attacked by the alkyl radical to afford the final product and regenerate the catalyst (Scheme 7.5).

Very recently, Nakamura *et al.* reported the remarkable activity of SIPr in the iron-mediated Kumada–Corriu cross-coupling of arylmagnesium bromides with aryl chlorides.[14] Thus, a system composed of $FeF_3{\cdot}3H_2O$ and SIPr·HCl proved efficient at relatively low catalyst loadings (3–5 mol%) and allowed for the coupling of both electron-rich and electron-poor aryl chlorides in good to excellent yields. It should be noted that cobalt and nickel fluorides exhibited a higher reactivity in the coupling of aryl bromides or iodides, and therefore nicely complemented the iron catalysts. On the basis of stoichiometric control experiments and DFT calculations, the outstanding catalytic effect of the fluoride counterions compared to other halides could be ascribed to the

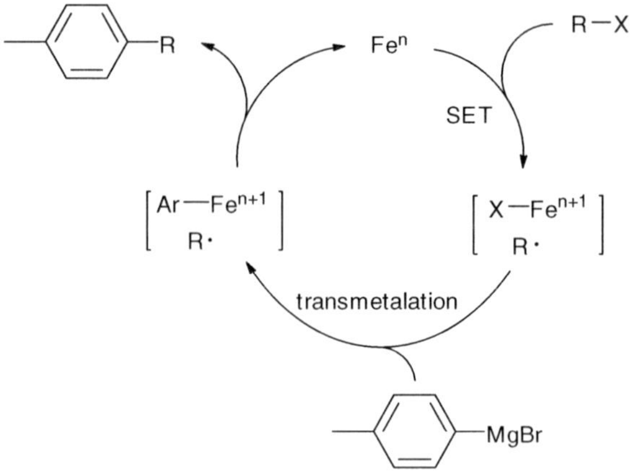

Scheme 7.5 Plausible mechanism for iron-catalysed Kumada cross-coupling.

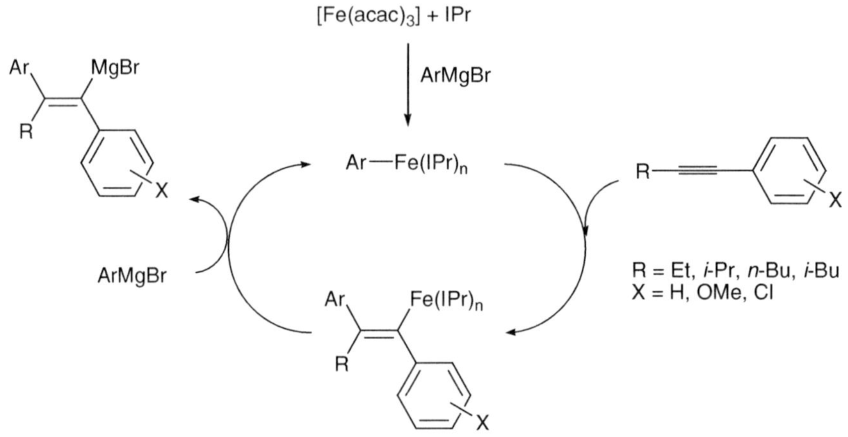

Scheme 7.6 Possible mechanism for the iron-catalysed arylmagnesiation of aryl(alkyl)acetylenes.

formation of a high-valent heteroleptic metalate [$ArMF_2$][MgBr] as a key intermediate in the catalytic cycle.

In the related field of C–C bond formation *via* carbometalation reactions, a series of arylmagnesium bromides were successfully employed by Shirakawa and Hayashi for the arylmagnesiation of aryl(alkyl)acetylenes in the presence of [$Fe(acac)_3$] (acac = acetylacetonato) and IPr (Scheme 7.6).[15] The alkenylmagnesium products obtained were further transformed into tetrasubstituted alkenes by subsequent treatment with electrophiles. As for the mechanism, coordination of the NHC ligand might stabilise low valent iron intermediates, allowing for the transmetalation of an alkenyliron intermediate with the arylmagnesium reagent to proceed.

Last but not least, Plietker and co-workers reported in 2008 an unprecedented dichotomy in allylic substitution reactions catalysed by NHC–Fe systems.[16] Hence, with $[Bu_4N][Fe(CO)_3(NO)]$ the regioselectivity of the addition of malonates onto allyl carbonates could be controlled by simply changing the NHC and the base (Equation (7.4)). Indeed, switching from SItBu to SIPr allowed for a remarkable inversion of the selectivity attributed to the possible intermediacy of a π-allyl complex instead of a σ-allyl one when the steric pressure of the nitrogen substituents on the NHC was moved away from the metal centre. A fast σ-π-σ isomerisation process took place with the SIPr ligand, which might set the stage for the future development of an iron-catalysed, asymmetric, dynamic–kinetic allylic substitution.

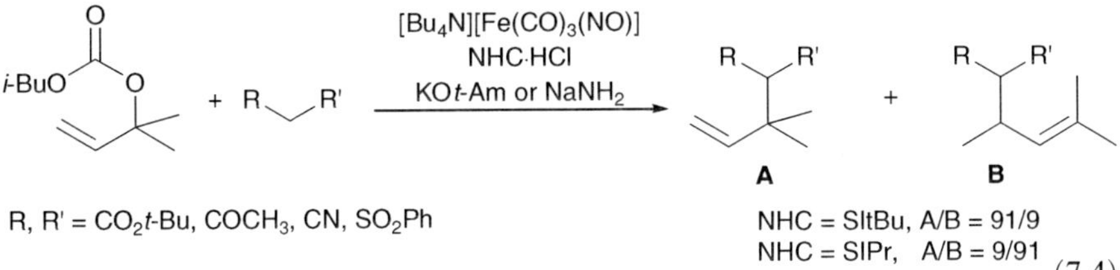

(7.4)

7.3 NHC–Ruthenium-catalysed Reactions

Ruthenium's supremacy in the carbene chemistry of Group 8 elements is a direct consequence of the tremendous interest raised by NHC–Ru complexes as catalysts for olefin metathesis.[17] Indeed, the synergy of a late transition metal tolerant of a wide variety of functional groups, together with a class of ligands whose physical and chemical properties are easily modulated to tailor the activity,[18,19] stability,[20,21] water-solubility,[22,23] recoverability,[24] or latency[25] of the resulting catalytic species translated into an unprecedented success story of modern synthetic chemistry. Yet, the ability of ruthenium complexes to promote carbon–carbon bond formation goes well beyond olefin metathesis.[26–28] Advances in these fields still remain in the shade of olefin metathesis, but they are likely to gain more visibility in a near future. Tandem reactions that combine a metathetical step with another ruthenium-promoted transformation in a single process have also benefited from the vast amount of knowledge gathered from studying and optimising in great depth individual catalytic cycles.[29–31]

7.3.1 Metathesis Reactions

7.3.1.1 Scope and Mechanism

Thanks to the development of well-defined molybdenum and ruthenium alkylidene catalysts initiated by Schrock[32] and Grubbs[33] in the late 1990s, alkene metathesis became a key methodology in organic synthesis[34] and in polymer chemistry.[35] Indeed, judicious combinations of C=C double bond formation and cleavage provided a whole gamut of valuable carbon skeletal rearrangements, which include *inter alia* cross-metathesis (CM), ring-opening cross-metathesis (ROCM), ring-opening metathesis polymerisation (ROMP), ring-closing metathesis (RCM), or acyclic diene metathesis polymerisation (ADMET).[36] The successful extension of these reactions to carbon–carbon triple bonds also afforded enyne[37] and alkyne metathesis,[38] while recent advances in asymmetric catalysis now render feasible the desymmetrisation of prochiral polyolefins or the kinetic resolution of racemates *via* alkene metathesis.[39]

Many researchers put forward proposals to explain how metathesis could take place,[40] but it was Hérisson and Chauvin who proposed that in the catalyst the metal was bound to the carbon through a double bond (metal alkylidene).[41] In the catalytic cycle of cross-metathesis, this active species first reacts with the olefin to form a four-membered ring (Scheme 7.7). This metallacyclobutane intermediate then cleaves, yielding ethylene and a new metal alkylidene, which reacts with a new alkene substrate to yield another metallacyclobutane. On decomposition in the forward direction, this second intermediate yields the internal alkene product and regenerates the initial metal alkylidene.

The first well-defined alkylidene complex based on ruthenium was synthesised by Grubbs *et al.* in 1992 using triphenylphosphine as an ancillary ligand.[42] Poorly active, it only polymerised highly strained cycloolefins, such as norbornene. However, a related initiator **9** containing two strongly basic tricylohexylphosphine ligands proved to be much more efficient.[43] Detailed mechanistic investigations suggested the formation of a highly active monophosphine intermediate (formally a 14-electron complex) that would further react with an alkene substrate to afford the propagating methylidene species involved in the Chauvin mechanism (see Scheme 7.7).[44] The synthesis of mixed complexes of type **9** where one phosphine was replaced by a NHC ligand led to

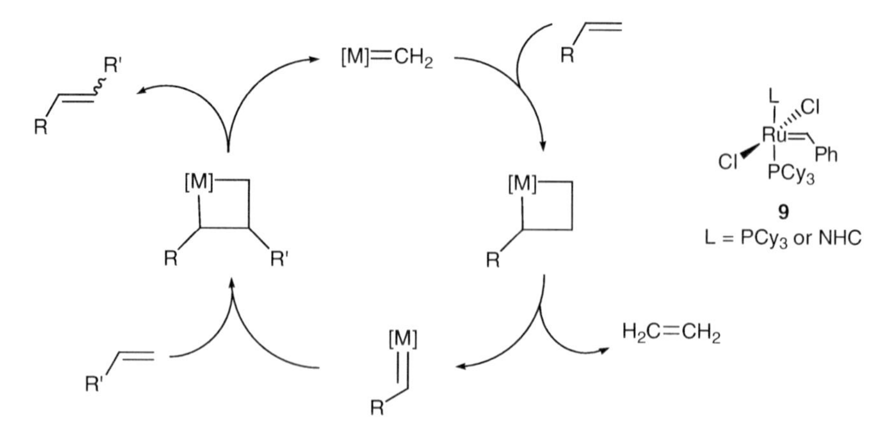

Scheme 7.7 Possible mechanism for alkene cross-metathesis.

second-generation catalysts with higher stability and increased activity compared to their diphosphine analogues.[45–47] Kinetic studies showed that these beneficial features could be assigned, respectively, to a reduced rate of phosphine dissociation from the initial 16-electron complex and to a much greater affinity of the highly unstable 14-electron species toward alkene substrates.[48] Several excellent reviews have been published over the last few years, that summarise these endeavours.[49] Hence, the following paragraphs aim only at providing a few significant examples of the latest research efforts in the field.

7.3.1.2 Benzylidene Catalysts

The second-generation Grubbs catalyst (**10**, Figure 7.2) was the first NHC–Ru olefin metathesis catalyst to gain widespread recognition, thanks to its high activity, good thermal stability, and commercial availability.[46] It remains one of the most efficient catalyst precursors for a wide range of metathesis reactions with the possible exception of ROMP, where a faster initiator, such as the third-generation Grubbs catalyst (**11**)[50] is usually preferred.[51]

Fogg *et al.* introduced chelating aryloxide (pseudohalide) ligands onto complexes **12** instead of the usual chloro substituents.[52] These catalysts

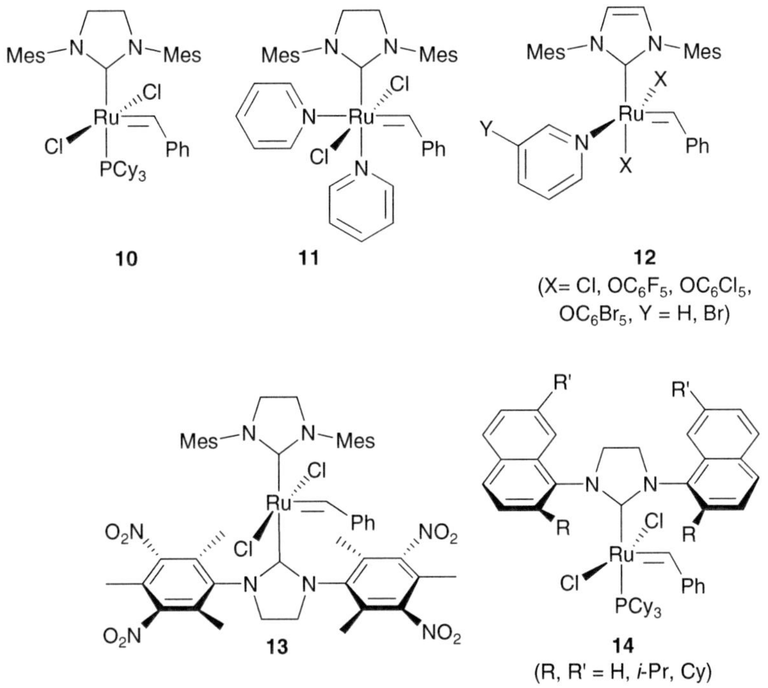

10

11

12
(X= Cl, OC_6F_5, OC_6Cl_5, OC_6Br_5, Y = H, Br)

13

14
(R, R' = H, *i*-Pr, Cy)

Figure 7.2 NHC–Ru catalysts bearing benzylidene ligands.

performed RCM with exceptionally high efficiency, due to enhanced lability of the substituted pyridine ligand. Moreover, they could be easily removed from the reaction media by one-run flash chromatography.

In another astute variation of parent compound **10**, Plenio and co-workers described the synthesis of **13**, which bore a tetranitrated NHC ligand designed to serve as leaving group in place of the customary phosphine.[53] Indeed, the electron-donating ability of this new carbene ligand was considerably reduced compared to SIMes and much closer to those of $P(i\text{-}Pr)_3$ or PEt_3, leading to a highly active bis(NHC)–Ru catalyst precursor for the RCM of tri- and tetra-substituted alkenes. This is in sharp contrast with earlier complexes of the type $[(NHC)_2RuCl_2(=CHPh)]$ bearing identical, strongly coordinated NHC ligands that displayed only mediocre activities in RCM or ROMP reactions.[18,54]

Alternatively, Dorta and co-workers used SINap ligands in order to improve the activity and stability of second-generation ruthenium metathesis catalysts (**14**).[55] Effects of NHC backbone substitution on imidazolinylidene-derived ligands were studied by Grubbs *et al.*,[56] who also replaced the imidazole-based N-heterocycles with their thiazole analogues.[57] Because these modifications were also applied to oxygen-chelated alkylidene catalysts, they are discussed in more details in the next Section.

7.3.1.3 Oxygen-chelated Alkylidene Catalysts

The stoichiometric reaction between a NHC–Ru complex bearing a metathetically active benzylidene fragment and a styrenyl ether afforded a distinct family of complexes with internal stabilisation due to chelation with the oxygen atom (Figure 7.3).[58] Accordingly, these complexes displayed a remarkable air and moisture tolerance, as well as resilience to silica gel chromatography, and they were often found more reliable catalysts than their simpler alkylidene analogues for the synthesis of pharmaceutical drugs on a multi-kilogram scale.[59] The lead compound in this series is the so-called second-generation Hoveyda–Grubbs catalyst **15**.[20,60] Numerous modifications of its properties originated from the groups of Blechert, Grela and Mauduit. For instance, we owe to Grela and Mauduit ammonium salt **16**,[61,62] and pyridinium salts **17**,[23] which could be used for metathesis reactions in conventional solvents, such as DCM or toluene, but also in alcohols, alcohol/water mixtures, or neat water, even in the presence of air. Moreover, ruthenium catalyst **16** was shown to behave as an inisurf (initiator and surfactant), thereby promoting metathesis under heterogeneous aqueous emulsion conditions.[62]

Catalysts **18** bearing amide or carbamate functional groups onto the aromatic ring of an isopropoxybenzylidene moiety were investigated by Nolan and Mauduit.[63] Introduction of a trifluoroacetamide group led to a significant enhancement of catalytic activity compared to **15**. Further modulation involved the replacement of the IMes ligand with its IPr, SIMes or SIPr congeners. When used in conjunction with the CF_3CONH group, SIPr afforded a catalyst precursor of remarkable activity and stability that could be recycled at the end of the reaction.

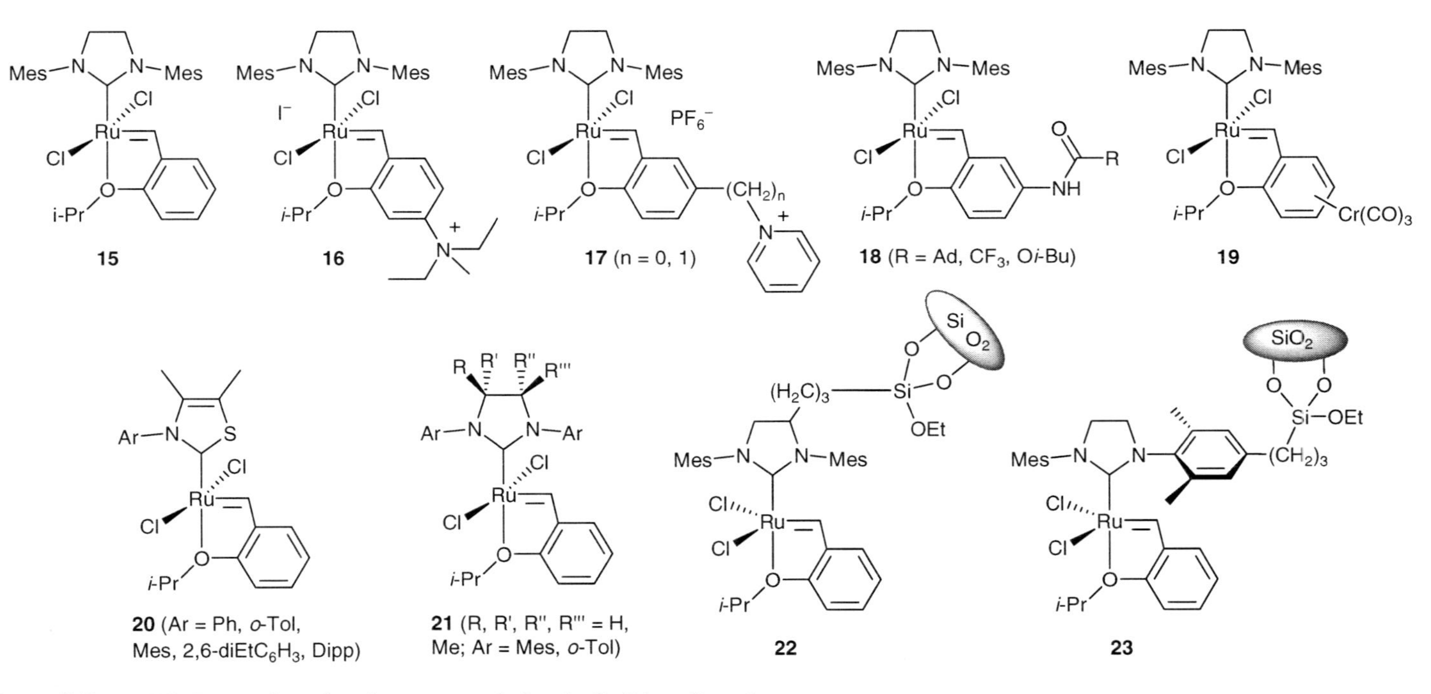

Figure 7.3 NHC–Ru catalysts bearing oxygen-chelated alkylidene ligands.

In 2008, Butenschön and co-workers reported an unusual variation on the second-generation Hoveyda–Grubbs catalyst, in which the 2-isopropoxybenzylidene ligand was coordinated to a highly electron-withdrawing tricarbonylchromium moiety (**19**).[64] In a series of RCM, CM and enyne metathesis reactions carried out to screen its catalytic properties, complex **19** matched or outperformed the results obtained with its parent **15** and related initiators.

In 2008, Grubbs and co-workers investigated a novel family of ruthenium complexes bearing a *N*-aryl thiazol-2-ylidene ligand.[57] Despite the decreased steric bulk of these ligands compared to imidazole-based NHCs, the resulting complexes **20** efficiently promoted standard RCM , ROMP and CM reactions, as well as the macrocyclic ring-closure of a 14-membered lactone. While removing steric bulk from the *ortho* positions in the *N*-aryl group of the thiazol-2-ylidene decreased their stability, substituents too bulky resulted in prolonged induction periods.

A subsequent study by Grubbs focused on ruthenium catalysts bearing imidazolin-2-ylidene ligands with varying degrees of backbone and *N*-aryl substitution (**21**).[56] These complexes showed greater resistance to decomposition through C–H activation of the *N*-aryl group, resulting in increased catalyst lifetimes, but the effects were subtle and sometimes contradictory.

Last but not least, two ruthenium isopropoxybenzylidene complexes bearing triethoxysilyl-functionalised NHC ligands were synthesised and grafted onto silica gel. The resulting solid-supported catalysts (**22** and **23**) were found efficient for a number of metathesis reactions.[65] They could be recycled a number of times with an eventual gradual decrease in activity, but most strikingly, they did not leach ruthenium under the experimental conditions adopted, a feature of utmost importance for the development of pharmaceutical applications.

7.3.1.4 Other Chelated Alkylidene Catalysts

Although fast initiation rates are highly desirable for many metathesis reactions, there are some applications where a delayed action of the catalyst is not an adverse feature. For instance, some industrial processes for ROMP, such as reaction injection moulding (RIM), require the handling or storage of a mixture of monomer and catalyst before polymerisation begins. Therefore, several types of latent initiators that can be triggered chemically, photochemically, or thermally were developed.[66] Most of them were based on alkylidene ligands strongly chelated to the metal centre *via* N or S atoms (Figure 7.4). Latency effects due to the O,S-bidentate Schiff base ligands in benzylidene complexes **24** were also reported by Verpoort and co-workers.[67] For these systems, addition of a controlled amount of hydrochloric acid or trichlorosilane to the reaction media successfully triggered the transformation of dormant pre-catalysts into active species for the ROMP of cycloalkenes.

In 2008–2009, Grela and co-workers disclosed the synthesis of NHC–Ru complexes bearing chelating sulfoxide ligands (**25**) that showed no catalytic activity in RCM or enyne metathesis reactions at room temperature, but became active upon heating to 40–110 °C.[68] Concomitantly, Lemcoff *et al.* prepared a closely related series of sulfur-chelated latent ruthenium alkene

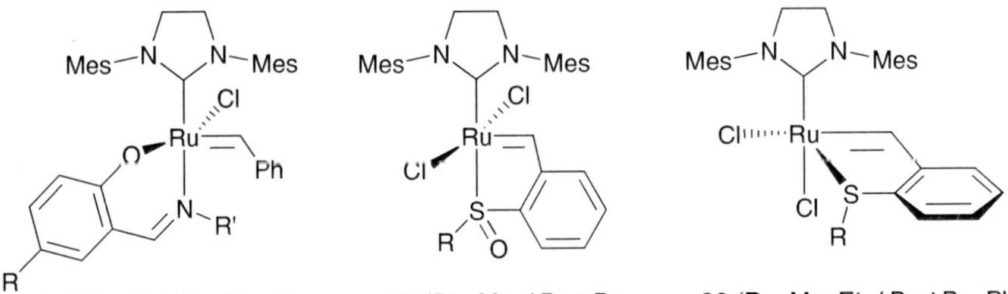

24 (R = H, NO_2; R' = Mes, Dipp, 2,6-diMe-4-BrC_6H_2)

25 (R = Me, *i*-Pr, *t*-Bu, Cy, Bn, Ph, *p*-$NO_2C_6H_4$)

26 (R = Me, Et, *i*-Pr, *t*-Bu, Ph)

Figure 7.4 NHC–Ru catalysts bearing nitrogen or sulfur-chelated alkylidene ligands.

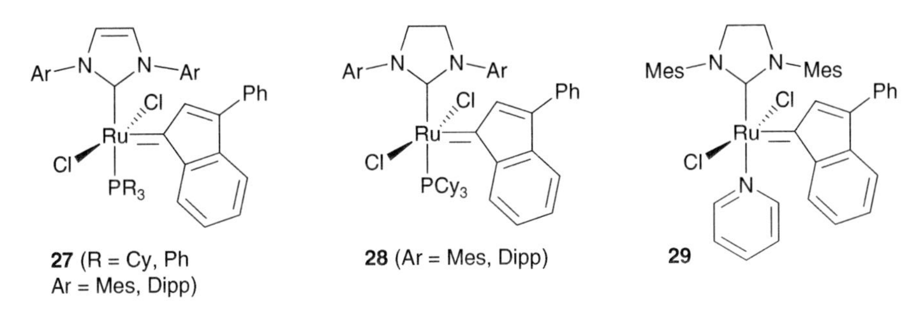

Figure 7.5 NHC–Ru catalysts bearing indenylidene ligands.

metathesis catalysts (**26**) that possessed an uncommon *cis*-dichloro arrangement and were mostly inactive at room temperature, but could be activated either thermally or photochemically.[69] Modifications of the size of remote substituents on the sulfur atom significantly affected the catalytic activity at different temperatures. More bulky substituents raised the activity at lower temperatures. Catalysts **26** were also stable in solution and retained their catalytic activity in RCM reactions even after being exposed to air for 2 weeks.

7.3.1.5 Indenylidene Catalysts

Over the past few years, ruthenium–indenylidene complexes bearing NHC ligands have emerged as robust and efficient catalyst precursors for olefin metathesis that nicely complement their benzylidene counterparts (Figure 7.5).[70] Indeed, they made possible reactions that were not promoted by many earlier Grubbs-type catalysts. In particular, they allowed the convenient synthesis of tri- and tetrasubstituted cycloalkenes, as well as RCM reactions involving highly substituted acrylates. They are also easier to synthesise, because the metathetically active 3-phenylindenylidene fragment is conveniently grafted onto ruthenium by coordination and rearrangement of the widely available propargyl alcohol 1,1-diphenyl-2-propyn-1-ol.

As with the benzylidene catalysts, second-generation ruthenium–indenylidene complexes result from the exchange of one phosphine ligand for a NHC within the coordination sphere of a diphosphine starting material. A substitution of this type was first reported by Nolan *et al.* in 1999 for the synthesis of complexes **27**. Surprisingly, it took almost 10 years for analogous mixed complexes bearing PCy_3 and the saturated SIMes and SIPr ligands (**28**) to appear in the literature.[71,72] The catalytic properties of all these species were evaluated in several types of metathesis reactions, including CM, RCM and ROMP.[73,74] Although high activities and selectivities were observed in many cases, it was difficult to anticipate the outcome of any specific reaction and to single out a 'one size fits all' catalyst that would perform best for any given substrate, a conclusion already formulated by Grela in a failed attempt to provide definite guidelines for the use of benzylidene, alkoxybenzylidene and indenylidene ruthenium olefin metathesis catalyst.[75]

A predictable move in the development of NHC–Ru complexes bearing indenylidene ligands was the preparation of the third-generation complex **29** reported by the groups of Slugovc[76] and Verpoort[71] in 2008. The first team established its X-ray crystallographic structure and demonstrated its aptitude to efficiently promote the controlled living ring-opening metathesis polymerisation of norbornene monomers,[76] while the second one probed its catalytic activity in diverse RCM, CM and ROMP reactions.[71,74]

7.3.1.6 Arene Catalysts

A distinctive approach for the advancement of metathesis catalysts relies on ruthenium–arene complexes either preformed or generated *in situ* from the $[RuCl_2(p\text{-cymene})]_2$ dimer and NHC ligands precursors (Figure 7.6).[77] Although these species did not initiate olefin metathesis *per se*, the required metal–alkylidene fragment was proposed to arise from the stoichiometric reaction between an alkene substrate or co-catalyst and a highly unsaturated ruthenium species generated *in situ* upon de-coordination of the arene ligand.[78] Indeed, NMR investigations showed that the active propagating species

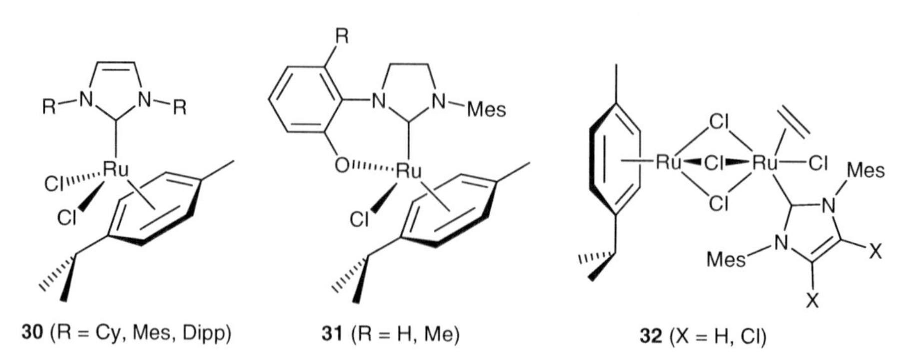

Figure 7.6 NHC–Ru catalysts bearing arene ligands.

involved in the self-metathesis of ethyl oleate obtained from ruthenium–arene catalyst precursors and trimethylsilyldiazomethane were probably identical to those derived from the first-generation Grubbs benzylidene complex.[79]

In 2006, Delaude *et al.* showed that imidazol(in)ium-2-carboxylates were convenient NHC precursors to produce NHC–Ru complexes *in situ* for the photoinduced ROMP of norbornene and cyclooctene in the presence of $[RuCl_2(p\text{-cymene})]_2$.[78] Subsequent work allowed the isolation of compounds **30** in pure form starting from the ruthenium dimer and three different $NHC{\cdot}CO_2$ adducts.[80] Alternatively, Fischmeister and Dixneuf used the free IMes carbene to synthesise the known $[(IMes)RuCl_2(p\text{-cymene})]$ complex, which was found to be an efficient pre-catalyst for the CM of various styrene derivatives and the RCM of a tetrasubstituted cycloolefin when activated with styrene.[81]

Chelated complexes **31** were reported by Ledoux *et al.* but they displayed very poor efficiencies in the ROMP of norbornene.[82] Attempts to open up the chelate by protonating the phenolic oxygen atom did not significantly improve the catalytic activity and caused instead a rapid decomposition of the transition metal active species.

In 2007, Delaude and Demonceau isolated homobimetallic ruthenium–arene complexes **32** in high yields upon heating a toluene solution of $[RuCl_2(p\text{-cymene})]_2$ with one equivalent of carbene ligand under ethylene atmosphere.[83] Coupling reactions with various styrene derivatives confirmed the outstanding aptitude of these two complexes to catalyse alkene metathesis. Unlike their monometallic counterparts **30**, they did not require the addition of a diazo compound nor visible light illumination to initiate the ROMP of norbornene or cyclooctene. When α,ω-dienes were tested in RCM reactions, cycloisomerisation products were also obtained in a non-selective way. However, addition of a terminal alkyne co-catalyst enhanced the metathetical activity, while completely repressing this side-reaction.

7.3.2 Non-metathesis Reactions

7.3.2.1 Introduction

Besides alkene or alkyne metathesis, a broad range of other non-metathetical reactions promoted by NHC–Ru complexes was reported in the literature.[17,31,84] Some of them were discovered serendipitously, as they constituted side reactions in metathesis catalysis. Other were deliberately investigated and optimised. Despite the usefulness of several of these processes, their significance has remained undervalued due to the huge appeal of olefin metathesis and related reactions.

Non-metathetical reactions catalysed by ruthenium–carbene complexes are multifaceted and cover a broad range of transformations, thanks in part to the large number of oxidation states and coordination geometries available for the metal centre,[85] in sharp contrast with other elements such as rhodium, palladium and platinum, which reluctantly form compounds with high oxidation states and have a strong preference for the square planar geometry. Several recent reviews gave a sense of the achievements already accomplished with recourse to NHC ligands for that matter.[17,31,84] Thus, the following paragraphs outline only recent developments involving non-metathetical, independent reactions involving NHC–Ru complexes. In view of their importance, oxidation and reduction processes (including hydrogenation reactions) are covered separately in Chapters 12 and 13, respectively.

7.3.2.2 Isomerisation Reactions

The use of the second-generation Grubbs catalyst **10** for olefin isomerisation was reported by Nishida and co-workers in 2002.[86] During an attempt to perform CM between an alkene substrate and trimethyl(vinyloxy)silane, the selective isomerisation of the terminal olefin to afford the corresponding propenyl species as a mixture of *E* and *Z* isomers was observed instead. Several other terminal alkenes were then reacted under the same experimental conditions, and the corresponding isomerisation products were obtained in moderate-to-excellent yields (Scheme 7.8). Slightly thereafter, Wagener and co-workers provided further evidence for a competition between self-metathesis and isomerisation when allyl derivatives were reacted in the presence of benzylidene complex **10**.[87] They also showed that this catalyst precursor promoted extensive isomerisation of both internal and terminal alkenes under typical ADMET conditions.[88] Subsequently, Nishida revealed that the actual catalyst responsible for the isomerisation was the ruthenium hydride complex **33** generated *in situ* from **10** and trimethyl(vinyloxy)silane.[28] This hydrido-carbonyl species had been previously isolated by Grubbs and Mol under various experimental conditions, but at that time its full potential as an isomerisation catalyst was not exploited.[18,89]

Building on these results, Hanessian and co-workers devised an efficient protocol for the isomerisation of terminal olefins with minimal self-dimerisation or cross-metathesis by employing methanol to generate hydride complex **33** *in situ* from its parent **10**.[90] The procedure was successfully applied to a variety of allylic compounds, including *O*- and *N*-allyl ethers, and cleanly afforded the corresponding propenyl species as *E*/*Z* isomeric mixtures, without any further isomerisation or conjugation in the cases of ketones, esters and lactams.

Another promising isomerisation process, the atom-economical conversion of allylic alcohols into aldehydes or ketones, was reported in 2006 by Fekete and Joó.[91] Reactions were carried out in aqueous/organic biphasic systems

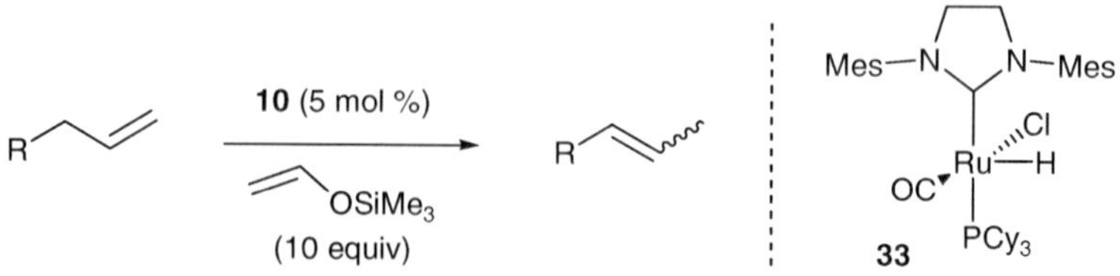

R
10 (5 mol %)
OSiMe$_3$
(10 equiv)
R
Mes—N N—Mes
Cl
Ru—H
OC
PCy$_3$
33

Scheme 7.8 NHC–Ru catalysed isomerisation of terminal olefins.

using a water-soluble NHC–Ru complex (**34**) (Equation (7.5)). Because hydrogen was needed to initiate the reaction, formation of saturated alcohols was also observed. This side reaction could be partially repressed by adding sodium chloride to the aqueous solution and by buffering its pH around 7. Altogether, the catalytic system proved moderately efficient, but could be reused up to four times upon recycling of the aqueous phase.

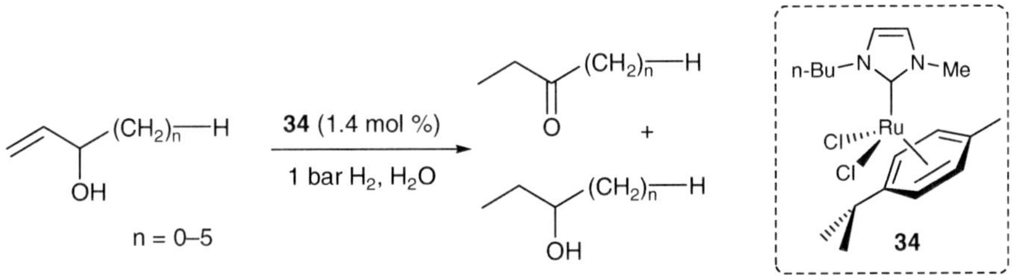

(7.5)

7.3.2.3 *Cycloisomerisation Reactions*

The RCM of 1,6-dienes into cyclopentenes is sometimes accompanied by a cycloisomerisation reaction to give cyclic products with an *exo*-methylene substituent, especially when ruthenium–arene complexes are used as catalyst precursors (see for instance compounds **32** in Section 7.3.1.6). In many cases, this side reaction was undesired and could be effectively suppressed by the addition of various co-catalysts. It is nevertheless possible to alter the reaction path to obtain the cycloisomerisation products with very high selectivities. Early examples of NHC–Ru complexes suitable for this task included the unstable cationic allenylidene complexes **36** prepared *in situ* from more robust chelated precursors **35** (Equation (7.6)).[92] Alternatively, a combination of IMes·HCl/Cs_2CO_3/[$RuCl_2$(*p*-cymene)]$_2$ also provided an efficient catalytic system.[93] A mechanism involving oxidative coupling of the 1,6-diene to a ruthenium(II) centre followed by β-elimination to generate a hydrido ruthenium(IV) intermediate and reductive elimination was proposed for the transformation (Scheme 7.9).[93]

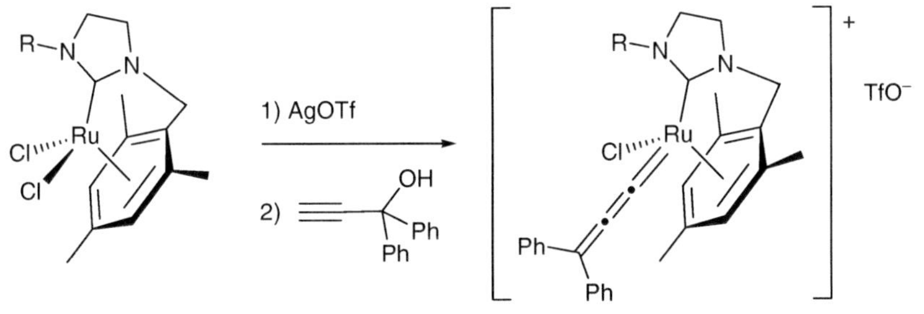

(7.6)

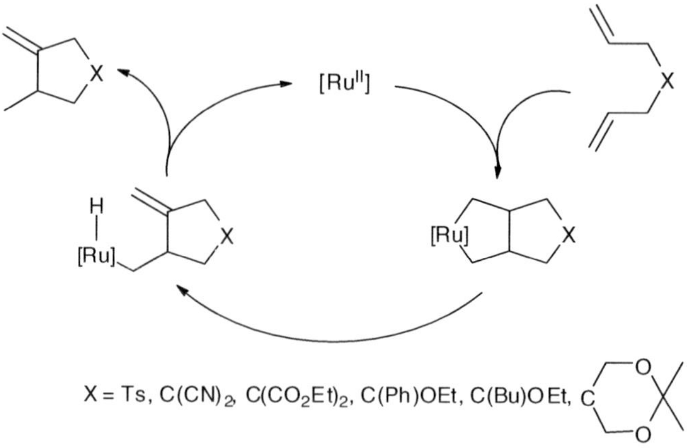

[Ru^{II}]

X = Ts, C(CN)$_2$, C(CO$_2$Et)$_2$, C(Ph)OEt, C(Bu)OEt, C

Scheme 7.9 Ruthenium-catalysed cycloisomerisation of 1,6-dienes.

Following this seminal work, Arisawa and Nishida investigated the behaviour of *N*-tosyl diallyl amine in the presence of the second-generation Grubbs initiator (**10**) and different catalyst modifiers.[28] Trimethyl(vinyloxy)silane afforded the best yields of cycloisomerised *versus* RCM product, allowing for the formation of diverse *exo*-methylene benzofurans and indolins *via* diene cycloisomerisation (Equation (7.7)). The method was also successfully applied to the synthesis of a novel indole alkaloid that displayed antifungal activity.[94] As mentioned previously (see Section 7.3.2.2), it is known that in the presence of the vinyl trimethylsilyl ether, compound **10** decomposed into hydrido-carbonyl complex **33**, which was believed to be the actual catalyst for the cycloisomerisation process.

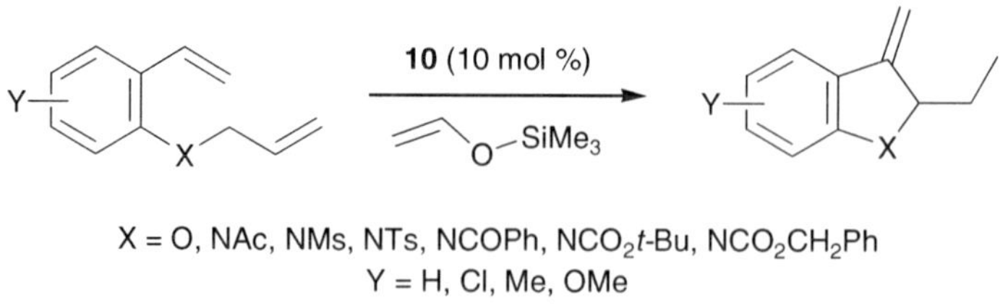

10 (10 mol %)

X = O, NAc, NMs, NTs, NCOPh, NCO$_2$*t*-Bu, NCO$_2$CH$_2$Ph
Y = H, Cl, Me, OMe

(7.7)

Mukai and co-workers examined the reactivity of 1,6-allenenes using **10**.[95] At high catalyst loadings, this system proved efficient for the cycloisomerisation of a number of allenes bearing a pendant allyl group into cyclohexene, tetrahydropyran or tetrahydropyridine derivatives (Scheme 7.10). The presence of heteroatoms was well tolerated, and the authors proposed a mechanism going through a ruthenacyclopentane.

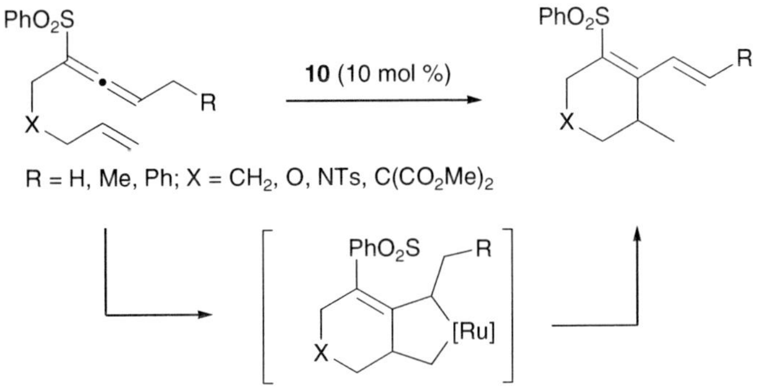

PhO$_2$S ... 10 (10 mol %) ... R = H, Me, Ph; X = CH_2, O, NTs, $C(CO_2Me)_2$ ... [Ru]

Scheme 7.10 Ruthenium-catalysed cycloisomerisation of allenes.

7.3.2.4 Oligomerisation and Polymerisation Reactions

Although ROMP or ADMET reactions hold a prominent position among polymerisation processes initiated by NHC–Ru complexes, other catalytic paths leading to macromolecular products were also investigated. The activity of compounds **30** and other similar monometallic [(NHC)$RuCl_2$(*p*-cymene)] complexes was tested in the atom transfer radical polymerisation (ATRP) of vinyl monomers by Delaude and Demonceau,[96] along with **32**.[83] These complexes led to the controlled polymerisation of methyl methacrylate at 85 °C (Equation (7.8)). Attempts to polymerise *n*-butyl acrylate and styrene turned out to be more challenging, because of difficulties to control the acrylate polymerisation and of competition with the self-metathesis of styrene.

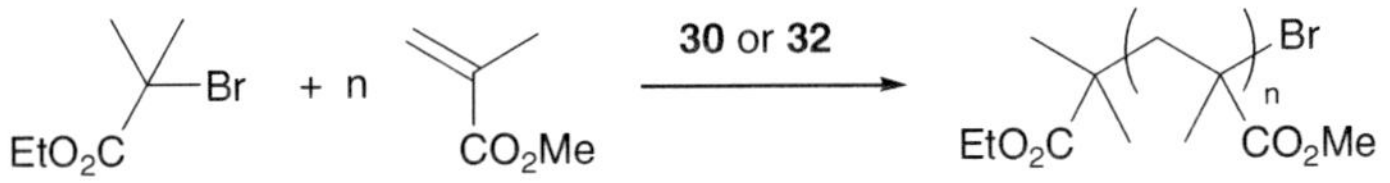

EtO_2C–C(CH_3)$_2$–Br + n CH_2=C(CH_3)CO_2Me —(**30** or **32**)→ EtO_2C–C(CH_3)$_2$–[CH_2–C(CH_3)(CO_2Me)]$_n$–Br (7.8)

The polymerisation of various mono- and disubstituted acetylene derivatives was investigated by Masuda and co-workers using the second-generation Hoveyda–Grubbs catalyst (**15**).[97] Owing to its excellent tolerance toward polar functional groups, this NHC–Ru complex was suitable for polymerising various alkyne monomers bearing silyl, silyloxy, ester, amide or carbamate groups that did not react with conventional tantalum initiators. Strikingly, polymerisation of *ortho*-substituted phenylacetylenes afforded macromolecular products that possessed high *trans* contents and took a helical conformation with predominantly one-handed screw sense when chiral side-chains were present.

The polymerisation of unsubstituted phenylacetylene into *trans*-polyphenylacetylene was achieved by Buchmeiser and co-workers using a related alkoxybenzylidene ruthenium initiator (**37**) bearing a six-membered NHC and pseudohalide ancillary ligands (Scheme 7.11).[98] Interestingly, when the ruthenium–arene complex **34** was employed as catalyst, only oligomers (pentamers

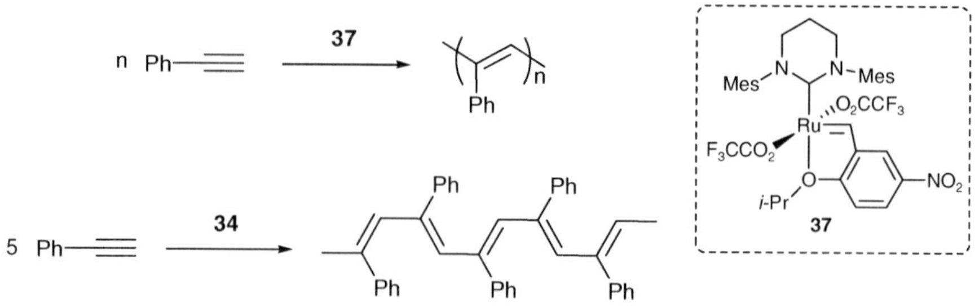

Scheme 7.11 NHC–Ru catalysed oligomerisation and polymerization of phenylacetylene.

mainly) were isolated.[99] Further analyses by NMR and mass spectrometry revealed, nevertheless, the non-negligible formation of linear chains terminated with a positively charged imidazolium end-group, a clear evidence for the de-coordination of the NHC rather than that of the *p*-cymene ligand during this catalytic process.

Non-classical NHCs were also studied in this context. The reader is referred to Chapter 5 for further details.

7.3.2.5 C–C and C–O Bond Forming Reactions

In order to complement earlier studies showing that *N*-alkyl-substituted benzimidazolylidene–ruthenium complexes were efficient catalysts for the regioselective alkylation of cinnamyl carbonate by dimethyl malonate or 1,3-diketones, as well as etherification of allylic halides by phenols,[100] Bruneau and co-workers prepared a wide range of imidazolinium **38**, tetrahydropyrimidinium **39**, and benzimidazolium salts **40** that were screened as NHC ligand precursors in various allylic substitution reactions (Equation (7.9)).[101] Unfortunately, linear *versus* branched selectivities were only modest, and no characterisation of the [(NHC)Ru(Cp*)] complexes assumed to take part in the reaction could be achieved, thereby preventing any further rational ligand modification that would have helped refine the catalytic system.

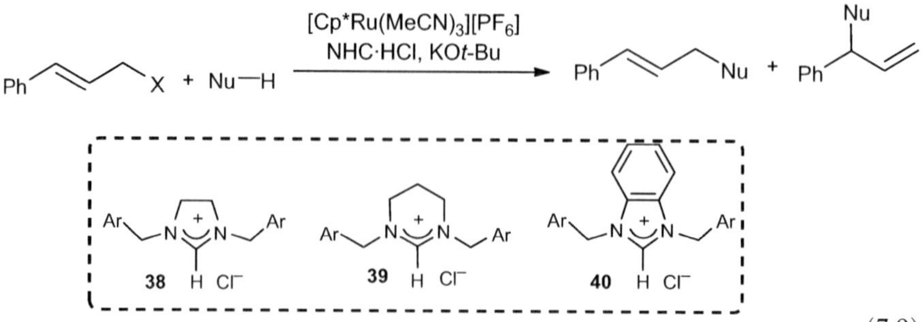

(7.9)

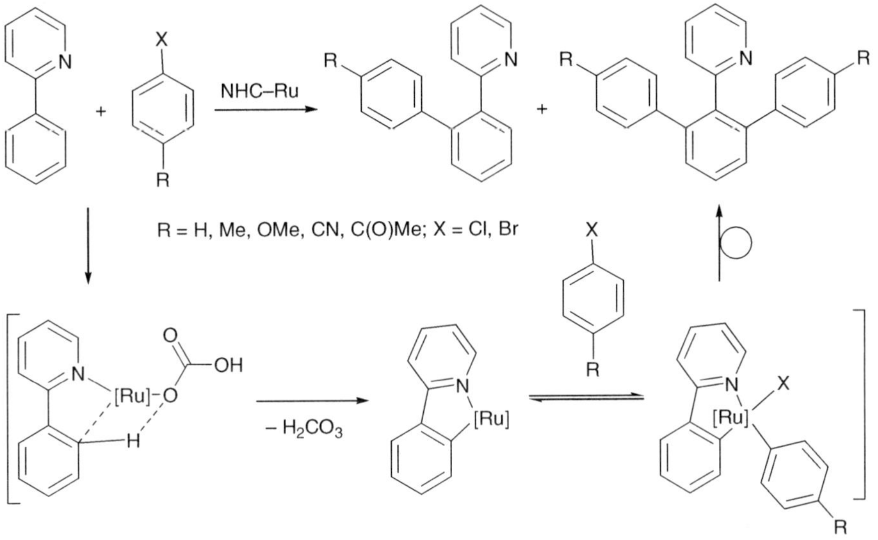

Scheme 7.12 Possible mechanism for the arylation of 2-phenylpyridine with aryl halides.

The potential of NHC–Ru complexes for C–H bond activation was also evidenced in the direct arylation of 2-phenylpyridine with aryl halides using either pre-formed or *in situ* generated catalysts (Scheme 7.12). Hence, Dixneuf and Maseras showed that a mixture of $[RuCl_2(p\text{-cymene})]_2/Cs_2CO_3/$**39** efficiently promoted the di-*ortho*-phenylation of 2-phenylpyridine with both electron-rich and electron-poor aryl bromides.[27] Based on DFT calculations, a cooperative metal/base proton abstraction was postulated for the initial step rather than an oxidative addition producing a Ru^{IV} species with a hydride ligand. The resulting *ortho*-metalated intermediate was then expected to undergo the reversible oxidative addition of the bromoarene coupling partner, followed by reductive elimination of the arylated product. Subsequent studies showed that various chelated NHC–ruthenium complexes prepared from $[RuCl_2(p\text{-cymene})]_2$ or $[RuCl_2(PPh_3)_3]$ were also active for this transformation, even when less reactive aryl chlorides were used instead of the corresponding bromides.[102,103]

7.3.2.6 Miscellaneous Reactions

Isolated reports mentioned the use of NHC–Ru species for the asymmetric Diels–Alder reaction of methacrolein and cyclopentadiene,[104] cyclopropanation of styrene with ethyl diazoacetate,[78] or dehydrative condensation of *N*-(2-pyridyl)benzimidazole and allyl alcohol.[105] Whittlesey and co-workers described the stoichiometric activation of C–F bonds in perfluoroaromatic substrates such as hexafluorobenzene, perfluorotoluene and pentafluoropyridine using the dihydrido complex $[(ICy)Ru(dppp)(CO)(H)_2]$ (**41**) (dppp = 1,3-bis(diphenylphosphino)propane).[106] Modification of the NHC nitrogen substituents from *N*-alkyl to

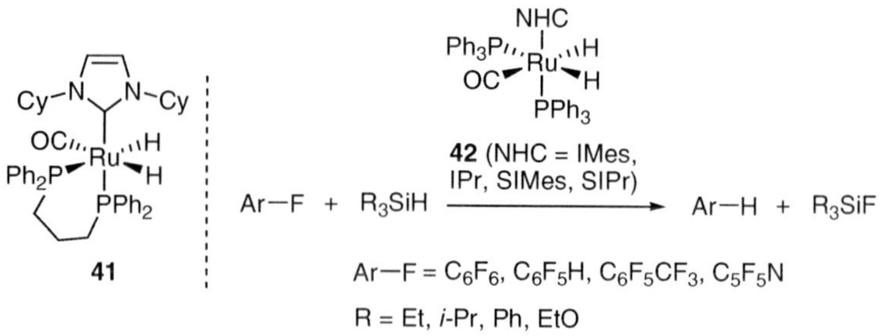

41

$$\text{Ar–F} + R_3SiH \xrightarrow{\textbf{42}\ (\text{NHC} = \text{IMes, IPr, SIMes, SIPr})} \text{Ar–H} + R_3SiF$$

Ar—F = C_6F_6, C_6F_5H, $C_6F_5CF_3$, C_5F_5N

R = Et, *i*-Pr, Ph, EtO

Scheme 7.13 NHC–Ru complexes for stoichiometric or catalytic C–F bond activation.

N-aryl groups and replacement of the chelating diphosphine with two triphenylphosphine ligands afforded complexes **42** that performed the catalytic hydrodefluorination of polyfluoroarenes with alkylsilanes with reasonable turnover number (up to 200) and frequency (up to 0.86 h^{-1}) (Scheme 7.13).[107]

7.3.3 Tandem Reactions

Tandem reactions refer to sequential transformations of a substrate *via* two or more mechanistically distinct processes that take place in a single vessel.[29] Because they generate less waste and minimise handling and work-up actions in multi-step syntheses, while significantly increasing molecular complexity, tandem protocols are often considered superior to step-wise procedures. Therefore, catalyst precursors that can be triggered to perform several types of chemical reactions are highly valuable commodities for organic synthesis. In this regard, NHC–ruthenium complexes, well-known for their alkene metathesis activities, proved highly suitable to function as pre-catalysts in joint alkene isomerisation, alkene hydrogenation, radical atom transfer reactions, and various other transformations.[30,31]

One of the most successful embodiment of this concept combined a metathetical step with the selective isomerisation of a terminal C=C double bond, and found applications in natural product synthesis.[108] As discussed in Section 7.3.2.2, it is indeed possible to convert the metathetically active benzylidene initiator **10** into hydrido complex **33**, simply by adding trimethyl(vinyloxy)silane[28,86] or methanol[90] to the reaction mixture, thereby triggering a consequent catalytic isomerisation process. Other methods for the decomposition of ruthenium metathesis catalysts for use in tandem with olefin isomerisation reactions include treatment with hydrogen,[109] sodium borohydride,[110] or sodium hydroxide in isopropanol.[111]

Snapper and co-workers investigated the enyne CM followed by cyclopropanation of the intermediate 1,3-diene produced with a diazoester (Scheme 7.14).[112] Because complex **10** did not promote cyclopropanation efficiently under the experimental conditions adopted, another related NHC–Ru alkylidene complex (**43**) was chosen as catalyst for the transformation. The whole three-component process gave access to vinyl cyclopropanes with satisfactory yields and selectivities.

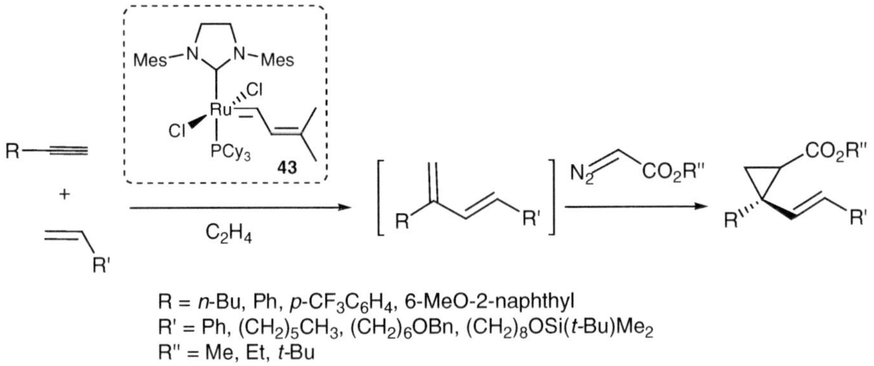
Mes
N
N
Mes
Cl
Ru
Cl
PCy3
43
R
+
R'
C2H4
R
R'
N2
CO2R''
CO2R''
R'''
R'
R = n-Bu, Ph, p-CF3C6H4, 6-MeO-2-naphthyl
R' = Ph, (CH2)5CH3, (CH2)6OBn, (CH2)8OSi(t-Bu)Me2
R'' = Me, Et, t-Bu

Scheme 7.14 NHC–Ru-catalysed tandem enyne CM/cyclopropanation.

7.4 NHC–Osmium-catalysed Reactions

Osmium is the least studied element of the iron triad and its main use consists in the preparation of stable models for reactive intermediates involved in reactions catalysed by ruthenium analogues.[113] Although NHC–Os compounds were first synthesised by Lappert in the late 1970s,[114] their catalytic properties were investigated only very recently.

The first report of NHC-containing osmium compounds acting as catalysts came from Esteruelas and co-workers in 2005.[115] Thus, cationic benzylidene complexes **44** were prepared by reaction of the corresponding 16-electron precursors [(NHC)OsCl(*p*-cymene)][OTf] (NHC = IMes or IPr) with phenyldiazomethane, and their potential as initiators for olefin metathesis was probed in the RCM of diethyl diallylmalonate, the ROMP of cyclooctene, and a variety of self- and cross-metathesis reactions (Equation (7.10)). Although they were not as efficient as standard ruthenium–benzylidene metathesis initiators, compounds **44** displayed, nevertheless, a fairly decent activity. More importantly, in addition to being the first NHC–Os catalytic application, this study constituted a rare example of osmium catalysed C–C bond formation.

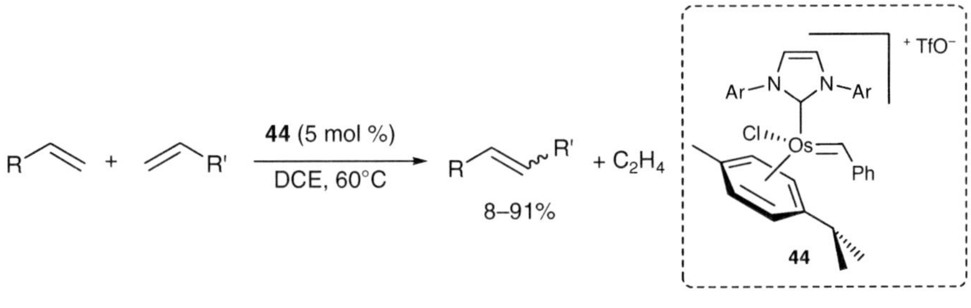
R
+
R'
44 (5 mol %)
DCE, 60°C
R
R'
+ C2H4
8–91%
Ar
N
N
Ar
+ TfO−
Cl
Os
Ph
44

(7.10)

In a subsequent work, Castarlenas and Esteruelas disclosed the synthesis of another cationic complex [(IPr)Os(OH)(*p*-cymene)][OTf] (**45**), which was an efficient pre-catalyst for transfer hydrogenation (Equation (7.11)).[116] Several aliphatic and aromatic aldehydes could be reduced in high yields and high turnover numbers, whereas under these same experimental conditions, acetophenone reacted only very slowly. Regarding the activation of the pre-catalyst, the authors provided strong NMR evidence that, in the presence of isopropanol, compound **45** was converted into a cationic acetone–hydride complex [(IPr)Os(H)(O=C(CH_3)$_2$)][OTf], which most likely acted as the active species for transfer hydrogenation.

RCHO + *i*-PrOH → (**45** (0.1–0.03 mol %), *i*-PrOH, 80°C) → RCH_2OH + acetone
72–98%
45: [(IPr)Os(OH)(*p*-cymene)]$^+$ TfO$^-$ (Dipp–N, N–Dipp, Os, OH)

(7.11)

7.5 Conclusions and Outlook

To date, catalytic applications of NHC–Fe systems are still scarce and do not reflect the full potential of this cheap, abundant and non-toxic metal that shows great promise for the formation of carbon–carbon bonds *via* cross-coupling reactions. Recourse to ill-defined species generated *in situ* and difficulties in understanding the intimate nature of reaction mechanisms do not ease the development of organo iron catalysis. Support from the related field of iron biocatalysis and recent discoveries in organometallic chemistry are expected to provide help and inspiration for further advances.

Contrastingly, the chemistry of NHC–Ru complexes has reached an unprecedented level of maturity during the past decade, thanks to the phenomenal development of olefin metathesis. Detailed mechanistic investigations and the in-depth characterisation of well-defined catalyst precursors have allowed the rational design and fine-tuning of ruthenium–alkylidene initiators with a degree of refinement rarely achieved in the history of homogeneous catalysis. Other synthetic pathways based on NHC–Ru promoters have begun to emerge, which benefit from the vast amount of knowledge already acquired for olefin metathesis.

Lastly, with only two reports to date, NHC–Os-based catalysis can hardly be considered anything other than a curiosity. The toxicity of osmium and the inertness of third-row transition metals are most likely to blame for this situation. It should be stressed, nevertheless, that the two transformations explored so far, olefin metathesis and transfer hydrogenation, are among the most important ones in modern organometallic catalysis.

References

1. (a) B. Plietker, ed. *Iron Catalysis in Organic Synthesis*, Wiley-VCH, Weinheim, 2008; (b) E. B. Bauer, *Curr. Org. Chem.*, 2008, **12**, 1341–1369; (c) C. Bolm, J. Legros, J. Le Paih and L. Zani, *Chem. Rev.*, 2004, **104**, 6217–6254.
2. B. D. Sherry and A. Fürstner, *Acc. Chem. Res.*, 2008, **41**, 1500–1511.
3. V. Lavallo and R. H. Grubbs, *Science*, 2009, **326**, 559–562.
4. (a) S. Jiang, J. Liu, Y. Shi, Z. Wang, B. Akermark and L. Sun, *Polyhedron*, 2007, **26**, 1499–1504; (b) D. Morvan, J.-F. Capon, F. Gloaguen, A. Le Goff, M. Marchivie, F. Michaud, P. Schollhammer, J. Talarmin, J.-J. Yaouanc, R. Pichon and N. Kervarec, *Organometallics*, 2007, **26**, 2042–2052; (c) T. Liu and M. Y. Darensbourg, *J. Am. Chem. Soc.*, 2007, **129**, 7008–7009; (d) C. M. Thomas, T. Liu, M. B. Hall and M. Y. Darensbourg, *Inorg. Chem.*, 2008, **47**, 7009–7024; (e) D. Morvan, J.-F. Capon, F. Gloaguen, F. Y. Pétillon, P. Schollhammer, J. Talarmin, J.-J. Yaouanc, F. Michaud and N. Kervarec, *J. Organomet. Chem.*, 2009, **694**, 2801–2807.
5. L.-C. Song, X. Luo, Y.-Z. Wang, B. Gai and Q.-M. Hu, *J. Organomet. Chem.*, 2009, **694**, 103–112.
6. C. Tard and C. J. Pickett, *Chem. Rev.*, 2009, **109**, 2245–2274.
7. J. Louie and R. H. Grubbs, *Chem. Commun.*, 2000, 1479–1480.
8. D. S. McGuinness, V. C. Gibson and J. W. Steed, *Organometallics*, 2004, **23**, 6288–6292.
9. M.-Z. Chen, H.-M. Sun, W.-F. Li, Z.-G. Wang, Q. Shen and Y. Zhang, *J. Organomet. Chem.*, 2006, **691**, 2489–2494.
10. G. Hilt, P. Bolze and I. Kieltsch, *Chem. Commun.*, 2005, 1996–1998.
11. (a) N. Saino, D. Kogure and S. Okamoto, *Org. Lett.*, 2005, **7**, 3065–3067; (b) N. Saino, D. Kogure, K. Kase and S. Okamoto, *J. Organomet. Chem.*, 2006, **691**, 3129–3136.
12. A. Rudolph and M. Lautens, *Angew. Chem., Int. Ed.*, 2009, **48**, 2656–2670.
13. R. B. Bedford, M. Betham, D. W. Bruce, A. A. Danopoulos, R. M. Frost and M. Hird, *J. Org. Chem.*, 2006, **71**, 1104–1110.
14. (a) T. Hatakeyama and M. Nakamura, *J. Am. Chem. Soc.*, 2007, **129**, 9844–9845; (b) T. Hatakeyama, S. Hashimoto, K. Ishizuka and M. Nakamura, *J. Am. Chem. Soc.*, 2009, **131**, 11949–11963.
15. T. Yamagami, R. Shintani, E. Shirakawa and T. Hayashi, *Org. Lett.*, 2007, **9**, 1045–1048.
16. B. Plietker, A. Dieskau, K. Möws and A. Jatsch, *Angew. Chem., Int. Ed.*, 2008, **47**, 198–201.
17. V. Dragutan, I. Dragutan, L. Delaude and A. Demonceau, *Coord. Chem. Rev.*, 2007, **251**, 765–794.

18. T. M. Trnka, J. P. Morgan, M. S. Sanford, T. E. Wilhelm, M. Scholl, T.-L. Choi, S. Ding, M. W. Day and R. H. Grubbs, *J. Am. Chem. Soc.*, 2003, **125**, 2546–2558.
19. (a) T. Ritter, M. W. Day and R. H. Grubbs, *J. Am. Chem. Soc.*, 2006, **128**, 11768–11769; (b) I. C. Stewart, C. J. Douglas and R. H. Grubbs, *Org. Lett.*, 2008, **10**, 441–444.
20. S. B. Garber, J. S. Kingsbury, B. L. Gray and A. H. Hoveyda, *J. Am. Chem. Soc.*, 2000, **122**, 8168–8179.
21. (a) H. Wakamatsu and S. Blechert, *Angew. Chem., Int. Ed*, 2002, **41**, 794–796; (b) K. Grela, S. Harutyunyan and A. Michrowska, *Angew. Chem., Int. Ed*, 2002, **41**, 4038–4040.
22. (a) S. H. Hong and R. H. Grubbs, *J. Am. Chem. Soc.*, 2006, **128**, 3508–3509; (b) D. Burtscher and K. Grela, *Angew. Chem., Int. Ed*, 2009, **48**, 442–454.
23. D. Rix, F. Caijo, I. Laurent, Ł. Gułajski, K. Grela and M. Mauduit, *Chem. Commun.*, 2007, 3771–3773.
24. (a) L. Jafarpour, M.-P. Heck, C. Baylon, H. M. Lee, C. Mioskowski and S. P. Nolan, *Organometallics*, 2002, **21**, 671–679; (b) L. Yang, M. Mayr, K. Wurst and M. R. Buchmeiser, *Chem.–Eur. J.*, 2004, **10**, 5761–5770; (c) A. Michrowska, K. Mennecke, U. Kunz, A. Kirschning and K. Grela, *J. Am. Chem. Soc.*, 2006, **128**, 13261–13267.
25. (a) T. Ung, A. Hejl, R. H. Grubbs and Y. Schrodi, *Organometallics*, 2004, **23**, 5399–5401; (b) A. Hejl, M. W. Day and R. H. Grubbs, *Organometallics*, 2006, **25**, 6149–6154; (c) N. Ledoux, R. Drozdzak, B. Allaert, A. Linden, P. Van Der Voort and F. Verpoort, *Dalton Trans.*, 2007, 5201–5210.
26. (a) B. M. Trost, M. U. Frederiksen and M. T. Rudd, *Angew. Chem., Int. Ed.*, 2005, **44**, 6630–6666; (b) B. M. Trost, F. D. Toste and A. B. Pinkerton, *Chem. Rev.*, 2001, **101**, 2067–2096; (c) A. Fürstner, *Angew. Chem., Int. Ed.*, 2000, **39**, 3012–3043.
27. I. Özdemir, S. Demir, B. Çetinkaya, C. Gourlaouen, F. Maseras, C. Bruneau and P. H. Dixneuf, *J. Am. Chem. Soc.*, 2008, **130**, 1156–1157.
28. M. Arisawa, Y. Terada, K. Takahashi, M. Nakagawa and A. Nishida, *J. Org. Chem.*, 2006, **71**, 4255–4261.
29. (a) D. E. Fogg, *Can. J. Chem.*, 2008, **86**, 931–941; (b) D. E. Fogg and E. N. dos Santos, *Coord. Chem. Rev.*, 2004, **248**, 2365–2379.
30. V. Dragutan and I. Dragutan, *J. Organomet. Chem.*, 2006, **691**, 5129–5147.
31. B. Alcaide, P. Almendros and A. Luna, *Chem. Rev.*, 2009, **109**, 3817–3858.
32. R. R. Schrock, *Angew. Chem., Int. Ed.*, 2006, **45**, 3748–3759.
33. R. H. Grubbs, *Angew. Chem., Int. Ed.*, 2006, **45**, 3760–3765.
34. (a) K. C. Nicolaou, P. G. Bulger and D. Sarlah, *Angew. Chem., Int. Ed*, 2005, **44**, 4490–4527; (b) A. H. Hoveyda and A. R. Zhugralin, *Nature*, 2007, **450**, 243–251.

35. (a) U. Frenzel and O. Nuyken, *J. Polym. Sci. A: Polym. Chem.*, 2002, **40**, 2895–2916; (b) C. Slugovc, *Macromol. Rapid Commun.*, 2004, **25**, 1283–1297.
36. R. H. Grubbs, ed. *Handbook of Metathesis*, Wiley-VCH, Weinheim, 2003.
37. (a) M. Mori, *Adv. Synth. Catal.*, 2007, **349**, 121–135; (b) S. T. Diver and A. J. Giessert, *Chem. Rev.*, 2004, **104**, 1317–1382.
38. (a) A. Fürstner and P. W. Davies, *Chem. Commun.*, 2005, 2307–2320; (b) W. Zhang and J. S. Moore, *Adv. Synth. Catal.*, 2007, **349**, 93–120.
39. (a) T. J. Seiders, D. W. Ward and R. H. Grubbs, *Org. Lett.*, 2001, **3**, 3225–3228; (b) J. J. Van Veldhuizen, D. G. Gillingham, S. B. Garber, O. Kataoka and A. H. Hoveyda, *J. Am. Chem. Soc.*, 2003, **125**, 12502–12508; (c) T. W. Funk, J. M. Berlin and R. H. Grubbs, *J. Am. Chem. Soc.*, 2006, **128**, 1840–1846; (d) S. J. Malcolmson, S. J. Meek, E. S. Sattely, R. R. Schrock and A. H. Hoveyda, *Nature*, 2008, **456**, 933–937.
40. A. M. Rouhi, *Chem. Eng. News*, 23 December 2002, 34–38.
41. (a) J.-L. Hérisson and Y. Chauvin, *Makromol. Chem.*, 1971, **141**, 161–176; (b) Y. Chauvin, *Angew. Chem., Int. Ed.*, 2006, **45**, 3741–3747.
42. S. T. Nguyen, L. K. Johnson, R. H. Grubbs and J. W. Ziller, *J. Am. Chem. Soc.*, 1992, **114**, 3974–3975.
43. (a) P. Schwab, M. B. France, J. W. Ziller and R. H. Grubbs, *Angew. Chem., Int. Ed. Engl.*, 1995, **34**, 2039–2041; (b) P. Schwab, R. H. Grubbs and J. W. Ziller, *J. Am. Chem. Soc.*, 1996, **118**, 100–110.
44. (a) E. L. Dias, S. T. Nguyen and R. H. Grubbs, *J. Am. Chem. Soc.*, 1997, **119**, 3887–3897; (b) M. Ulman and R. H. Grubbs, *J. Org. Chem.*, 1999, **64**, 7202–7207.
45. T. Weskamp, F. J. Kohl, W. Hieringer, D. Gleich and W. A. Herrmann, *Angew. Chem., Int. Ed.*, 1999, **38**, 2416–2419.
46. M. Scholl, S. Ding, C. W. Lee and R. H. Grubbs, *Org. Lett.*, 1999, **1**, 953–956.
47. J. Huang, E. D. Stevens, S. P. Nolan and J. L. Petersen, *J. Am. Chem. Soc.*, 1999, **121**, 2674–2678.
48. (a) M. S. Sanford, J. A. Love and R. H. Grubbs, *J. Am. Chem. Soc.*, 2001, **123**, 6543–6554; (b) J. A. Love, M. S. Sanford, M. W. Day and R. H. Grubbs, *J. Am. Chem. Soc.*, 2003, **125**, 10103–10109.
49. (a) C. Samojłowicz, M. Bieniek and K. Grela, *Chem. Rev.*, 2009, **109**, 3708–3742; (b) P. H. Deshmukh and S. Blechert, *Dalton Trans.*, 2007, 2479–2491; (c) E. Despagnet-Ayoub and T. Ritter, in *N-Heterocyclic Carbenes in Transition Metal Catalysis*, ed. F. Glorius, Springer, Berlin, 2007, *Topics in Organometallic Chemistry*, vol. 21, pp. 193–218; (d) S. Beligny and S. Blechert, in *N-Heterocyclic Carbenes in Synthesis*, ed. S. P. Nolan, Wiley-VCH, Weinheim, 2006, pp. 1–25.
50. J. A. Love, J. P. Morgan, T. M. Trnak and R. H. Grubbs, *Angew. Chem., Int. Ed.*, 2002, **41**, 4035–4037.
51. T.-L. Choi and R. H. Grubbs, *Angew. Chem., Int. Ed.*, 2003, **42**, 1743–1746.

52. (a) J. C. Conrad, H. H. Parnas, J. L. Snelgrove and D. E. Fogg, *J. Am. Chem. Soc.*, 2005, **127**, 11882–11883; (b) J. C. Conrad, K. D. Camm and D. E. Fogg, *Inorg. Chim. Acta*, 2006, **359**, 1967–1973; (c) S. Monfette and D. E. Fogg, *Organometallics*, 2006, **25**, 1940–1944.
53. T. Vorfalt, S. Leuthäußer and H. Plenio, *Angew. Chem., Int. Ed.*, 2009, **48**, 5191–5194.
54. T. Weskamp, W. C. Schattenmann, M. Spiegler and W. A. Herrmann, *Angew. Chem., Int. Ed.*, 1998, **37**, 2490–2493.
55. (a) L. Vieille-Petit, X. Luan, M. Gatti, S. Blumentritt, A. Linden, H. Clavier, S. P. Nolan and R. Dorta, *Chem. Commun.*, 2009, 3783–3785; (b) M. Gatti, L. Vieille-Petit, X. Luan, R. Mariz, E. Drinkel, A. Linden and R. Dorta, *J. Am. Chem. Soc.*, 2009, **131**, 9498–9499.
56. K. M. Kuhn, J.-B. Bourg, C. K. Chung, S. C. Virgil and R. H. Grubbs, *J. Am. Chem. Soc.*, 2009, **131**, 5313–5320.
57. G. C. Vougioukalakis and R. H. Grubbs, *J. Am. Chem. Soc.*, 2008, **130**, 2234–2245.
58. A. H. Hoveyda, D. G. Gillingham, J. J. Van Veldhuizen, O. Kataoka, S. B. Garber, J. S. Kingsbury and J. P. A. Harrity, *Org. Biomol. Chem.*, 2004, **2**, 8–23.
59. H. Wang, H. Matsuhashi, B. D. Doan, S. N. Goodman, X. Ouyang and W. M. Clark Jr, *Tetrahedron*, 2009, **65**, 6291–6303.
60. S. Gessler, S. Randl and S. Blechert, *Tetrahedron Lett.*, 2000, **41**, 9973–9976.
61. A. Michrowska, L. Gulajski, Z. Kaczmarska, K. Mennecke, A. Kirschning and K. Grela, *Green Chem.*, 2006, **8**, 685–688.
62. L. Gulajski, A. Michrowska, J. Naroznik, Z. Kaczmarska, L. Rupnicki and K. Grela, *ChemSusChem*, 2008, **1**, 103–109.
63. (a) D. Rix, F. Caijo, I. Laurent, F. Boeda, H. Clavier, S. P. Nolan and M. Mauduit, *J. Org. Chem.*, 2008, **73**, 4225–4228; (b) H. Clavier, F. Caijo, E. Borré, D. Rix, F. Boeda, S. P. Nolan and M. Mauduit, *Eur. J. Org. Chem.*, 2009, 4254–4265.
64. N. Vinokurov, J. R. Garabatos-Perera, Z. Zhao-Karger, M. Wiebcke and H. Butenschön, *Organometallics*, 2008, **27**, 1878–1886.
65. D. P. Allen, M. M. Van Wingerden and R. H. Grubbs, *Org. Lett.*, 2009, **11**, 1261–1264.
66. (a) S. Monsaert, A. Lozano Vila, R. Drozdzak, P. Van Der Voort and F. Verpoort, *Chem. Soc. Rev.*, 2009, **38**, 3360–3372; (b) A. Szadkowska and K. Grela, *Curr. Org. Chem.*, 2008, **12**, 1631–1647.
67. (a) B. Allaert, N. Dieltiens, N. Ledoux, C. Vercaemst, P. Van Der Voort, C. V. Stevens, A. Linden and F. Verpoort, *J. Mol. Catal. A: Chem.*, 2006, **260**, 221–226; (b) N. Ledoux, B. Allaert, D. Schaubroeck, S. Monsaert, R. Drozdzak, P. Van Der Voort and F. Verpoort, *J. Organomet. Chem.*, 2006, **691**, 5482–5486.
68. A. Szadkowska, A. Makal, K. Woźniak, R. Kadyrov and K. Grela, *Organometallics*, 2009, **28**, 2693–2700.
69. (a) A. Ben-Asuly, E. Tzur, C. E. Diesendruck, M. Sigalov, I. Goldberg and N. G. Lemcoff, *Organometallics*, 2008, **27**, 811–813; (b) T. Kost,

M. Sigalov, I. Goldberg, A. Ben-Asuly and N. G. Lemcoff, *J. Organomet. Chem.*, 2008, **693**, 2200–2203; (c) C. E. Diesendruck, Y. Vidavsky, A. Ben-Asuly and N. G. Lemcoff, *J. Polym. Sci. A: Polym. Chem.*, 2009, **47**, 4209–4213; (d) A. Ben-Asuly, A. Aharoni, C. E. Diesendruck, Y. Vidavsky, I. Goldberg, B. F. Straub and N. G. Lemcoff, *Organometallics*, 2009, **28**, 4652–4655.
70. (a) F. Boeda, H. Clavier and S. P. Nolan, *Chem. Commun.*, 2008, 2726–2740; (b) V. Dragutan, I. Dragutan and F. Verpoort, *Platinum Metals Rev.*, 2005, **49**, 33–40.
71. S. Monsaert, R. Drozdzak, V. Dragutan, I. Dragutan and F. Verpoort, *Eur. J. Inorg. Chem.*, 2008, 432–440.
72. H. Clavier, C. A. Urbina-Blanco and S. P. Nolan, *Organometallics*, 2009, **28**, 2848–2854.
73. (a) H. Clavier and S. P. Nolan, *Chem.–Eur. J.*, 2007, **13**, 8029–8036; (b) F. Boeda, X. Bantreil, H. Clavier and S. P. Nolan, *Adv. Synth. Catal.*, 2008, **350**, 2959–2966.
74. S. Monsaert, E. De Canck, R. Drozdzak, P. Van Der Voort, F. Verpoort, J. C. Martins and P. M. S. Hendrickx, *Eur. J. Org. Chem.*, 2009, 655–665.
75. M. Bieniek, A. Michrowska, D. L. Usanov and K. Grela, *Chem.–Eur. J.*, 2008, **14**, 806–818.
76. D. Burtscher, C. Lexer, K. Mereiter, R. Winde, R. Karch and C. Slugovc, *J. Polym. Sci. A: Polym. Chem.*, 2008, **46**, 4630–4635.
77. L. Delaude, A. Demonceau and A. F. Noels, *Curr. Org. Chem.*, 2006, **10**, 203–215.
78. A. Tudose, A. Demonceau and A. F. Noels, *J. Organomet. Chem.*, 2006, **691**, 5356–5365.
79. M. Ahr, C. Thieuleux, C. Copéret, B. Fenet and J.-M. Basset, *Adv. Synth. Catal.*, 2007, **349**, 1587–1591.
80. L. Delaude, X. Sauvage, A. Demonceau and J. Wouters, *Organometallics*, 2009, **28**, 4056–4064.
81. C. Lo, R. Cariou, C. Fischmeister and P. H. Dixneuf, *Adv. Synth. Catal.*, 2007, **349**, 546–550.
82. N. Ledoux, B. Allaert and F. Verpoort, *Eur. J. Inorg. Chem.*, 2007, 5578–5583.
83. X. Sauvage, Y. Borguet, A. F. Noels, L. Delaude and A. Demonceau, *Adv. Synth. Catal.*, 2007, **349**, 255–265.
84. (a) S. Díez-González, N. Marion and S. P. Nolan, *Chem. Rev.*, 2009, **109**, 3612–3676; (b) S. Burling, B. M. Paine and M. K. Whittlesey, in *N-Heterocyclic Carbenes in Synthesis*, ed. S. P. Nolan, Wiley-VCH, Weinheim, 2006, pp. 27–53.
85. (a) S.-I. Murahashi, ed. *Ruthenium in Organic Synthesis*, Wiley-VCH, Weinheim, 2005; (b) C. Bruneau and P. H. Dixneuf, ed. *Ruthenium Catalysts and Fine Chemistry*, Springer, Berlin, 2004.
86. M. Arisawa, Y. Terada, M. Nakagawa and A. Nishida, *Angew. Chem., Int. Ed.*, 2002, **41**, 473–4734.
87. J. C. Sworen, J. H. Pawlow, W. Case, J. Lever and K. B. Wagener, *J. Mol. Catal. A: Chem.*, 2003, **194**, 69–78.

88. S. E. Lehman, J. E. Schwendeman, P. M. O'Donnell and K. B. Wagener, *Inorg. Chim. Acta*, 2003, **345**, 190–198.
89. M. B. Dinger and J. C. Mol, *Eur. J. Inorg. Chem.*, 2003, 2827–2833.
90. S. Hanessian, S. Giroux and A. Larsson, *Org. Lett.*, 2006, **8**, 5481–5484.
91. M. Fekete and F. Joó, *Catal. Commun.*, 2006, **7**, 783–786.
92. B. Çetinkaya, S. Demir, I. Özdemir, L. Toupet, D. Sémeril, C. Bruneau and P. H. Dixneuf, *New J. Chem.*, 2001, **25**, 519–521.
93. D. Sémeril, C. Bruneau and P. H. Dixneuf, *Helv. Chim. Acta*, 2001, **84**, 3335–3341.
94. Y. Terada, M. Arisawa and A. Nishida, *J. Org. Chem.*, 2005, **71**, 1269–1272.
95. C. Mukai and R. Itoh, *Tetrahedron Lett.*, 2006, **47**, 3971–3974.
96. (a) F. Simal, D. Jan, L. Delaude, A. Demonceau, M.-R. Spirlet and A. F. Noels, *Can. J. Chem.*, 2001, **79**, 529–535; (b) L. Delaude, S. Delfosse, A. Richel, A. Demonceau and A. F. Noels, *Chem. Commun.*, 2003, 1526–1527; (c) A. Richel, S. Delfosse, C. Cremasco, L. Delaude, A. Demonceau and A. F. Noels, *Tetrahedron Lett.*, 2003, **44**, 6011–6015.
97. (a) T. Katsumata, M. Shiotsuki, S. Kuroki, I. Ando and T. Masuda, *Polym. J.*, 2005, **37**, 608–616; (b) T. Katsumata, M. Shiotsuki and T. Masuda, *Macromol. Chem. Phys.*, 2006, **207**, 1244–1252; (c) T. Katsumata, M. Shiotsuki, F. Sanda, X. Sauvage, L. Delaude and T. Masuda, *Macromol. Chem. Phys.*, 2009, **210**, 1891–1902.
98. Y. Zhang, D. Wang, K. Wurst and M. R. Buchmeiser, *J. Organomet. Chem.*, 2005, **690**, 5728–5735.
99. P. Csabai, F. Joó, A. M. Trzeciak and J. J. Ziółkowski, *J. Organomet. Chem.*, 2006, **691**, 3371–3376.
100. N. Gürbüz, I. Özdemir, B. Çetinkaya, J.-L. Renaud, B. Demerseman and C. Bruneau, *Tetrahedron Lett.*, 2006, **47**, 535–538.
101. S. Yaşar, I. Özdemir, B. Çetinkaya, J.-L. Renaud and C. Bruneau, *Eur. J. Org. Chem.*, 2008, 2142–2149.
102. S. Yaşar, Ö. Doğan, I. Özdemir and B. Çetinkaya, *Appl. Organomet. Chem.*, 2008, **22**, 314–318.
103. I. Özdemir, S. Demir, N. Gürbüz, B. Çetinkaya, L. Toupet, C. Bruneau and P. H. Dixneuf, *Eur. J. Inorg. Chem.*, 2009, 1942–1949.
104. J. W. Faller and P. P. Fontaine, *J. Organomet. Chem.*, 2006, **691**, 5798–5803.
105. K. Araki, S. Kuwata and T. Ikariya, *Organometallics*, 2008, **27**, 2176–2178.
106. S. P. Reade, A. L. Acton, M. F. Mahon, T. A. Martin and M. K. Whittlesey, *Eur. J. Inorg. Chem.*, 2009, 1774–1785.
107. S. P. Reade, M. F. Mahon and M. K. Whittlesey, *J. Am. Chem. Soc.*, 2009, **131**, 1847–1861.
108. T. J. Donohoe, T. J. C. O'Riordan and C. P. Rosa, *Angew. Chem., Int. Ed.*, 2009, **48**, 1014–1017.
109. A. E. Sutton, B. A. Seigal, D. F. Finnegan and M. L. Snapper, *J. Am. Chem. Soc.*, 2002, **124**, 13390–13391.

110. V. Böhrsch and S. Blechert, *Chem. Commun.*, 2006, 1968–1970.
111. B. Schmidt and A. Biernat, *Synlett*, 2007, 2375–2378.
112. R. P. Murelli, S. Catalán, M. P. Gannon and M. L. Snapper, *Tetrahedron Lett.*, 2008, **49**, 5714–5717.
113. M. A. Esteruelas and A. M. Lopez, *Organometallics*, 2005, **24**, 3584–3613.
114. (a) P. B. Hitchcock, M. F. Lappert and P. L. Pye, *J. Chem. Soc., Dalton Trans.*, 1978, 826–836; (b) M. F. Lappert and P. L. Pye, *J. Chem. Soc., Dalton Trans.*, 1978, 837–844.
115. R. Castarlenas, M. A. Esteruelas and E. Oñate, *Organometallics*, 2005, **24**, 4343–4346.
116. R. Castarlenas, M. A. Esteruelas and E. Oñate, *Organometallics*, 2008, **27**, 3240–3247.

CHAPTER 8

NHC–Cobalt, Rhodium and Iridium Complexes in Catalysis

VINCENT CÉSAR,[a,*] LUTZ H. GADE[b,*] AND STÉPHANE BELLEMIN-LAPONNAZ[c,*]

[a] Laboratoire de Chimie de Coordination du CNRS, 205 route de Narbonne, 31077 Toulouse Cedex 4, France; [b] Anorganisch-Chemisches Institut, Universität Heidelberg, Im Neuenheimer Feld 270, 69120 Heidelberg, Germany; [c] Institut de Physique et Chimie des Matériaux de Strasbourg, CNRS-Université de Strasbourg, 23 Rue du Loess, 67034 Strasbourg, France

8.1 Introduction

This Chapter is devoted to the development of cobalt-, rhodium- and iridium-based catalysts that contain N-heterocyclic carbene (NHC) ligands. It will cover their most relevant catalytic applications, along with stoichiometric model reactions, except for catalytic oxidations and reductions such as hydrogenations, hydrosilylations and hydroborations, which are treated in detail in Chapters 12 and 13. Since the NHC chemistry of Group 9 metals is one of the most developed areas in this field, this overview will only cover the chemistry of 'classical' NHCs, namely cyclic diaminocarbenes. Chapter 5 reviews the chemistry of the 'non-classical' NHCs, to which the reader is referred. Finally, this Chapter does not pretend to be exhaustive and further details may be found in previous overviews.[1,2]

8.2 NHC–Cobalt Complexes

The N-heterocyclic carbene complexes of cobalt have been less studied than their heavier analogues, rhodium and iridium, even though NHC–cobalt complexes have been known for more than 3 decades. Despite the fact that their

RSC Catalysis Series No. 6
N-Heterocyclic Carbenes: From Laboratory Curiosities to Efficient Synthetic Tools
Edited by Silvia Díez-González
© Royal Society of Chemistry 2011
Published by the Royal Society of Chemistry, www.rsc.org

first preparation dates back to the 1970s,[3,4] applications of NHC–cobalt catalysts only emerged recently, but this field is growing at a rapid pace.

8.2.1 Stoichiometric Activation of Small Molecules

Although most studies related to the activation of small molecules concern stoichiometric transformations, this topic is relevant to the development of new catalysts and catalytic reactions. Notable achievements in this field were due to Meyer and co-workers using the potentially tetradentate tris-NHC ligand tris[2-(3-arylimidazol-2-ylidene)ethyl]amine ($TIMEN^{Ar}$).[5] Reaction of $TIMEN^{Ar}$ with $[Co(PPh_3)_3Cl]$ yielded the coordinatively unsaturated cobalt(I) complex $[(TIMEN^{Ar})Co]Cl$ $\mathbf{1^{Ar}}$, with a high-spin ground state. The coordinated $TIMEN^{Ar}$ ligand wrapped itself around the cobalt centre, thus protecting it, but leaving sufficient space for additional ligand binding (Scheme 8.1).[6]

Thus, complex $\mathbf{1^{Xyl}}$ reacted with CO by displacement of the weakly coordinated chloro ligand to form the cationic cobalt(I) complex $[(TIMEN^{Xyl})Co(CO)]^+$ $\mathbf{2^{Xyl}}$ and with benzyl chloride or chlorinated solvents (dichloromethane or chloroform) in a one-electron oxidation, yielding the blue-coloured, paramagnetic cobalt(II) complex $\mathbf{3^{Xyl}}$. Reaction of $\mathbf{1^{Xyl}}$ with molecular dioxygen produced the cobalt(III) complex $\mathbf{4^{Xyl}}$ containing a side-on bound peroxo ligand and capable of transferring an oxygen atom to organic electrophiles such as benzoyl chloride or benzylidene malodinitrile. Furthermore, complex $\mathbf{1^{Ar}}$ reacted with aryl azides at low temperature (−35 °C) to form cobalt(III) imido complexes $\mathbf{5^{Ar}}$.[7] The imido ligand is highly electrophilic and inserted into a cobalt–carbene bond at room temperature to form a cobalt(I)–imine species which in turn readily disproportionated, to produce the cobalt(II)–imine complex $\mathbf{6^{Ar}}$ and metallic cobalt(0).

Clearly, these studies established interesting reactivity patterns that may lead to the development of new catalytic reactions in the future.

8.2.2 Cobalt-catalyzed Cyclizations

The first application of NHC–cobalt catalysts was reported in 2003 by Gibson and Loch.[8] Several bimetallic complexes of the type $[L^1\text{-}Co(CO)_3\text{-}Co(CO)_3\text{-}L^2]$ **7**–**9** were synthesized in low to moderate yields (20–35%) by displacement of CO from the complex $[Co_2(CO)_8]$. The competitive formation of the disproportionation product $[(NHC)_2Co(CO)_3]^+[Co(CO)_4]^-$ in THF was found to be a limiting factor in the syntheses of these species.[9] Their catalytic activity was tested in the Pauson–Khand reaction (PKR) using a standard substrate (Scheme 8.2). NHC-containing catalysts **7**–**9** displayed lower activity than their phosphine analogue $[(Ph_3P)Co_2(CO)_7]$ as well as the non-substituted reference system $[Co_2(CO)_8]$. Subsequently, a chiral IPr-containing Nicholas-type complex, $[(\mu\text{-alkyne})Co_2(CO)_6]$, was reported as a stoichiometric reagent in a diastereoselective PKR with good diastereoselectivity (92% *de*).[10]

In 2005, Okamoto and co-workers reported that the combination of $CoCl_2$ (1–3 mol%) with 2 equivalents of IPr ligand and zinc powder (10 mol%, as

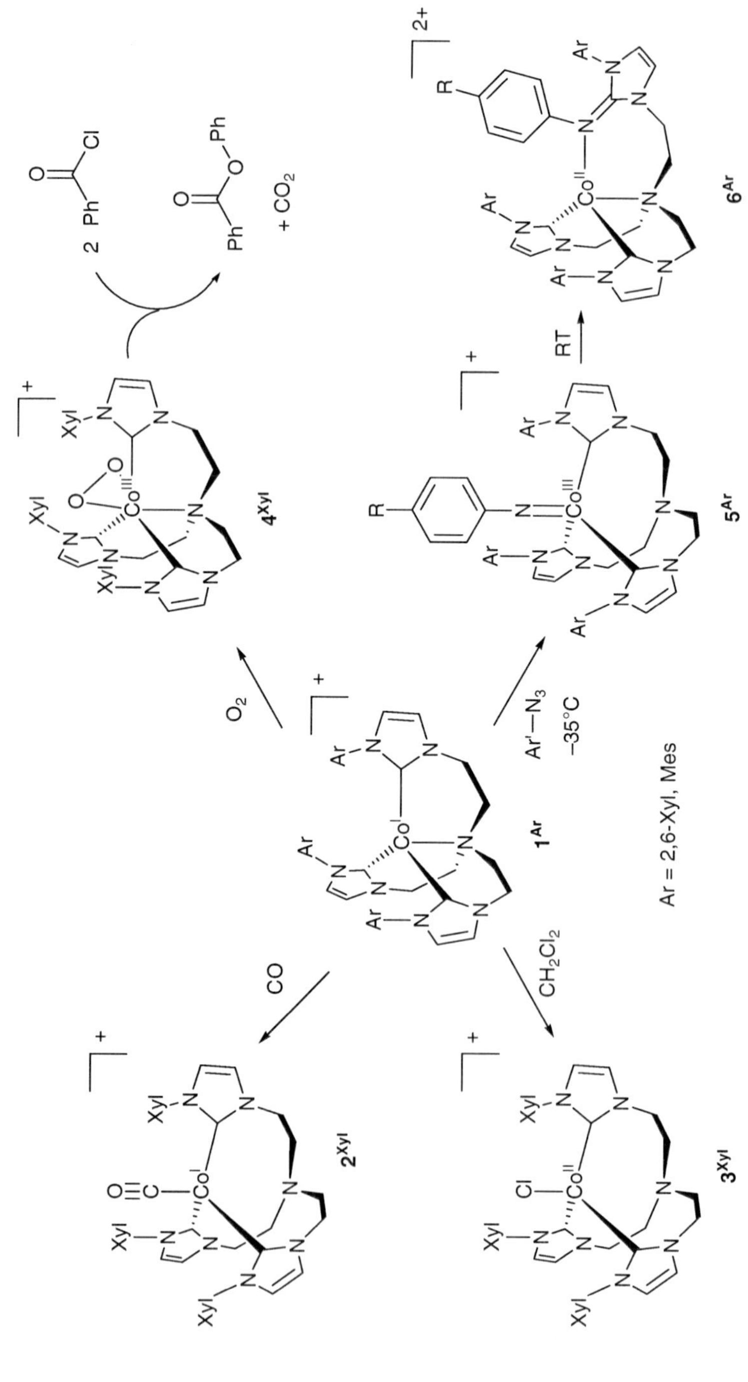

Scheme 8.1 Reactivity of TIMEN^{Ar}-containing cobalt(I) complex $\mathbf{1^{Ar}}$ towards small molecules (Xyl = xylyl; Mes = mesityl).

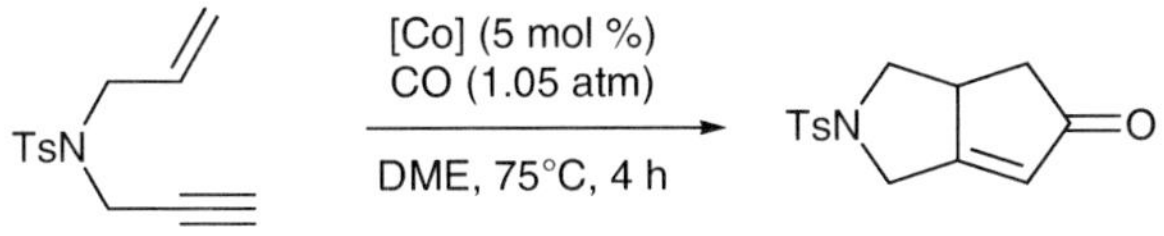

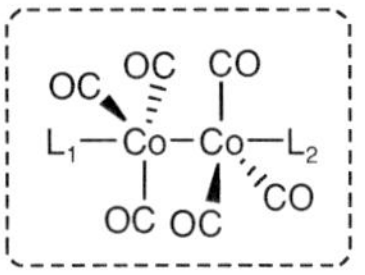

[Co]	L_1	L_2	yield (%)
7	IMes	CO	73
8	IMes	PPh_3	40
9	SIMes	PPh_3	59
$[(PPh_3)Co_2(CO)_7]$	PPh_3	CO	97
$[Co_2(CO)_8]$	CO	CO	87

Scheme 8.2 Activity of cobalt(0) dimers in a Pauson–Khand reaction.

reducing agent for the pre-catalyst) allowed the intramolecular cyclotrimerization of triynes (Equation (8.1)).[11] A low valent $[(IPr)_2Co^0]$ complex was proposed to be the active catalytic species in this transformation.

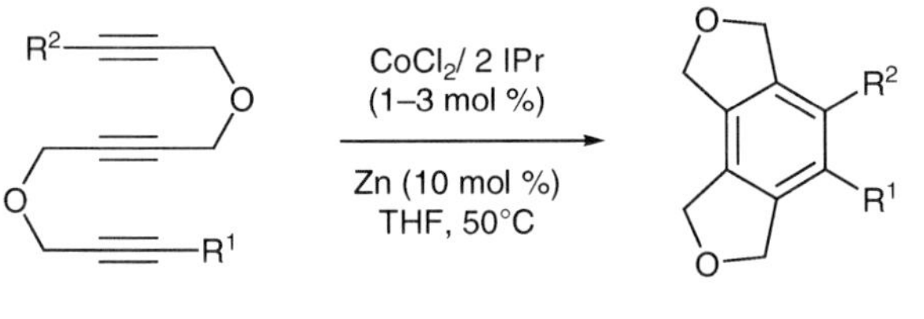

(8.1)

In continuation of their work on the [2+2+2]-cycloaddition of enediynes,[12] Gandon, Aubert and Malacria discovered that replacement of PPh_3 by IPr in their previously reported $CoI_2/PPh_3/Mn$ reagent greatly improved the performance of this system (Equation (8.2)).[13] Whereas a stoichiometric amount of CoI_2 was still required, IPr could be used catalytically (6 mol%) with 1 equivalent of manganese, whereas previously 2 equivalents of PPh_3 and 10 equivalents of Mn were generally needed to obtain acceptable yields. This system proved to be tolerant to functional groups and cyclization occurred diastereospecifically with substrates that had terminal substituents on the olefin.

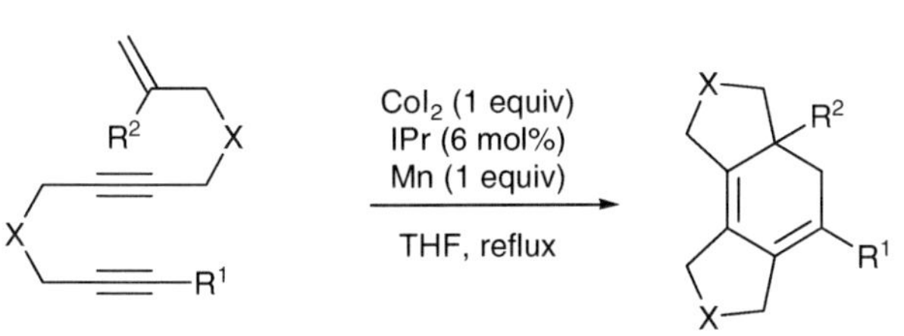

(8.2)

X = O, NTs, $C(COOEt)_2$
R_1 = H, Me, Ph, Bu, CO_2Me
R_2 = H, Me, Ph

8.2.3 Activation of Carbon–Halogen Bonds

Yorimitsu and Oshima exploited the capacity of cobalt to act as a one-electron reductant towards carbon–halogen bonds in several very interesting organic reactions. In a first report, the sequential cyclization/cross-coupling reaction of several 6-halo-1-hexene derivatives with Grignard reagents was described (Equation (8.3)).[14] A series of alkyl iodides and bromides cyclized in 5-*exo*-dig fashion *via* Co-mediated generation of a carbon radical, which was generated by one-electron reduction of the C–X bond. The trapping of the exocyclic radical by the cobalt centre and subsequent reductive elimination were proposed to be the key steps in this transformation.

Ts N I
$CoCl_2$ (5 mol %)
SIEt·HCl (5 mol %)
RMe_2SiCH_2MgCl (3 equiv)
Dioxane, RT
R = allyl
Ts N $SiMe_2R$
81%
Cl^- N + N H
SIEt·HCl

(8.3)

The best catalytic system resulted from the combination of $CoCl_2$ (5 mol%) and the imidazolinium salt SIEt·HCl (5 mol%) as NHC ligand precursor, and was found to outperform the catalyst containing IPr, IMes and SIMes. This methodology was further applied to the synthesis of diverse 1,3-diols, starting from iodovinylsilyloxy derivatives and subsequently performing a one-pot Tamao–Fleming oxidation of the cyclized product (Equation (8.4)).[15]

O Si I
+
$PhMe_2SiCH_2MgCl$
$CoCl_2$
SIEt·HCl
Dioxane
O Si $SiMe_2Ph$
KF, $KHCO_3$
H_2O_2 aq (30%)
MeOH/THF
(1:1)
51% (2 steps)
(50/50 mixture)
OH OH $SiMe_2Ph$

(8.4)

Modifying their previous catalytic system, Yorimitsu and Oshima succeeded recently in developing a remarkable catalytic dehydrohalogenation of alkyl halides.[16] In this case, IMes was found to be superior to other NHCs, phosphines and amines as ancillary ligand, whilst the bulky dimethylphenylsilylmethylmagnesium chloride proved to be the best co-reagent. Primary and secondary alkyl iodides and bromides appeared to be suitable substrates for this reaction but not the corresponding chlorides and tosylates. The main advantage of this process was its high regioselectivity favouring the formation of the less substituted alkene, unlike the corresponding base-promoted elimination. The proposed mechanism involved a single-electron reduction of the

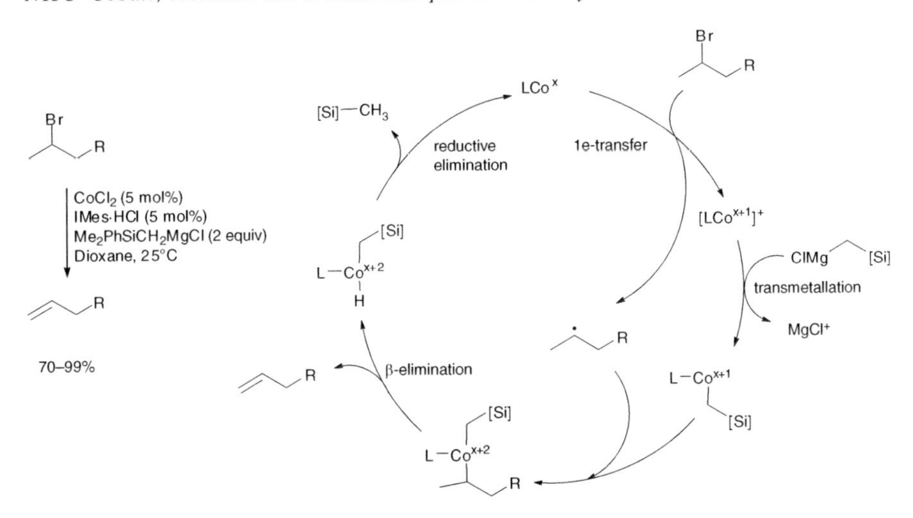

Scheme 8.3 Dehydrohalogenation of alkyl bromides and proposed catalytic cycle.

alkyl halide to generate the carbon radical, transmetalation and subsequent trapping of the radical, followed by *β*-hydride elimination (Scheme 8.3).

Finally, two reports on the Kumada–Corriu cross-coupling of aryl halogenides with Grignard reagents appeared.[17] Most of the aryl bromides were readily converted; however, chloroarenes remained poor substrates reacting either sluggishly or being totally unreactive. Although these preliminary results could not compete with the 'classical' Ni- and Pd-catalyzed Kumada–Corriu cross-coupling, they constituted the proof-of-concept that NHC–Co complexes are potential catalysts for C–C cross-coupling. The low cost of cobalt salts should encourage increasing research efforts in this field.

8.2.4 Miscellaneous Reactions

8.2.4.1 Isomerization of 1-Alkenes

During the course of their work on the dehydrohalogenation of alkyl halides (see Section 8.2.3), Yorimitsu and Oshima observed the isomerization of the 1-alkene product to (*E*)-2-alkene upon prolonged heating. From this starting point, they recently reported an optimized and highly selective cobalt-catalyzed isomerization of 1-alkenes to (*E*)-2-alkenes.[18] A cobalt(0) complex obtained by combining $CoCl_2$, IMes·HCl and $PhMe_2SiCH_2MgCl$ was proposed as active species in this process. This synthetic method was applied in the stereoselective formation of (*E*)-2-crotylsilanes (Equation (8.5)) and (*E*)-1-propenylsilanes.

$$\text{CH}_2{=}\text{CHCH}_2\text{SiR}_3 \xrightarrow[\text{Dioxane, 50 °C}]{\text{CoCl}_2\text{ (5 mol \%)},\ \text{IMes·HCl (5 mol \%)},\ \text{Me}_2\text{PhSiCH}_2\text{MgCl (0.5 equiv)}} (E)\text{-CH}_3\text{CH}{=}\text{CHCH}_2\text{SiR}_3 \quad \text{51–77\%} \tag{8.5}$$

8.2.4.2 *Hydroformylation of Alkenes*

The hydroformylation of alkenes is an important reaction in industry and the largest homogeneous catalytic process in terms of volume.[19] It is industrially so important because the resulting aldehydes are easily converted into many secondary products that are used as specialty chemicals (detergents, fragrances···). The catalyst precursors are typically dinuclear complexes $[(R_3P)Co(CO)_3]_2$ which are converted to the hydride catalysts $[(R_3P)Co(CO)_3H]$ under the reaction conditions.[20] Unfortunately, the NHC-based dinuclear complexes $[(IMes)Co(CO)_3]_2$ were found to be totally inactive under the chosen conditions (170 °C, 60 bar),[9b] mainly because of the formation of the imidazolium salt $[IMes{\cdot}H]^+$ by reductive elimination of IMes and the *in situ* generated hydride.[21] In contrast, Llewellyn *et al.* reported promising results employing milder conditions (50 °C, 8 bar) with complex $[(IMes)CoH(CO)_3]$ as catalyst (Equation (8.6)).[22] The unusually high selectivity for the branched aldehyde product 2-methyloctanal was proposed to arise from a [(NHC)Co]-catalyzed isomerization of the terminal alkene to an internal alkene before hydroformylation, although this aspect certainly requires further mechanistic studies.

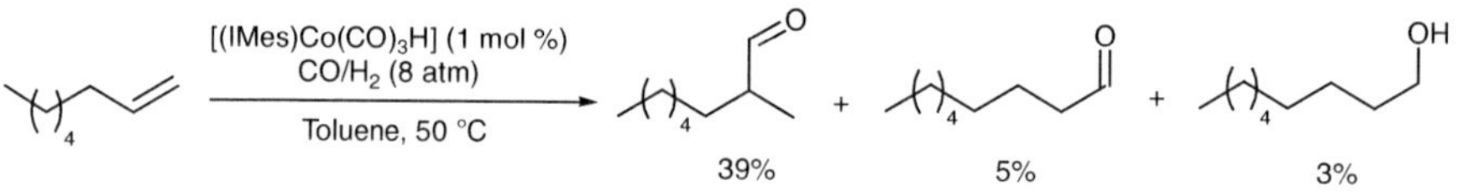

(8.6)

To summarize, applications of NHC–cobalt complexes in homogeneous catalysis only emerged a few years ago, with emphasis on the comparison between NHC and phosphine ligands in known cobalt catalysis such as the Pauson–Khand or hydroformylation reactions. However, it appears that the introduction of NHCs in cobalt catalysts may lead to interesting and previously unknown patterns of reactivity. For this reason and in view of the accelerating research activity since 2007/2008, a growing interest in NHC–cobalt catalysis is expected.

8.3 NHC–Rhodium Complexes

The first efficient catalytic applications of N-heterocyclic carbene ligands with late transition metals were reported for rhodium.[23] Since the mid 1990s, NHC–rhodium chemistry has been extensively studied and may be comparable to palladium or nickel in terms of the number of catalytic applications. Outside the context of catalysis, NHC–Rh complexes have featured prominently in studies probing the electronic properties of NHCs, usually quantified by comparison of the μ_{CO} frequencies in the IR spectra of $[LRh(CO)_2Cl]$ complexes.[24]

8.3.1 Arylation of Carbonyl and Related Compounds with Organoboron Reagents

The study of rhodium-catalyzed additions of organoboron reagents to carbonyl compounds and related substrates such as α,β-unsaturated ketones or esters is an active field of research mainly employing phosphorus or diene-based ligands.[25] Given the analogy between NHC and phosphine ligands, it is not surprising that this field has been one of the most intensely investigated in NHC–rhodium catalysis (along with hydrosilylations and hydrogenations reviewed in Chapter 13). This section will focus on three classes of substrates that require different reaction conditions and catalyst compositions.

8.3.1.1 Synthesis of Diarylmethanol Derivatives from Aldehydes

Diarylmethanol units are key structural elements in a range of pharmaceutically active compounds,[25b] and can be obtained by the rhodium-catalyzed addition of organoboronic acids to aldehydes, a reaction first discovered by Miyaura and co-workers in 1998.[26] Soon thereafter, Fürstner *et al.* greatly improved this catalytic method in terms of activity (5–8 times faster than with phosphine-based systems) and scope using monodentate NHC ligands (Scheme 8.4).[27] The optimized catalytic system was composed of the imidazolium chloride IPr·HCl (1 mol%) and the simplest possible rhodium source, $RhCl_3{\cdot}3H_2O$ (1 mol%), and allowed for the arylation of numerous aryl and alkyl aldehydes in moderate to excellent yields (52–99%). Vinylboronic acids were also suitable substrates and led to the formation of the corresponding allylic alcohols.

Subsequently, a great number of NHC–rhodium complexes were reported to be active catalysts in this addition reaction. These systems fall into two representative categories. First, rhodium(I) complexes of the general type [(NHC)RhCl(COD)] (COD = 1,5-cyclooctadiene) were found to be active pre-catalysts for the addition of arylboronic acids to benzaldehydes (1 mol% [Rh], 80 °C). In these complexes, the NHC ligand was always monodentate with the heterocyclic ring being either a

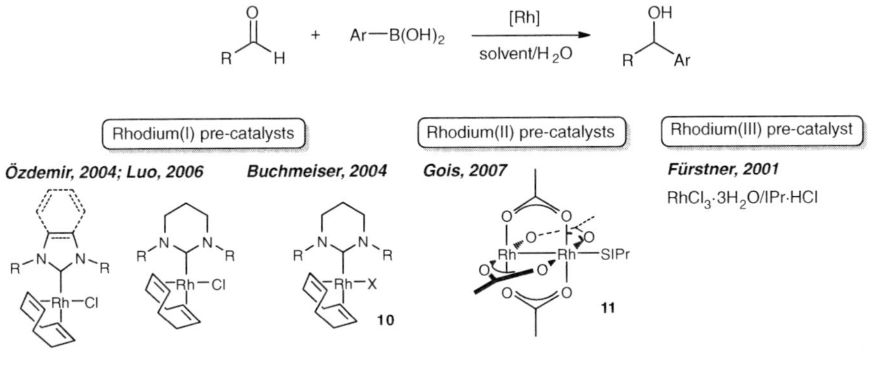

+ Ar—B(OH)$_2$ —[Rh], solvent/H_2O→

Rhodium(I) pre-catalysts

Özdemir, 2004; Luo, 2006

Buchmeiser, 2004

10

Rhodium(II) pre-catalysts

Gois, 2007

11

Rhodium(III) pre-catalyst

Fürstner, 2001

$RhCl_3{\cdot}3H_2O$/IPr·HCl

Scheme 8.4 Addition of organoboronic acids to aldehydes catalysed by NHC–rhodium catalysts.

five-membered ring such as imidazol-2-ylidene,[28] imidazolin-2-ylidene,[29] benzimidazol-2-ylidene[30] or perhydrobenzimidazol-2-ylidene.[31] Alternatively, six-membered ring NHCs, such as tetrahydropyrimidin-2-ylidene, with different substitution patterns on the nitrogen atoms, were employed.[32] Among them, complex **10** (with R = *i*-Pr, X = CF_3COO) was found to be the most active to date with TONs of up to 1230 (Scheme 8.4).

In 2007, Gois and co-workers reported another important family of catalysts for this reaction, derived from rhodium(II) 'paddlewheel' complexes $[Rh(OAc)_2]_2$ whose axial coordination sites were mono- or double-substituted with NHCs.[33] Four complexes were tested and the best results were obtained with the mono-substituted complex **11** possessing the saturated carbene SIPr as axial ligand (Scheme 8.4). An unusual mechanism emerged from experimental mechanistic studies combined with DFT modelling (Scheme 8.5).[34] The metal centre apparently did not participate directly in the aryl transfer to the aldehyde, but the binding of boronic acid to the second free axial coordination site may facilitate the formation of the borate complex **B** by addition of a molecule of solvent (an alcohol) to the boron atom. Interaction with the aldehyde would lead to intermediate **C**. The subsequent transfer of the aryl moiety to the aldehyde *via* a transition state $\mathbf{TS_{CD}}$, stabilized by hydrogen bonding would be the key reaction step. Since the NHC ligand was located at the opposite axial position of the molecule, it had a profound effect on the reactive site by

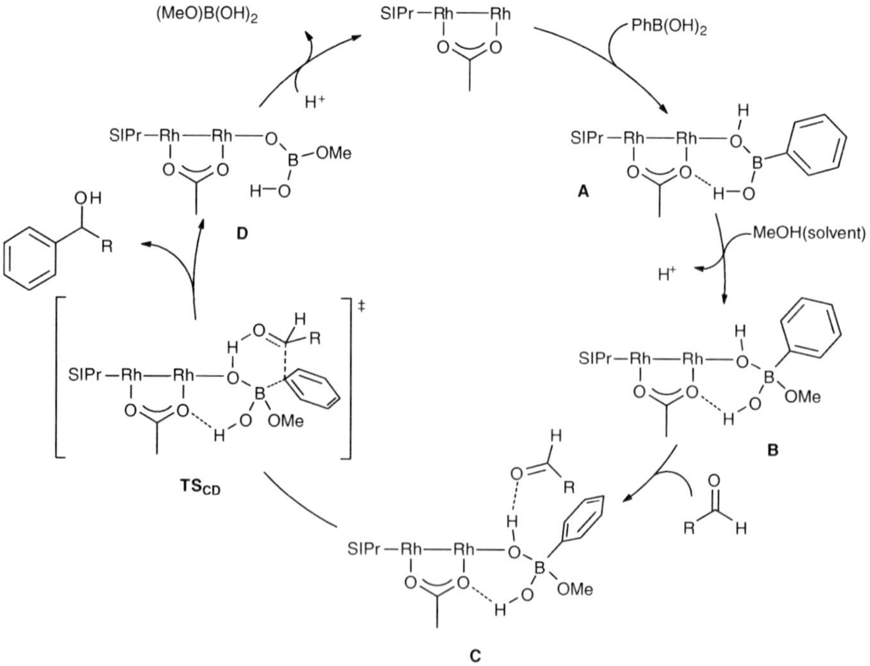

Scheme 8.5 Mechanistic proposal for the arylation of aldehydes using complex **11** (the three other acetato ligands are omitted for clarity).

adjusting the electronic properties of the catalyst and by the steric shielding of the second coordination site.

Since this reaction produces secondary alcohols, an asymmetric version using a chiral NHC ligand is conceivable.[35] However, only one report by Bolm and co-workers has appeared to date, in which a catalyst bearing an enantiopure paracyclophane-containing NHC was employed, but only gave low enantioselectivities (up to 38% *ee*).[36]

8.3.1.2 1,4-Addition to α,β-Unsaturated Compounds

Interestingly and in contrast to the vast majority of rhodium-based catalysts,[37] addition of organo-boronic acids to conjugated enals using complex **11** as catalyst resulted exclusively in the formation of 1,2-addition products.[34] With the exception of this particular system, NHC-based catalysts generally produced the 1,4-addition products, and this transformation was widely studied as reference reaction in the evaluation of the stereodirecting potential of new chiral NHC scaffolds (Equation (8.7)). Among the systems studied, the C_2-symmetric, monodentate imidazolinylidene derived from **12**, with very bulky chiral paracyclophanes as *N*-substituents, gave the 1,4-addition products in high enantiomeric excesses with cyclic enones (83–98% *ee*) and good *ee*'s with acyclic substrates (73–91%).[38] Complex **13**, containing a chiral bidentate phosphine–NHC ligand, also furnished good enantioselectivities for cyclic substrates (67–94% *ee*) and a remarkable 99% *ee* for one acyclic α,β-unsaturated ester.[39]

+ Ph—B(OH)$_2$ —[Rh], solvent/H_2O→

12	75%, 95% *ee*
13	91%, 94% *ee*

[Rh(acac)(C_2H_4)]$_2$ / **12**; **13** (8.7)

8.3.1.3 Synthesis of Secondary Amines from Imines

Despite the considerable synthetic potential of this reaction, only two NHC-containing catalysts have been reported to date. The *in situ* generated complex from the reaction between [RhCl(COD)]$_2$ and a NHC–silver ligand transfer reagent was active in the arylation of an *N*-phosphinoyl aldimine with phenylboronic acid, but unfortunately only one catalytic test was performed (Equation (8.8)).[40] In a more systematic study, Suzuki and Sato screened various azolium salts, among which the system $[RhCl(COD)]_2$/IAd·HCl proved to be the most active catalyst for the arylation of a series of *N*-sulfonyl and *N*-phosphinoyl arylimines.[41]

$$\text{ArCH=NR} + \text{Ph–B(OH)}_2 \xrightarrow{[\text{Rh}]} \text{ArCH(Ph)–NHR} \qquad (8.8)$$

R = *p*-TolSO$_2^-$, Ph$_2$P(=O)$^-$

[Rh] = $[RhCl(COD)]_2$/ dinuclear Ag–NHC complex (Mes, Ph, Ag—Ag)

8.3.2 Hydroformylations

Since the 1970s, most hydroformylation protocols have relied on catalysts based on phosphine/rhodium systems, such as $[(Ph_3P)_3RhH(CO)]$. A range of NHC–Rh complexes were also tested for this reaction, and first promising results have been reported.[42] Figure 8.1 displays some representative NHC–Rh complexes tested in hydroformylations, the first examples having been reported by Crudden and co-workers in 2000 (**14** and **15**).[43] Complex **14** was capable of hydroformylating various styrene derivatives in high yield and high selectivity favouring the branched product (Equation (8.9)). However, its activity was very low (up to 8 turnovers h^{-1} under the conditions studied). Similar activities were established for the dimetallic Rh complex **16** reported by Peris and co-workers, which displayed TOFs up to 29 h^{-1}.[44] The most active NHC-based catalyst was complex **17**, with a tetrazole-based

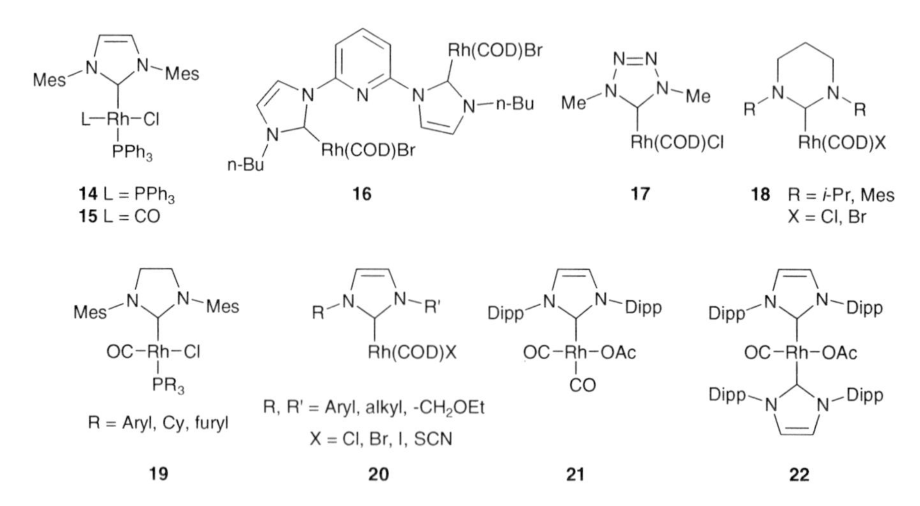

Figure 8.1 Selection of active NHC–Rh catalysts in the hydroformylation of alkenes.

CO/H$_2$
14 (1 mol %)
87% conv.
linear : branched = 3 / 97

(8.9)

NHC, with TOFs of up to 3500 h^{-1}.[45] Tetrahydropyrimidinylidene-based systems like **18** also gave high TOFs of around 1500 h^{-1}; however, olefin isomerization occurred concomitantly with hydroformylation, leading to a loss of branched/linear ratio as the reaction proceeded.[46] This turned out to be a general feature of most carbene complexes studied in this context.

Of note, substitution of unsaturated NHC ligands by their saturated analogues, such as **19**, doubled the activity but left the branched/linear ratios unaffected.[47] An increase in activity was also observed upon addition of one equivalent of triethylamine with no appreciable loss of selectivity. This increase in activity might be attributed to a base-assisted generation of the active catalyst, which is thought to be a rhodium hydride species. Many other rhodium complexes of the general formula [(NHC)Rh(COD)X] **20** were tested and usually displayed reasonable activity.[48,49]

Finally, the most promising hydroformylation catalysts, **21** and **22**, were halide-free complexes and contain the bulky IPr ligand.[50] High regioselectivities of up to 40/1 in favor of the branched isomer were observed in the hydroformylation of terminal olefins, with TOFs of up to 725 h^{-1}. Most importantly, the reaction occurred without concomitant isomerization, which was attributed to the steric bulk of the IPr ligand.

8.3.3 Rhodium-catalyzed Cyclizations

[(NHC)Rh(COD)Cl] complexes are active catalysts for various cyclizations involving alkenes and alkynes. In 2005, Evans *et al.* showed that [(IMes)Rh(COD)Cl] catalyzed the diastereoselective [4 + 2 + 2] carbocyclization of 1,6-enynes in the presence of 1,3-butadiene and silver triflate (Equation (8.10)).[51] Classical phosphine/rhodium systems generally displayed lower activities. For example, the use of the electron-rich tricyclohexylphosphine gave only 30% yield, demonstrating that strong σ-donation is crucial for the catalytic turnover in this carbocyclization.

[Rh]
AgOTf
1,3-butadiene

>19 : 1

[(IMes)Rh(COD)Cl]	75%
[RhCl(COE)$_2$]$_2$ + 2 PCy$_3$	30%

(8.10)

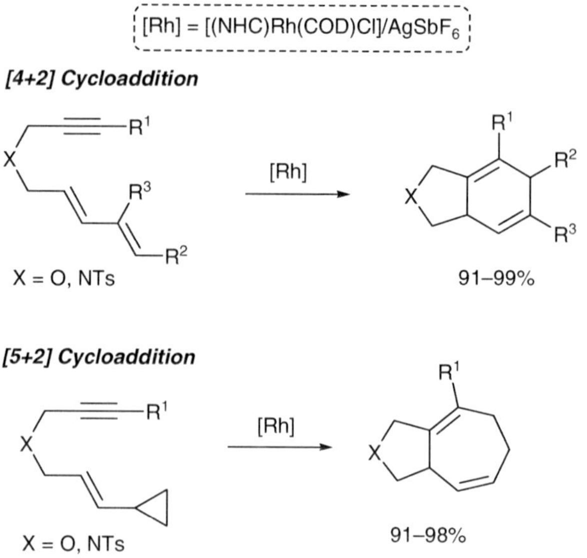

$[Rh] = [(NHC)Rh(COD)Cl]/AgSbF_6$

[4+2] Cycloaddition

X = O, NTs

[Rh]

91–99%

[5+2] Cycloaddition

X = O, NTs

[Rh]

91–98%

Scheme 8.6 [(NHC)Rh]-catalyzed [4 + 2] and [5 + 2] cycloaddition reactions.

This type of complexes also induced efficiently the intra- and intermolecular [4 + 2] and intramolecular [5 + 2] cycloaddition reactions shown in Scheme 8.6.[52] However, these catalysts were not effective for the intermolecular [5 + 2] cycloaddition reaction. The $[(IPr)Rh(COD)Cl]/AgSbF_6$ system had several attractive features, such as mild reaction conditions (15–20 °C) combined with high yields (up to 99%) and high turnover numbers (up to 1900).[53]

A quinoline-tethered NHC–rhodium complex was used as catalyst for the [3 + 2] cycloaddition of diphenylcyclopropenone and internal alkynes (Equation (8.11)).[54] In the presence of 2 mol% of **23**, the desired pentenone derivatives were isolated in up to 86% yield. Also, this catalyst tolerated functional substituents such as ketone, ester or hydroxyl groups.

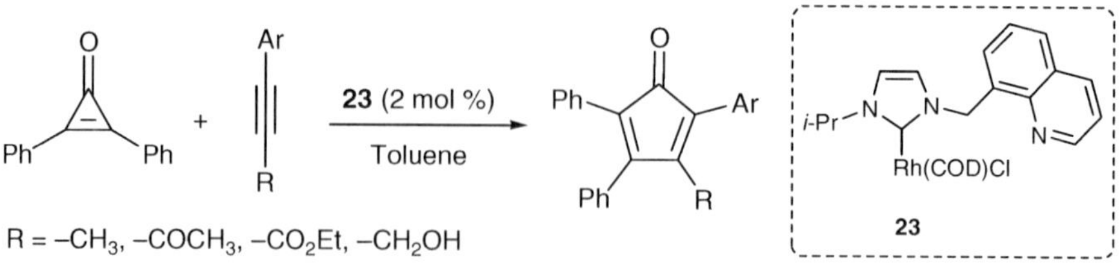

23 (2 mol %), Toluene

$R = -CH_3, -COCH_3, -CO_2Et, -CH_2OH$

23

(8.11)

Dinuclear rhodium(II) complexes were tested as catalysts for intramolecular C–H insertions of diazo-acetamides. In this study, the reactivity of axially monocoordinated $[(NHC)Rh_2(OAc)_4]$ complexes was found to be different from that of the parent [$Rh_2(OAc)_4$] complex.[55] More importantly, unexpected decarboxylation was observed in certain cases when using these NHC–Rh catalysts (Equation (8.12)).

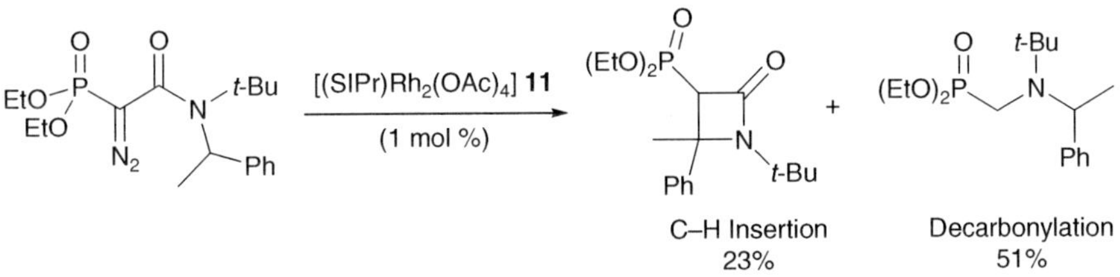

[(SIPr)$Rh_2(OAc)_4$] **11** (1 mol %)

C–H Insertion 23%

Decarbonylation 51%

(8.12)

8.3.4 Miscellaneous Reactions

NHC–rhodium complexes have been employed in a diverse range of chemical transformations of which only selected recent examples will be highlighted in this section.[56]

The hydroamination of alkenes and alkynes provides a highly atom economical method for the preparation of substituted amines and imines. Despite substantial efforts and recent progress, the development of a generally applicable functional group-tolerant catalyst for this reaction remains a challenge.[57] An interesting example was reported by using Rh combined with a bidentate NHC ligand (Equation (8.13)).[58] Complex **24** catalyzed the intramolecular hydroamination of aminoalkynes; however, the turnover rates remained modest with values up to 50 h^{-1}.

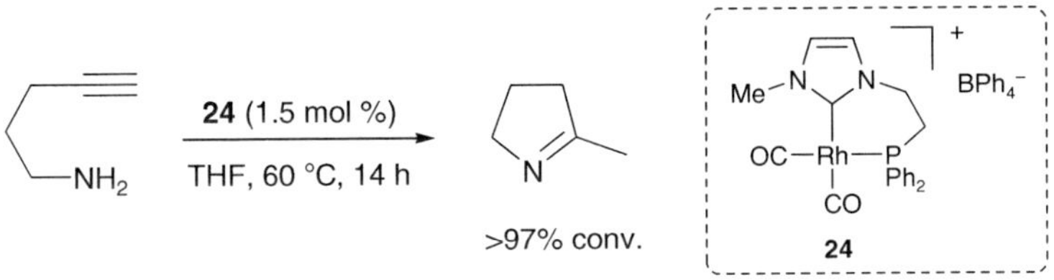

24 (1.5 mol %), THF, 60 °C, 14 h

>97% conv.

(8.13)

Chloroesterification of alkynes by chloroformates produces chloroacrylate esters, which are valuable intermediates in organic synthesis. Chung and coworkers investigated the [(NHC)Rh]-catalyzed chloroesterification of alkynes and enynes (Equation (8.14)).[59] The reaction was studied using [(NHC)Rh(COD)Cl] (NHC = IPr or IMes) as catalyst. At 100 °C and 1 mol% of catalyst loading, the chloroesterification product was isolated in 83% yield whereas only 18% yield was obtained with [(Ph_3P)Rh(COD)Cl]. Interestingly, only conjugated terminal alkynes (*i.e.* 1,3-enynes or phenylacetylenes) were suitable substrates for this catalytic reaction, whereas alkyl acetylenes gave poor yields and internal alkynes were found to be unreactive.

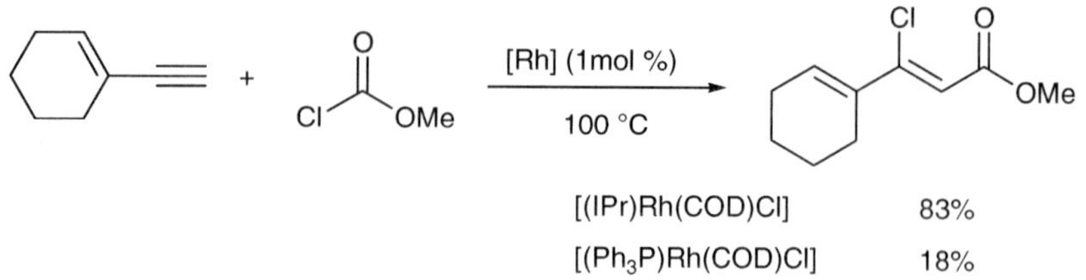

[Rh] (1mol %)
100 °C

[(IPr)Rh(COD)Cl]	83%
[(Ph_3P)Rh(COD)Cl]	18%

(8.14)

A NHC–rhodium complex displayed an intriguing reactivity towards cyclobutanone derivatives. With complex **25**, a cyclobutanone having an additional aldehyde function underwent chemoselective decarbonylation of the ketone moiety whilst the aldehyde carbonyl group remained intact, giving the corresponding cyclopropane in 82% yield (Scheme 8.7).[60] On the other hand, no chemoselectivity was observed using $[Rh(COD)Cl]_2$/dppp (dppp = 1,3-bis(diphenylphosphino)propane), and both aldehyde and ketone units were decarbonylated to afford 1-isopropenyl-4-propoxymethylbenzene in 84% yield. The use of the NHC ligand therefore provides a rare example of preferential activation of a C–C bond over a C–H bond.

Other applications of NHC-based rhodium catalysts include the cyclization of acetylene carboxylic acids,[61] cyclopropanation of olefins with diazo compounds,[62] or aryl–aryl cross-coupling combined with dynamic kinetic resolution with the help of a lipase or Beckmann rearrangement.[63] Thus, the chemistry of NHC–Rh catalysts is rich and varied, and we expect new catalysts and patterns of reactivity in the years to come. These will provide new tools for synthetic organic chemistry.

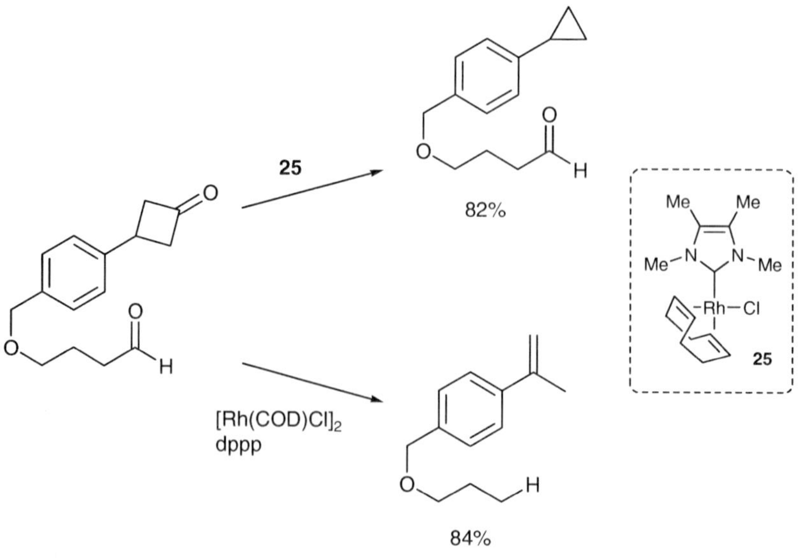

Scheme 8.7 Decarbonylation reactions.

8.4 NHC–Iridium Complexes

8.4.1 *C*-, *N*- and *O*-Alkylations

Alcohols constitute one of the most basic and abundant classes of organic compounds and have a wide variety of uses in synthetic chemistry. They generally react as nucleophiles at the oxygen atom, but the OH groups themselves are poor nucleofuges, and the alcohol function needs to be transformed into a better leaving group such as a halogenide or sulfonate to allow a nucleophilic substitution at the OH-bearing carbon. To overcome these limitations, the 'borrowing hydrogen' strategy exploits the wider range of reactions available to the carbonyl functionality.[64] In particular, the latter are better electrophiles and good nucleophiles (as the enol/enolate). This methodology is at the basis of the [(NHC)Ir]-catalyzed C and N alkylations using primary alcohols and amines.

Since the iridium(III) complex $[(Cp^*)IrCl_2]_2$ (Cp^* = pentamethylcyclopentadienyl) is an active catalyst for the β-alkylation of secondary alcohols with primary alcohols,[65] a series of iridium(III) complexes **26**–**28** bearing a Cp* unit tethered to an imidazolylidene was synthesized (Equation (8.15)).[66,67] These complexes displayed similar activities in the β-alkylation of secondary alcohols with primary alcohols as electrophiles (Equation (8.15)), and surpassed the performance of their parent compound $[(Cp^*)IrCl_2]_2$. Control of the reaction time was found to be crucial to avoid the undesirable dehydrogenation of the product (see Section 8.4.2 for further details). The sequence of catalytic reaction steps was thought to involve the oxidation of both alcohols and the formation of an iridium hydride species. Base-promoted cross-aldolization and elimination to form the α-enone and hydrogenation of the C=C and C=O bonds to regenerate an iridium–alkoxide species would complete the cycle.

26–28 (1 mol %), KOH (1 equiv), Toluene, 110 °C; 82–100%

26a (X = Cl), **26b** (X = OTf); **27a** (R = Ph), **27b** (R = H); **28**

(8.15)

Peris and co-workers reported the benzylation of arenes using complex **26b** as catalyst (0.1 or 1 mol%).[68] Different benzylating agents such as benzyl alcohols, benzyl ethers or styrene derivatives were used in combination with toluene, xylenes, phenol and anisole (Equation (8.16)). Catalyst **26b** displayed high activities, except for a few cases, but poor selectivities for both the *ortho* and *para* isomers were observed (40:60 to 10:90). Although a mixture of *ortho*/*N*-alkylated/*para* products was obtained when anilines were used as substrates, the results compared well with those previously reported which made use of Brønstedt acids or metal complexes as catalysts.

R1 OR2 OR R + arene —26b (0.1–1 mol %), 110 °C→ R1 arene 32–95% OR arene R 30–95% (8.16)

Alternatively, the cross-coupling between two alcohols to provide unsymmetrical ethers was catalyzed by 1 mol% of the bis-triflate complex **26b**.[69] Excellent conversions (generally > 95%) as well as good selectivities for the unsymmetrical ether were generally obtained. An example using allyl alcohol is displayed in Equation (8.17).

OH + HO —26a (1 mol %), AgOTf (3 mol %), 110 °C→ Ph O + Ph O Ph + O
93 % conv.
85/8/15 mixture
(8.17)

The (speculative) mechanism may first involve oxidative addition of an OH function to provide a Ir^{V}–H species which would behave as a Brønstedt acid towards the second alcohol and activate it for an nucleophilic attack at the C–O carbon atom. Displacement of water and release of the ether from the coordination sphere would regenerate the catalyst and close the cycle.

For details on the [(NHC)Ir]-catalyzed *N*-alkylation processes, the reader is referred to Chapter 12.

8.4.2 Hydroamination of Alkenes

The iridium complex **29**, bearing a 'pincer' ligand, was an efficient catalyst for the hydroamination/cyclization of *secondary* amines as shown in Equation (8.18).[70] Remarkably, both Rh and Ir catalysts were found to be air and water stable, and no appreciable loss of catalytic activity was observed when carrying out the reaction in water as solvent. For example, the desired reaction product was observed in more than 98% yield (as observed by ^{1}H NMR) using catalyst

29. The authors examined the scope of the system and found that the cyclized products were selectively formed in high yields. In contrast, primary amines were not converted into the corresponding N-heterocycles but gave the internal alkene isomer as the only detectible product.

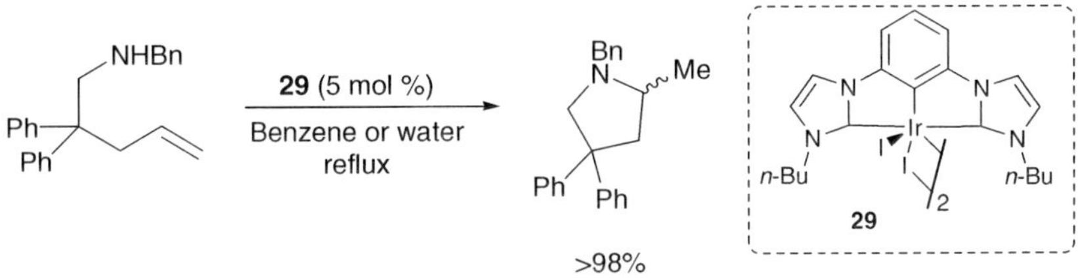

(8.18)

8.4.3 Miscellaneous Reactions

During their study on the β-alkylation of alcohols, Peris *et al.* observed an oxidative dehydrogenation of the final alcohol to the corresponding ketone under the reaction conditions (toluene, reflux).[66a] This result was interpreted in the light of the previous report by Fujita and Yamaguchi where a [Ir(Cp*)] complex was efficient for the oxidation of alcohol in an 'oxidant-free' environment (with release of H_2),[71] and demonstrated that [(NHC)Ir] complexes are good candidates for this promising reaction.

As already discussed for rhodium, the addition of acid chlorides to alkynes should lead to the formation of chloro-substituted unsaturated ketones. However, the carbonyl group is often lost during the catalytic process by decarbonylation.[72] Tsuji and collaborators found that [(IPr)Ir(COD)Cl] successfully catalyzed the addition of aromatic acid chlorides to terminal alkynes *without* loss of the carbonyl functionality (Equation (8.19)).[73] Under optimized conditions, the reaction of benzoyl chloride and phenylacetylene in the presence of 5 mol% of Ir catalyst (90 °C) yielded the product in 91% yield with high *Z*-selectivity (99/1). In contrast, when the IPr ligand was changed to a phosphine ligand, the addition occurred with decarbonylation. These opposite reactivities are a testimony to the complementary nature of phosphine and the NHC ligands in catalysis. At higher temperature (110 °C), terminal alkynes bearing a methylene unit adjacent to the triple bond afforded the corresponding 2,5-disubstituted furans in moderate to good yields (Equation (8.20)).

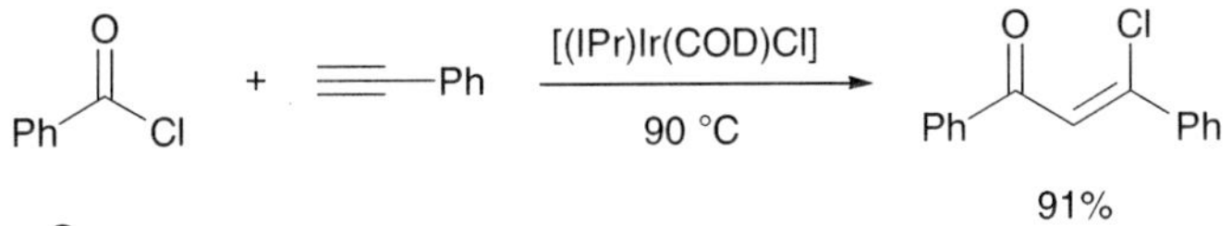

(8.19)

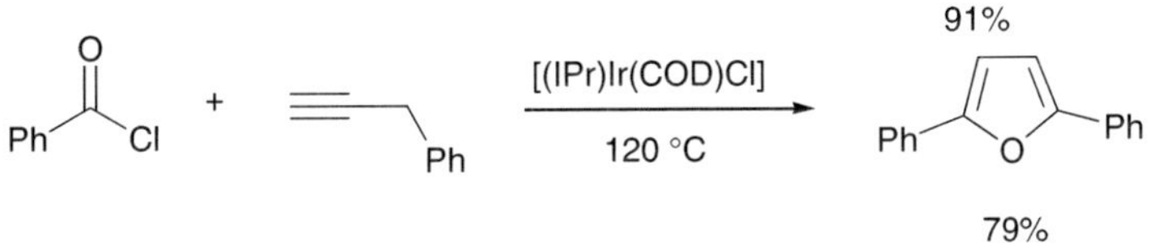

(8.20)

Along with their high activity in the 'borrowing hydrogen' methodology (see Section 8.4.1), complexes of general composition [(NHC)Ir(Cp*)] (NHC = mono- or bidentate) were also reported as active catalysts in various other catalytic transformations. For example, some [(NHC)Ir(Cp*)Cl_2] derivatives were efficiently employed for H/D exchange reactions in challenging substrates such as diethyl ether, THF or isopropanol (Equation (8.21)).[74]

$$H_3C\text{–}CH_2\text{–}O\text{–}CH_2\text{–}CH_3 \xrightarrow[CD_3OD,\ 100\ ^\circ C]{\mathbf{26a}\ (2\ mol\ \%),\ AgOTf\ (excess)} D_3C\text{–}CD_2\text{–}O\text{–}CD_2\text{–}CD_3 \quad (8.21)$$

deuterium incorporation >99%

Alternatively, they were also found to be good racemization catalysts.[75] Iridium complexes, but also rhodium compounds, catalyzed the racemization step in the enzymatic dynamic kinetic resolution of secondary alcohols, and excellent results were reported for alkyl–aryl as well as dialkyl secondary alcohols. Finally, picolyl and pyridine functionalized N-heterocyclic carbene iridium complexes [(C^N)Ir(Cp*)Cl]Cl were moderately active catalysts for the polymerization of norbornene in the presence of methylaluminoxane as co-catalyst.[76]

In summary, NHC-containing iridium complexes have shown remarkable activities in various reactions and are akin to rhodium-based systems. Most importantly, such Ir complexes are frequently stable towards air and water. Thus, the development of catalytic applications with these easy-to-handle complexes is expected to continue at a rapid pace.

8.5 Conclusions

Since the first application in homogeneous catalysis in the mid 1990s, N-heterocyclic carbenes have been established as crucial components for the development of Group 9 metal-based catalysts. In many cases, these have displayed better activities and selectivities than their analogs containing more classical ligands such as phosphines. It is clear from this overview of Group 9 metal chemistry that applications will continue to emerge rapidly. Commercially available NHCs (*i.e.* IMes, IPr, SIMes, SIPr) have played a key role in this development. The molecular shape of the NHCs is readily modified by variation of N-substituents and the structure of the backbone, and this provides many opportunities for development of new catalytic reactions in the future.

References

1. For monographs, see: (a) F. Glorius (ed.), *N-Heterocyclic Carbenes in Transition Metal Catalysis*, Topics in Organometallic Chemistry, vol. 21, Springer, Berlin, 2007; (b) S. P. Nolan (ed.) *N-Heterocyclic Carbenes in Synthesis*, Wiley-VCH, Weinheim, 2006.

2. For the most recent reviews including Group 9 chemistry, see: (a) S. Díez-González, N. Marion and S. P. Nolan, *Chem. Rev.*, 2009, **109**, 3612–3676; (b) M. Poyatos, J. A. Mata and E. Peris, *Chem. Rev.*, 2009, **109**, 3677–3707; (c) F. E. Hahn and M. C. Jahnke, *Angew. Chem., Int. Ed*, 2008, **47**, 3122–3172.
3. For the first publication on NHC–cobalt complexes, see: M. F. Lappert and P. L. Pye, *J. Chem. Soc., Dalton Trans.*, 1977, 2172–2180.
4. A. W. Coleman, P. B. Hitchcock, M. F. Lappert, R. K. Maskell and J. H. Müller, *J. Organomet. Chem.*, 1985, **296**, 173–196, and references therein.
5. X. Yu, I. Castro-Rodriguez and K. Meyer, *J. Am. Chem. Soc.*, 2004, **126**, 13464–13473.
6. For a mini review on this ligand, see: X. Hu and K. Meyer, *J. Organomet. Chem.*, 2005, **690**, 5474–5484.
7. X. Hu and K. Meyer, *J. Am. Chem. Soc.*, 2004, **126**, 16322–16323.
8. S. E. Gibson, C. Johnstone, J. A. Loch, J. W. Steed and A. Stevenazzi, *Organometallics*, 2003, **22**, 5374–5377.
9. (a) H. van Rensburg, R. P. Tooze, D. F. Foster and A. M. Z. Slawin, *Inorg. Chem.*, 2004, **43**, 2468–2470; (b) H. van Rensburg, R. P. Tooze, D. F. Foster and S. Otto, *Inorg. Chem.*, 2007, **46**, 1963–1965.
10. A. M. Poulton, S. D. R. Christie, R. Fryatt, S. H. Dale and M. R. J. Elsegood, *Synlett*, 2004, 2103–2106.
11. N. Saino, D. Kogure and S. Okamoto, *Org. Lett.*, 2005, **7**, 3065–3067.
12. F. Slowinski, C. Aubert and M. Malacria, *Adv. Synth. Catal.*, 2001, **343**, 64–67.
13. A. Geny, S. Gaudrel, F. Slowinski, M. Amatore, G. Chouraqui, M. Malacria, C. Aubert and V. Gandon, *Adv. Synth. Catal.*, 2009, **351**, 271–275.
14. H. Someya, H. Ohmiya, H. Yorimitsu and K. Oshima, *Org. Lett.*, 2007, **9**, 1565–1567.
15. H. Someya, H. Ohmiya, H. Yorimitsu and K. Oshima, *Tetrahedron*, 2007, **63**, 8609–8618.
16. T. Kobayashi, H. Oshimiya, H. Yorimitsu and K. Oshima, *J. Am. Chem. Soc.*, 2008, **130**, 11276–11277.
17. (a) H. Hamaguchi, M. Uemura, H. Yasui, H. Yorimitsu and K. Oshima, *Chem. Lett.*, 2008, **37**, 1178–1179; (b) Z. Xi, B. Liu and W. Chen, *Dalton Trans.*, 2009, 7008–7014.
18. T. Kobayashi, H. Yorimitsu and K. Oshima, *Chem. Asian J.*, 2009, **4**, 1078–1083.
19. (a) B. Cornils and W. A. Herrmann, (ed.), *Applied Homogeneous Catalysis with Organometallic Compounds*, Wiley-VCH, Weinheim, 2002; (b) P. W. N. M. van Leeuwen and C. Claver, (ed.), *Rhodium Catalyzed Hydroformylation*, Kluwer Academic Press, Dordrecht, 2000.
20. For reviews, see: (a) F. Hebrard and P. Kalck, *Chem. Rev.*, 2009, **109**, 4272–4282; (b) C. Dwyer, H. Assumption, J. Coetzee, C. Crause, L. Damoense and M. Kirk, *Coord. Chem. Rev.*, 2004, **248**, 653–669.
21. For a review on NHC as reactive ligands, see: K. J. Cavell, *Dalton Trans.*, 2008, 6676–6685.

22. S. A. Llewellyn, M. L. H. Green and A. R. Cowley, *Dalton Trans.*, 2006, 4164–4168.
23. M. F. Lappert and R. K. Maskell, *J. Organomet. Chem.*, 1984, **264**, 217–228.
24. See for some recent examples: (a) M. Alcarazo, S. J. Roseblade, A. R. Cowley, R. Fernández, J. M. Brown and J. M. Lassaletta, *J. Am. Chem. Soc.*, 2005, **127**, 3290–3291; (b) A. Fürstner, M. Alcarazo, H. Krause and C. W. Lehmann, *J. Am. Chem. Soc.*, 2007, **129**, 12676–12677; (c) D. M. Khramov, E. L. Rosen, V. M. Lynch and C. W. Bielawski, *Angew. Chem., Int. Ed.*, 2008, **47**, 2267–2270; (d) J. Kobayashi, S.-Y. Nakafuji, A. Yatabe and T. Kawashima, *Chem. Commun.*, 2008, 6233–6235.
25. For recent reviews, see: (a) C. Defieber, H. Grützmacher and E. M. Carreira, *Angew. Chem., Int. Ed.*, 2008, **47**, 4482–4502; (b) F. Schmidt, R. T. Stemmler, J. Rudolph and C. Bolm, *Chem. Soc. Rev.*, 2006, **35**, 454–470; (c) T. Hayashi and K. Yamasaki, *Chem. Rev.*, 2003, **103**, 2829–2844; (d) K. Fagnou and M. Lautens, *Chem. Rev.*, 2003, **103**, 169–196.
26. M. Sakai, M. Ueda and N. Miyaura, *Angew. Chem., Int. Ed.*, 1998, **37**, 3279–3281.
27. A. Fürstner and H. Krause, *Adv. Synth. Catal.*, 2001, **343**, 343–350.
28. (a) C. Yan, X. Zeng, W. Zhang and M. Luo, *J. Organomet. Chem.*, 2006, **691**, 3391–3396; (b) J. Chen, X. Zhang, Q. Feng and M. Luo, *J. Organomet. Chem.*, 2006, **691**, 470–474.
29. (a) I. Özdemir, S. Demir and B. Çetinkaya, *J. Mol. Catal. A: Chem.*, 2004, **215**, 45–48; (b) M. Yigit, I. Özdemir, E. Çetinkaya and B. Çetinkaya, *Heteroat. Chem.*, 2005, **16**, 461–465; (c) R. Kilincarslan, M. Yigit, I. Özdemir, E. Çetinkaya and B. Çetinkaya, *J. Heterocycl. Chem.*, 2007, **44**, 69–73.
30. (a) H. Türkmen, S. Denizalti, I. Özdemir, E. Çetinkaya and B. Çetinkaya, *J. Organomet. Chem.*, 2008, **693**, 425–434; (b) I. Özdemir, N. Gürbüz, Y. Gök, B. Çetinkaya and E. Çetinkaya, *Transition Met. Chem.*, 2005, **30**, 367–371.
31. M. Yigit and I. Özdemir, *Transition Met. Chem.*, 2007, **32**, 536–540.
32. (a) I. Özdemir, S. Demir, B. Cetinkaya and E. Cetinkaya, *J. Organomet. Chem.*, 2005, **690**, 5849–5855; (b) N. Imlinger, M. Mayr, D. Wang, K. Wurst and M. R. Buchmeiser, *Adv. Synth. Catal.*, 2004, **346**, 1836–1843.
33. P. M. P. Gois, A. F. Trindade, L. F. Veiros, V. André, M. T. Duarte, C. A. M. Afonso, S. Caddick and F. G. N. Cloke, *Angew. Chem., Int. Ed*, 2007, **46**, 5750–5753.
34. A. F. Trindade, P. M. P. Gois, L. F. Veiros, V. André, M. T. Duarte, C. A. M. Afonso, S. Caddick and F. G. N. Cloke, *J. Org. Chem.*, 2008, **73**, 4076–4086.
35. For reviews on chiral NHC ligands, see: (a) L. H. Gade and S. Bellemin-Laponnaz, in *Topics in Organometallic Chemistry*, ed. F. Glorius, Springer, Berlin, 2007, pp. 117–157; (b) V. César, S. Bellemin-Laponnaz and L. H. Gade, *Chem. Soc. Rev.*, 2004, **33**, 619–636.
36. T. Focken, J. Rudolph and C. Bolm, *Synthesis*, 2005, 429–436.

37. For a review, see: T. Hayashi and K. Yamasaki, *Chem. Rev.*, 2003, **103**, 2829–2844.
38. Y. Ma, C. Song, C. Ma, Z. Sun, Q. Chai and M. B. Andrus, *Angew. Chem., Int. Ed.*, 2003, **42**, 5871–5874.
39. J.-M. Becht, E. Bappert and G. Helmchen, *Adv. Synth. Catal.*, 2005, **347**, 1495–1498.
40. C. Y. Legault, C. Kendall and A. B. Charette, *Chem. Commun.*, 2005, 3826–3828.
41. A. Bakar Mr., Y. Suzuki and M. Sato, *Chem. Pharm. Bull.*, 2008, **56**, 973–976.
42. For a review, see: J. M. Praetorius and C. M. Crudden, *Dalton Trans.*, 2008, 4079–4094.
43. A. C. Chen, L. Ren, A. Decken and C. M. Crudden, *Organometallics*, 2000, **19**, 3459–3461.
44. M. Poyatos, P. Uriz, J. A. Mata, C. Claver, E. Fernandez and E. Peris, *Organometallics*, 2003, **22**, 440–444.
45. M. Bortenschlager, J. Schültz, D. von Preysing, O. Nuyken, W. A. Herrmann and R. Weberskirch, *J. Organomet. Chem.*, 2005, **690**, 6233–6237.
46. M. Bortenschlager, M. Mayr, O. Nuyken and M. R. Buchmeiser, *J. Mol. Catal. A: Chem.*, 2005, **233**, 67–71.
47. A. C. Chen, D. P. Allen, C. M. Crudden and A. Decken, *Can. J. Chem.*, 2005, **83**, 943–957.
48. (a) M. Ahmed, C. Buch, L. Routaboul, R. Jackstell, H. Klein, A. Spannenberg and M. Beller, *Chem.–Eur. J.*, 2007, **13**, 1594–1601; (b) W. Gil, A. M. Trzeciak and J. J. Ziółkowski, *Organometallics*, 2008, **27**, 4131–4138.
49. Other NHC-related systems were also tested, see: (a) A. Neveling, G. R. Julius, S. Cronje, C. Esterhuysen and H. G. Raubenheimer, *Dalton Trans.*, 2005, 181–192; (b) S. Dastgir, K. S. Coleman, A. R. Cowley and M. L. H. Green, *Organometallics*, 2006, **25**, 300–306; (c) M. Green, C. L. McMullin, G. J. P. Morton, A. G. Orpen, D. F. Wass and R. L. Wingad, *Organometallics*, 2009, **28**, 1476–1479; (d) S. P. Downing, P. J. Pogorzelec, A. A. Danopoulos and D. J. Cole-Hamilton, *Eur. J. Inorg. Chem.*, 2009, 1816–1824; (e) M. S. Jeletic, M. T. Jan, I. Ghiviriga, K. A. Abboud and A. S. Veige, *Dalton Trans.*, 2009, 2764–2776; For supported NHC–Rh, see; (f) M. T. Zarka, M. Bortenschlager, K. Wurst, O. Nuyken and R. Weberskirch, *Organometallics*, 2004, **23**, 4817–4820; (g) W. Gil, K. Boczon, A. M. Trzeciak, J. J. Ziółkowski, E. Garcia-Verdugo, S. V. Luis and V. Sans, *J. Mol. Catal. A: Chem.*, 2009, **309**, 131–136.
50. J. M. Praetorius, M. W. Kotyk, J. D. Webb, R. Y. Wang and C. M. Crudden, *Organometallics*, 2007, **26**, 1057–1061.
51. (a) P. A. Evans, E. W. Baum, A. N. Fazal and M. Pink, *Chem. Commun.*, 2005, 63–65; (b) M.-H. Baik, E. W. Baum, M. C. Burland and P. A. Evans, *J. Am. Chem. Soc.*, 2005, **127**, 1602–1603.
52. S. I. Lee, S. Y. Park, J. H. Park, I. G. Jung, S. Y. Choi, Y. K. Chung and B. Y. Lee, *J. Org. Chem.*, 2006, **71**, 91–96.

53. A rhodium/*N*-alkoxyimidazolylidene system was also reported as a good catalyst for these [5+2] and [4+2] cycloaddition reactions, see: F. J. Gómez, N. E. Kamber, N. M. Deschamps, A. P. Cole, P. A. Wender and R. M. Waymouth, *Organometallics*, 2007, **26**, 4541–4545.
54. H. M. Peng, R. D. Webster and X. Li, *Organometallics*, 2008, **27**, 4484–4493.
55. L. F. R. Gomes, A. F. Trindade, N. R. Candeias, P. M. P. Gois and C. A. M. Afonso, *Tetrahedron Lett.*, 2008, **49**, 7372–7375.
56. For other reviews covering a larger scope of applications for NHC–Rh complexes, see reference 2.
57. For a review, see: T. E. Müller, K. C. Hultzsch, M. Yus, F. Foubelo, and M. Tada, *Chem. Rev.*, 2008, **108**, 3795–3892.
58. L. D. Field, B. A. Messerle, K. Q. Vuong and P. Turner, *Organometallics*, 2005, **24**, 4242–4250.
59. J. Y. Baek, S. I. Lee, S. H. Sim and Y. K. Chung, *Synlett*, 2008, 551–554.
60. T. Matsuda, M. Shigeno and M. Murakami, *Chem. Lett.*, 2006, **35**, 288–289.
61. (a) E. Mas-Marzá, E. Peris, I. Castro-Rodríguez and K. Meyer, *Organometallics*, 2005, **24**, 3158–3162; (b) E. Mas-Marzá, M. Sanaú and E. Peris, *Inorg. Chem.*, 2005, **44**, 9961–9967; (c) M. Viciano, E. Mas-Marzá, M. Sanaú and E. Peris, *Organometallics*, 2006, **25**, 3063–3069.
62. M. L. Rosenberg, A. Krivokapic and M. Tilset, *Org. Lett.*, 2009, **11**, 547–550.
63. M. Kim, J. Lee, H.-Y. Lee and S. Chang, *Adv. Synth. Catal.*, 2009, **351**, 1807–1812.
64. For review, see: (a) T. D. Nixon, M. K. Whittlesey and J. M. J. Williams, *Dalton Trans.*, 2009, 753–762; (b) M. H. S. A. Hamid, P. A. Slatford and J. M. J. Williams, *Adv. Synth. Catal.*, 2007, **349**, 1555–1575; (c) G. Guillena, D. J. Ramon and M. Yus, *Angew. Chem., Int. Ed.*, 2007, **46**, 2358–2364.
65. K.-I. Fujita, C. Asai, T. Yamaguchi, F. Hanasaka and R. Yamaguchi, *Org. Lett.*, 2005, **7**, 4017–4019.
66. (a) A. Pontes da Costa, M. Viciano, M. Sanaú, S. Merino, J. Tejeda, E. Peris and B. Royo, *Organometallics*, 2008, **27**, 1305–1309; (b) A. Pontes da Costa, M. Sanaú, E. Peris and B. Peris, *Dalton Trans.*, 2009, **1**, 6960–6966.
67. D. Gnanamgari, E. L. O. Sauer, N. D. Schley, C. Butler, C. D. Incarvito and R. H. Crabtree, *Organometallics*, 2009, **28**, 321–325.
68. A. Prades, R. Corberán, M. Poyatos and E. Peris, *Chem.–Eur. J.*, 2009, **15**, 4610–4613.
69. A. Prades, R. Corberán, M. Poyatos and E. Peris, *Chem.–Eur. J.*, 2008, **14**, 11474–114479.
70. E. B. Bauer, G. T. S. Andavan, T. K. Hollis, R. J. Rubio, J. Cho, G. R. Kuchenbeiser, T. R. Helgert, C. S. Letko and F. S. Tham, *Org. Lett.*, 2008, **10**, 1175–1178.
71. K. Fujita, N. Tanino and R. Yamaguchi, *Org. Lett.*, 2007, **9**, 109–111.
72. K. Kokubo, K. Matsumasa, M. Miura and M. Nomura, *J. Org. Chem.*, 1996, **61**, 6941–6946.

73. T. Iwai, T. Fujihara, J. Terao and Y. Tsuji, *J. Am. Chem. Soc.*, 2009, **131**, 6668–6669.
74. (a) R. Corberán, M. Sanaú and E. Peris, *J. Am. Chem. Soc.*, 2006, **128**, 3974–3979; See also (b) J. A. Brown, S. Irvine, A. R. Kennedy, W. J. Kerr, S. Andersson and G. N. Nilsson, *Chem. Commun.*, 2008, 1115–1117.
75. A. C. Marr, C. L. Pollock and G. C. Saunders, *Organometallics*, 2007, **26**, 3283–3285.
76. X.-Q. Xiao and G.-X. Jin, *J. Organomet. Chem.*, 2008, **693**, 3363–3368.

CHAPTER 9

NHC–Palladium Complexes in Catalysis

ADRIEN T. NORMAND[a] AND KINGSLEY J. CAVELL[b,*]

[a] Department of Chemistry, University of Ottawa, 10 Marie Curie Street, K1N 6N5, Ottawa, ON, Canada; [b] School of Chemistry, Cardiff University, 51 Park Place, CF10 3AT, Cardiff, UK

9.1 Introduction

The application of N-heterocyclic carbenes (NHCs) as ligands in catalysis has grown exponentially since the isolation of the first free NHCs in the early 1990s. Interest in these species as ligands grew initially from their use as alternatives to the ubiquitous phosphines. It has since been found that NHCs offer unique structures, reactivity and ligand properties, quite distinct from those of phosphines, and the field has grown from there. The field of NHC chemistry has passed the first exploratory/developmental phase and could now be classified as a mature discipline; nowhere is this more evident than in the area of palladium-based catalysis. In the following pages, we will appraise the application of NHC–Pd complexes in catalysis, illustrating recent trends and developments in this exciting and vital topic since 2004. Numerous reviews and book chapters have been written covering literature prior to 2004, and a selection of these is provided.[1]

9.2 C–C Bond Formation

Carbon–carbon cross-coupling reactions dominate catalytic applications of NHC–Pd complexes, demonstrating the importance of this class of reaction and the unique success of Pd compounds in these transformations.[1a,d] In the

RSC Catalysis Series No. 6
N-Heterocyclic Carbenes: From Laboratory Curiosities to Efficient Synthetic Tools
Edited by Silvia Díez-González
© Royal Society of Chemistry 2011
Published by the Royal Society of Chemistry, www.rsc.org

past five years, there has been a substantial body of work published on cross-coupling reactions catalysed by NHC–Pd complexes; therefore, to provide some degree of structure to this chapter reactions have been broadly grouped under Heck, Suzuki and "other" types of reaction categories. Not surprisingly there is overlap between sections, and many publications report more than one class of reaction; however, the layout does provide a coherent structure.

9.2.1 Mizoroki–Heck Coupling and Related Chemistry

The Mizoroki–Heck reaction (Equation (9.1)) is a valuable synthetic tool for the catalytic formation of a range of aryl-olefins, and other related building blocks, widely used in the pharmaceutical and fine chemicals industries. Many Pd-based compounds are known to catalyse the reaction to varying extents. In the past 10–12 years, NHC-based systems were added to the list, and seemed to provide some advantages with respect to other catalysts.

$$\text{R-C}_6\text{H}_4\text{-X} + \text{CH}_2{=}\text{CHR} \xrightarrow{[\text{Pd}]} \text{R-C}_6\text{H}_4\text{-CH}{=}\text{CH-R} \qquad (9.1)$$

9.2.1.1 Imidazol-2-ylidene Type NHC Ligands

Several studies reported the application of "established" five-membered ring (imidazol-2-ylidene type) NHC ligands in Heck chemistry. Whilst most authors generally focus on catalytic activity, Jutand and co-workers examined the mechanism of oxidative addition of aryl halide to NHC–Pd^0 complexes.[2] Complexes $[Pd^0L_2]$ (L = saturated NHC) were generated electrochemically and tested *in situ* for oxidative addition. They proposed that either an associative or dissociative mechanism could operate, depending on steric and electronic properties of the NHC ligand.[1a,3]

Several studies using mixed ligand NHC complexes of Pd were reported. Herrmann and co-workers investigated the performance of well-defined phospha-palladacycle NHC complexes **1**–**4** in Heck coupling (Figure 9.1).[4] The coupling of styrene with a variety of aryl halides (bromides and chlorides) yielded mixed results. TONs (72 h) in several examples were high, although conversions, particularly for aryl chlorides were modest.

Ying and co-workers investigated a NHC–palladacycle pre-catalyst **5**, a variant of complex **4**, in the Heck reaction of a large range of functionalised aryl bromides and iodides with a variety of olefins (several sterically demanding and deactivated substrates were investigated).[5] Low catalyst loadings were used and high TONs were achieved over long reaction times (164 h).

A significant number of studies were directed at palladium complexes of chelating bridged bis-NHCs and their performance as catalysts in coupling reactions; the main focus was on varying the linker; however, several studies also investigated the impact of the *N*-substituent. Following earlier studies exploring xylyl linked bis-NHCs, Baker and co-workers tested Pd complexes **6**

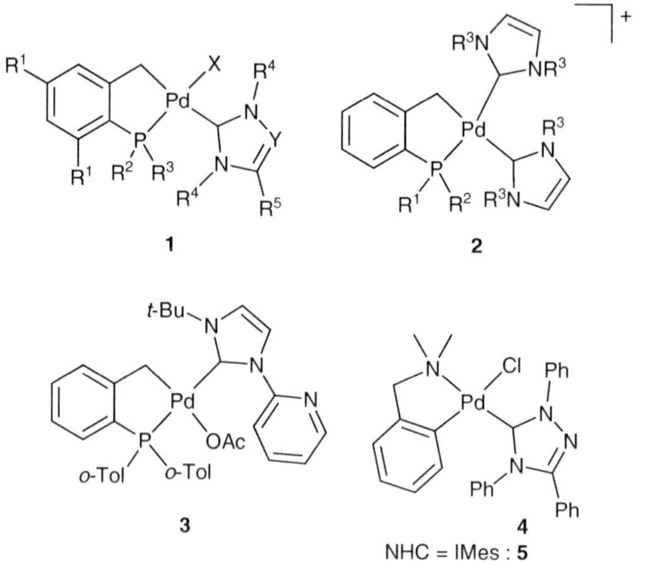

Figure 9.1 Mixed ligand NHC–Pd^{II} complexes.

and **7** in Heck and Suzuki coupling reactions with iodobenzene and 4-bromotoluene respectively as substrates (Figure 9.2).[6] At low catalyst loadings (0.002–0.0002 mol%), high temperature (140 °C) and long reaction times, high TONs and reasonable conversions were obtained. Herrmann and co-workers applied complex **8** in Heck arylation with aryl bromides and chlorides, but its catalytic performance was modest.[7] Similarly, Biffis and co-workers prepared Pd complexes of bis-NHC chelating ligands with methyl and phenyl linkers and investigated their performance in the Heck reaction.[8] Strassner and co-workers investigated PdX_2 complexes of bis-NHC chelating ligands with varying chain length linkers and *N*-substituents.[9] Complexes with the Et bridged ligand gave the best results for these reactions.

Fused ring NHC ligands were also investigated in Heck and Suzuki coupling,[10] although the catalytic performance of **9** in the Heck reaction was not exceptional (Figure 9.2).[10b] A bis-chelate tetracarbene Pd^{II} complex **10** was also synthesised and its efficiency in Heck chemistry investigated.[11] With high catalyst loadings and temperatures (1.2 mol% [Pd], 140 °C) substantial conversions of aryl bromides were obtained. Importantly, the catalyst could be reused in six consecutive runs without any apparent diminution in performance, and after the initial run the induction period was greatly reduced. A series of ionic liquids (ILs) based on mixed imidazolium–triazolium salts were as solvent and ligand in Pd-catalysed Heck coupling.[12] Catalysts were generated *in situ*, using $PdCl_2$ as the Pd source, or pre-formed complexes such as **11** were dissolved in the IL (Figure 9.2). Catalytic conversions were good for aryl iodides, but poor for bromides and chlorides. Importantly, the IL solvent and catalyst could be recycled three times without loss in performance.

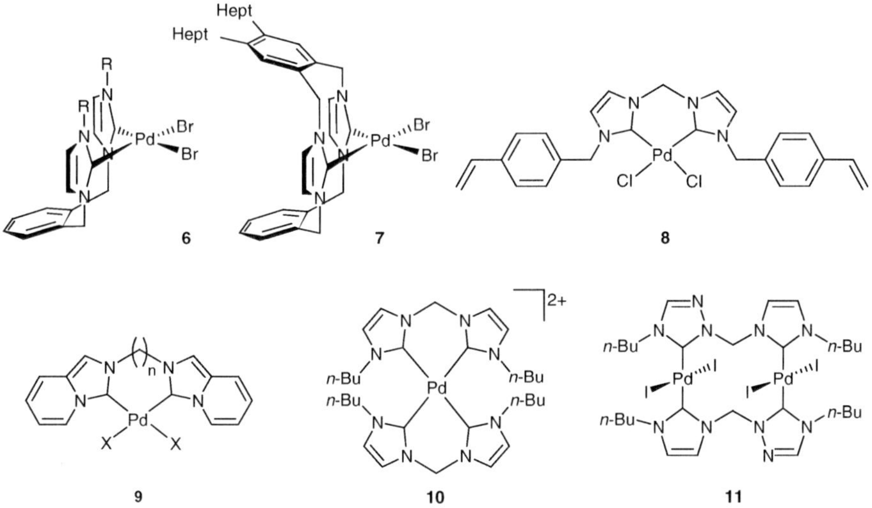

Figure 9.2 Pd^{II} catalysts with chelating di-NHCs.

9.2.1.2 Functionalised Imidazol-2-ylidene NHC Ligands

Functionalised NHCs were also widely studied as catalysts in Heck coupling and in recent years a range of different functionalities were employed.[11] NHCs with *N*-donor substituents, pyrazolyl- (**12**),[13] pyrimidyl- (**13–16**),[14] pyrazyl- (**17**),[15] picolyl-[14a,16] and pyridyl-,[17] predominated (Figure 9.3). Whilst interesting new complexes were prepared, catalytic performances were generally unspectacular as most of them were only effective with the more reactive aryl iodides and bromides. Shreeve and co-workers used pre-formed pyrazolyl-functionalised complexes in IL solvent for Heck and Suzuki coupling. Performances were generally modest; however, they were able to recycle the catalytic system without loss in performance.[13] Comparison of complexes **15** and **16** in Heck coupling indicated that **16** consistently gave the best results.[13b] These systems appeared to be very stable and were also tested for C–H activation of methane under quite severe conditions, and no formation of Pd black was observed. Mechanistic studies on pre-formed Pd complexes of pyridyl-, picolyl-, diphenylphosphinoethyl- and diphenphosphinomethyl-functionalised NHCs provided a number of valuable insights.[16] Whereas the NHC–Pd bond appeared very stable, at reaction temperatures the *N*-donor ligand was labile and dissociated, providing a vacant site for reaction. Dimeric, trimeric species and soluble Pd colloids, stabilised by the NHC ligand, appeared to be involved during the formation of the active species.

Chen and co-workers investigated bis-chelating *N*-donor functionalised complex **18** (Figure 9.3) in simple Heck coupling with aryl iodides and bromides. This catalyst was also active in one-pot sequential Heck–Suzuki, Heck–Heck and Heck–Sonogashira coupling reactions of aryl dihalides, affording asymmetrically substituted arenes in very good yields.[17] Palladium complexes

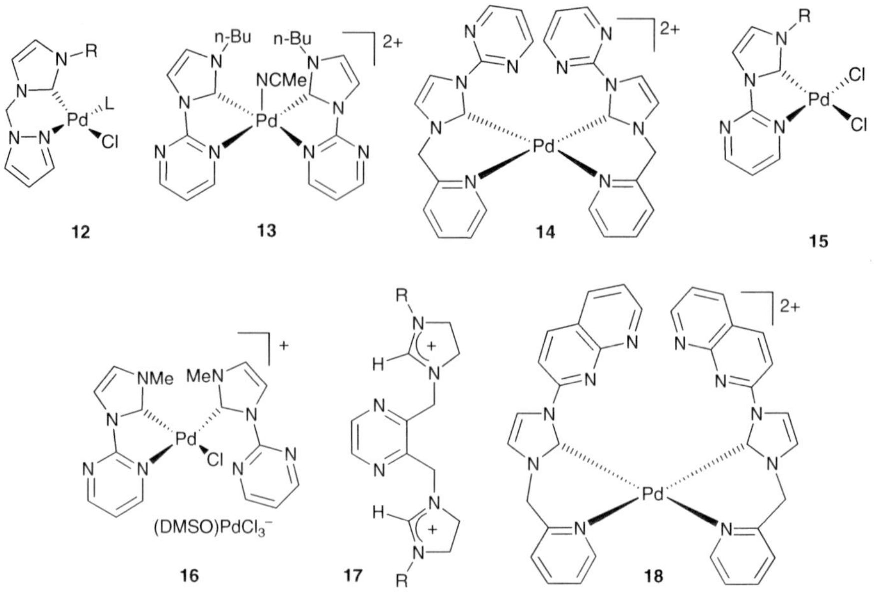

Figure 9.3 *N*-Donor functionalised NHC–Pd^{II} complexes.

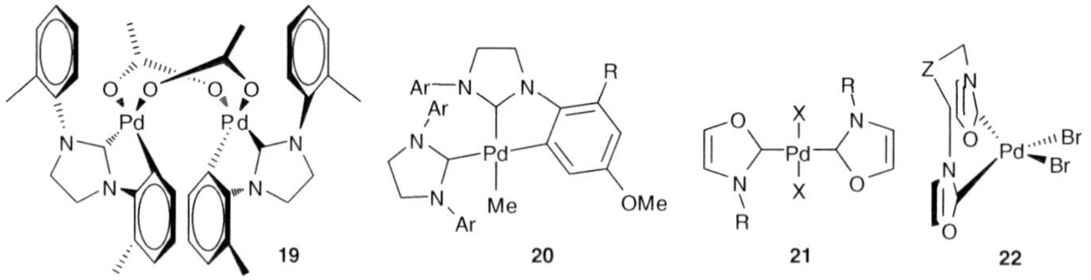

Figure 9.4 Further NHC–Pd^{II} complexes.

of several ether-functionalised NHCs were also investigated in Heck and Suzuki coupling.[18]

A number of cationic complexes of pincer type ligands with pyridyl-,[19] pycolyl-,[19] and amino-[20] bridges, and neutral complexes with ether[21] bridges were also investigated as catalysts. The pyridine-bridged complexes were generally found to be stable at high temperatures, and in consequence rather poor catalysts for Heck coupling.[19] The amino-bridged systems, investigated in hydroamination and Heck coupling, proved to be better catalysts, possibly because of more facile dissociation of the weak amino- donor during catalysis, and modest yields of product were achieved with low catalyst loadings (0.1–0.0001 mol%).[20]

Cyclometalated–NHC complexes of Pd **19** and **20** (Figure 9.4) were tested in Heck coupling of *n*-butyl acrylate with several aryl bromides and chlorides. At high temperatures and after long reaction times (140–160 °C, 18 h), complex **20** gave good conversions with 4-$CH_3COC_6H_4Cl$.[22]

Alternatives to the more traditional imidazole-based NHC structures were described. In particular, complexes of monodentate and chelating oxazol-2-ylidene ligands **21** and **22**[23] led to good conversions, but only with activated substrates (Figure 9.4). For the activity of other non-classical NHC-based catalysts, the reader is referred to Chapter 5.

9.2.1.3 Benzimidazol-2-ylidene and Benzothiazol-2-ylidene NHC Ligands

Palladium complexes of benzimidazol-2-ylidene NHCs were widely studied and applied to various C–C coupling processes. Hahn and co-workers, important contributors in this area, applied a wide variety of benzannulated NHCs to diverse catalytic processes.[1c] Pyridine-bridged and xylene-bridged bis-benzimidazol-2-ylidene NHCs **23** were tested in Heck couplings of styrene derivatives (Figure 9.5).[24] Pyridine-bridged ligands were tested *in situ* in air using 0.5 mol% $PdCl_2$ at 115 °C.[24a] Conversions and selectivities were reasonable and the catalytic activity of neutral xylene–NHC complexes was found to be generally lower than systems based on the pyridine bridged pincers.[24b]

The Hahn group also investigated the catalytic behaviour of phosphine- and picoline-functionalised benzimidazol-2-ylidene-Pd complexes, **24**–**26**.[25] Complexes **24** (R = Et, *n*-Pr or *n*-Bu) and **25** catalysed the coupling of 4-bromobenzaldehyde with styrene in good yield (1 mol% [Pd], 110 °C, 2 h), but **25**

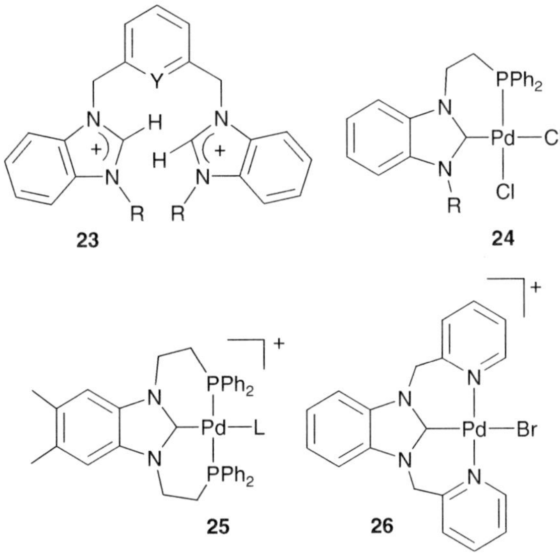

Figure 9.5 Pd^{II}–benzimidazol-2-ylidene chelate and pincer complexes.

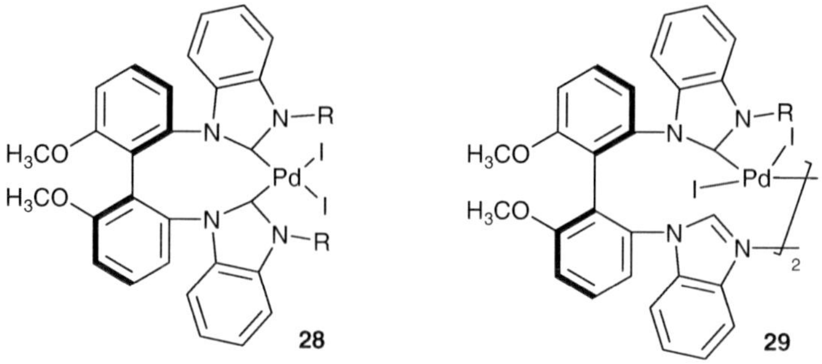

Figure 9.6 Biphenyl-based benzimidazol-2-ylidene–Pd^{II} complexes.

consistently led to lower yields. In contrast, pincer complex **26** gave the best yields in the coupling of a range of aryl bromides with several different olefins.[25b] Overall catalytic performance was modest (1 mol% [Pd], 110 °C) and high conversions were only achieved after 24 h, although the reactions could be carried out in air.

Chiral complexes **27** were tested in oxidative Heck-type reactions of boronic acids with acyclic alkenes (Equation (9.2)).[26] Product yields were modest; however, enantioselectivities were excellent (90–98% *ee*). Other examples of functionalised benzimidazol-2-ylidene–Pd complexes include **28** and **29** (Figure 9.6), which required high temperatures and long reaction times to afford reasonable conversions.[27]

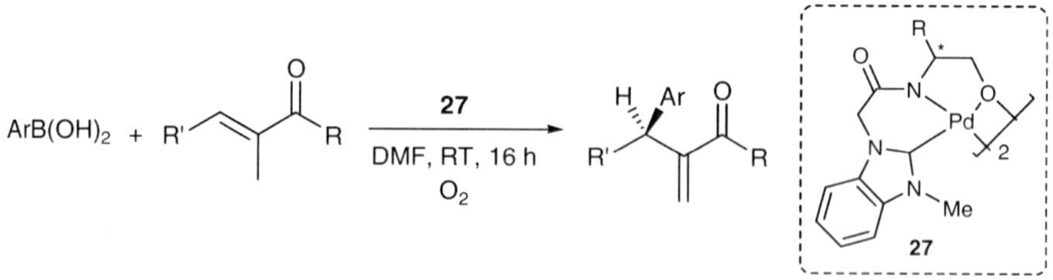

(9.2)

Zou and co-workers reported Heck arylation of a variety of aryl bromides and chlorides using catalysts formed *in situ* from several Pd compounds and *N*,*N*′-disubstituted benzimidazolium salts.[28] High catalyst loadings and long reaction times were required to obtain satisfactory yields from aryl chlorides. Baker and co-workers also investigated 5,6-dibutoxy- and 4,7-dibutoxy-substituted benzimidazol-2-ylidenes as ligands in Pd-catalysed Mizoroki–Heck and Suzuki–Miyaura reactions.[29] High TONs and TOFs were obtained at low catalyst loadings for the coupling of iodobenzene with *n*-butyl acrylate, but poor results were obtained for bromobenzene. Huynh and co-workers also investigated Heck coupling using a variety of NHC–Pd complexes also based on benzimidazole.[30] In particular, a comparison of the catalytic behaviour of

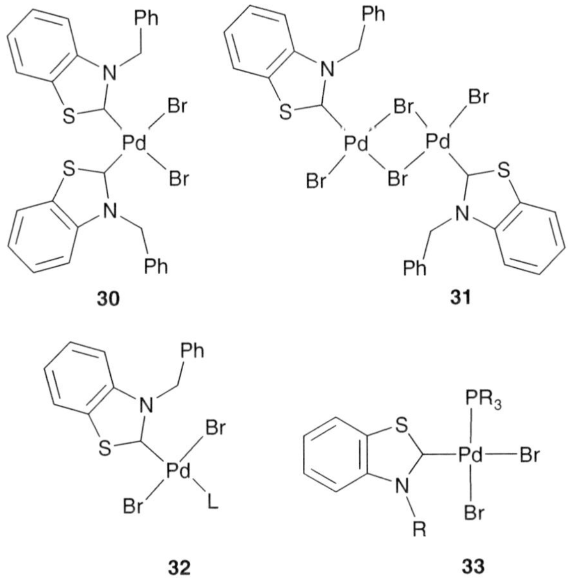

Figure 9.7 Benzothiazol-2-ylidene Pd^{II} complexes.

trans and *cis* (benzimidazol-2-ylidene)palladium(II) diiodides showed that formation of the catalytically active species occurred more rapidly from the *cis* species.

A naphthylene-annulated NHC was also investigated as a ligand in Heck coupling reactions.[31] The NHC salt was tested *in situ* with $[Pd(OAc)_2]$ and as a pre-formed complex and even if high catalyst loadings were used (3 mol%), conversions remained low.

Palladium complexes of *N*-benzylbenzothiazol-2-ylidene **30–33** (Figure 9.7) were tested in coupling reactions (Heck and Suzuki–Miyaura).[32] Relatively low temperatures were used (60–110 °C), but in general catalytic performances were unexceptional.

9.2.2 Suzuki–Miyaura Cross-coupling

The Suzuki–Miyaura cross-coupling reaction involves the coupling of an aryl- or vinyl-boronic acid with aryl, vinyl or alkyl halides, commonly catalysed by a Pd complex (Equation (9.3)).

$$R^1\text{—}B(OH)_2 + R^2\text{—}X \xrightarrow[\text{Base}]{[Pd]} R^1\text{—}R^2 \qquad (9.3)$$

R^1 = aryl, vinyl
R^2 = aryl, vinyl, alkyl
X = halide, triflate, etc...

9.2.2.1 Imidazol-2-ylidene NHC Ligands

As for the Heck reaction, the use of NHCs as ligands in the Suzuki reaction was extensively studied and ligand types largely mirror those described above, although catalytic performance may vary between reactions. For example, complexes **30**–**33** (see Figure 9.7), gave rise to more effective catalysts in Suzuki coupling, particularly for the conversion of chlorobenzaldehyde with arylboronic acids.[32b,c]

A series of phosphine and imidazolyl-based NHC complexes of phosphite and phosphinite palladacycles **34** were tested in Suzuki coupling of alkyl[33] and aryl bromide[34] with phenylboronic acid (Figure 9.8). Activity (using 1 mol% catalyst) was acceptable for activated aryl bromides, but lower for deactivated substrates and essentially non-existent for aryl chlorides and more sterically demanding compounds. In two closely related and important studies, Nolan and co-workers investigated the performance of pre-formed [(NHC)Pd(allyl)Cl] complexes **35** as catalysts for room temperature Suzuki–Miyaura coupling and Buchwald–Hartwig amination reactions.[35] These catalysts are among the best reported to date, and excellent results were generally obtained where the allyl scaffold contained a substituent in the terminal position, even with sterically hindered substrates. Impressive conversions for unactivated aryl chlorides were obtained with short reaction times and using catalyst loadings as low as 50 ppm. It appeared the allyl group provided an effective route to the monoligated [(NHC)Pd^0] active species. The catalysts were also very effective with aryl bromides and triflates. Complexes **36**, which generate C_2-symmetric (rac) and C_s-symmetric (meso) atropisomers, proved to be excellent catalysts for several catalytic reactions such as Suzuki coupling, and Ru-catalysed metathesis.[36] In particular, the Pd complexes out-performed related **35**[35] in

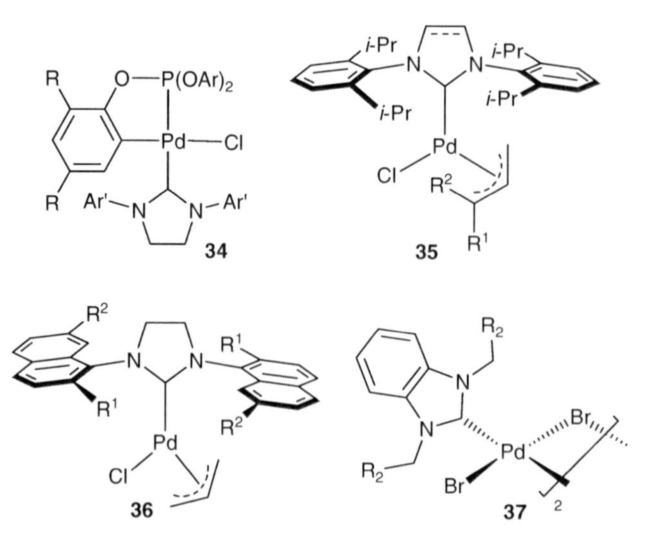

Figure 9.8 NHC–Pd^{II} complexes for Suzuki reactions.

Suzuki coupling, and excellent results were obtained even with aryl chlorides and sterically encumbered arylboronic acids (Figure 9.8). The Nolan group also investigated the application of simple pre-formed $[(NHC)Pd(OAc)_2]$ complexes (NHC = IMes, IPr) in Suzuki coupling.[37] Sterically hindered aryl and activated alkyl chlorides, bearing β-hydrogens, were successfully coupled (1 mol% [Pd], 70–100% yield, 0.5–7 h reaction times).

A variety of complexes of benzimidazole-based NHCs were also investigated in Suzuki coupling reactions. Huynh and co-workers studied dimeric complexes **37** as pre-catalysts in Suzuki coupling reactions (Figure 9.8).[38] Good activities were obtained for the cross-coupling of 4-bromobenzaldehyde with phenylboronic acid, in water as solvent. However, the catalysts were less effective for coupling aryl chlorides.

Other studies investigating the application of Pd complexes of imidazolyl-based NHCs in Suzuki-type coupling, including carbonylative cross-coupling,[39] and homocoupling of arylboronic acids,[40] were reported.[41] Unsaturated and saturated NHCs, and pre-formed or *in situ*-generated catalysts were studied in Suzuki coupling (and amination) reactions; complexes $[(NHC)_2Pd^0]$ (NHC = IPr, SIPr, ItBu, SItBu) were compared with *in situ* generated catalyst systems.[41a] *In situ* catalysts formed from saturated and bulky ligands consistently performed better, probably because of more facile generation of the active $[(NHC)Pd^0]$ active species.

Shi and co-workers extended their studies on novel Pd complexes of bridging and chelating benzimidazol-2-ylidenes[27] to coupling reactions. Dinuclear **38** and chelate **39** demonstrated moderate performance in Suzuki coupling of aryl bromides and chlorides, but low activities for the Heck reaction with aryl bromides and iodides (Figure 9.9).[42] Cyclometalated **40** also showed limited performance (long reaction times, high catalyst loading and modest yields) in Suzuki coupling.[43] Pd complexes of bidentate NHCs containing anthracene **41** and xanthenes **42** linkers were tested by Saito and co-workers in Suzuki and Heck coupling with unexceptional results.[44] Somewhat better results were achieved in Suzuki coupling of aryl bromides with complex **43**; however, the complex was not effective with aryl chlorides.[45]

9.2.2.2 *Functionalised Imidazol-2-ylidene NHC Ligands*

The synthesis of Pd complexes containing functionalised NHC ligands is a very active area of study. Many and varied attachments were added *via* the nitrogen of the NHC. These included monodentate[46] and chelating[47] ligand systems. A number of complexes with monodentate ligands such as **44**[46a] and **45**,[46e] gave rise to modest catalytic performances for coupling reactions (Figure 9.10). For chelating ligand systems like **46** and **47**, the functional group commonly contained a *N*-donor (heterocyclic-N, amido or imino donors), although *O*-[47c,f,k,l] and *P*-[47d,m] donors were also reported. Unfortunately, chelating ligands, whether used *in situ* or in pre-formed complexes, tended to lead to low activity in coupling reactions, particularly with aryl chlorides as substrates.

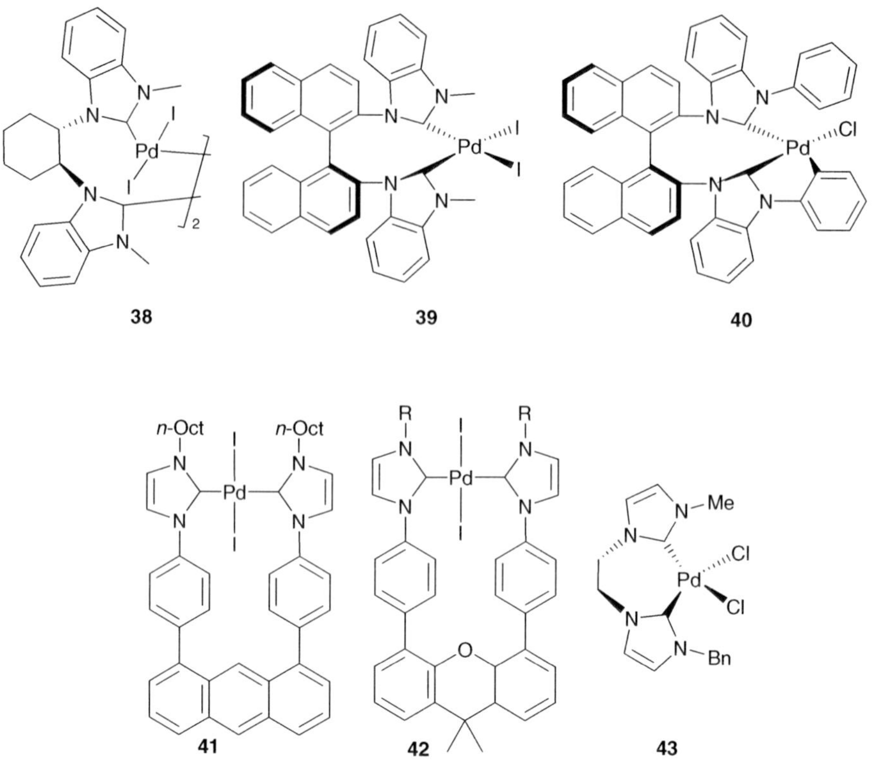

Figure 9.9 Chelating NHC–Pd^{II} complexes.

Further functionalised complexes[48] include complex **48** (Figure 9.10), an unusual bimacrocyclic compound that showed interesting performance with aryl chlorides (0.2 mol% [Pd], 60 °C, 2 h, 83–92% conversion).[18b] Catalyst **49** only gave high conversions (in 2 h) with activated aryl bromides and high catalyst loadings were required to achieve reasonable yields with aryl chlorides.[47g]

With pincer complexes **50**, acceptable conversions were obtained for activated and deactivated aryl bromides with phenylboronic acid in air.[49] Complex **51** proved to be particularly interesting for the coupling of aryl bromides with arylboronic acids.[50] Essentially complete conversions were obtained with 0.1 mol% catalyst, at 100 °C in 2 h using water as solvent. High conversions and extremely high TONs (up to 10^9) and TOFs (1.66×10^6) were claimed with catalyst loadings as low as 10^{-7} mol%, and the catalysts could be separated from the products and reused up to five times without loss in performance. The authors argued against the formation of Pd nanoparticles, although an investigation of this possibility was not reported.[46c]

NHC scaffolds other than imidazole-based were also investigated. Complex **52** ($R^1 = CH_3$; $R^2 = p\text{-}CH_3C_6H_4$) was tested, for coupling of phenylboronic acid

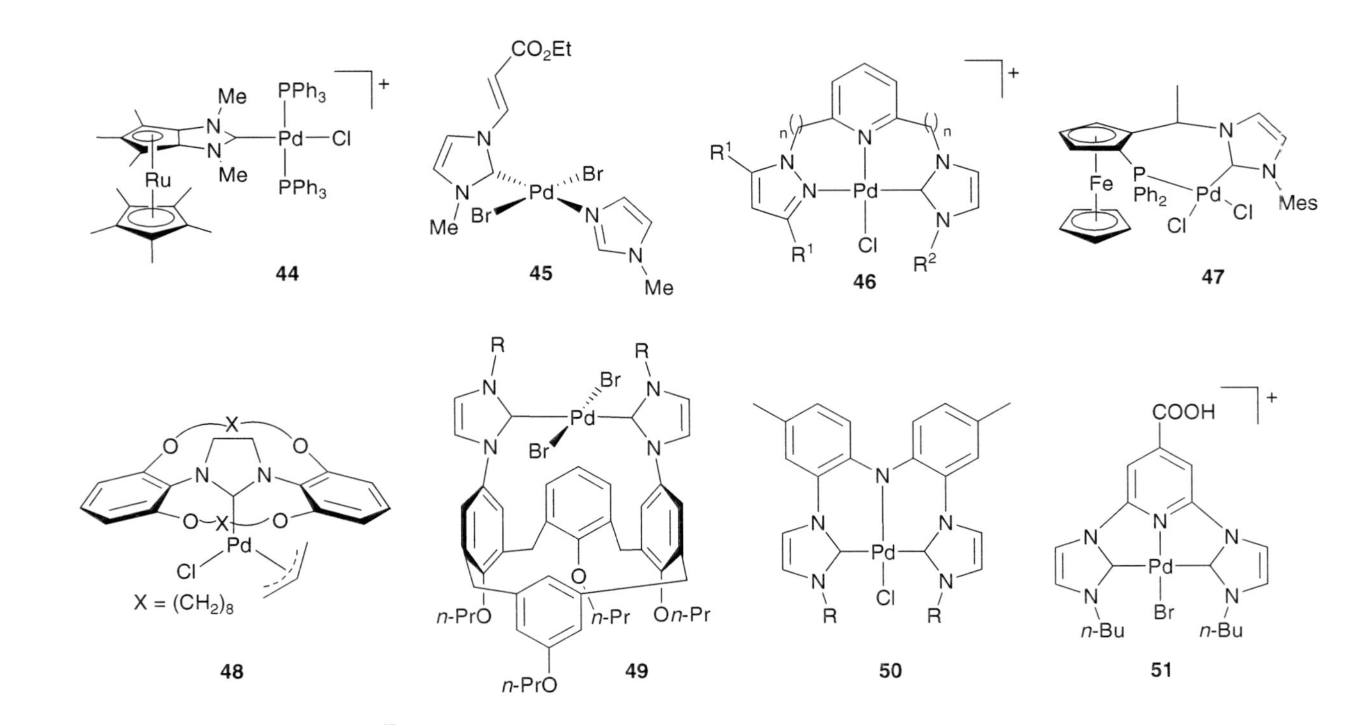

Figure 9.10 Functionalised NHC–Pd^{II} complexes.

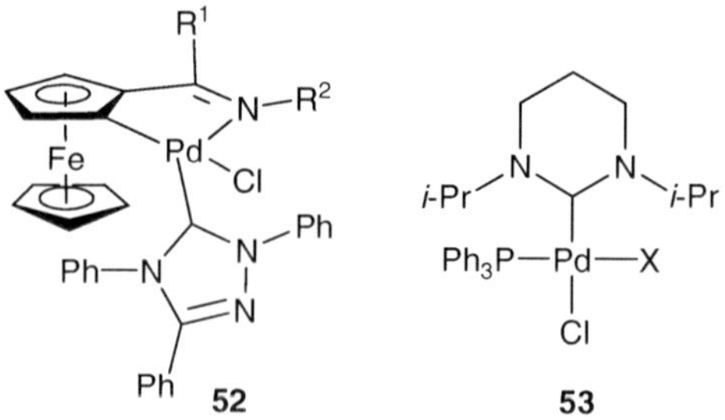

Figure 9.11 Non-traditional NHC–Pd^{II} complexes.

with a large range of aryl bromides, including deactivated and sterically hindered aryl bromides (Figure 9.11). Using low catalyst loadings (0.01–0.001 mol%), and generally short reaction times, high product yields were obtained.[51] Complex **53** was also found to be active for coupling of phenylboronic acid with a range of aryl bromides and aryl chlorides. TONs of up to 10^6 were achieved with low catalyst loadings for activated aryl bromides. However, high temperatures (130 °C) and long reaction times (14 h) were necessary.[52] For the activity of non-classical NHCs in this context, the reader is referred to Chapter 5.

9.2.3 Sonogashira Coupling

For the coupling aryl and alkyl bromides with alkynes, Sonogashira reaction, a small number of papers using NHC–Pd systems as catalysts was recently published. Glorius and co-workers developed bisoxazoline-based NHC ligands (IBox), which generated effective catalysts for coupling unactivated alkyl bromides with alkyl substituted alkynes (**54** and **55**, Figure 9.12).[53] These results represented the first examples of unactivated secondary alkyl halides (bromides and iodides) being successfully coupled in Sonogashira reactions. Ghosh and co-workers employed *trans* and *cis* isomers of keto- and imido-functionalised NHCs as ligands in **56** and **57** in amine-free Sonogashira coupling in air using a mixed aqueous solvent.[54] High yields of coupled product were obtained for the reaction of aryl iodides with aryl- and alkyl-acetylenes. Complex **58**, with a tridentate phenanthroline-functionalised NHC was also used as a ligand in Cu-free Sonogashira coupling in water.[15b]

9.2.4 Application of the PEPPSI Protocol in Coupling Reactions

A notable recent development in coupling reactions is the so-called PEPPSI (**p**yridine **e**nhanced **p**re-catalyst **p**reparation, **s**tabilisation and **i**nitiation) approach to catalyst design. The design principle was based on the need for facile generation of a monoligated [LPd^0] species, thought to be the active component in cross-coupling reactions. Organ and co-workers synthesised a

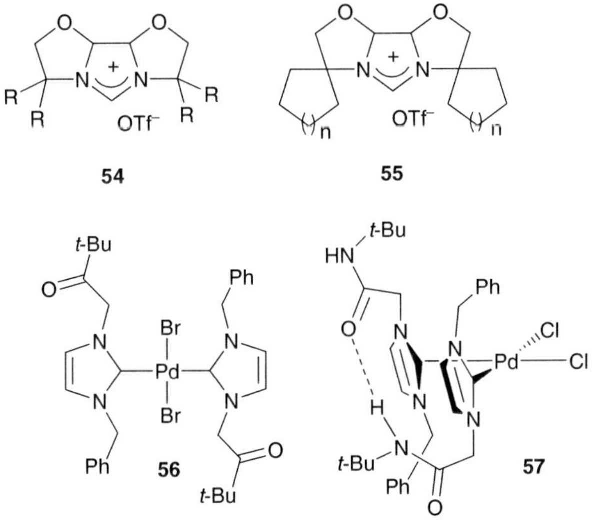

Figure 9.12 NHC ligands and Pd^{II} complexes used in Sonogashira coupling.

variety of Pd complexes of type **58**, based on the concept that the pyridine ligand is readily lost/displaced, under reaction conditions, to give the required active species (Equation (9.4)). Such species were successfully tested in a variety of coupling reactions: alkyl–alkyl Suzuki and Negishi,[55] Suzuki–Miyaura,[55,56] and Kumada–Tamao–Corriu[57] reactions, using a wide variety of substrates and yielding an array of coupling products. The pre-catalysts were very effective under mild conditions (low temperature and short reaction times) and sterically hindered RCl compounds were effectively coupled, although in some cases high catalyst loadings (up to 10 mol%)[56] were used. The PEPPSI–IPr catalyst was also effectively applied to Suzuki–Miyaura and Buchwald–Hartwig coupling to generate benzo[*b*]thiophenes of interest to the pharmaceutical industry.[58]

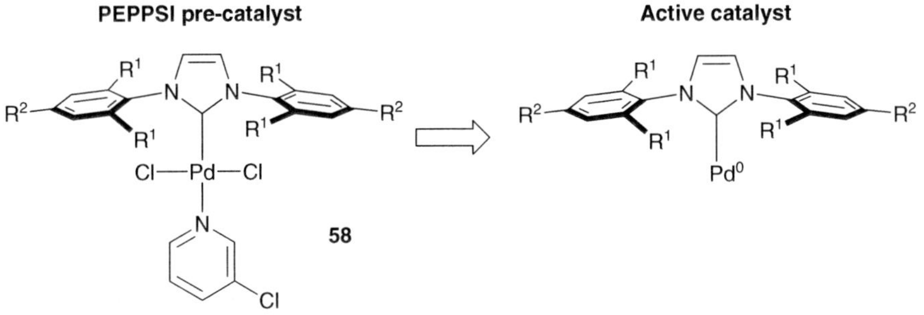

(9.4)

Ghosh and co-workers synthesised a number of modified versions of the PEPPSI pre-catalysts, employing pyridine rather than 3-chloropyridine and applying a range of different NHCs—saturated and unsaturated versions—and triazole-based NHCs. The complexes were successfully applied to Suzuki–Miyaura coupling[59] and in fluoride-free Hiyama, and copper- and amine-free Sonogashira coupling (in air and aqueous medium).[60] The PEPPSI pre-catalyst was immobilised on polyvinylpyridine, the polymer-bound pyridine replacing the 3-chloropyridine of the original complex. However, the catalyst activation was based on loss of pyridine ligand, the "supported" species were converted to an inactive form after one use.[61] The PEPPSI protocol was also applied to carbonylative Suzuki–Miyaura and carbonylative Negishi coupling using a variety of substrate giving products in moderate to high yields.[39b]

9.2.5 Immobilised Catalysts for Coupling Reactions

Immobilisation of homogenous catalysts is an important way of facilitating product and catalyst separation and hence highly attractive to industry. Commonly used supports are either organic polymers or inorganic supports such as silica or related materials. Because of the inherently strong NHC–M bond, and the unique relationship between NHCs and imidazolium salts, immobilisation in imidazolium-based ionic liquid solvents offers a particularly interesting approach to forming stable catalysts.

Recent publications reported the use of metal particles or organic polymers as supports for Pd cross-coupling catalysts. System **59** (Figure 9.13) gave interesting results in Suzuki coupling: conversions of 85–99%, for sterically demanding and deactivated aryl chlorides with phenylboronic acid, were obtained in 6 min at 100 °C (MW) with 0.5 mol% catalyst.[62] Mayr and Buchmeiser reported the immobilisation of a bis(tetrahydropyrimidin-2-ylidene)–Pd complex on Merrifield and Wang resins **60**, and on a cross-linked poly(norborn-2-ene)-polymer **61** in Heck type coupling reactions.[63] The significant variations in activity for different supports indicated that the reactions proceeded *via* supported species. Weberskirch and co-workers used amphiphilic, water-soluble copolymers with pendant NHC–Pd complexes as catalysts for Heck and Suzuki coupling in water, **62**.[64] Notably, **62** represents the best system reported to date for Suzuki water as solvent.[64a]

Lee and co-workers undertook a series of investigations using polymer-supported NHC–Pd complexes in various cross-coupling reactions (Heck, Suzuki and copper-free Sonogashira). Selected polymer supports including surface grafted polystyrene,[65] and amphiphilic polyethylene glycol (PS-PEG-NHC) on Merrifield resin.[66] Good conversions (91–97%) were obtained in Suzuki cross-coupling for aryl iodides and bromides with phenyl boronic acid at 50 °C with 1 mol% [Pd].[65a] The catalyst could be recycled up to 10 times without loss in performance. The PS-PEG-NHC catalyst system (2 mol% [Pd], 50 °C, aryl iodides and bromides with $PhB(OH)_2$) gave reasonable conversions in neat water as solvent; five recycling sequences were undertaken with some loss of catalyst performance.[66] Other supports such as polynorbornene[67] and "organic silica"[68] were also explored, but silica remains the preferred inorganic support for these transformations. In several studies organosilane linkers were used to attach a NHC–Pd complex to silica

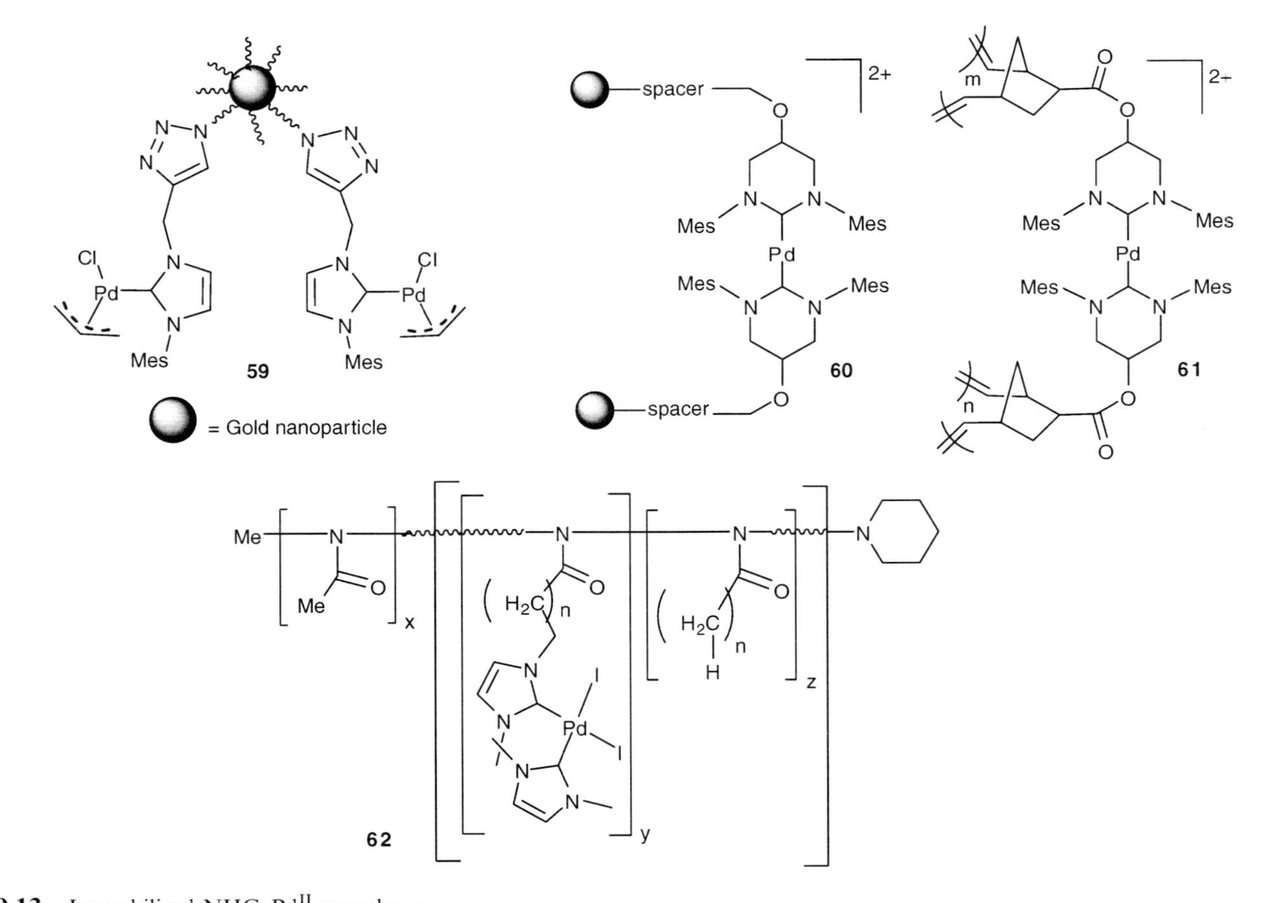

Figure 9.13 Immobilised NHC–Pd^{II} complexes.

particles, and the resulting systems tested in Suzuki[69] and Heck[70] coupling with aryl iodides and bromides. Alternatively, Karimi and Enders used a novel approach to anchor a NHC–Pd/IL matrix on silica,[71] and careful examination of the system after one catalytic run revealed the presence of Pd nanoparticles (10–40 nm) surrounded by an IL layer on the surface of the silica. A second study by Artok and co-workers also revealed that the initial NHC–Pd complex acted as a pre-catalyst for highly active Pd-nanoparticles.[72] Silica nanoparticles were used to support a NHC–Pd complex in Suzuki, and Heck coupling reactions with aryl iodides and bromides (but not chlorides) under relatively mild conditions.[73]

Seddon and co-workers investigated the Heck arylation of 2-methylprop-2-en-1-ol in several solvents including ILs with a *trans*-bis(benzothiazol-2-ylidene)Pd complex as catalyst precursor.[74] The best results were achieved in IL solvents; however, it is possible that nanoparticles are formed during the reaction. Finally, Dyson and co-workers investigated functionalised IL solvents (nitrile-functionalised imidazolium salts,[75] and ether/polyether-functionalised imidazolium and pyridinium based Ils[76]) in Pd-catalysed Suzuki couplings. In the latter report the authors were notably able to rule out the *in situ* formation of NHC–Pd during catalysis.

9.2.6 Pd–Allyl Mediated C–C Bond Formation

Allylic alkylation, also known as the Tsuji–Trost reaction, operates *via* a unique mechanism that exploits the electrophilicity of π-allyl Pd complexes. It is a versatile transformation in asymmetric synthesis,[77] and new catalysts are generally tested in this benchmark reaction. The investigation of functionalised NHC ligands containing electronically dissimilar groups has met limited success. Actually, allylic alkylation is one of the rare transformations in which phosphines still outperform NHCs.

In 2006, Fernández reported S-functionalised NHC complexes **63** and **64**, the latter featuring an original *N*-pyrrolidino substituent (Figure 9.14).[78] Good activities (up to 97% yield with 5 mol% [Pd]) and enantioselectivities up to 82 % *ee* were obtained at room temperature. Wang reported an interesting template of stereoselective monoligating NHC ligands based on the natural antitumour drug podophyllotoxin. Salt **65** used *in situ* with [Pd(OAc)$_2$] gave high activity (93% yield at 5 mol% [Pd]) and good enantioselectivity (87% *ee*).[79] Despite its originality, this class of NHCs lacked modularity, which is inconvenient for further catalyst development.

In 2007, Roland reported a series of simple NHC–Ag complexes, which proved to be quite active when used in conjunction with a Pd precursor. Interestingly, the use of PPh$_3$ as additive was required in order to get activity; however, an asymmetric version was not investigated.[80]

In direct contrast to what is observed for η^3-allyl complexes, the allyl ligand in η^1 compounds is nucleophilic. This was exploited in the asymmetric Pd-catalysed nucleophilic addition of allylic reagents to aldehydes (also known as umpolung allylation of aldehydes). There are only three recent studies with NHC–Pd catalysts, by Shi,[81] and Jarvo,[82] but activities and/or enantioselectivities were only modest.

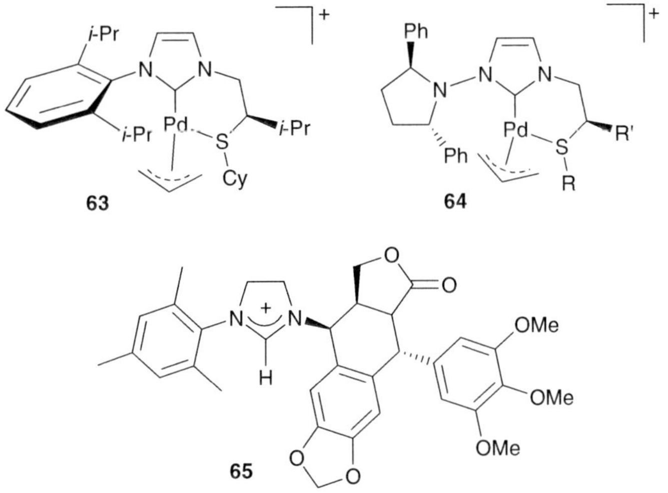

Figure 9.14 NHC–Pd^{II} complexes for asymmetric allylic alkylation.

9.2.7 Direct Arylation by C–H Functionalisation

During the current decade, reactions based on direct C–H functionalisation, and in particular Pd-catalysed direct arylation, have witnessed extraordinary progress.[83] The use of NHC–Pd catalysts (especially well-defined complexes) in these transformations is much less documented than ligandless systems or those using commercially available phosphines *in situ*.

In 2005, Sames reported the use of several well-defined NHC–Pd complexes in the direct arylation of protected heterocycles with aryl halides.[84] In particular, complex **66** gave moderate to excellent activities at low catalyst loading across a wide range of substrates. Noteworthy, only the C2 position was arylated under these conditions (Equation (9.5)). Although the NHC in **89** was monoligating, the authors did not look for evidence of hemilabile behaviour under catalytic conditions. Another interesting question is the potential for hydrogen bond interactions in the second coordination sphere of Pd between the oxygen and the acidic proton of the indole substrate.

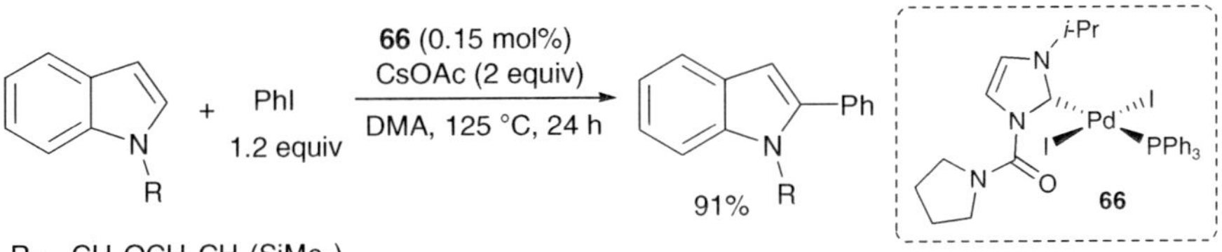

R= $-CH_2OCH_2CH_2(SiMe_3)$

(9.5)

Sanford later reported the direct arylation of *N*-methylindole with hypervalent iodine reagents using the simple Pd complex [(IMes)Pd(OAc)$_2$I].

Although the reaction was performed at room temperature and is mechanistically interesting (strong evidence was provided for high valent Pd intermediates), the use of the Ar_2I^+ cation as arylating agent severely limited the applicability of this methodology, especially on large scale.[85] Fagnou reported the intramolecular direct arylation of aryl chlorides catalysed by a series of well-defined complexes. Catalyst **67** enabled convenient formation of a range of tricyclic compounds, notably the important class of carbazoles (Equation (9.6)).[86] The authors found evidence of reductive elimination of the NHC ligand to give 2-arylimidazolium salts. As a result, the addition of NHC precursor IPr·HCl to the reaction led to increased catalyst activity.

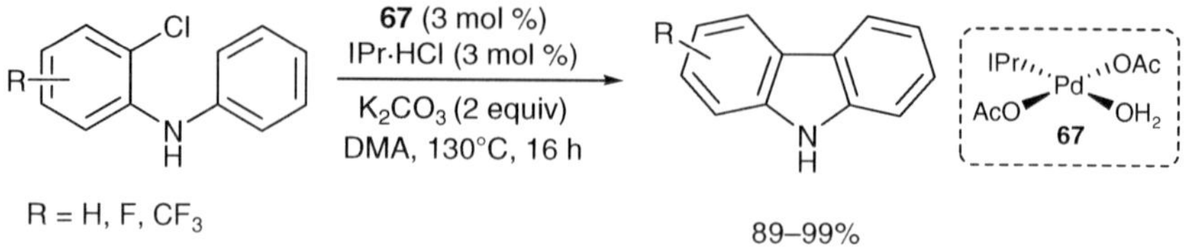

(9.6)

Finally, Greaney recently reported the use of PEPPSI–IPr in the direct arylation of oxazoles, although the scope of this reaction was limited.[87]

9.2.8 α-Carbonyl Arylation

The acidity of protons in á position of carbonyl compounds can be exploited in Pd-catalysed ketone arylation. This transformation is atom-economical, and is usually performed under mild conditions although strong bases are often employed.

Monoligating NHCs proved extremely successful in ketone arylation. The groups of Nolan[88] and Bertrand[89] reported catalysts **68** and **69**–**71**, respectively (Figure 9.15). Whilst complex **68** was based on commercially available IPr and had a rather unremarkable structure, Bertrand's catalysts were the first reported complexes of so-called cyclic alkyl(amino)carbenes, or CAAC. These ligands display unique steric properties and they are also stronger σ-donors

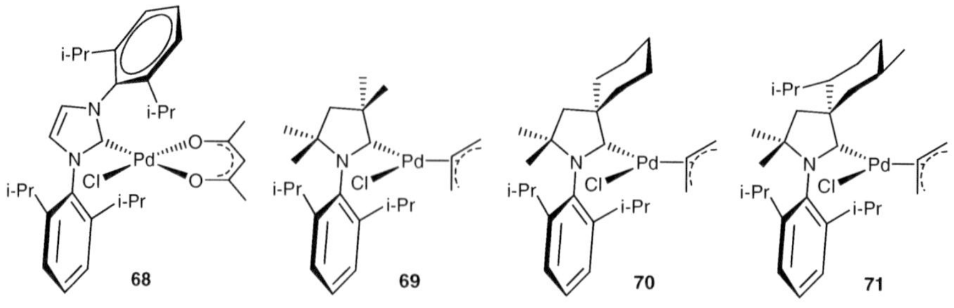

Figure 9.15 NHC–Pd^{II} catalysts used in ketone arylation.

than diaminocarbenes.[90] **71** was considerably more active than **68** and **69–70** when phenyl chloride was used as substrate. However, due to its rigid conformation, **71** was completely inactive in the reaction of propiophenone with bulky aryl halides, whereas **68** and **70** performed well, probably due to the more flexible substituents on the NHCs.

Amide arylation is closely related to ketone arylation. In an elegant study reported in 2008, Dorta described the application of atropoisomeric napthyl-substituted NHC ligands to the intramolecular asymmetric arylation of amides.[91] Using catalyst **72** (Equation (9.7)), the authors obtained excellent yields (>89%) and up to 88% *ee*. The two other atropoisomers of **72** (not depicted) gave lower enantioselectivities, which was ascribed to the matching/mismatching of chirality in the Pd intermediate prior to reductive elimination. Glorius reported the application of IBox ligand **73** to the same reaction.[92] The selectivities observed were sometimes excellent (up to 97% *ee*), and aryl bromides and chlorides were successfully coupled in good to excellent yields (up to 99%). The bulky isopropyl groups in **73** apparently locked the NHC in the depicted conformation, and as a result the ligand exerted extremely strong and asymmetric steric pressure on the Pd fragment, hence the observed enantioselectivity.

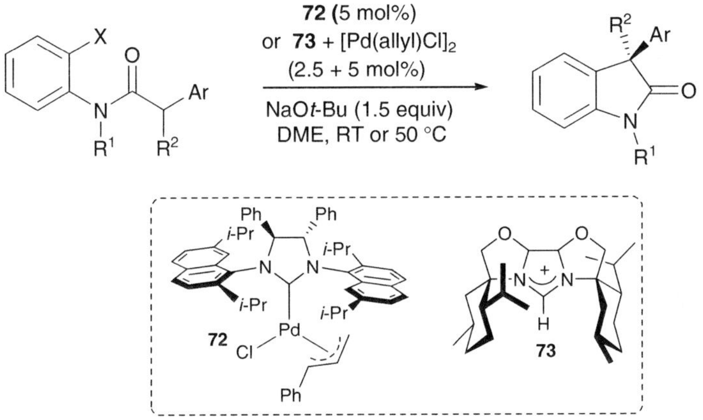

72 (5 mol%)
or **73** + $[Pd(allyl)Cl]_2$
(2.5 + 5 mol%)
NaO*t*-Bu (1.5 equiv)
DME, RT or 50 °C

(9.7)

9.2.9 Polymerisation Reactions

A limited number of studies focussed on the development of NHC–Pd catalysts for polymerisation reactions. The α-ketone arylation reactions, described in Section 9.2.8, can be applied to the polycondensation of haloarylketones. Thus, Matsubara reported the polymerisation of bromo- or chloro-propiophenone catalysed by mixtures of $[Pd(OAc)_2]$ and IPr·HCl or IMes·HCl (3 mol% [Pd]). The former system was more active and the use of excess imidazolium salts resulted in decreased activity, due to the formation of bis-NHC Pd species.[93] Chung reported the application of Nolan-type Pd-allyl catalysts (see Figure 9.8) to the polymerisation of ester-functionalised norbornene monomers.[94] These substrates are typically more difficult to polymerize than non-functionalised monomers, and such catalysts efficiently catalysed the polymerisation of simple norbornene with 0.002 mol% [Pd] ,while the methyl ester-containing monomer required 1 mol% [Pd]. Catalyst activation was effected by chloride abstraction with silver salts.

9.2.10 Telomerisation of Dienes

The telomerisation of dienes is a powerful methodology for the atom-economical construction of bulk chemicals from cheap feedstock such as butadiene and alcohols, amines and even CO_2.[95] Beller reported the telomerisation of 1,3-butadiene with diols (*e.g.* ethylene glycol) using IMes-based systems at low catalyst loading (0.001 mol% [Pd]).[96] Following up on this report, a study looking at the scaling-up (up to 400 g scale) of this methodology was reported and in particular, excellent chemoselectivity was observed with *N*-methylaminoethanol.[97]

9.2.11 Miscellaneous C–C Bond-forming Reactions

A number of isolated, less systematic reports describing catalytic applications of NHC–Pd complexes appeared, sometimes with remarkable results. For example, Shi reported the use of **74** in the asymmetric arylation of *N*-tosylimines with arylboronic acids (Equation (9.8)).[98] Using 3 mol% **74** under mild conditions, the authors obtained mostly excellent yields and high *ee*s for a broad range of substrates.

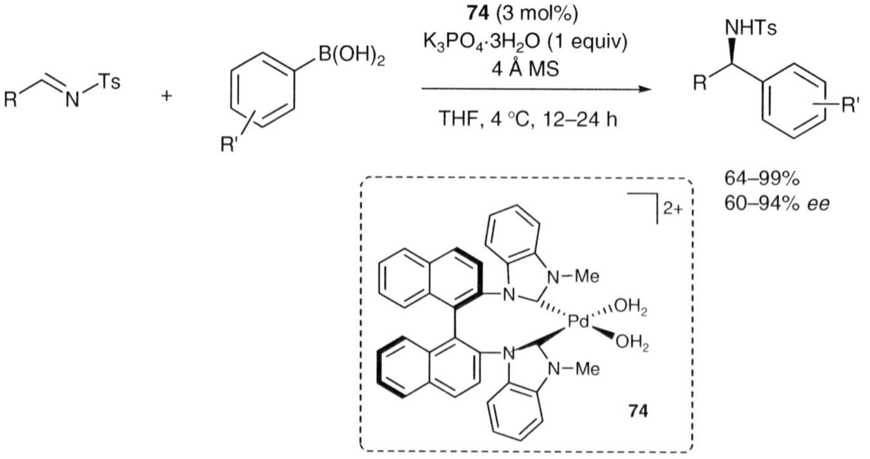

R–CH=N–Ts + R'–C_6H_4–$B(OH)_2$ → (**74** (3 mol%), $K_3PO_4{\cdot}3H_2O$ (1 equiv), 4 Å MS; THF, 4 °C, 12–24 h) → R–CH(NHTs)–C_6H_4–R'

64–99%
60–94% *ee*

74 ($[Pd(NHC)_2(OH_2)_2]^{2+}$; N–Me, Pd, OH_2)

(9.8)

In 2008, Sigman reported the hydroarylation of styrene derivatives with phenylboronic esters under oxygen.[99] Using 0.75 mol% *trans*-$[(SIPr)PdCl_2]_2$ and 6 mol% (–)-sparteine as a co-catalyst, the authors obtained good to very high yields of hydroarylation product for a variety of substrates. However, the reaction was plagued with decomposition pathways affecting the phenylboronate partner, generating biphenyl and phenol by-products. Other transformations of interest catalysed by NHC–Pd catalysts include the asymmetric 1,4-conjugate addition of carbon nucleophiles to α,β-unsaturated ketones,[26,100] alkyne hydroarylation,[101] diphenylacetylene arylation,[102] iodobenzene methoxycarbonylation,[103] cycloisomerisation,[104] and Friedel–Crafts alkylation of indole.[43]

9.3 C–N Bond Formation

9.3.1 Buchwald–Hartwig Aryl Amination

Aryl amination, also known as Buchwald–Hartwig coupling, is one of the most important transformations in organic synthesis, due to its versatility and the ubiquitous nature of nitrogen-containing molecules in therapeutically active compounds.[105] There have been numerous studies on catalytic applications of NHC–Pd complexes to aryl amination, sometimes with impressive results.

The Pd-allyl NHC complexes described in Section 9.2.2 were successfully applied to Buchwald–Hartwig coupling. A 1:1 Pd/NHC ratio, enforced by the well-defined nature of these catalysts, appeared to play a crucial role in achieving high efficiency (see Figure 9.8). In 2002, Nolan reported the synthesis and catalytic activity of simple [(NHC)Pd(π-allyl)Cl] complexes in cross-coupling reactions, including aryl amination.[106] By modifying the simple allyl ligand to cinnamyl as in **35a** (Figure 9.16), much improved activity was observed even for bulky and/or deactivated substrates.[35a] The performance of **48b** in the benchmark coupling of morpholine with 4-chlorotoluene and 2-chloropyridine was excellent and compared favourably with that of phosphine ligands.[35] Considering the essentially non-optimised nature of the supporting NHC in **35a**, modified Pd-allyl complexes are an extremely successful paradigm for pre-catalyst design. Similar complexes to **35a** were reported by Dorta, in which SIPr was replaced by naphthyl-substituted NHCs (Figure 9.16).[36,107] Among these, **36a** and **36d** displayed high activity in aryl amination. Direct comparison with Nolan's system revealed identical performances of **35a**, **36a** and **36d**. Of note, **36b** and **36c** also performed well with heteroaryl chlorides. Finally, Caddick

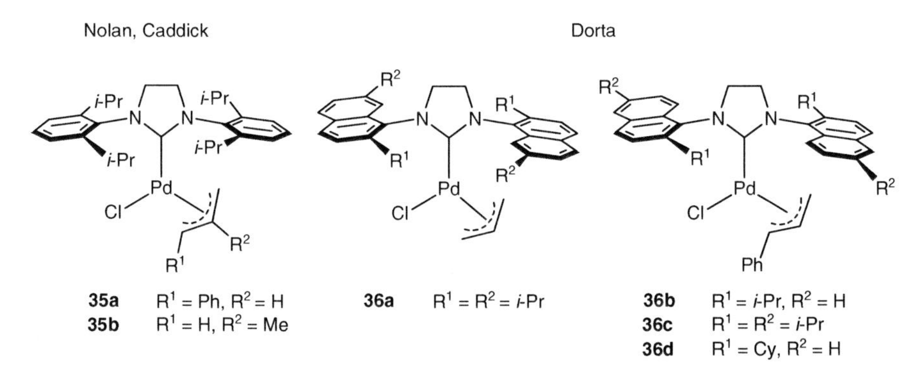

Figure 9.16 Pd^{II}-allyl NHC catalysts for aryl amination.

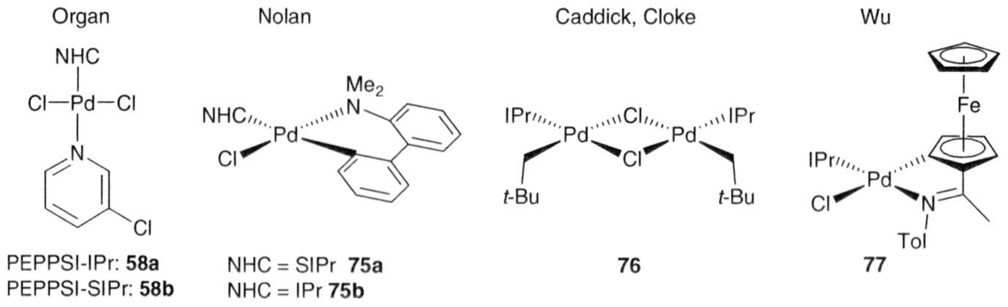

Organ Nolan Caddick, Cloke Wu

PEPPSI-IPr: **58a**
PEPPSI-SIPr: **58b**

NHC = SIPr **75a**
NHC = IPr **75b**

76

77

Figure 9.17 IPr and SIPr containing Pd catalysts for aryl amination.

recently reported complex **35b**, containing a methylallyl ligand, but its activity was well below that observed with the cinnamyl systems.[108]

Another successful monoligated Pd^0 precursor design is the PEPPSI family of catalysts developed by Organ (see Section 9.2.4). In 2008, Organ reported a study on aryl amination catalysed by PEPPSI complexes.[109] Amongst the various catalysts tested, those containing IPr and SIPr (**58a** and **58b**, Figure 9.17) gave the best results,[110] although their activity was much lower than that of Pd-cinnamyl. Interestingly, for electron-deficient aryl chlorides (*e.g.* 2-chloropyridine) and sufficiently reactive amines (*e.g.* morpholine), amine deprotonation could be effected by much weaker bases than KO*t*-Bu, *e.g.* Cs_2CO_3. This is a crucial advantage over Pd-cinnamyl catalysts for base-sensitive substrates, since for these catalysts the presence of *tert*-butoxide was required to generate the catalytically active species.[35] Other monoligated Pd^0 precursors were reported by Nolan (**75b**),[88bc,111] Caddick and Cloke (**76**),[40] and Wu (**77**).[112] The observed catalytic activities were generally high, especially for **74b**. Interestingly, IPr analogue **75a** was previously reported to have considerably lower activity, although experimental conditions were slightly different.[113]

Although monoligated Pd^0 is the dominant paradigm in NHC–Pd catalyst development for Buchwald–Hartwig amination, other approaches were investigated with some success.[1j] Shi reported the use of a (ferrocenyl)phosphine-functionalised NHC, but its activity was low, giving credit to the monodentate approach.[114]

9.3.2 Allylic Amination

Allylic amination is a variant of the reaction described in Section 9.2.6 where the nucleophilic partner is an amine. Visentin studied the amination of diphenylpropenylacetate with benzylamine using [(NHC)Pd(η^3-allyl)]CF_3CO_2, where NHC was a (ferrocenyl)phosphine-functionalised NHC.[115] The poor activity (TON = 10) observed with this catalyst compared to analogous P,N-ligands was ascribed to the highly electron donating nature of the NHC, which would decrease the electrophilicity of the Pd-allyl intermediate. The authors also explained the low enantioselectivity (5% *ee*) by the similar electronic properties of NHCs and phosphines, which caused the amine to attack at either position of the allyl ligand. In contrast, Roland reported the catalytic activity of [(NHC)Pd(η^3-allyl)Cl] **35** with mono-coordinating NHCs.[116] As observed with allylic alkylation,[80] this catalyst was active only in the presence of PPh_3. Depending on the conditions (nucleophile, use of anhydrous or aqueous biphasic conditions, *etc.*), TONs of 8–20 were observed. Clearly, these contrasting results point to a gap in the current knowledge of NHC and phosphine chemistry. Improvement of catalytic performances will only stem from a better understanding of fundamental ligand properties.

9.3.3 Miscellaneous C–N Bond-forming Reactions

In 2006, Stahl reported the activity of **78a** and *rac*-**78b** in the intramolecular amination of olefins (Equation (9.9)). A variety of substituted heterocycles were generated in fair to very good yields.[117] Interesting features of this reaction are the use of atmospheric O_2 as oxidant and the potential for an asymmetric version.

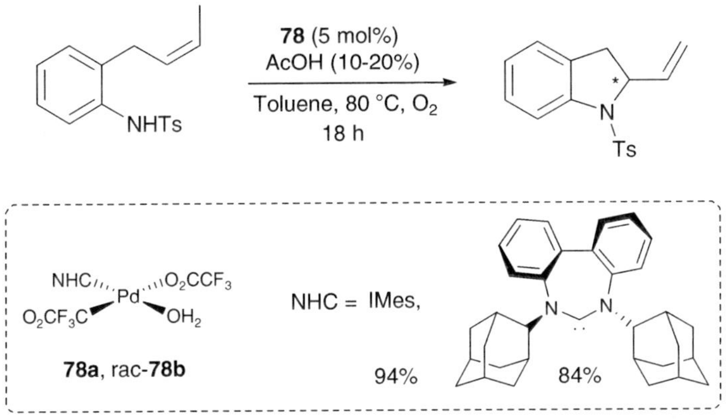

(9.9)

Slaughter recently reported a modular entry into chiral bis(ADC) complexes (ADC = acyclic diaminocarbene) starting from the corresponding Pd isocyanides and chiral diamines (Equation (9.10)).[118] Complex **79** gave modest activity and enantioselectivity in Aza–Claisen rearrangement; however, this new family of complexes has great potential for asymmetric catalysis, owing to the large variety of available chiral diamines. Finally, Xia reported the application of a number of simple NHC–Pd complexes to the synthesis of ureas from the corresponding amines and CO.[119] High activities were observed (TOF up to $9900\,h^{-1}$).

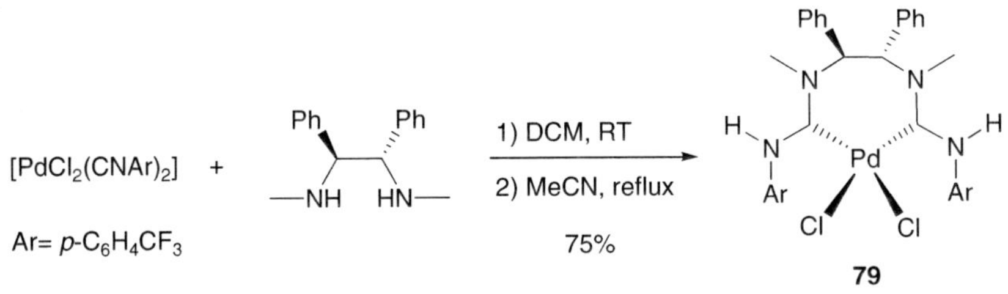

(9.10)

9.4 Other Transformations

Details on oxidation and reduction reactions mediated by [(NHC)Pd] complexes can be found in Chapters 12 and 13, respectively. Further reports of interest disclosed in recent years include diboration of alkenes catalysed by a pincer complex,[120] deuteration of C–H bonds with an N,O-functionalised NHC complex,[121] and an intriguing Suzuki-type reaction of $[FeI(Cp)(CO)_2]$ with arylboronic acids.[122]

9.5 Conclusions and Outlook

The number of reports on the design and application of NHCs as ligands has expanded dramatically in the last 15 years. The motivation for this exponential growth in publications is principally the application of NHCs in catalysis, and to a large extent the focus has been on reactions catalysed by NHC–Pd complexes. A substantial variety of NHC ligands (with different structural motifs and additional functionalities), and many different Pd complexes have been synthesised and tested in numerous catalytic reactions. Several processes involving NHC–Pd complexes appear to be approaching commercial exploitation. However, in comparison to the range of studies carried out on phosphines and their complexes, the field of NHC chemistry is in its infancy. Many exciting opportunities still exist, limited only by the imagination and synthetic skills of the chemists involved. Rather than simply treat NHCs as pseudo phophines, the prime focus should be on gaining a fundamental understanding of the behaviour and influences of NHCs, allowing rational design and the logical application of these exciting ligands. Nonetheless, there is no doubt another quantum leap in knowledge and catalyst development will occur in the next 15 years.

References and Notes

1. (a) S. Díez-González, N. Marion and S. P. Nolan, *Chem. Rev.*, 2009, **109**, 3612–3676; (b) N. Marion and S. P. Nolan, *Acc. Chem. Res.*, 2008, **41**, 1440–1449; (c) A. T. Normand and K. J. Cavell, *Eur. J. Inorg. Chem.*, 2008, 2781–2800; (d) F. E. Hahn and M. C. Jahnke, *Angew. Chem., Int. Ed.*, 2008, **47**, 3122–3172; (e) E. A. B. Kantchev, C. J. O'Brien and M. G. Organ, *Angew. Chem., Int. Ed.*, 2007, **46**, 2768–2813; (f) J. A. Mata, M. Poyatos and E. Peris, *Coord. Chem. Rev.*, 2007, **251**, 841–859; (g) L. H. Gade and S. Bellemin-Laponnaz, *Coord. Chem. Rev.*, 2007, **251**, 718–725; (h) R. E. Douthwaite, *Coord. Chem. Rev.*, 2007, **251**, 702–717; (i) D. Pugh and A. A. Danopoulos, *Coord. Chem. Rev.*, 2007, **251**, 610–641; (j) K. J. Cavell and D. S. McGuinness, in *Comprehensive Organometallic Chemistry III*, ed. R. H. Crabtree, D. M. P. Mingos and A. J. Canty, Elsevier, Amsterdam, 2007, pp. 197–268; (k) D. J. Nielsen and K. J. Cavell, in *N-Heterocyclic Carbenes in Synthesis*, ed. S. P. Nolan, Wiley-VCH Verlag, Weinheim, 2006, pp. 73–102; (l) V. César,

S. Bellemin-Laponnaz and L. H. Gade, *Chem. Soc. Rev.*, 2004, **33**, 619–636; (m) K. J. Cavell and D. S. McGuinness, *Coord. Chem. Rev.*, 2004, **248**, 671–679; (n) W. A. Herrmann, *Angew. Chem., Int. Ed.*, 2002, **41**, 1290–1309.

2. S. Roland, P. Mangeney and A. Jutand, *Synlett*, 2006, 3088–3094.
3. A. K. de K. Lewis, S. Caddick, F. G. N. Cloke, N. C. Billingham, P. B. Hitchcock and J. Leonard, *J. Am. Chem. Soc.*, 2003, **125**, 10066–10073.
4. G. D. Frey, J. Schutz, E. Herdtweck and W. A. Herrmann, *Organometallics*, 2005, **24**, 4416–4426.
5. (a) E. A. B. Kantchev, G.-R. Peh, C. Zhang and J. Y. Ying, *Org. Lett.*, 2008, **10**, 3949–3952; (b) G.-R. Peh, E. A. B. Kantchev, C. Zhang and J. Y. Ying, *Org. Biomol. Chem.*, 2009, **7**, 2110–2119.
6. M. V. Baker, D. H. Brown, P. V. Simpson, B. W. Skelton, A. H. White and C. C. Williams, *J. Organomet. Chem.*, 2006, **691**, 5845–5855.
7. T. Scherg, S. K. Schneider, G. D. Frey, J. Schwarz, E. Herdtweck and W. A. Herrmann, *Synlett*, 2006, 2894–2907.
8. C. Tubaro, A. Biffis, C. Gonzato, M. Zecca and M. Basato, *J. Mol. Catal. A: Chem.*, 2006, **248**, 93–98.
9. (a) S. Ahrens, A. Zeller, M. Taige and T. Strassner, *Organometallics*, 2006, **25**, 5409–5415; (b) M. A. Taige, A. Zeller, S. Ahrens, S. Goutal, E. Herdtweck and T. Strassner, *J. Organomet. Chem.*, 2007, **692**, 1519–1529.
10. (a) C. Burstein, C. W. Lehmann and F. Glorius, *Tetrahedron*, 2005, **61**, 6207–6217; (b) M. Nonnenmacher, D. Kunz, F. Rominger and T. Oeser, *J. Organomet. Chem.*, 2007, **692**, 2554–2563.
11. C.-S. Lee, S. Pal, W.-S. Yang, W.-S. Hwang and I. J. B. Lin, *J. Mol. Catal. A: Chem.*, 2008, **280**, 115–121.
12. (a) C.-M. Jin, B. Twamley and J. M. Shreeve, *Organometallics*, 2005, **24**, 3020–3023; (b) R. Wang, C.-M. Jin, B. Twamley and J. M. Shreeve, *Inorg. Chem.*, 2006, **45**, 6396–6403.
13. (a) R. Wang, B. Twamley and J. M. Shreeve, *J. Org. Chem.*, 2006, **71**, 426–429; (b) R. Wang, Z. Zeng, B. Twamley, M. M. Piekarski and J. M. Shreeve, *Eur. J. Org. Chem.*, 2007, 655–661.
14. (a) J. Ye, W. Chen and D. Wang, *Dalton Trans.*, 2008, 4015–4022; (b) D. Meyer, M. A. Taige, A. Zeller, K. Hohlfeld, S. Ahrens and T. Strassner, *Organometallics*, 2009, **28**, 2142–2149.
15. (a) J. Ye, X. Zhang, W. Chen and S. Shimada, *Organometallics*, 2008, **27**, 4166–4172; (b) S. Gu and W. Chen, *Organometallics*, 2009, **28**, 909–914.
16. S. G. Fiddy, J. Evans, T. Neisius, M. A. Newton, N. Tsoureas, A. A. D. Tulloch and A. A. Danopoulos, *Chem.–Eur. J.*, 2007, **13**, 3652–3659.
17. (a) X. Zhang, A. Liu and W. Chen, *Org. Lett.*, 2008, **10**, 3849–3852; (b) X. Zhang, Z. Xi, A. Liu and W. Chen, *Organometallics*, 2008, **27**, 4401–4406.
18. (a) W.-H. Yang, C.-S. Lee, S. Pal, Y.-N. Chen, W.-S. Hwang, I. J. B. Lin and J.-C. Wang, *J. Organomet. Chem.*, 2008, **693**, 3729–3740; (b) O.

Winkelmann, C. Näther and U. Lüning, *J. Organomet. Chem.*, 2008, **693**, 2784–2788.
19. D. J. Nielsen, K. J. Cavell, B. W. Skelton and A. H. White, *Inorg. Chim. Acta*, 2006, **359**, 1855–1869.
20. J. Houghton, G. Dyson, R. E. Douthwaite, A. C. Whitwood and B. M. Kariuki, *Dalton Trans.*, 2007, 3065–3073.
21. D. J. Nielsen, K. J. Cavell, B. W. Skelton and A. H. White, *Organometallics*, 2006, **25**, 4850–4856.
22. N. Stylianides, A. A. Danopoulos, D. Pugh, F. Hancock and A. Zanotti-Gerosa, *Organometallics*, 2007, **26**, 5627–5635.
23. C. Tubaro, A. Biffis, M. Basato, F. Benetollo, K. J. Cavell and L. Ooi, *Organometallics*, 2005, **24**, 4153–4158.
24. (a) F. E. Hahn, M. C. Jahnke, V. Gomez-Benitez, D. Morales-Morales and T. Pape, *Organometallics*, 2005, **24**, 6458–6463; (b) F. E. Hahn, M. C. Jahnke and T. Pape, *Organometallics*, 2007, **26**, 150–154.
25. (a) F. E. Hahn, M. C. Jahnke and T. Pape, *Organometallics*, 2006, **25**, 5927–5936; (b) M. C. Jahnke, T. Pape and F. E. Hahn, *Eur. J. Inorg. Chem.*, 2009, **20**, 1960–1969.
26. S. Sakaguchi, K. S. Yoo, J. O'Neill, J. H. Lee, T. Stewart and K. W. Jung, *Angew. Chem., Int. Ed.*, 2008, **47**, 9326–9329.
27. (a) L.-j. Liu, F. Wang and M. Shi, *Eur. J. Inorg. Chem.*, 2009, **20**, 1723–1728See also; (b) T. Zhang, S. Liu, M. Shi and M. Zhao, *Synthesis*, 2008, 2819–2824.
28. G. Zou, W. Huang, Y. Xiao and J. Tang, *New. J. Chem.*, 2006, **30**, 803–809.
29. M. V. Baker, D. H. Brown, P. V. Simpson, B. W. Skelton and A. H. White, *Eur. J. Inorg. Chem.*, 2009, 1977–1988.
30. (a) H. V. Huynh, J. H. H. Ho, T. C. Neo and L. L. Koh, *J. Organomet. Chem.*, 2005, **690**, 3854–3860; (b) H. V. Huynh, T. C. Neo and G. K. Tan, *Organometallics*, 2006, **25**, 1298–1302; (c) Y. Han, H. V. Huynh and L. L. Koh, *J. Organomet. Chem.*, 2007, **692**, 3606–3613.
31. H. Türkmen, O. Sahin, O. Büyükgüngör and B. Çetinkaya, *Eur. J. Inorg. Chem.*, 2006, **20**, 4915–4921.
32. (a) S. K. Yen, L. L. Koh, F. E. Hahn, H. V. Huynh and T. S. A. Hor, *Organometallics*, 2006, **25**, 5105–5112; (b) S. K. Yen, L. L. Koh, H. V. Huynh and T. S. A. Hor, *Dalton Trans.*, 2007, 3952–3958; (c) S. K. Yen, L. L. Koh, H. V. Huynh and T. S. A. Hor, *Dalton Trans.*, 2008, 699–706.
33. R. B. Bedford, M. Betham, S. J. Coles, R. M. Frost and M. B. Hursthouse, *Tetrahedron*, 2005, **61**, 9663–9669.
34. R. B. Bedford, M. Betham, M. E. Blake, R. M. Frost, P. N. Horton, M. B. Hursthouse and R.-M. López-Nicolás, *Dalton Trans.*, 2005, 2774–2779.
35. (a) N. Marion, O. Navarro, J. Mei, E. D. Stevens, N. M. Scott and S. P. Nolan, *J. Am. Chem. Soc.*, 2006, **128**, 4101–4111; (b) O. Navarro, N. Marion, J. Mei and S. P. Nolan, *Chem.–Eur. J.*, 2006, **12**, 5142–5148. See also; (c) B. H. Lipshutz, T. B. Petersen and A. R. Abela, *Org. Lett.*, 2008, **10**, 1333–1336.

36. X. Luan, R. Mariz, M. Gatti, C. Costabile, A. Poater, L. Cavallo, A. Linden and R. Dorta, *J. Am. Chem. Soc.*, 2008, **130**, 6848–6858.
37. R. Singh, M. S. Viciu, N. Kramareva, O. Navarro and S. P. Nolan, *Org. Lett.*, 2005, **7**, 1829–1832.
38. (a) H. V. Huynh, Y. Han, J. H. H. Ho and G. K. Tan, *Organometallics*, 2006, **25**, 3267–3274; (b) Y. Han, Y.-T. Hong and H. V. Huynh, *J. Organomet. Chem.*, 2008, **693**, 3159–3165.
39. (a) S. Zheng, L. Xu and C. Xia, *App. Organomet. Chem.*, 2007, **21**, 772–776; (b) B. M. O'Keefe, N. Simmons and S. F. Martin, *Org. Lett.*, 2008, **10**, 5301–5304.
40. O. Esposito, P. M. P. Gois, A. K. de K. Lewis, S. Caddick, F. G. N. Cloke and P. B. Hitchcock, *Organometallics*, 2008, **27**, 6411–6418.
41. (a) K. Arentsen, S. Caddick and F. G. N. Cloke, *Tetrahedron*, 2005, **61**, 9710–9715; (b) H. Türkmen and B. Çetinkaya, *J. Organomet. Chem.*, 2006, **691**, 3749–3759; (c) Y. Yamamoto, *Synlett*, 2007, 1913–1916.
42. (a) M. Shi and H.-x. Qian, *Tetrahedron*, 2005, **61**, 4949–4955; (b) Q. Xu, W.-L. Duan, Z.-Y. Lei, Z.-B. Zhu and M. Shi, *Tetrahedron*, 2005, **61**, 11225–11229.
43. Z. Liu, T. Zhang and M. Shi, *Organometallics*, 2008, **27**, 2668–2671.
44. S. Saito, H. Yamaguchi, H. Muto and T. Makino, *Tetrahedron Lett.*, 2007, **48**, 7498–7501.
45. T. A. P. Paulose, J. A. Olson, J. W. Quail and S. R. Foley, *J. Organomet. Chem.*, 2008, **693**, 3405–3410.
46. (a) A. J. Arduengo III, D. Tapu and W. J. Marshall, *J. Am. Chem. Soc.*, 2005, **127**, 16400–16401; (b) C. Fliedel, A. Maisse-François and S. Bellemin-Laponnaz, *Inorg. Chim. Acta*, 2007, **360**, 143–148; (c) L. Ray, M. M. Shaikh and P. Ghosh, *Organometallics*, 2007, **26**, 958–964; (d) J.-M. Suisse, L. Douce, S. Bellemin-Laponnaz, A. Maisse-François, R. Welter, Y. Miyake and Y. Shimizu, *Eur. J. Inorg. Chem.*, 2007, 3899–3905; (e) C. S. Linninger, E. Herdtweck, S. D. Hoffmann, W. A. Herrmann and F. E. Kühn, *J. Mol. Struct.*, 2008, **890**, 192–197; (f) C. Fliedel, G. Schnee and P. Braunstein, *Dalton Trans.*, 2009, 2474–2476.
47. (a) P. L. Chiu, C.-L. Lai, C.-F. Chang, C.-H. Hu and H. M. Lee, *Organometallics*, 2005, **24**, 6169–6178; (b) F. Zeng and Z. Yu, *J. Org. Chem.*, 2006, **71**, 5274–5281; (c) T. Chen, J. Gao and M. Shi, *Tetrahedron*, 2006, **62**, 6289–6294; (d) J.-C. Shi, P.-Y. Yang, Q. Tong, Y. Wu and Y. Peng, *J. Mol. Catal. A: Chem.*, 2006, **259**, 7–10; (e) C.-Y. Wang, Y.-H. Liu, S.-M. Peng, J.-T. Chen and S.-T. Liu, *J. Organomet. Chem.*, 2007, **692**, 3976–3983; (f) H. Ren, P. Yao, S. Xu, H. Song and B. Wang, *J. Organomet. Chem.*, 2007, **692**, 2092–2098; (g) I. Dinarès, C. Garcia de Miguel, M. Font-Bardia, X. Solans and E. Alcalde, *Organometallics*, 2007, **26**, 5125–5128; (h) C.-Y. Liao, K.-T. Chan, J.-Y. Zeng, C.-H. Hu, C.-Y. Tu and H. M. Lee, *Organometallics*, 2007, **26**, 1692–1702; (i) F. Li, S. Bai and T. S. A. Hor, *Organometallics*, 2008, **27**, 672–677; (j) K. A. Netland, A. Krivokapic, M. Schröder, K. Boldt, F. Lundvall and M. Tilset, *J. Organomet. Chem.*, 2008, **693**, 3703–3710; (k) T. Zhang,

W. Wang, X. Gu and M. Shi, *Organometallics*, 2008, **27**, 753–757; (l) M. Ulusoy, O. Sahin, O. Büyükgüngör and B. Çetinkaya, *J. Organomet. Chem.*, 2008, **693**, 1895–1902; (m) H. Willms, W. Frank and C. Ganter, *Chem.–Eur. J.*, 2008, **14**, 2719–2729.
48. M. Meise and R. Haag, *ChemSusChem*, 2008, **1**, 637–642.
49. W. Wei, Y. Qin, M. Luo, P. Xia and M. S. Wong, *Organometallics*, 2008, **27**, 2268–2272.
50. F. Churruca, R. SanMartin, B. Inés, I. Tellitu and E. Domínguez, *Adv. Synth. Catal.*, 2006, **348**, 1836–1840.
51. J.-Y. Li, A. J. Yu, Y.-J. Wu, Y. Zhu, C.-X. Du and H.-W. Yang, *Polyhedron*, 2007, **26**, 2629–2637.
52. S. K. Schneider, W. A. Herrmann and E. Herdtweck, *J. Mol. Catal. A: Chem.*, 2006, **245**, 248–254.
53. G. Altenhoff, S. Würtz and F. Glorius, *Tetrahedron Lett.*, 2006, **47**, 2925–2928.
54. L. Ray, Barman, S. Mobin, M. Shaikh and P. Ghosh, *Chem.–Eur. J.*, 2008, **14**, 6646–6655.
55. (a) C. J. O'Brien, E. A. B. Kantchev, C. Valente, N. Hadei, G. A. Chass, A. Lough, A. C. Hopkinson and M. G. Organ, *Chem.–Eur. J.*, 2006, **12**, 4743–4748; (b) M. G. Organ, S. Avola, I. Dubovyk, N. Hadei, E. A. B. Kantchev, C. J. O'Brien and C. Valente, *Chem.–Eur. J.*, 2006, **12**, 4749–4755; (c) C. Valente, S. Baglione, D. Candito, C. J. O'Brien and M. G. Organ, *Chem. Commun.*, 2008, 735–737.
56. M. G. Organ, S. Çalimsiz, M. Sayah, K. H. Hoi and A. J. Lough, *Angew. Chem., Int. Ed.*, 2009, **48**, 2383–2387.
57. M. G. Organ, M. Abdel-Hadi, S. Avola, N. Hadei, J. Nasielski, C. J. O'Brien and C. Valente, *Chem.–Eur. J.*, 2007, **13**, 150–157.
58. G. Lamanna and S. Menichetti, *Adv. Synth. Catal.*, 2007, **349**, 2188–2194.
59. L. Ray, M. M. Shaikh and P. Ghosh, *Dalton Trans.*, 2007, 4546–4555.
60. C. Dash, M. M. Shaikh and P. Ghosh, *Eur. J. Inorg. Chem.*, 2009, 1608–1618.
61. K. Mennecke and A. Kirschning, *Synthesis*, 2008, 3267–3272.
62. W. J. Sommer and M. Weck, *Langmuir*, 2007, **23**, 11991–11995.
63. M. Mayr and M. R. Buchmeiser, *Macromol. Rapid Commun.*, 2004, **25**, 231–236.
64. (a) D. Schönfelder, O. Nuyken and R. Weberskirch, *J. Organomet. Chem.*, 2005, **690**, 4648–4655; (b) D. Schönfelder, K. Fischer, M. Schmidt, O. Nuyken and R. Weberskirch, *Macromolecules*, 2005, **38**, 254–262.
65. (a) J.-H. Kim, J.-W. Kim, M. Shokouhimehr and Y.-S. Lee, *J. Org. Chem.*, 2005, **70**, 6714–6720; (b) D.-H. Lee, J.-H. Kim, B.-H. Jun, H. Kang, J. Park and Y.-S. Lee, *Org. Lett.*, 2008, **10**, 1609–1612.
66. J.-W. Kim, J.-H. Kim, D.-H. Lee and Y.-S. Lee, *Tetrahedron Lett.*, 2006, **47**, 4745–4748.
67. W. J. Sommer and M. Weck, *Adv. Synth. Catal.*, 2006, **348**, 2101–2113.
68. V. Polshettiwar and R. S. Varma, *Tetrahedron*, 2008, **64**, 4637–4643.

69. (a) S.-M. Lee, H.-J. Yoon, J.-H. Kim, W.-J. Chung and Y.-S. Lee, *Pure Appl. Chem.*, 2007, **79**, 1553–1559; (b) H. Qiu, S. M. Sarkar, D.-H. Lee and M.-J. Jin, *Green Chem.*, 2008, **10**, 37–40.
70. V. Polshettiwar, P. Hesemann and J. J. E. Moreau, *Tetrahedron Lett.*, 2007, **48**, 5363–5366.
71. B. Karimi and D. Enders, *Org. Lett.*, 2006, **8**, 1237–1240.
72. Ö. AksIn, H. Türkmen, L. Artok, B. Çetinkaya, C. Ni, O. Büyükgüngör and E. Özkal, *J. Organomet. Chem.*, 2006, **691**, 3027–3036.
73. S. Tandukar and A. Sen, *J. Mol. Catal. A: Chem.*, 2007, **268**, 112–119.
74. (a) S. A. Forsyth, H. Q. N. Gunaratne, C. Hardacre, A. McKeown, D. W. Rooney and K. R. Seddon, *J. Mol. Catal. A: Chem.*, 2005, **231**, 61–66. For a review, see; (b) V. Calò, A. Nacci and A. Monopoli, *J. Organomet. Chem.*, 2005, **690**, 5458–5466.
75. Z. Fei, D. Zhao, D. Pieraccini, W. H. Ang, T. J. Geldbach, R. Scopelliti, C. Chiappe and P. J. Dyson, *Organometallics*, 2007, **26**, 1588–1598.
76. X. Yang, Z. Fei, T. J. Geldbach, A. D. Phillips, C. G. Hartinger, Y. Li and P. J. Dyson, *Organometallics*, 2008, **27**, 3971–3977.
77. B. M. Trost and M. L. Crawley, *Chem. Rev.*, 2003, **103**, 2921–2944.
78. A. Ros, D. Monge, M. Alcarazo, E. Álvarez, J. M. Lassaletta and R. Fernández, *Organometallics*, 2006, **25**, 6039–6046.
79. S.-J. Li, J.-H. Zhong and Y.-G. Wang, *Tetrahedron: Asymmetry*, 2006, **17**, 1650–1654.
80. A. Flahaut, S. Roland and P. Mangeney, *J. Organomet. Chem.*, 2007, **692**, 5754–5762.
81. (a) T. Zhang, M. Shi and M. Zhao, *Tetrahedron*, 2008, **64**, 2412–2418; (b) W. Wang, T. Zhang and M. Shi, *Organometallics*, 2009, **28**, 2640–2642.
82. N. T. Barczak, R. E. Grote and E. R. Jarvo, *Organometallics*, 2007, **26**, 4863–4865.
83. (a) J. A. Ellman, *Science*, 2007, **316**, 1131–1132; (b) K. Dogula and D. Sames, *Science*, 2006, **312**, 67–72; (c) V. Ritleng, C. Sirlin and M. Pfeffer, *Chem. Rev.*, 2002, **102**, 1731–1770.
84. B. B. Touré, B. S. Lane and D. Sames, *Org. Lett.*, 2006, **8**, 1979–1982.
85. N. R. Deprez, D. Kalyani, A. Krause and M. S. Sanford, *J. Am. Chem. Soc.*, 2006, **128**, 4972–4973.
86. L.-C. Campeau, P. Thansandote and K. Fagnou, *Org. Lett.*, 2005, **7**, 1857–1860.
87. E. F. Flegeau, M. E. Popkin and M. F. Greaney, *Org. Lett.*, 2008, **10**, 2717–2720.
88. (a) O. Navarro, N. Marion, N. M. Scott, J. González, D. Amoroso, A. Bell and S. P. Nolan, *Tetrahedron*, 2005, **61**, 9716–9722; (b) N. Marion, E. C. Ecarnot, O. Navarro, D. Amoroso, A. Bell and S. P. Nolan, *J. Org. Chem.*, 2006, **71**, 3816–3821; (c) N. Marion, P. de Frémont, I. M. Puijk, E. C. Ecarnot, D. Amoroso, A. Bell and S. P. Nolan, *Adv. Synth. Catal.*, 2007, **349**, 2380–2384.

89. V. Lavallo, Y. Canac, C. Präsang, B. Donnadieu and G. Bertrand, *Angew. Chem., Int. Ed.*, 2005, **44**, 5705–5709.
90. For further details, see Chapter 5.
91. X. Luan, R. Mariz, C. Robert, M. Gatti, S. Blumentritt, A. Linden and R. Dorta, *Org. Lett.*, 2008, **10**, 5569–5572.
92. S. Würtz, C. Lohre, R. Fröhlich, K. Bergander and F. Glorius, *J. Am. Chem. Soc.*, 2009, **131**, 8344–8345.
93. K. Matsubara, H. Okazaki and M. Senju, *J. Organomet. Chem.*, 2006, **691**, 3693–3699.
94. (a) I. G. Jung, J. S. Young, K. C. Dong, M. S. Sung-Ho and C. S. U. Son, *J. Polym. Sci., Part A: Polym. Chem.*, 2007, **45**, 3042–3052; (b) I. G. Jung, Y. T. Lee, S. Y. Choi, D. S. Choi, Y. K. Kang and Y. K. Chung, *J. Organomet. Chem.*, 2009, **694**, 297–303.
95. For a general review on telomerisation, see: (a) A. Behr, M. Becker, T. Beckmann, L. Johnen, J. Leschinski and S. Reyer, *Angew. Chem., Int. Ed.*, 2009, **48**, 3598–3614. For reviews on [(NHC)Pd] systems, see: (b) N. D. Clement, L. Routaboul, A. Grotevendt, R. Jackstell and M. Beller, *Chem.–Eur. J.*, 2008, **14**, 7408–7420. See also ref. 1k.
96. A. Grotevendt, R. Jackstell, D. Michalik, M. Gomez and M. Beller, *ChemSusChem*, 2009, **2**, 63–70.
97. R. Jackstell, A. Grotevendt, M. G. Andreu and M. Beller, *Org. Process Res. Dev.*, 2009, **13**, 349–353.
98. G.-N. Ma, T. Zhang and M. Shi, *Org. Lett.*, 2009, **11**, 875–878.
99. Y. Iwai, K. M. Gligorich and M. S. Sigman, *Angew. Chem., Int. Ed.*, 2008, **47**, 3219–3222.
100. K. J. Cavell, M. C. Elliott, D. J. Nielsen and J. S. Paine, *Dalton Trans.*, 2006, 4922–4925.
101. (a) A. Biffis, C. Tubaro, G. Buscemi and M. Basato, *Adv. Synth. Catal.*, 2008, **350**, 189–196; (b) G. Buscemi, A. Biffis, C. Tubaro and M. Basato, *Catal. Today*, 2009, **140**, 84–89.
102. C.-C. Ho, S. Chatterjee, T.-L. Wu, K.-T. Chan, Y.-W. Chang, T.-H. Hsiao and H. M. Lee, *Organometallics*, 2009, **28**, 2837–2847.
103. W. Zawartka, A. M. Trzeciak, J. J. Ziókowski, T. Lis and J. P. Z. Ciunik, *Adv. Synth. Catal.*, 2006, **348**, 1689–1698.
104. (a) Y.-J. Song, I. G. Jung, H. Lee, Y. T. Lee, Y. K. Chung and H.-Y. Jang, *Tetrahedron Lett.*, 2007, **48**, 6142–6146; (b) Y. Yang and X. Huang, *Synlett*, 2008, 1366–1370.
105. (a) S. L. Buchwald, C. Mauger, G. Mignani and U. Scholz, *Adv. Synth. Cat.*, 2006, **348**, 23–39; (b) J. F. Hartwig, *Acc. Chem. Res.*, 2008, **41**, 1534–1544.
106. M. S. Viciu, R. F. Germaneau, O. Navarro-Fernandez, E. D. Stevens and S. P. Nolan, *Organometallics*, 2002, **21**, 5470–5472.
107. L. Vieille-Petit, X. Luan, R. Mariz, S. Blumentritt, A. Linden and R. Dorta, *Eur. J. Inorg. Chem.*, 2009, 1861–1870.
108. M. J. Cawley, F. G. N. Cloke, R. J. Fitzmaurice, S. E. Pearson, J. S. Scott and S. Caddick, *Org. Biomol. Chem.*, 2008, **6**, 2820–2825.

109. M. G. Organ, M. Abdel-Hadi, S. Avola, I. Dubovyk, N. Hadei, E. A. B. Kantchev, C. J. O'Brien, M. Sayah and C. Valente, *Chem.–Eur. J.*, 2008, **14**, 2443–2452.
110. Catalysts containing less bulky NHCs such as IMes did not perform well.
111. J. Broggi, H. Clavier and S. P. Nolan, *Organometallics*, 2008, **27**, 5525–5531.
112. J. Li, M. Cui, A. Yu and Y. Wu, *J. Organomet. Chem.*, 2007, **692**, 3732–3742.
113. M. S. Viciu, R. A. Kelly III, E. D. Stevens, F. Naud, M. Studer and S. P. Nolan, *Org. Lett.*, 2003, **5**, 1479–1482.
114. J.-c. Shi, P. Yang, Q. Tong and L. Jia, *Dalton Trans.*, 2008, 938–945.
115. F. Visentin and A. Togni, *Organometallics*, 2007, **26**, 3746–3754.
116. S. Roland, W. Cotet and P. Mangeney, *Eur. J. Inorg. Chem.*, 2009, 1796–1805.
117. (a) M. M. Rogers, J. E. Wendlandt, I. A. Guzei and S. S. Stahl, *Org. Lett.*, 2006, **8**, 2257–2260; (b) G. Liu and S. S. Stahl, *J. Am. Chem. Soc.*, 2007, **129**, 6328–6335.
118. (a) Y. A. Wanniarachchi and L. M. Slaughter, *Chem. Commun.*, 2007, 3294–3296; (b) Y. A. Wanniarachchi, Y. Kogiso and L. M. Slaughter, *Organometallics*, 2008, **27**, 21–24.
119. (a) S. Zheng, X. Peng, J. Liu, W. Sun and C. Xia, *Helv. Chim. Acta*, 2007, **90**, 1471–1476; (b) S.-Z. Zheng, X.-G. Peng, J.-M. Liu, W. Sun and C.-G. Xia, *Chin. J. Chem.*, 2007, **25**, 1065–1068.
120. V. Lillo, E. Mas-Marzá, A. M. Segarra, J. J. Carbó, C. Bo, E. Peris and E. Fernandez, *Chem. Commun.*, 2007, 3380–3382.
121. J. H. Lee, K. S. Yoo, C. P. Park, Janet, M. Olsen, S. Sakaguchi, G. K. S. Prakash, T. Mathew and K. W. Jung, *Adv. Synth. Catal.*, 2009, **351**, 563–568.
122. Y. Asada, S. Yasuda, H. Yorimitsu and K. Oshima, *Organometallics*, 2008, **27**, 6050–6052.

CHAPTER 10

NHC–Nickel and Platinum Complexes in Catalysis

YVES FORT* AND CORINNE COMOY

Laboratoire SRSMC-SOR, UMR CNRS-UHP 7565, Nancy Université, BP 70239, Bd des Aiguillettes, 54506 Vandoeuvre-les-Nancy, France

10.1 Introduction

Even if in the last decades coordinated Pd species have gained the leading part in catalytic processes, Ni and Pt catalysts must not be considered only as a pale image of their Group 10 congener. Ni or Pt might often be less efficient in carbon–carbon bond formations, multi-step syntheses or enantioselective processes, but they work better in specific reactions such as hydrocyanation or hydrosilylation, respectively. Then, a warned organic chemist actually recognizes Ni and (in a smaller extent) Pt as good complements of Pd.

In 1991, a new card set appeared in organometallic chemistry with the first isolation by Arduengo *et al.* of a free stable crystalline imidazol-2-ylidene.[1] This pioneering work was quickly and efficiently completed by Hermann *et al.*, who described the striking similarity of electron-rich organophosphanes PR_3 and these N-heterocyclic carbenes (NHCs) in terms of their metal coordination chemistry.[2] NHCs are generally considered neutral, two electrons donor ligands with low π-back-bonding tendency. However, Radius and Bickelhaupt recently showed that π-interactions were to be considered in the case of electron-rich nickel complexes.[3] They demonstrated that π-interactions accounted for at least 10% and up to 40%, depending on the nature of the metal (Ni > Pd > Pt), the NHC substitution and the electronic nature of other ligands of the complex. Also, Cavallo and co-workers reviewed the understanding of

RSC Catalysis Series No. 6
N-Heterocyclic Carbenes: From Laboratory Curiosities to Efficient Synthetic Tools
Edited by Silvia Díez-González
© Royal Society of Chemistry 2011
Published by the Royal Society of Chemistry, www.rsc.org

the NHC–M bond obtained by the application of advanced computational techniques.[4] Following the abundant work describing the remarkable catalytic properties of NHC–Pd species, Ni (and Pt) mediated processes also gained popularity in the use of NHCs as ligands. This Chapter summarizes the recent impact of NHC ligands in broadening the field of Ni-and Pt-catalyzed reactions.

10.2 Preparation and Properties of NHC–Ni^0 Complexes

The first example of NHC–Ni^0 systems was reported by Arduengo *et al.* in 1994.[5] Typically the addition of an appropriate ligand to [$Ni(COD)_2$] (COD = 1,5-cyclooctadiene) or another nickel(0) precursor was used to generate catalytically active species.

Following the initial report, a plethora of complexes were described both by varying/mixing NHC ligands, and using other Ni precursors.[6] The more recent publications described the use of bi- or tridentate[7] or pincer[8] as well as asymmetric[9] NHC ligands.

In this context, there are several documented examples of discrepancies in reactivity. For example, the reactions of IPr and ItBu with various nickel sources have drastically different outcomes. The reaction of IPr with [$Ni(COD)_2$] was reported to form [$(IPr)_2Ni$] cleanly,[10] whereas the reaction of ItBu produced several products, but not [$(ItBu)_2Ni$].[11]

During their studies on the aerobic oxidation of π-allylchloro (NHC)nickel(II) complexes, Sigman and co-workers demonstrated that conformational restriction around NHC–metal bonds was relevant to catalysis and should be considered in terms of catalyst design and for rationalization of observed reactivity patterns.[12] In particular, it appeared that a monodentate ligand with a lesser volume within the square plane was more likely to promote β-hydride elimination.[13] The issue of conformational flexibility *versus* rigidity was then expected to directly impact the catalytic behaviour of NHC–Ni complexes.

In an alternative preparation, [(NHC)Ni^0] complexes could be prepared from nickel(II) salts through *in situ* reduction in the presence of the ligand. This approach initially developed by Fort *et al.* conducted under safety conditions to stable [$(NHC)_2Ni^0$] species that could be directly used for reduction, amination or coupling reactions.[14] Transmission electron microscopy of the obtained mixture revealed a homogeneous distribution of uniformly sized amorphous and sub-nanometrical (less than 0.5 nm) Ni particles. The NMR analysis of the reaction medium revealed a spectra comparable to the reported spectra for [$(IPr)_2Ni^0$] prepared from [$Ni(COD)_2$]. This procedure was recently adapted by Matsubara to prepare free [$(IPr)_2Ni^0$] and [$(IPr)Ni^{II}(acac)$] (acac = acetylacetonato) complexes in the absence of NaO*t*-Bu.[15] This later structural study also showed that the [$(NHC)_2Ni$] systems displayed anagostic Ni–H(methyl) interactions, stabilizing this 14e species.

10.3 Preparation and Properties of NHC–Ni^{II} Complexes

Some years after the first report of NHC–Ni synthesis by Arduengo, Hermann *et al.* described the synthesis of stable NHC–Ni^{II} complexes *via* phosphine displacement.[16] Alternatively, the same complexes were obtained by *in situ* deprotonation of the corresponding azolium salts in the presence of [Ni(OAc)$_2$]. NHC–Ni^{II} syntheses were also regarded as a suitable starting point to study oxidative addition of organic halides, bond activation, carbon–carbon couplings or the Heck reaction either by theoretical methods or crystallography.[6a,17]

Chen and co-workers proposed a new efficient and direct route to NHC–Ni^{II} complexes by reacting commercially available nickel powder with bis(pyridylimidazoliumyl)methane dihexafluorophosphate in the presence of Ag$_2$O (Equation (10.1)).[18] The same complex could also be obtained from reaction of the imidazolium salt and the corresponding metal powders under air.

Ag$_2$O/Ni powder
or
Ni powder/air

2+

94%

(10.1)

10.4 Dehalogenation and Dehydrogenation Mediated by NHC–Ni Complexes

For related hydrogenation reactions, the reader is referred to Chapter 13.

10.4.1 Dehalogenation Mediated by NHC–Ni Complexes

The reduction of chlorinated arenes into arenes represents an important chemical transformation due to deleterious environmental and health impact these reluctant compounds,[19] and nickel reagents have shown high efficiencies in this reaction compared to other metals.[20] Fort and co-workers reported the first use of NHC–Ni for reduction of C–Cl bonds.[21] They showed that dehalogenation of aryl halides was efficiently performed in refluxing THF using a catalytic combination composed of Ni^0/IMes/*i*-PrONa. IMes was the most effective NHC, compared to IPr, SIPr, SIMes, ITol, SITol or ItBu, for the dehalogenation of functionalized aryl chlorides, bromides, iodides and polyhalogenated (hetero)aromatic compounds (Equation (10.2)).

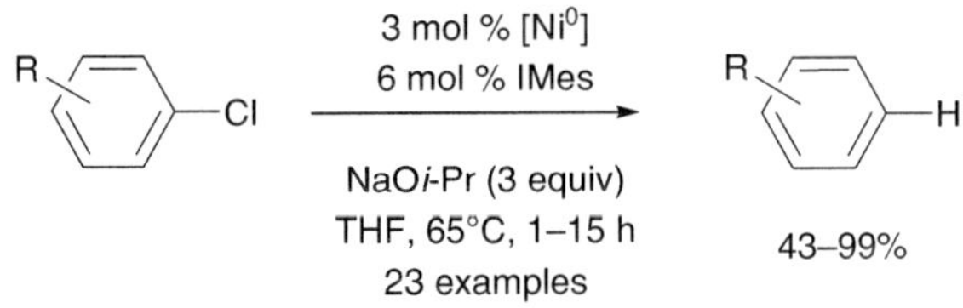
R
Cl
3 mol % [Ni⁰]
6 mol % IMes
NaO*i*-Pr (3 equiv)
THF, 65°C, 1–15 h
23 examples
R
H
43–99%

(10.2)

Both electron-poor and electron-rich aryl chlorides were cleanly dehalogenated in good to excellent yields in short reaction times. The catalyst was only slightly sensitive to the steric hindrance of the starting material. Deuterium incorporation experiments performed with $(CD_3)_2CDONa$ confirmed that sodium isopropoxide was the main hydrogen donor in these reductions. The same authors showed that carbon–fluoride bonds were efficiently cleaved or reduced by Ni^0/IMes (1:1) catalyst associated to β-hydrogen-containing alkoxide used in excess.[22] For example, the 1-fluoronaphtalene was defluorinated at 100 °C within 3 h. Except strongly coordinating 2-fluoropyridine, various substituted aromatic and heteroaromatic fluorides were defluorinated in good to excellent yields under mild conditions. Easily handled and of low cost, the proposed system constitutes one of the best known reagents of defluorination. These results also demonstrated the usefulness of NHC in activation of C–X bonds since $[Ni(bpy)_2]$ or $[Ni(PPh_3)_4]$ did not react.

Two years later, Schaub and Radius showed that dimeric complex $[(IiPr_2)_4Ni_2(COD)]$ also exhibited an interesting reactivity towards C–F bonds.[23] This catalyst activated the carbon–fluorine bond of hexafluorobenzene producing a new complex result of oxidative addition (Equation (10.3)). However, this reaction only worked with strongly activated substrates (*i.e.* hexafluorobenzene, octafluoronaphtalene).

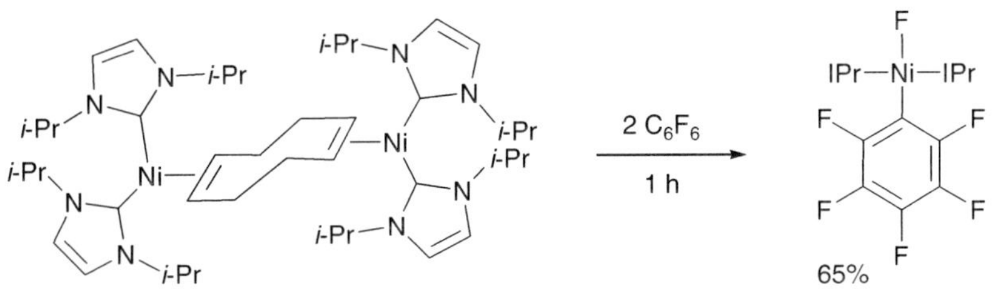
N—*i*-Pr
i-Pr
N
i-Pr
Ni
N
N
i-Pr
i-Pr—N
N
Ni
i-Pr
i-Pr
N
i-Pr—N
2 C₆F₆
1 h
F
IPr—Ni—IPr
F
F
F
F
F
65%

(10.3)

By a combined experimental and theoretical study on C–F bond activation by *in situ* generated $[(IiPr)_2Ni]$, Radius and co-workers also showed that the reaction was very fast, regio- and chemoselective.[24] For example, the activation of octafluorotoluene was only observed at the *para* position to afford a *trans*-$[(IiPr_2)_2Ni(F)(4\text{-}(CF_3)C_6F_4)]$ complex. Mechanistic studies on this activation led to the detection in solution of intermediates with the aromatic system coordinated in a η^2-fashion. For the C–F activation step, the authors suggested a pre-coordination of the aromatic system prior to a concerted oxidative addition that might be facilitated by the nucleophilicity of the [(IiPr)$_2$Ni] fragment.

Alternatively [(IiPr)$_2$Ni(F)(C$_6$F$_5$)] was isolated from the reaction of [Ni(COD)$_2$] with IiPr (2 equiv) and hexafluorobenzene.[25] Square-planar (pentafluorophenyl)nickel complexes were obtained by a systematic derivatization of this pre-formed complex. Fluorido ligand was replaced by hydrido, halogenido (Cl, I), trifluoromethanesulfonato, cyanido, alkyl, aryl, selenolato, and thiolato ligands affording the corresponding *trans*-complexes in moderate to good yields (23–85%) (Equation (10.4)).

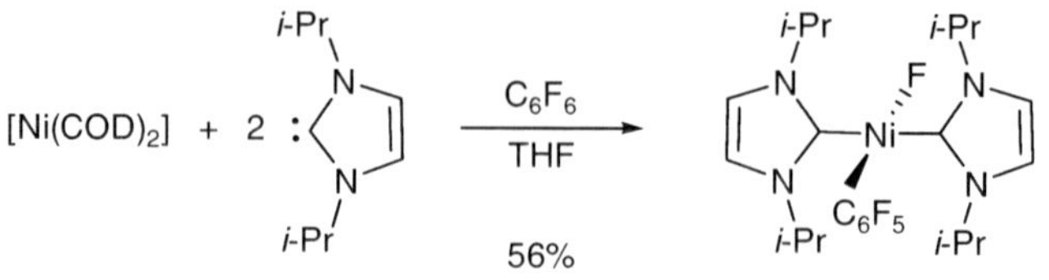

(10.4)

10.4.2 Dehydrogenation Mediated by NHC–Ni Complexes

In the last few years, ammonia–borane (AB) has attracted increasing attention as a very promising chemical hydrogen storage material. AB is solid at room temperature and presents a potential capacity of 19.6 wt % H$_2$.[26]

In 2007, Baker *et al.* reported the first example of a homogeneous NHC–Ni catalyst active in dehydrogenation of AB.[27] Enders' NHC, TPT, exhibited higher activity than IPr or IMes and the dehydrogenation then produced 18 wt % H$_2$. The TPT/Ni system consumed AB twice faster than Ru and four times faster than Rh systems. The authors postulated a mechanistic pathway involving initial formation of a σ-complex, followed by B–H bond activation and β-H elimination from the N–H (Scheme 10.1).

A year later, Yang and Hall reported a theoretical study of the reaction of AB with [(TPT)$_2$Ni].[28] They notably proposed that the dehydrogenation of AB would begin with an unusual proton transfer from nitrogen to the metal-bound carbene carbon, instead of B–H or N–H bond activation. This new C–H bond would be then activated by the metal, transfer the H to Ni, then form the H$_2$ molecule by transferring another H from B to Ni, rather than evolving by β-H transfer. This reaction pathway would better explain the importance of NHC ligands in the catalytic process.

Paul and co-workers completed this first study and demonstrated the role of free NHC in the catalytic dehydrogenation of AB.[29] They concluded that the

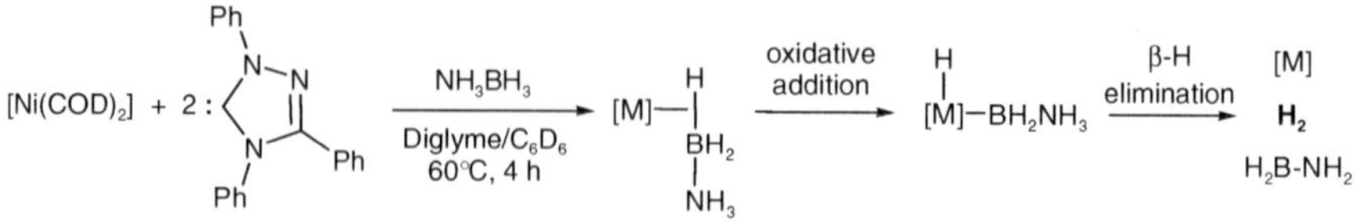

Scheme 10.1 [(NHC)Ni]-catalyzed AB dehydrogenation.

ability of the NHC–Ni system to generate more than two equivalents of H_2 from AB is partly accounted for by the assistance of free NHC in breaking down AB oligomers that may not react quickly with the Ni catalysts due to steric hindrance.

10.5 Activation/Cleavage of C–C, C–S or C–CN Bonds Mediated by NHC–Ni Complexes

During their study on activation of C–F bond, Schaub and Radius noted an interesting C–C bond activation of biphenylene.[23] They next showed that this activation was obtained with various dimeric complexes $[(NHC)_4Ni_2(COD)]$.[30]

The insertion of phenylacetylene into the strained C–C bond was completed by reaction with $[(IiPr)_4Ni_2(COD)]$ at 80 °C (Equation (10.5)). The reaction probably proceeded *via* the insertion of Ni into the C–C bond of biphenylene leading to a $[(NHC)_2Ni(2,2'\text{-biphenyl})]$ intermediate. Of note, a $[NiL_n(2,2'\text{-biphenyl})]$ intermediate had already been postulated for the desulfurization of dibenzophiophenes.[31] The activation of phenylacetylene as diphenylacetylene complexes $[(NHC)_2Ni(\eta^2\text{-}C_2Ph_2)]$ could be confirmed by X-ray analysis.

biphenylene (2, 2') + Ph–≡–Ph → [(IiPr)₄Ni₂(COD)] (3 mol %), Toluene, 80°C, 91% → 9,10-diphenylphenanthrene (Ph, Ph) (10.5)

Radius and co-workers also described the first example of C–S bond cleavage of alkyl- and aryl sulfoxides induced by the same complex $[(IiPrMe)_4\text{-}Ni_2(COD)]$.[32] For example, the *trans*-addition was obtained from DMSO as substrate in toluene (76% yield). The chemical behavior of such compounds should be of interest since Ni-complexes are common desulfurization agents as well as implicated in bioorganic chemistry.

It is well know that nickel(0) complexes play a crucial role in the commercial synthesis of adiponitrile (AdN), the major nylon-6,6 precursor.[33] In the global process, the isomerization of the branched 2-methyl-3-butenenitrile (2M3BN) to the linear 3-pentenenitrile (3PN) is a key step.[34] Such isomerization is obtained through the C–CN bond activation involving a Ni^{II} intermediate.

In this context, Radius studied the activation of organonitriles with NHC–Ni complex.[35] They showed that $[(IiPr)_4Ni_2(COD)]$ led to the efficient addition and C–CN cleavage of aromatic and aliphatic nitriles. This reaction irreversibly proceeded *via* η^2-coordination of organonitriles, followed by formation of *trans* aryl cyanide complexes (Scheme 10.2). They applied this reaction to acetonitrile or trimethylsilylnitrile as well as adiponitrile, but with the latter, some by-reactions were observed.

Concomitantly, García and co-workers studied the catalytic isomerization of 2M3BN in the presence of $[(NHC)_2Ni]$.[36] They showed that the catalytic

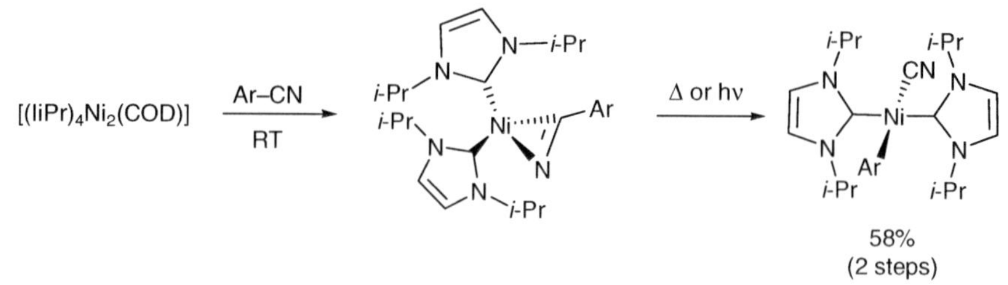

Scheme 10.2 [(NHC)Ni]-mediated activation of organonotriles.

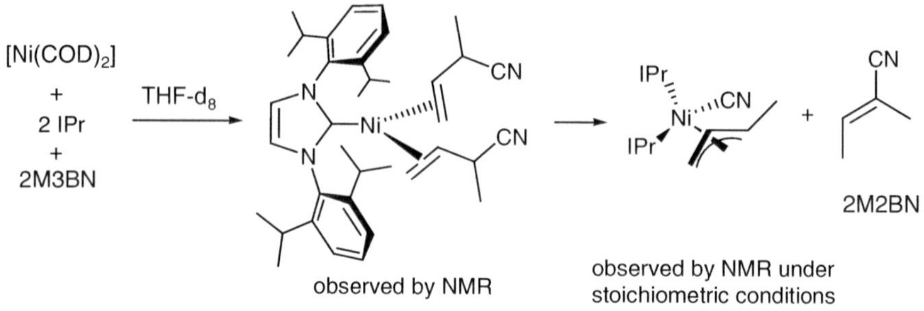

Scheme 10.3 [(NHC)Ni]-catalyzed isomerization of organonotriles.

activity strongly depended on the nature of NHC, bulky IPr affording the most active catalyst. They also obtained evidence of C–CN bond cleavage by NMR analysis under stoichiometric conditions (Scheme 10.3). The isomerization to *E*-2-methyl-2-butenenitrile (2M2BN; $Z/E = 46{:}53$) was then nearly quantitatively obtained in 15 min at room temperature. It appeared that the C–H/C–CN cleavage selectivities were increased with an excess of substrate and the isomerization of 2M3BN into 3PN was not achieved whatever the conditions used. The authors finally concluded that the use of NHC ligands did not appear to present any advantage over classical P-donor ligands.

10.6 Aryl Amination, Aryl Thiolation and Hydrothiolation Mediated by NHC–Ni Complexes

10.6.1 Aryl Amination Mediated by NHC–Ni Complexes

Along with cross-coupling reactions (see Section 10.7), direct functionalization of aromatic or vinylic halides plays an important role in biologically active substances as well as in the design of new molecular materials. In this context, the synthesis of aromatic amines has been the area of intensive research.

However, only few reports explored nickel-catalyzed amination reactions,[37] and the use of NHC allowed a significant improvement in these reactions.

Fort and co-workers first reported that [(IPr)$_2$Ni] (formed *in situ* from [Ni(acac)$_2$], SIPr·HCl, and base) allowed for the efficient amination of various aryl chlorides under mild conditions (Equation (10.6)).[38] Both electron-poor and electron-rich substituents were supported as well as pyridinic core. A great variety of anilines and secondary amines were easily introduced in good to excellent yields. Primary amines also reacted in moderate yields with activated aryl chlorides. The catalytic cycle was presumably similar to that postulated for the palladium-catalyzed amination of aryl halides using NHC–Pd systems.[39]

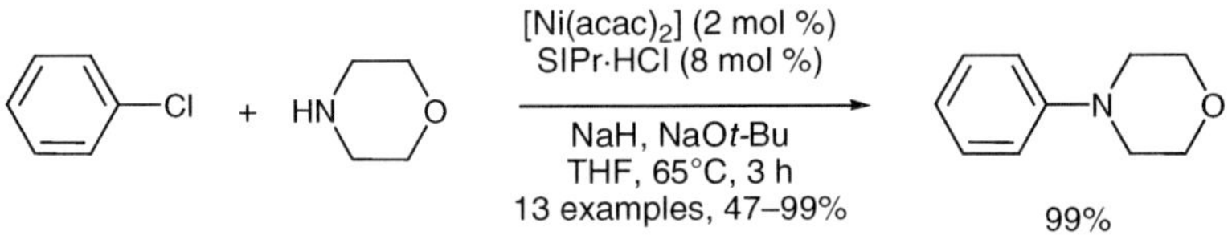

Cl + HN O → [Ni(acac)$_2$] (2 mol %), SIPr·HCl (8 mol %); NaH, NaO*t*-Bu, THF, 65°C, 3 h; 13 examples, 47–99% → N O 99% (10.6)

This aryl amination protocol was next extended to aromatic diamines. In this case, IPr was found as the best ligand and *N*,*N*′-diarylation of aromatic diamines was obtained in high yields in 1,4-dioxane at 100 °C.[40] Diamines with spacers such as biphenyl, phenoxyphenyl or benzylbenzene were also efficiently arylated. Interestingly, the selective mono-amination of unsymmetrical diamines was obtained and therefore a two-step procedure for the synthesis of unsymmetrical *N*,*N*′-diaryl aromatic diamines could be developed.

These nickel-catalyzed aryl aminations found application in molecular materials with the preparation of 1,3,5-tris(4-aminophenyl)benzenes or highly functionalized naphtidines (Figure 10.1), which possess interesting electronic properties potentially useful for molecular switches.[41]

Alternatively, the formation of five-, six- and seven-membered rings, such as indoline, quinoline, indolizidine and benzazepine derivatives obtained with [(SIPr)$_2$Ni] by intramolecular amination of appropriated chloro-aromatic amines (Equation (10.7)).[42]

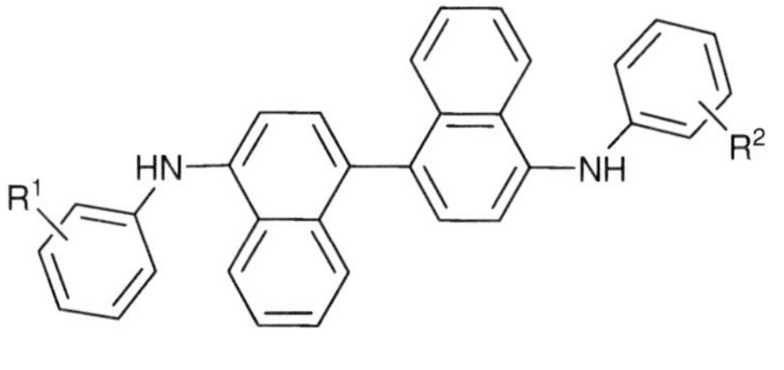

HN NH R^1 R^2

Figure 10.1 Naphtidines prepared *via* [(NHC)Ni]-catalyzed aryl amination.

$$\text{2-Cl-C}_6\text{H}_4\text{-X-(CH}_2)_n\text{-NHR} \xrightarrow[\text{NaH, NaO}t\text{-Bu; Dioxane, 100°C; 12 examples, 44–95\%}]{[\text{Ni(acac)}_2]\ (2\ \text{mol \%}),\ \text{SIPr·HCl}\ (2\ \text{mol \%})} \text{N-R heterocycle};\quad n = 1\text{–}3;\ X = CH_2, O \tag{10.7}$$

Also, Gao and Yang described the efficient amination of aryltosylates. The catalytic system used for these reactions was constituted from $[Ni(PPh_3)_2(1\text{-naphtyl})(Cl)]$ and IPr·HCl in a 1:1 ratio (Equation (10.8)).[43] The catalyst system was ineffective for the reaction without the use of a NHC as ligand and it was postulated that the active species might be a Ni^0 complex coligated with IPr and PPh_3. The reaction was limited to electron-neutral and weakly electron-withdrawing tosylates.

$$\text{Ar—OTs} + \text{H—N}R^1R^2 \xrightarrow[\text{NaO}t\text{-Bu, dioxane, 110°C; 14 examples, 24–96\%}]{[\text{Ni}^{\text{II}}]\ (5\ \text{mol \%}),\ \text{IPr·HCl}\ (5\ \text{mol \%})} \text{Ar—N}R^1R^2 \qquad [\text{Ni}^{\text{II}}] = \text{Naph—Ni(PPh}_3)_2\text{—X} \tag{10.8}$$

10.6.2 Aryl Thiolation Induced by NHC–Ni Complexes

In 2007, Zhang *et al.* reported the first example of C–S coupling (also called arylthiolation) induced by a NHC–Ni complex, including the reactions of some electron-poor aryl chlorides and bromides.[44] Surprisingly, the best NHC for this reaction was IBn that was found far more reactive than IPr (Equation (10.9)). It is worth noting that bis-carbenic ligands derivating from IBn were found slightly more effective but less synthetically accessible.

$$\text{R-C}_6\text{H}_4\text{-Br} + \text{HSR'} \xrightarrow[\text{NaO}t\text{-Bu, DMF, 110°C; 13 examples, 78–99\%}]{[(\text{IBn})_2\text{Ni}]\ (3\ \text{mol \%})} \text{R-C}_6\text{H}_4\text{-SR'} \tag{10.9}$$

10.6.3 Hydrothiolation of Alkynes Mediated by NHC–Ni Complexes

The sulfur–hydrogen bond addition to alkynes represents a convenient access to vinyl sulfides. Nolan, Beletskaya and co-workers reported a new selective hydrothiolation process induced by homogeneous NHC–Ni complexes.[45] They showed that [(IMes)Ni(Cp)Cl] was effective in catalyzing the reaction of terminal alkynes and various aryl thiols (Equation (10.10)). The process was found highly selective and the classical transfer of two thioaryl groups did not take place. In addition, the reaction was highly regioselective producing the Markovnikov-type adducts as major products.

$$R^1\text{—}\equiv \; + \; HS\text{–}C_6H_4R^2 \xrightarrow[Et_3N,\ 80°C,\ 5\ h]{[(IMes)Ni(Cp)Cl]\ (1\ mol\ \%)} CH_2{=}C(R^1)\text{–}S\text{–}C_6H_4R^2 \qquad (10.10)$$

7 examples, 61–87%
regioselectivity 7:1 to 28:1

From crystallographic data, the authors proposed a catalytic cycle, based on an alkyne insertion into the Ni–S bond and protonolysis of the Ni–C bond by trapping with arylthiol.

10.7 NHC–Ni-catalyzed Cross-coupling Reactions

Transition metal-catalyzed cross-coupling reactions of organic electrophiles with organometallic reagents are a powerful synthetic tool for carbon–carbon bond formation. These reactions replaced the classical Ullman reaction that often exhibited a lack of generality.[46] Even though Pd catalysts are the most popular in C–C cross-coupling reactions, the use of catalysts based on nickel has the advantages of lower cost and easier removal from the final products. Furthermore, some of the Ni^0-based catalysts have been shown to be more effective than their Pd counterparts.[47] The recent introduction of NHCs in nickel catalysis conducted to a real breakthrough since it allowed the preparation of stable Ni^0 complexes with increased catalytic activities.

10.7.1 NHC–Ni-catalyzed Corriu–Kumada Cross-couplings

The nickel-catalyzed reaction of a Grignard reagent with an aryl halide, widely known as Corriu–Kumada,[48] constituted the first example of efficient cross-coupling reaction of aryl halides, long before the most popular Migita–Stille coupling. However, the classical protocol was not properly optimized and recent improvements were achieved using NHC–Ni complexes.

Following the first use of NHC–Pd by Nolan and co-workers with aryl chlorides,[49] Hermann disclosed the first Kumada cross-couplings using a NHC–Ni catalyst at room temperature. Compared to NHC–Pd, NHC–Ni-catalyzed reactions displayed a better group tolerance and lower reaction temperatures.[50] A first screening study showed that the air stable IMes or IPr are the most efficient ligands for this methodology. The catalyst was prepared from an equimolar amount (3 mol%) of $[Ni(acac)_2]$ and imidazolium salt in the presence of an excess of Grignard reagent. Under these conditions, efficient and selective cross-couplings were obtained from various brominated or chlorinated aryl halides and arylmagnesium compounds (Equation (10.11)).

$$\text{R}^1\text{-C}_5\text{H}_3\text{E-Cl} + \text{BrMg-C}_6\text{H}_4\text{-R}^2 \xrightarrow[\text{THF, RT, 18 h}]{[\text{Ni(acac)}_2]\ (3\ \text{mol}\ \%),\ \text{IMes or IPr}\ (3\ \text{mol}\ \%\ \textit{in situ})} \text{R}^1\text{-C}_5\text{H}_3\text{E-C}_6\text{H}_4\text{-R}^2 \quad \text{18 examples, 38–99\%} \tag{10.11}$$

As classically observed in coupling reactions, electron-poor arenes were more reactive. Also, steric hindrance and *ortho*-substitution were not really deleterious, as generally observed in nickel catalysis.

A year later, Hermann and co-workers showed that these Corriu–Kumada cross-coupling were efficiently obtained under NHC–Ni catalysis with aryl fluorides as starting materials.[51] They showed that an *in situ* generated species worked similarly if not better than a pre-formed $[(NHC)_2Ni]$ complex. This observation suggested that the catalytic active species would be a zero-valent nickel coordinated with only one N-heterocyclic carbene ligand. The formation of a 12-electrons complex would be evidently favored in an *in situ* process. Both catalysts were active with electron-rich or electron-poor fluoroarenes as well as with congested organometallic species (Equation (10.12)).

$$\text{R}^1\text{-C}_6\text{H}_4\text{-F} + \text{BrMg-C}_6\text{H}_4\text{-R}^2 \xrightarrow[\text{THF, RT, 18 h}]{[\text{Ni(acac)}_2]\ (5\ \text{mol}\ \%),\ \text{IPr}\ (5\ \text{mol}\ \%\ \textit{in situ})} \text{R}^1\text{-C}_6\text{H}_4\text{-C}_6\text{H}_4\text{-R}^2 \quad \text{20 examples, 23–95\%} \tag{10.12}$$

A polar pathway through oxidative addition was proposed in this highly selective transformation, in contrast with the classical radical pathways described for $NiCl_2$-catalyzed reactions that produced higher amounts of ter-phenyl by-products.

In 2005, Fürstner and co-workers introduced an alternative pathway for the preparation of catalytic precursors based on the oxidative addition of 2-chloroimidazolium salts in place of unhalogenated derivatives.[52] The use of $[Ni(COD)_2]$ in combination with PPh_3 (2 equiv) in THF at ambient temperature led to clean conversions and allowed formation of cationic mono-diaminocarbene nickel complexes in good to excellent yields (Figure 10.2).

These novel mixed nickel carbene complexes were found to effect Kumada cross-couplings of *p*-methoxyphenylmagnesium bromide with chloro- or bromobenzene, and 2-chloropyridine, but accompanied with significant amounts of 4,4′-dimethoxybiphenyl (15–25%).

Matsubara *et al.* investigated $[(IPr)NiCl_2(PPh_3)]$.[53] This monocarbene complex showed higher catalytic activity (TON > 400) in the Grignard coupling than the corresponding bis-carbene and bis-phosphine complexes at room

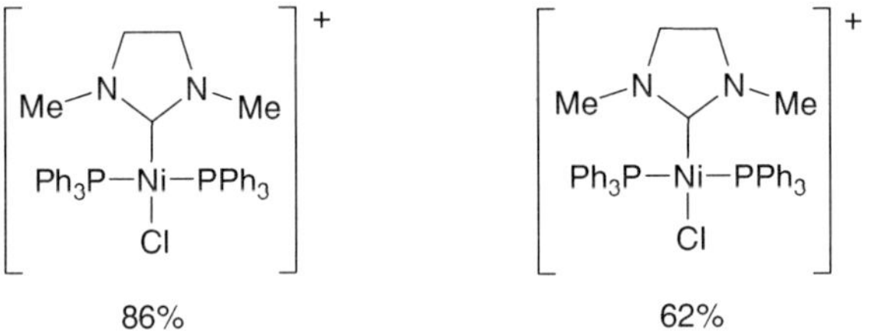

Figure 10.2 [(NHC)Ni] complexes prepared through oxidative addition.

temperature (TON = 240 and 40, respectively). However, these complexes were only efficient for cross-couplings of aryl iodides and bromides.

From these results, the Matsubara group postulated that the high activity of mixed phosphine/NHC Ni^{II} complex might be related to the PPh_3 decoordination under mild conditions, to form *in situ* a coordinatively unsaturated nickel centre. This hypothesis conducted them to develop a new interesting system of α-arylation of ketones in the presence of strong base (*i.e.* NaO*t*-Bu).[54] Under optimized conditions various ethyl aryl ketones were arylated in good yields (Equation (10.13)). Of note, only acetophenone gave aldol condensation products. Under identical conditions, aminations of aryl bromides were carried out in moderate to good yields.

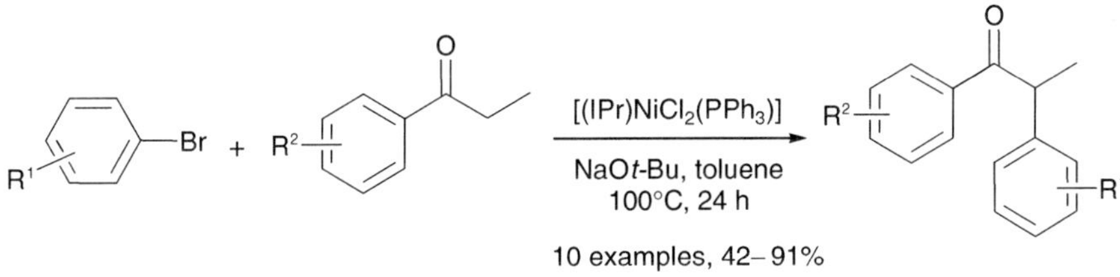

(10.13)

Successively Hiroya[55] and Chen[56] reported that Ni^{II} complexes containing pyridine-functionalized bis(NHC) ligands were highly efficient catalysts for coupling reactions of aryl chlorides, vinyl chlorides and heteroaryl chlorides with aromatic Grignard reagents at ambient temperature. The high catalytic activities of these multidentate NHC-containing nickel complexes were probably due to the strong stabilization effect of such NHC ligands on the catalytically active. Of note, Herrmann and co-workers previously reported mixed pyridyl- and quinolinylidene nickel carbene complexes for Kumada couplings.[57]

10.7.2 NHC–Ni-catalyzed Organomanganese Cross-couplings

Compared to Grignard compounds, organomanganese reagents combine high reactivity and increased chemoselectivity. Schneider and co-workers reported that *in situ* generated [$(IPr)_2Ni$] efficiently catalyzed the cross-coupling of functionalized aryl bromides with organomanganese reagents under mild conditions (0 °C to room temperature).[58] Both electron-deficient and electron-rich aryl bromides were coupled in good to excellent yields. The reaction was also applied to heteroaryl bromides and to neutral or electron-deficient aryl chlorides (Equation (10.14)).

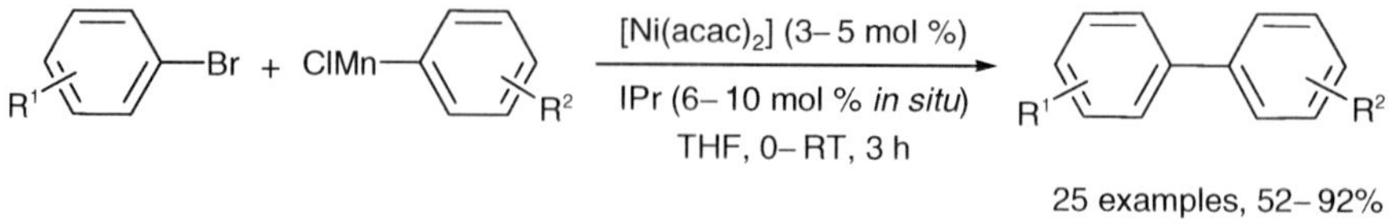

R^1-C$_6$H$_4$-Br + ClMn-C$_6$H$_4$-R^2 → [Ni(acac)$_2$] (3–5 mol %), IPr (6–10 mol % *in situ*), THF, 0–RT, 3 h → R^1-C$_6$H$_4$-C$_6$H$_4$-R^2

25 examples, 52–92%

(10.14)

10.7.3 NHC–Ni-catalyzed Suzuki–Miyaura and Negishi Cross-couplings

In 1999, Cavell and co-workers reported that the Suzuki coupling reaction could be catalyzed by NHC–Ni complexes, although not as efficiently as observed with Pd.[6a] The introduction of a bulky NHC on the catalyst precursor led to a significant increase in activity, although not as pronounced as in the Pd case. In fact, Blakey and MacMillan described the first efficient Suzuki–Miyaura cross-coupling reaction mediated by NHC–Ni in 2003.[59] The original reaction used aryltrimethyl ammonium triflates as coupling partners and in the presence of CsF, the authors showed that IMes–Ni^0 [1:1] led to the efficient preparation of various biphenyl derivatives (Equation (10.15)).

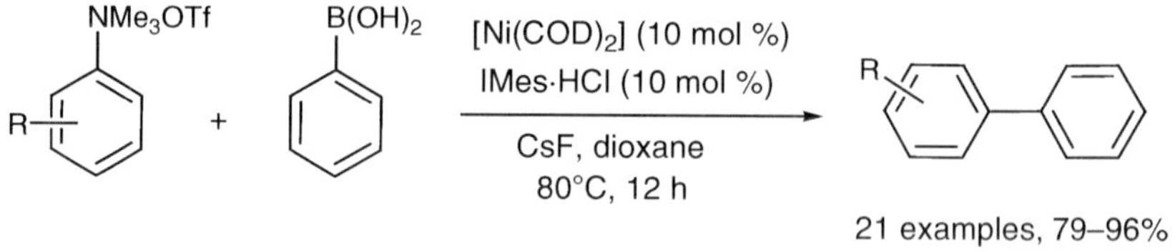

R-C$_6$H$_4$-NMe$_3$OTf + C$_6$H$_5$-B(OH)$_2$ → [Ni(COD)$_2$] (10 mol %), IMes·HCl (10 mol %), CsF, dioxane, 80°C, 12 h → R-C$_6$H$_4$-C$_6$H$_5$

21 examples, 79–96%

(10.15)

Radius and co-workers next reported examples of Suzuki-type cross-coupling reactions of perfluorinated arenas (*i.e.* octafluorotoluene or perfluorobiphenyl) (Equation (10.16)).[60] Again, [$(IiPr)_4Ni_2(COD)$] was used to activate the C–F bond (see Section 10.4.2).

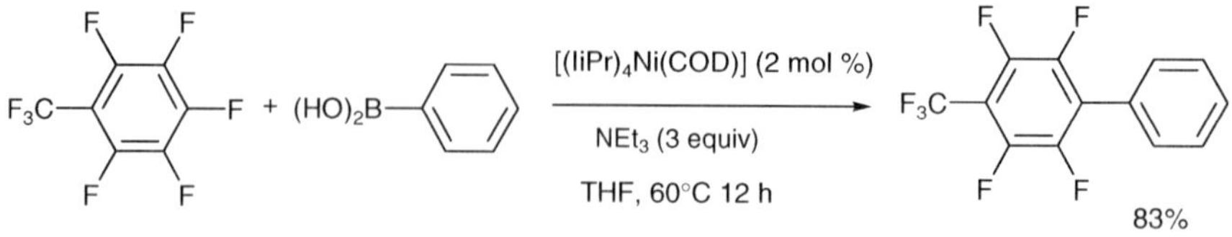

F_3C-C$_6$F$_4$-F + (HO)$_2$B-C$_6$H$_5$ → [(IiPr)$_4$Ni(COD)] (2 mol %), NEt$_3$ (3 equiv), THF, 60°C 12 h → F_3C-C$_6$F$_4$-C$_6$H$_5$

83%

(10.16)

In 2004, Liu and Robbins demonstrated that a Ni/SIPr [1:1] system allowed the synthesis of 6-arylpurine 2′-deoxynucleosides and nucleosides from 6-(imidazol-1-yl)- and 6-(1,2,4-triazol-4-yl)purine derivatives by a Suzuki coupling (Equation (10.17)).[61] These reactions were not observed with classical $[Pd(PPh_3)_4]$ while $[Ni(dppp)Cl_2]$ (dppp = 1,3-bis(diphenylphosphino)propane) only led to 30% of desired product.

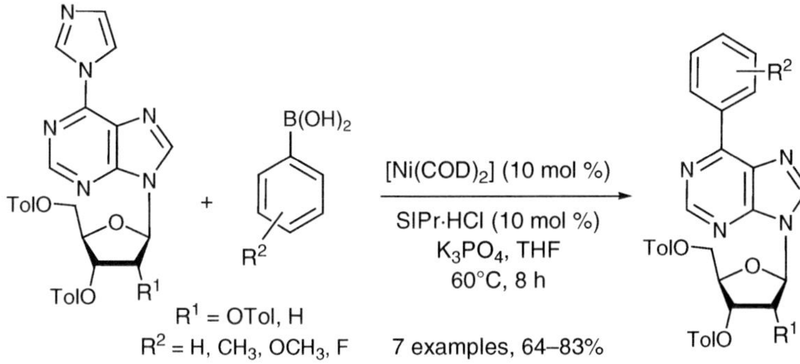

$[Ni(COD)_2]$ (10 mol %); SIPr·HCl (10 mol %); K_3PO_4, THF; 60°C, 8 h

R^1 = OTol, H
R^2 = H, CH_3, OCH_3, F

7 examples, 64–83%

(10.17)

Two other contributions from Hiroya[55] and Chen[62], respectively, reported the use of pyridine-functionalized NHC, supported or not, for Suzuki–Miyaura cross-coupling reactions.

Finally, the efficient Negishi coupling of unactivated aryl, heterocyclic, vinyl chlorides as well as aryl dichlorides catalyzed by binuclear and mononuclear NHC–Ni complexes was recently described by Chen and co-workers (Equation (10.18)).[63] Elaborated heteroarene-functionalized NHC were found to be highly effective for the preparation biaryls and terphenyls in good to excellent yields under mild conditions. Notably, the binuclear nickel catalysts showed higher activities than mononuclear analogues, probably because of a bimetallic cooperative effect.

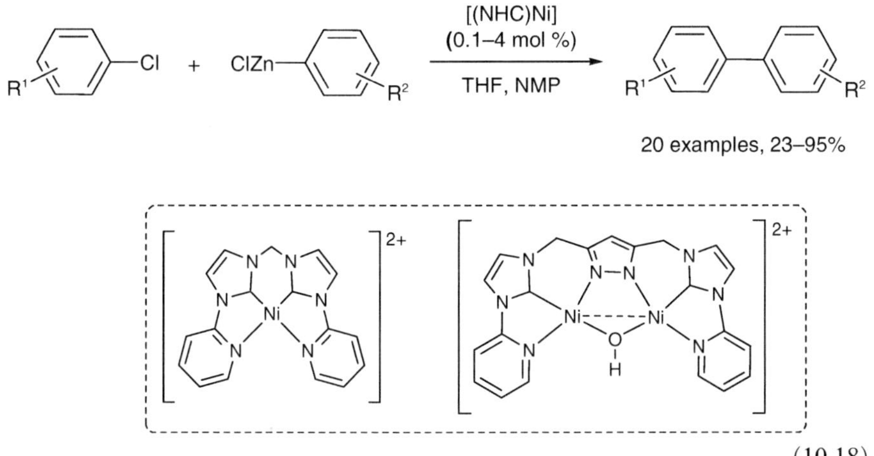

[(NHC)Ni] (0.1–4 mol %); THF, NMP

20 examples, 23–95%

(10.18)

10.8 Synthesis of Heterocyclic and Polycyclic Compounds by Cycloaddition Reactions

Due to its numerous applications in life sciences or molecular materials, one-pot construction of poly(hetero)cyclic units constitutes a synthetic challenge in organic chemistry. The main objective is to obtain cycloadditions and rearrangements of unsaturated substrates under mild conditions (low or room temperature, atmospheric pressure and short reaction times) in order to support various functionalities or enantioselectivities. This goal can be reached under transition metal catalysis and NHC–Ni systems appeared as efficient and versatile tools for the access to substituted nitrogen- and oxygen-containing heterocycles. A major contribution to this specific field came from the Louie group, which described various cycloadditions of diynes or enynes to unsaturated derivatives such as carbon dioxide, ketones or aldehydes, isocyanates or nitriles producing numerous pyrones, pyridones, pyrans or pyridines.[64]

10.8.1 Pyrones from Diynes and Carbon Dioxide

In a first report, Louie *et al.* showed that a [$(IPr)_2Ni$] system led to efficient, selective and general synthesis of pyrones from diynes and CO_2 under ambient pressure and low reaction temperature (Scheme 10.4).[65]

Except for terminal diynes, the classical diyne oligomerization obtained in the presence of a phosphine ligand[66] was suppressed under these mild reaction. Two possible intermediates were proposed for this process: the first one would be issue of an initial [2 + 2] cycloaddition of CO_2 and one acetylenic moiety, while the second one would arise form the insertion of another alkyne to form a metallalactone (Scheme 10.4).

This reaction was sensitive to steric hindrance since *t*-Bu and TMS (TMS = trimethylsilyl) disubstituted diynes showed low reactivity. On the other hand, the regioselectivity of the reaction was strongly dependent on the

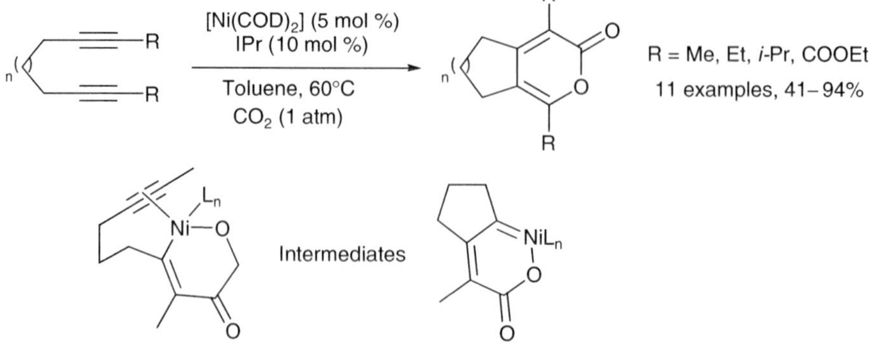

Scheme 10.4 [(NHC)Ni]-catalyzed preparation of pyrones.

substrate substitution. For example, the reaction of unsymmetrical diynes could in certain cases produce only one pyrone isomer (Equation (10.19)).

EtOOC COOEt
Me R
$[Ni(COD)_2]$ (5 mol %)
IPr (10 mol %)
CO_2 (1 atm)
Toluene, 60°C
EtOOC COOEt
R Me O O
EtOOC COOEt
Me R O O
R = *t*-Bu 64%
R = TMS 83%
100 / 0

(10.19)

10.8.2 Pyridones or Pyrimidine-diones from Diynes or Alkynes and Isocyanates

Louie and co-workers later used their methodology for the cycloaddition of diynes with isocyanates.[67] In this case, SIPr was superior to IPr while isocyanates were found more reactive than carbon dioxide. Also of interest, it appeared that a 1:1 Ni/SIPr ratio was equally efficient to a 1:2 ratio. Except for some particular cases, the preparation of a large variety of highly substituted 2-pyridones were then efficiently obtained in the presence of 3 mol% catalyst under very mild conditions (Equation (10.20)).

R^1 R^1 n
$[Ni(COD)_2]$ (3 mol %)
SIPr (3 mol %)
R^2NCO (1 equiv)
Toluene, RT, 30 min
n N R^2 O R
R^1 = TMS, Me, *t*-Bu, CO_2Et
R^2 = Ph, Cy, Et, Pr
15 examples, 58–91%

(10.20)

Some interesting points emerged from this systematic study: (i) steric hindrance on the diynes only had a little influence on the catalytic process; (ii) when starting from unsymmetrical diynes, larger substituents were introduced in the 3-position of the formed pyridone; (iii) under dilute conditions, unsymmetrical diynes were converted into pyridones without telomerization to aromatic by-products; and (iv) aryl- as well as alkyl-isocyanates were reactive; however, electron-poor arylisocyanate required a higher reaction temperature (80 °C).

Surprisingly, isocyanurates were not produced during this reaction while they could be obtained nearly quantitatively in the presence of 0.3 mol% of free SIPr.[68] This result illustrated the strong coordination properties of SIPr to Ni, which allowed for the absence of free SIPr in the reaction media under the employed reaction conditions.

Finally, the high efficiency of SIPr–Ni systems was applied to a three-component cycloadditions and the reaction of 3-hexyne and 2 equivalents of phenyl isocyanate afforded 2-pyridone in excellent isolated yield (Equation (10.21)).

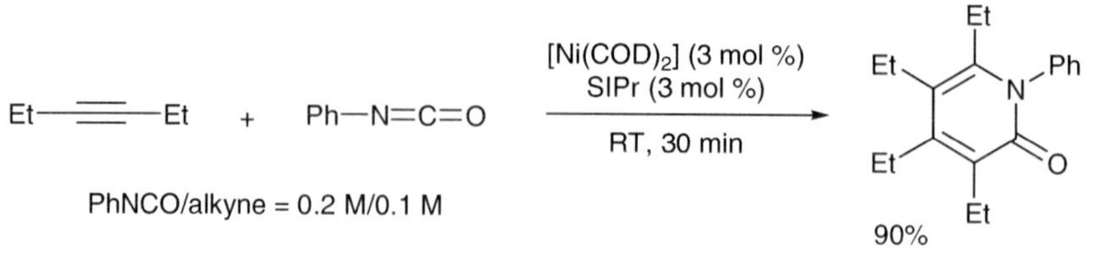

Et—≡—Et + Ph—N=C=O → ([Ni(COD)$_2$] (3 mol %), SIPr (3 mol %), RT, 30 min) 90%

PhNCO/alkyne = 0.2 M/0.1 M

(10.21)

In a more detailed study,[69] Duong and Louie showed that the combination of Ni and IPr is slightly more efficient than SIPr. More importantly, they also showed that the reaction outcome was strongly dependent on isocyanate concentration as well as on alkyne substitution. When isocyanate concentrations were high, pyrimidine-diones were produced in good yields from unsymmetrically hindered alkynes by a double insertion of isocyanates (Equation (10.22)).

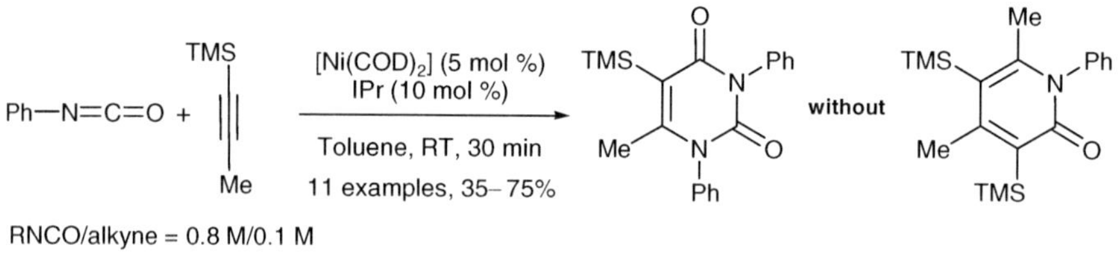

Ph—N=C=O + TMS≡Me → ([Ni(COD)$_2$] (5 mol %), IPr (10 mol %), Toluene, RT, 30 min, 11 examples, 35–75%) without

RNCO/alkyne = 0.8 M/0.1 M

(10.22)

10.8.3 Pyridines from Diynes and Nitriles

A Ni/SIPr (1:2) combination catalyzed the cycloaddition of diynes and nitriles into pyridines.[70] Aryl, heteroaryl as well as alkyl nitriles were reactive under ambient conditions (Equation (10.23)). In contrast with aryl isocyanates (*vide supra*), the reaction was not dependent on the electronic demand on the aryl nitrile since both electron-rich and electron-poor substrates gave similar results. Finally, and similarly to isocyanates reactions, the larger substituent was introduced in the 3-position of the pyridine ring.

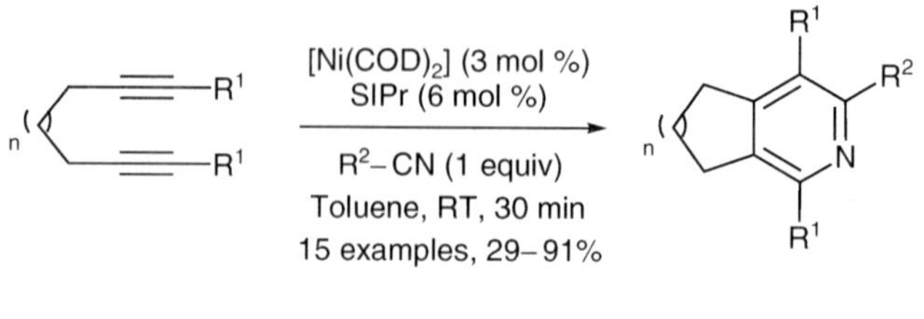

→ ([Ni(COD)$_2$] (3 mol %), SIPr (6 mol %), R^2–CN (1 equiv), Toluene, RT, 30 min, 15 examples, 29–91%)

(10.23)

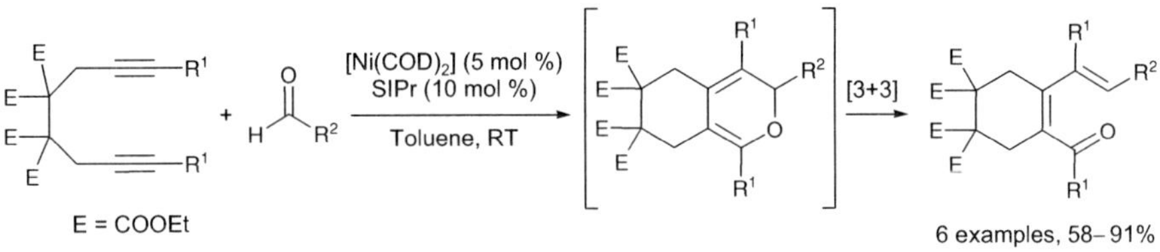

Scheme 10.5 [(NHC)Ni]-catalyzed cycloaddition of diynes and aldehydes.

10.8.4 Pyrans from Unsaturated Hydrocarbons and Carbonyl Substrates

The cycloaddition of diynes and benzaldehyde or propanal could be obtained at room temperature with 5 mol% of Ni/SIPr (1:2).[71] In these reactions, the replacement of classical phosphines by SIPr produced an increase of reactivity and pyranic derivatives were obtained in good yields (Scheme 10.5).

Unactivated ketones such as cyclohexanone produced spiropyrans in good yields (Equation (10.24)). The authors envisioned that this surprising reactivity was probably due to the ability of the NHC ligand to enhance C–O bond-forming reductive elimination.

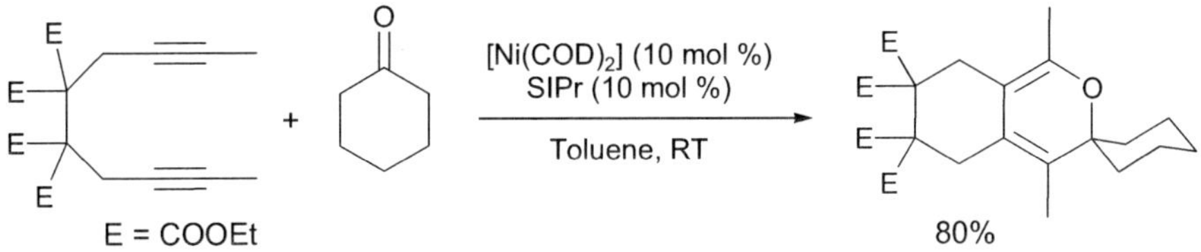

(10.24)

It is worth noting that [2 + 2 + 2] cycloadditions of imines and alkynes have not been reported in the presence of NHC–Ni systems. These reactions seem to fail since Ogoshi *et al.* noted that the reaction of 2-butyne and *N*-benzenesulfonylbenzaldimine did not proceed in the presence of a NHC–Ni catalyst (IMes or SIMes).[72] This result contrasted with the results obtained by the same authors for the reaction catalyzed by Ni/phosphine systems.

10.9 Cycloaddition of Alkynes to Unsaturated Derivatives

[2 + 2 + 2] Cycloadditions constitute a remarkable tool for the synthesis of 1,3-cyclohexadiene derivatives and related heterocyclic compounds.[73] Surprisingly, no cyclotrimerization of alkynes (Reppe reaction)[74] has been reported to date with NHC–Ni complexes.

Sato and co-workers described the nickel-catalyzed [2 + 2 + 2] cycloaddition of arynes and an unactivated alkene allowing for the original preparation of

9,10-dihydrophenanthrene derivatives resulting from the reaction of two arynes with an unsaturated part of a α,ω-diene (Equation (10.25)).[75] SIMes and IMes appeared as the best ligands, and interestingly, formation of the expected tetrahydronaphtalene derivatives was not observed. Also noteworthy is the fact that a mono-alkene did not react under these conditions, suggesting that the unreacted alkene moiety was necessary for the cycloaddition to take place.

MeO_2C, MeO_2C + TMS, TfO, OMe, OMe

$[Ni(COD)_2]$ (10 mol %)
SIMes (10 mol %)
CsF (6 equiv)
MeCN, 50°C, 6 h

MeO_2C, MeO_2C, OMe, OMe, OMe, OMe
68%

(10.25)

10.10 Ni-catalyzed Isomerization of Vinylcyclopropanes and Derivatives

Zuo and Louie discovered that the combination of Ni^0 with a sterically hindered NHC ligand catalyzed the isomerization of a variety of activated or unactivated vinylcyclopropanes under mild conditions to afford the corresponding cyclopentenes in very good yields (Equation (10.26)).[76] This process also allowed for the efficient preparation of cyclopentene possessing a bicyclic framework, useful for potential biological applications.

R^1, R^2

$[Ni(COD)_2]$ (1 mol %)
IPr (2 mol %)
Toluene or hexanes
RT, 1–12 h

R^1, R^2
6 examples, 92–96%

(10.26)

Zuo and Louie also found that NHC–Ni complexes catalyzed the rearrangement of cyclopropylen-yne derivatives.[77] However, the selectivity of the reaction was strongly dependent on the NHC–Ni catalyst as well as on the substrate substitution. With a SIPr/Ni 1:1 system, a hindered vinylcyclopropylene-yne gave the isomerized seven-membered ring as the sole product while the corresponding methyl derivative afforded a disubstituted tetrahydrofuran derivative. The authors then described a successful and general access to the later by using the ItBu/Ni 1:1 system (Scheme 10.6).

Furthermore, the Louie group reported the extension of IPr–Ni catalysis to the rearrangement of vinyl aziridines and aziridinylenynes (Equation (10.27)).[78] The rearrangement of vinylaziridines to vinyl imines or 1,3-butadienylamines could be rationalized by the formation of a key π-allyl Ni complex conducting

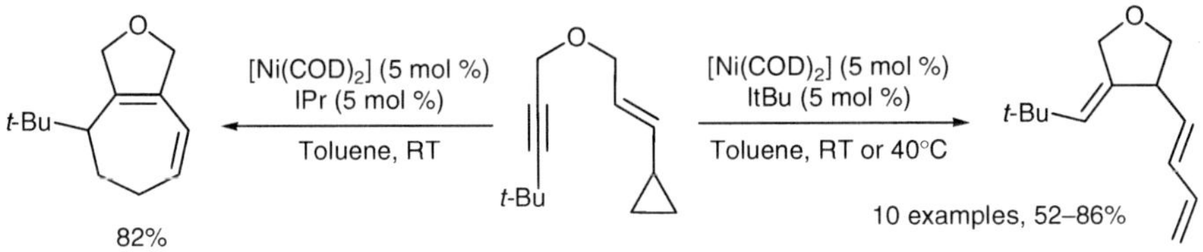

t-Bu
82%
$[Ni(COD)_2]$ (5 mol %)
IPr (5 mol %)
Toluene, RT
t-Bu
$[Ni(COD)_2]$ (5 mol %)
ItBu (5 mol %)
Toluene, RT or 40°C
t-Bu
10 examples, 52–86%

Scheme 10.6 Reactivity of cyclopropylenynes with [(NHC)Ni].

to a metalazetidine or metalapiperidine intermediate. As previously observed with the corresponding cyclopropane derivatives, aziridinylenynes led to three products whose ratio were dependent on which the NHC ligand was employed. The same trends were found since azepines were formed with IPr/Ni catalyst while ItBu/Ni produced conjugated unsaturated imines (Equation (10.28)).

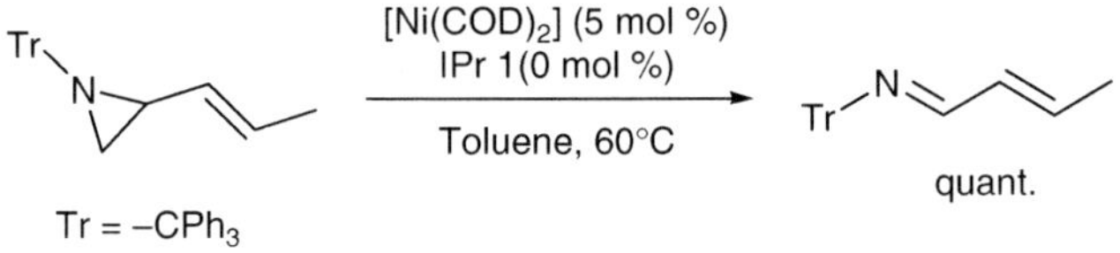

Tr
N
Tr = $-CPh_3$
$[Ni(COD)_2]$ (5 mol %)
IPr 1(0 mol %)
Toluene, 60°C
Tr N
quant.
(10.27)

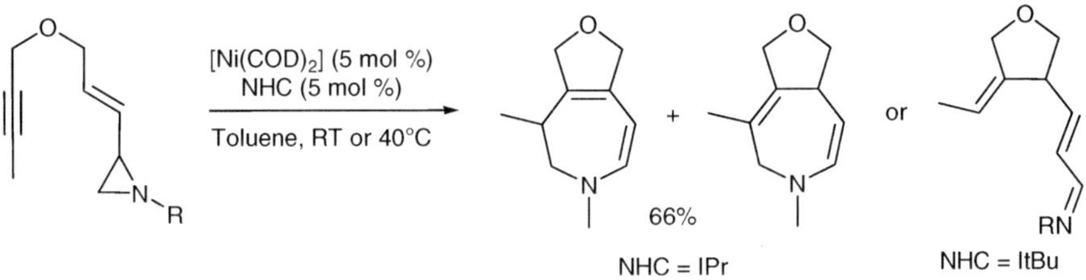

N
R
$[Ni(COD)_2]$ (5 mol %)
NHC (5 mol %)
Toluene, RT or 40°C
N
66%
NHC = IPr
+
N
or
RN
NHC = ItBu
(10.28)

10.11 Multi-component Reactions with Aldehydes and Ketones

Multi-component reactions open a direct access to poly-functional derivatives.[79] For instance, reactions of unsaturated substrates and aldehydes in the presence of Et_3SiH have opened a rapid access to allylic, homoallylic or ω-unsaturated alcohols and NHC–Ni^0 complexes were recently reported to be efficient catalysts for such oxidative coupling. Mori and co-workers described the first example of such condensation in 2001.[80] They showed that an *in situ* generated IPr–Ni^0 1:1 complex led to the efficient and stereoselective formation of (*Z*)-homoallylic silyl alcohols (Equation (10.29)) while classical Ni^0–PPh_3 catalyst produced only (*E*)-coupling products. Of note, IPr/Ni^0 1:2 and IMes/Ni^0 1:2 were less efficient in this reaction. These first results were extended to silylated dienes.[81] Interestingly, it was showed that a mixed complex [(IiPr)-$Ni(PPh_3)$] stereoselectively produced (*Z*)-allylsilanes while $[Ni(PPh_3)_4]$ led only to (*E*)-isomeric products.

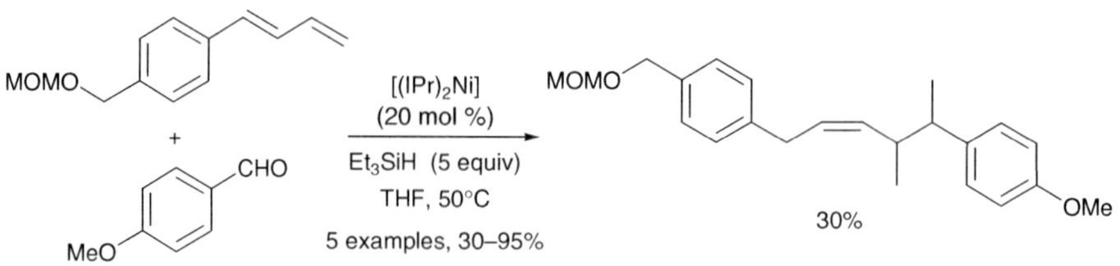

(10.29)

The asymmetric version of the reaction was described in 2007 by using a chiral carbene bearing 1-(2,4,6-trimethylphenyl)-propyl groups on the nitrogen.[82] Various coupling products were produced in good yields with high regio-, diastereo- (*anti* selective in the case of the internal 1,3-diene), and enantioselectivities (up to 97% *ee*).

Saito *et al.* published an interesting extension dealing with multicomponent couplings of ynamides, aldehydes, and silanes leading to functionalized enamides. They showed that IMes–Ni stereoselectively catalyzed such coupling reactions to give the corresponding γ-silyloxyenamide derivatives (Scheme 10.7).[83] The intervention of an oxanickelacycle was proposed to explain the high stereoselectivity observed.

On the other hand, Jamison and co-workers focused on the development of reactions involving allenes as source of chirality. They described the enantio- and regioselective coupling of various allenes, aldehydes and silanes catalyzed by [(IPr)$_2$Ni] to produce (*Z*)-silylated allyl alcohols (Equation (10.30)).[84]

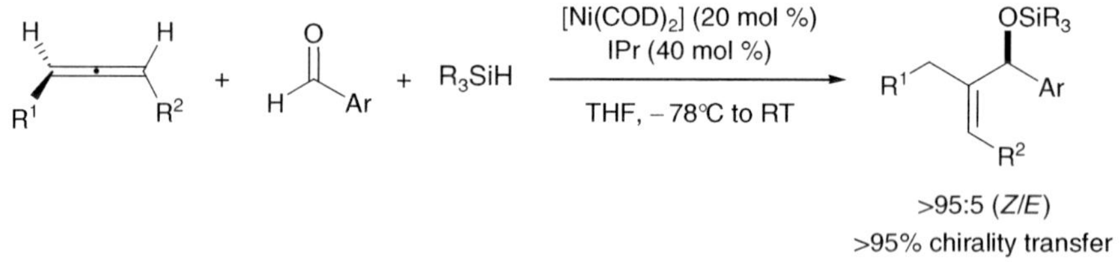

(10.30)

The reaction was extended to the coupling of alkenes and isocyanates to provide α,β-acrylamides.[85] The most active catalyst was [(IPr)Ni] that allowed a good regioselectivity at the 2-position of the olefin (Equation (10.31)) and deprotection of the obtained products afforded the corresponding primary amides.

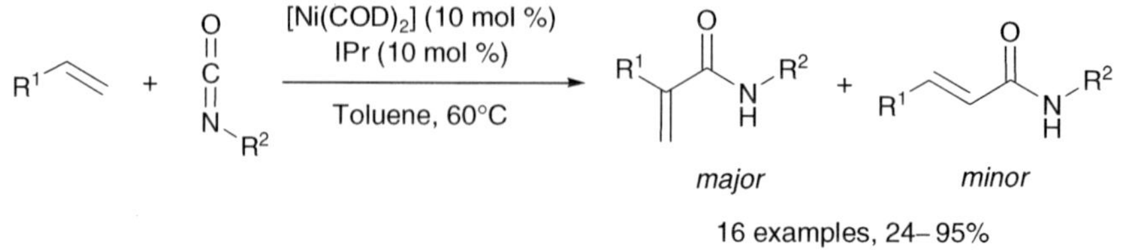

(10.31)

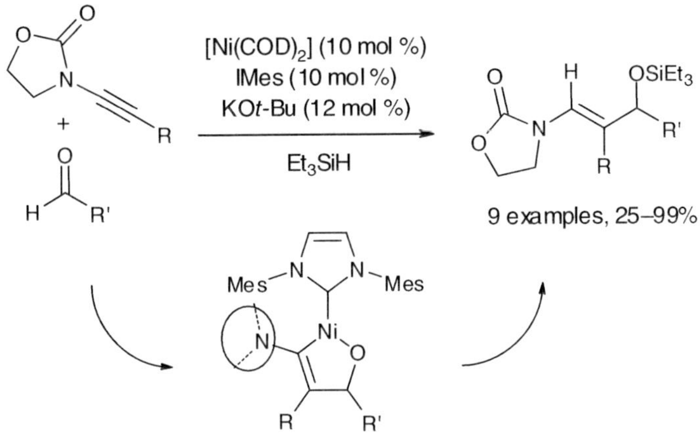

Scheme 10.7 [(NHC)Ni]-mediated synthesis of γ-silyloxyenamide.

More interestingly, the same authors used their procedure for highly selective couplings of alkenes and aldehydes (Equation (10.32)).[86] The catalytic system combined a strong σ donor IPr and a strong π acceptor $P(OPh)_3$. This combination was proposed to produce a synergistic relationship by reducing the electron density at the coordinatively unsaturated Ni centre, which would accelerate the elimination step in the catalytic cycle.

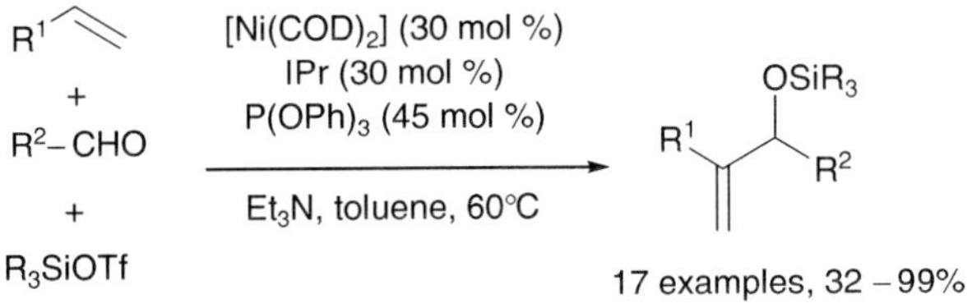

(10.32)

Montgomery and co-workers examined this three-component reaction from a mechanistic point of view (Scheme 10.8).[87] Using a combination of cross-over experiments between Et_3SiD and Pr_3SiH, they demonstrated that catalyst formulations involving PBu_3 and IMes proceed largely by fundamentally different pathways. As a lack of cross-over with IMes was observed, it was concluded that the R_3Si and H units in the product were mainly from a single molecule of hydrosilane. Intermediates **I** or **II** appeared the most probable in the catalytic cycle. Also, since PBu_3 gave cross-over between Et_3SiD and Pr_3SiH, it was postulated that the trialkylsilane would react at the end of the reaction with a transient allyl-nickalate.

Subsequently, Montgomery described an interesting and efficient method for the hydrosilylation of alkynes involving NHC–Ni as catalyst (see Chapter 13 for further details),[88] and the diastereoselective additions of alkynes to α-silyoxyaldehydes.[89] Under optimized conditions, this reaction efficiently produced *anti*-1,2-diols (Scheme 10.9). The use of a chiral NHC in this

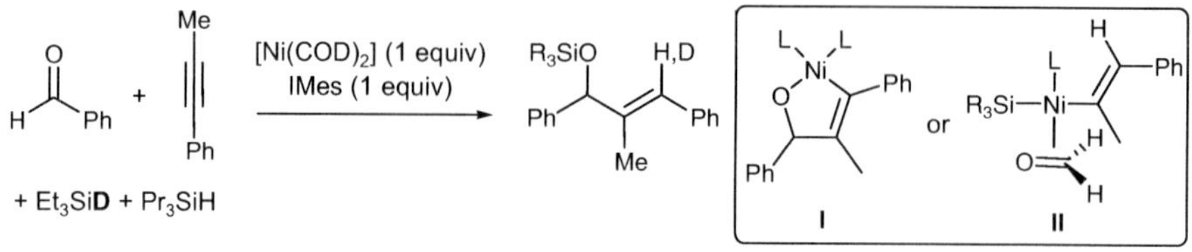

Scheme 10.8 Mechanistic studies on three-component reactions.

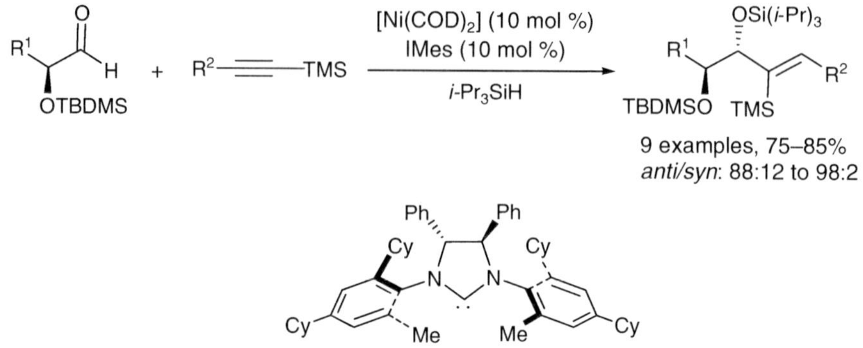

Scheme 10.9 [(NHC)Ni]-mediated additions of alkynes to aldehydes.

three-component reaction allowed for the production of allylic alcohols in good yields and moderate enantioselectivities (65–85% *ee*) (Scheme 10.9).[90]

Finally, Murakami *et al.* reported in 2005 a concise and efficient access to substituted cyclohexenones from cyclobutanones and alkynes.[91] While phosphine ligands were generally employed to facilitate the reaction, the authors demonstrated that IPr was an effective ligand and led to the selective ring-expansion affording cyclohexenones (Scheme 10.10).

These results could be explained by an initial oxidative coupling between the carbonyl and the alkyne to afford a nickelapentenacycle that would undergo subsequent β-carbon elimination to produce a seven-membered nickel-acycle. In the absence of IPr ligand, a β-hydride elimination competed with the β-carbon one. With unsymmetrical alkynes the smaller substituent was mainly introduced α to the carbonyl group.

10.12 Dimerization, Oligomerization and Polymerization Mediated by NHC–Ni Complexes

NHC–Ni complexes displayed moderate activities compared to literature data in the polymerization of alkenes.[92] In addition, they showed low selectivity

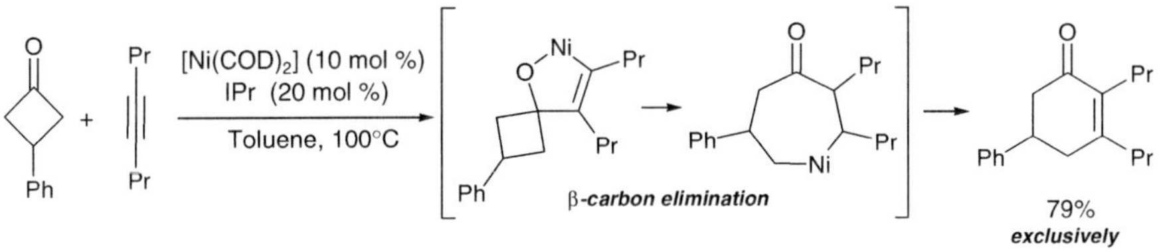

Scheme 10.10 [(NHC)Ni]-mediated ring expansion reactions.

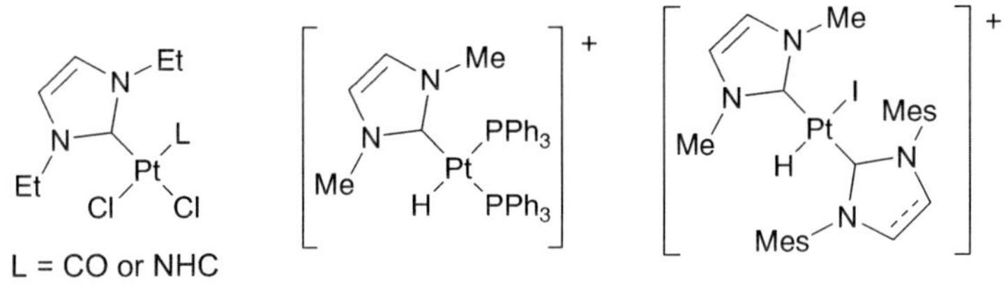

Figure 10.3 [(NHC)Pt] complexes.

since higher molecular weight distributions were observed. Similar results were obtained for dimerization of ethylene.[93] In many cases, the decomposition of the catalyst (increased by temperature) competed with the dimerization or oligomerization processes. Interestingly, Wasserscheid, Cavell and co-workers showed that this decomposition could be partially or totally inhibited in ionic liquids.[94]

10.13 Syntheses of NHC–Pt Complexes

While the NHC–Ni-mediated catalysis has attracted a considerable interest in organic chemistry for several decades, the development of complexes NHC–Pt is much more recent. Several examples of NHC–Pt complexes are shown in Figure 10.3.

In 1998, Liu *et al.* reported the preparation of carbene carbonyl- or biscarbene-Pt^{II} complexes using a transfer reaction between W-complex and $[(PhCN)_2PtCl_2]$.[95] On the other hand, the Cavell group proposed an oxidative addition of imidazolium salt to Pt^0 complex to yield Pt^{II}-hydrido/NHC complexes.[96]

In the last few years, an increasing interest in the elaboration of new NHC–Pt families of complexes has arisen. This includes Pt-oxazoline/NHC complexes,[97] combining the heterocyclic ligand moiety as *N*-substituent of NHC unit, or mixed Pt NHC/phosphine complexes developed by Rourke (Figure 10.4).[98] The Cavell group described a 1,3-diazepan-2-ylidene carbene and its coordination with Pt^0.[99] On the other hand, Steinborn reported the synthesis and the characterization of Pt^{IV} and Pt^{II} complexes with monodentate *N,N*-, *N,O*- or

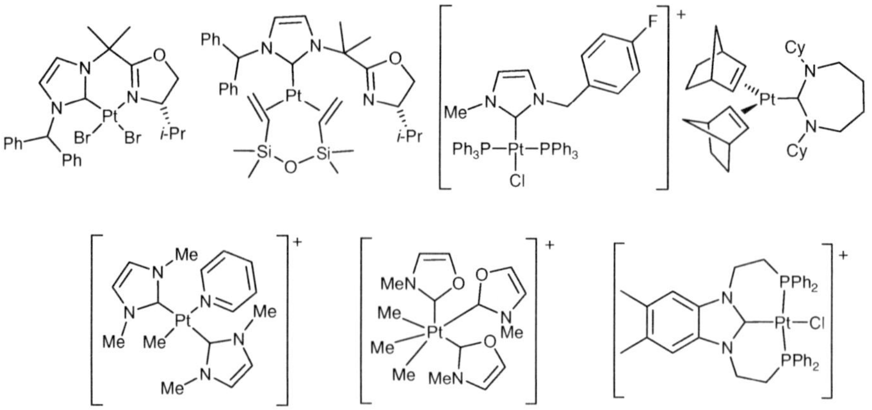

Figure 10.4 Recent families of [(NHC)Pt] complexes.

N,S-NHC ligands.[100] Additionally, Hahn and co-workers disclosed the preparation of Pt complexes bearing phosphine/benzimidazol-2-ylidene as ligands.[101]

Certainly, the most popular application of [(NHC)Pt] complexes to date is the hydrosilylation reactions.[102] However, these are covered in Chapter 13 instead.

10.14 Cycloaddition Reactions Mediated by NHC–Pt Complexes

Reductive cyclization of diynes or enynes was catalyzed by NHC–Pt complexes in the presence of H_2 and $SnCl_2$ as activator (Equation (10.33)) to afford an interesting pathway to 2,5-dihydrofurans, pyrroles or cyclopentenes in moderate to good yields (50–88%).[103]

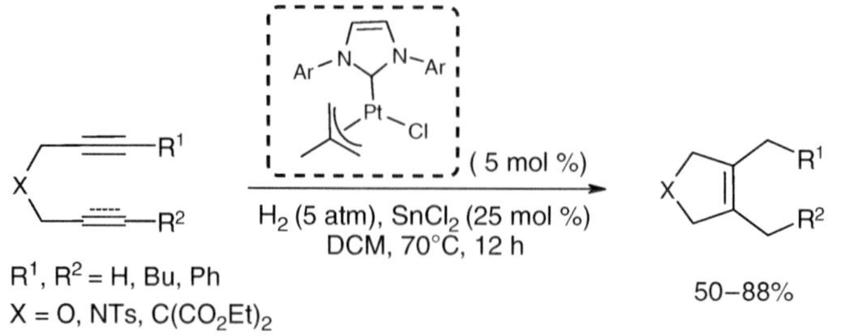

(5 mol %)

H_2 (5 atm), $SnCl_2$ (25 mol %)
DCM, 70°C, 12 h

R^1, R^2 = H, Bu, Ph
X = O, NTs, $C(CO_2Et)_2$

50–88%

(10.33)

10.15 Cycloisomerization Reactions

In 2009, Marinetti and co-workers reported the preparation and structural data of NHC–Pt^{II} complexes and their catalytic activity in model 1,6-enyne cycloisomerization reactions.[104] The elaboration of square-planar Pt^{II} complexes bearing symmetrical or unsymmetrical chiral diphosphines was described in a sequence involving an iodine oxidative addition to a NHC–Pt^0 complex and subsequent complexation of a chiral chelating diphosphine.

In this context, the synthesis of a new family of Pt^{II} six-membered metal-acyclic NHC complexes was reported.[104b] This new platinacyclic complex was then used in an enantioselective 1,6-enyne cycloisomerization to afford, under mild conditions, the expected fused azabicycles in very high enantiomeric excesses and good to excellent yields (Equation (10.34)).

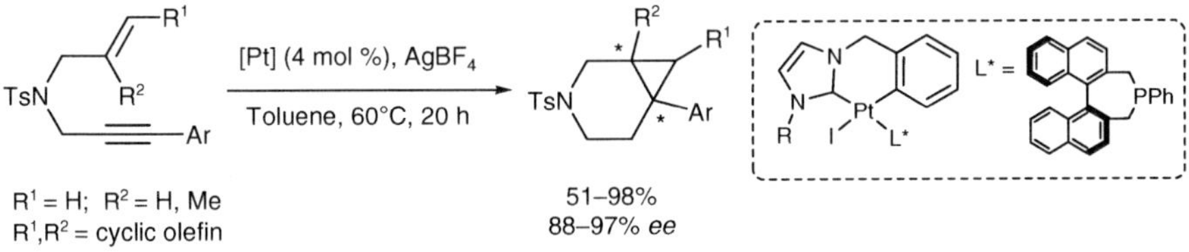

[Pt] (4 mol %), $AgBF_4$ / Toluene, 60°C, 20 h

R^1 = H; R^2 = H, Me
R^1,R^2 = cyclic olefin

51–98%
88–97% *ee*

(10.34)

10.16 Catalytic B–B and B–H Addition Reactions

Recently, Fernández and Mata reported the use of mixed Pt^0-siloxane/NHC complexes as active catalytic species in 1,2-diboration of alkenes or alkynes (Equation (10.35)).[105]

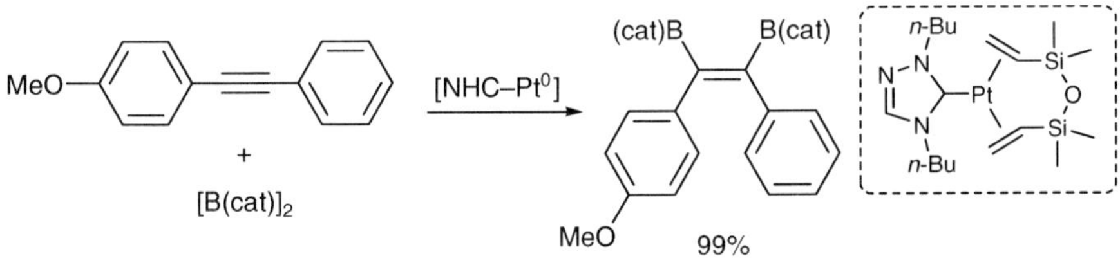

$[B(cat)]_2$ + MeO–C6H4–C≡C–Ph → [NHC–Pt^0] → (cat)B, B(cat) alkene, 99%

(10.35)

On the other hand, regioselective H–B addition reaction to vinylarenes and alkynes mediated by this NHC–Pt complex was also reported (Equation (10.36)).[106] The reaction, carried out under mild conditions, afforded preferentially the corresponding branched products with moderate to good regioselectivity (branched/linear = 90/10).

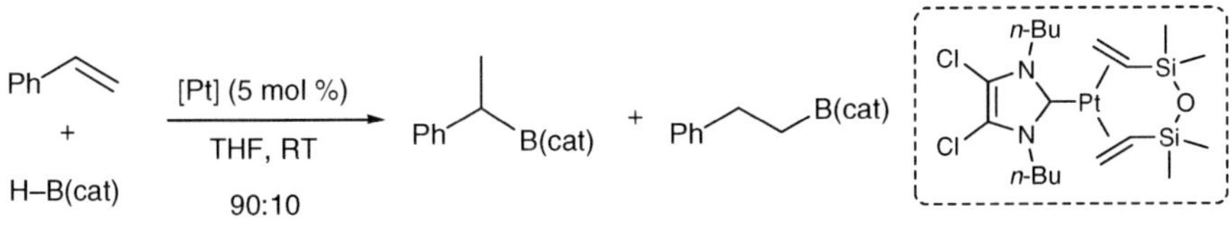

Ph–CH=CH2 + H–B(cat) → [Pt] (5 mol %) / THF, RT → Ph–CH(CH3)–B(cat) + Ph–CH2CH2–B(cat) (90:10)

(10.36)

The regioselectivity appeared to be dependent on the electron richness of the starting material. At least an example of tandem H–B addition/Suzuki–Miyaura coupling was explored but it provided the expected product without any regioselectivity and in moderate yield.

10.17 Conclusion and Outlook

In the last decade, NHC–Ni complexes have become ubiquitous reagents in catalysis. As shown in this Chapter, sterically hindered NHC ligands have led to improved reactivity and selectivity compared to classical Ni-phosphine reagents. The activation of C–F bond is probably the best example of such improvements. Even if (S)IPr and (S)IMes appear to be the most commonly used NHCs with Ni, the design of new bi- or tridentate ligands as well as asymmetric units appears as a promising area of research. The future of NHC–Ni complexes will probably be sunny.

A similar fate might be predicted for NHC–Pt analogues even if the reactions explored so far remain limited. Regioselective and/or asymmetric hydroboration as well as hydrosilylation processes are the two most promising transformations catalyzed by NHC–Pt.

References

1. (a) A. J. Arduengo III, R. L. Harlow and M. Kline, *J. Am. Chem. Soc.*, 1991, **113**, 361–363. (b) A. J. Arduengo III, M. Kline, J. C. Calabrese and F. Davidson, *J. Am. Chem. Soc.*, 1991, **113**, 9704–9705. (c) A. J. Arduengo III, H. V. R. Dias, R. L. Harlow and M. Kline, *J. Am. Chem. Soc.*, 1992, **114**, 5530–5534.
2. (a) W. A. Herrmann, D. Mihalios, K. Öfele, P. Kiprof and F. Belmedjahed, *Chem. Ber.*, 1992, **125**, 1795–1799. (b) K. Öfele, W. A. Herrmann, D. Mihalios, M. Elison, E. Herdtweck, W. Scherer and J. Mink, *J. Organomet. Chem.*, 1993, **459**, 177–184. (c) W. A. Herrmann, K. Öfele, M. Elison, F. E. Kühn and P. W. Roesky, *J. Organomet. Chem.*, 1994, **480**, C7–C9. For a review, see: (d) W. A. Herrmann, *Angew. Chem., Int. Ed.*, 2002, **39**, 1291–1309.
3. (a) U. Radius and F. M. Bickelbaupt, *Organometallics*, 2008, **27**, 3410–3414. (b) U. Radius and F. M. Bickelhaupt, *Coord. Chem. Rev.*, 2009, **253**, 678–686.
4. H. Jacobsen, A. Correa, A. Poater, C. Costabile and L. Cavallo, *Coord. Chem. Rev.*, 2009, **253**, 687–703.
5. A. J. Arduengo III, S. F. Gamper, J. C. Calabrese and F. Davidson, *J. Am. Chem. Soc.*, 1994, **116**, 4391–4394.
6. (a) D. S. McGuinness, K. J. Cavell, B. W. Skelton and A. H. White, *Organometallics*, 1999, **18**, 1596–1605. (b) J. A. Chamizo and J. Morgado, *Transition Met. Chem. (Dordrecht, Neth)*., 2000, **25**, 161–165. (c) C. D. Abernethy, A. H. Cowley and R. A. Jones, *J. Organomet. Chem.*,

2000, **596**, 3–5. (d) R. Dorta, E. D. Stevens, C. D. Hoff and S. P. Nolan, *J. Am. Chem. Soc.*, 2003, **125**, 10490–10491. (e) A. A. Danopoulos and D. Pugh, *Dalton Trans.*, 2008, 30–31.

7. (a) R. E. Douthwaite, M. L. H. Green, P. J. Silcock and P. T. Gomes, *Organometallics*, 2001, **20**, 2611–2615. (b) X. Hu, I. Castro-Rodriguez and K. Meyer, *Chem. Commun.*, 2004, 2164–2165. (c) C. C. Lee, W. C. Ke, K. T. Chan, C. L. Lai, C. H. Hu and H. M. Lee, *Chem.–Eur. J.*, 2007, **13**, 582–591. (d) C. Y. Liao, K. T. Chan, Y. C. Chang, C. Y. Chen, C. Y. Tu, C. H. Hu and H. M. Lee, *Organometallics*, 2007, **26**, 5826–5833. (e) S. K. Schneider, C. F. Rentzsch, A. Krueger, H. G. Raubenheimer and W. A. Herrmann, *J. Mol. Catal. A: Chem.*, 2007, **265**, 50–58.
8. (a) S. Winston, N. Stylianides, A. A. D. Tulloch, J. A. Wright and A. A. Danopoulos, *Polyhedron*, 2004, **23**, 2813–2820. (b) K. Inamoto, J.-I. Kuroda, K. Hiroya, Y. Noda, M. Watanabe and T. Sakamoto, *Organometallics*, 2006, **25**, 3095–3098. (c) D. Pugh, A. Boyle and A. A. Danopoulos, *Dalton Trans.*, 2008, 1087–1094.
9. (a) W. A. Herrmann, L. J. Goossen, C. Koecher and G. R. J. Artus, *Angew. Chem., Int. Ed. Engl.*, 1997, **35**, 2805–2807. (b) D. S. Clyne, J. Jin, E. Genest, J. C. Gallucci and T. V. RajanBabu, *Org. Lett.*, 2000, **2**, 1125–1128. (c) M. V. Baker, B. W. Skelton, A. H. White and C. C. Williams, *J. Chem. Soc., Dalton Trans.*, 2001, 111–120. (d) M. V. Baker, B. W. Skelton, A. H. White and C. C. Williams, *Organometallics*, 2002, **21**, 2674–2678.
10. Of note, this reaction was also proposed to be reversible, see: (a) V. P. W. Böhm, C. W. K. Gstöttmayr, T. Weskamp and W. A. Herrmann, *Angew. Chem., Int. Ed*, 2001, **40**, 3387–3389; (b) J. Louie, J. E. Gibby, M. V. Farnworth and T. N. Tekavec, *J. Am. Chem. Soc.*, 2002, **124**, 15188–15189.
11. S. Caddick, F. G. N. Cloke, P. B. Hitchcock and A. K. D. K. Lewis, *Angew. Chem., Int. Ed.*, 2004, **43**, 5824–5827.
12. (a) B. R. Dible, M. S. Sigman and A. M. Arif, *Inorg. Chem.*, 2005, **44**, 3774–3776. (b) B. R. Dible and M. S. Sigman, *Inorg. Chem.*, 2006, **45**, 8430–8441.
13. G. Zuo and J. Louie, *J. Am. Chem. Soc.*, 2005, **127**, 5798–5799.
14. B. Gradel, E. Brenner, R. Schneider and Y. Fort, *Tetrahedron Lett.*, 2001, **42**, 5689–5692.
15. K. Matsubara, S. Miyazaki, Y. Koga, Y. Nibu, T. Hashimura and T. Matsumoto, *Organometallics*, 2008, **27**, 6020–6024.
16. (a) W. A. Herrmann, G. Gerstberger and M. Spiegler, *Organometallics*, 1997, **16**, 2209–2212. (b) W. A. Herrmann, J. Schwarz, M. G. Gardiner and M. Spiegler, *J. Organomet. Chem.*, 1999, **575**, 80–86.
17. For recent contributions, see: (a) S. K. Schneider, G. R. Julius, C. Loschen, H. G. Raubenheimer, G. Frenking and W. A. Herrmann, *Dalton Trans.*, 2006, 1226–1233; (b) K. Matsubara, S. Miyazaki, Y. Koga, Y. Nibu, T. Hashimura and T. Matsumoto, *Organometallics*, 2008, **27**, 6020–6024; (c) X. Zhang, B. Liu, A. Liu, W. Xie and W. Chen,

Organometallics, 2009, **28**, 1336–1349; (d) L.-Z. Xie, H.-M. Sun, D.-M. Hu, Z.-H. Liu, Q. Shen and Y. Zhang, *Polyhedron*, 2009, **28**, 2585–2590; (e) T. Steinke, B. K. Shaw, H. Jong, B. O. Patrick, M. D. Fryzuk and J. C. Green, *J. Am. Chem. Soc.*, 2009, **131**, 10461–10466; (f) R. Jothibasu and H. V. Huynh, *Organometallics*, 2009, **28**, 2505–2511.

18. B. Liu, Q. Xia and W. Chen, *Angew. Chem., Int. Ed.*, 2009, **48**, 5513–5516.
19. (a) M. Hudlicky, in *Comprehensive Organic Synthesis*, ed. B. M. Trost and I. Fleming, Pergamon, Oxford, 1991, vol. 8, pp. 895–923. (b) R. A. Hites, *Acc. Chem. Res.*, 1990, **23**, 194–201. (c) A. R. Pinder, *Synthesis*, 1980, 425–452.
20. (a) F. Massicot, R. Schneider, Y. Fort, S. Illy-Cherrey and O. Tillement, *Tetrahedron*, 2000, **56**, 4765–4768. (b) X. Jurvilliers, R. Schneider, Y. Fort and J. Ghanbaja, *Appl. Organomet. Chem.*, 2001, **15**, 744–748.
21. C. Desmarets, S. Kuhl, R. Schneider and Y. Fort, *Organometallics*, 2002, **21**, 1554–1559.
22. S. Kuhl, R. Schneider and Y. Fort, *Adv. Synth. Catal.*, 2003, **345**, 341–344.
23. T. Schaub and U. Radius, *Chem.–Eur. J.*, 2005, **11**, 5024–5030.
24. T. Schaub, P. Fischer, A. Steffen, T. Braun, U. Radius and A. Mix, *J. Am. Chem. Soc.*, 2008, **130**, 9304–9317.
25. T. Schaub, M. Backes and U. Radius, *Eur. J. Inorg. Chem.*, 2008, 2680–2690.
26. (a) C. W. Hamilton, R. T. Baker, A. Staubitz and I. Manners, *Chem. Soc. Rev.*, 2009, **38**, 279–293. (b) F. H. Stephens, V. Pons and R. T. Baker, *Dalton Trans.*, 2007, 2613–2626.
27. R. J. K. Keaton, J. M. Blacquiere and R. T. Baker, *J. Am. Chem. Soc.*, 2007, **129**, 1844–1845.
28. (a) X. Z. Yang and M. B. Hall, *J. Am. Chem. Soc.*, 2008, **130**, 1798–1799. (b) X. Z. Yang and M. B. Hall, *J. Organomet. Chem.*, 2009, **694**, 2831–2838.
29. (a) P. M. Zimmerman, A. Paul, Z. Zhang and C. B. Musgrave, *Angew. Chem., Int. Ed.*, 2009, **48**, 2201–2205. (b) P. M. Zimmerman, A. Paul and C. B. Musgrave, *Inorg. Chem.*, 2009, **48**, 5418–5433.
30. T. Schaub, M. Backes and U. Radius, *Organometallics*, 2006, **25**, 4196–4206.
31. S. Becker, Y. Fort and P. Caubere, *J. Org. Chem.*, 1990, **55**, 6194–6198.
32. T. Schaub, M. Backes and U. Radius, *Chem. Commun.*, 2007, 2037–2039.
33. P. W. N. M. van Leeuwen, *Homogeneous Catalysis: Understanding the Art.*, Kluwer, Dordrecht, 2004, pp. 229–233.
34. R. J. McKinney, in *Homogeneous Catalysis*, ed. G. W. Parshall and S. D. Ittel, Wiley, New York, 1992, pp. 42–50.
35. T. Schaub, C. Doering and U. Radius, *Dalton Trans.*, 2007, 1993–2002.
36. A. Acosta-Ramírez, D. Morales-Morales, J. M. Serrano-Becerra, A. Arévalo, W. D. Jones and J. J. García, *J. Mol. Catal. A: Chim.*, 2008, **288**, 14–18.

37. (a) J. P. Wolfe and S. L. Buchwald, *J. Am. Chem. Soc.*, 1997, **119**, 6054–6058. (b) C. Bolm, J. P. Hildebrand and J. Rudolph, *Synthesis*, 2000, 911–913. (c) B. H. Lipshutz and H. Ueda, *Angew. Chem., Int. Ed.*, 2000, **39**, 4492–4494.
38. (a) B. Gradel, E. Brenner, R. Schneider and Y. Fort, *Tetrahedron Lett.*, 2001, **42**, 5689–5692. (b) C. Desmarets, R. Schneider and Y. Fort, *J. Org. Chem.*, 2002, **67**, 3029–3036.
39. J. Huang, G. Grasa and S. P. Nolan, *Org. Lett.*, 1999, **1**, 1307–1309.
40. S. Kuhl, Y. Fort and R. Schneider, *J. Organomet. Chem.*, 2005, **690**, 6169–6177.
41. (a) C. Desmarets, R. Schneider, Y. Fort and A. Walcarius, *J. Chem. Soc., Perkin Trans. 2*, 2002, 1844–1849. (b) C. Desmarets, R. Schneider, Y. Fort and A. Walcarius, *Inorg. Chem. Commun.*, 2003, **6**, 278–280. (c) C. Desmarets, B. Champagne, A. Walcarius, C. Bellouard, R. Omar-Amrani, A. Ahajji, Y. Fort and R. Schneider, *J. Org. Chem.*, 2006, **71**, 1351–1361.
42. R. Omar-Amrani, A. Thomas, E. Brenner, R. Schneider and Y. Fort, *Org. Lett.*, 2003, **5**, 2311–2314.
43. C.-Y. Gao and L.-M. Yang, *J. Org. Chem.*, 2008, **73**, 1624–1627.
44. Y. G. Zhang, K. C. Ngeow and J. Y. Ying, *Org. Lett.*, 2007, **9**, 3495–3498.
45. D. A. Malyshev, N. M. Scott, N. Marion, E. D. Stevens, V. P. Ananikov, I. P. Beletskaya and S. P. Nolan, *Organometallics*, 2006, **25**, 4462–4470.
46. For the only example of NHC–Ni in Ullmann coupling, see: H. V. Huynh, L. R. Wong and P. S. Ng, *Organometallics*, 2008, **27**, 2231–2237.
47. (a) S. B. Blakey and D. W. C. MacMillan, *J. Am. Chem. Soc.*, 2003, **125**, 6046–6047. (b) J. Liu and M. J. Robins, *Org. Lett.*, 2004, **6**, 3421–3423.
48. (a) K. Tamao, K. Sumitani and M. Kumada, *J. Am. Chem. Soc.*, 1972, **94**, 4374–4376. (b) R. J. P. Corriu and J. P. Masse, *J. Chem. Soc., Chem. Commun.*, 1972, 144a. (c) M. Kumada, *Pure Appl. Chem.*, 1980, **52**, 669–679. (d) T. Takahashi and K. I. Kanno in *Modern Organonickel Chemistry*, ed. Y. Tamaru, Wiley-VCH, Weinheim, 2005.
49. J. Huang and S. P. Nolan, *J. Am. Chem. Soc.*, 1999, **121**, 9889–9890.
50. V. P. W. Böhm, T. Weskamp, C. W. K. Gstöttmayr and W. A. Herrmann, *Angew. Chem., Int. Ed.*, 2000, **39**, 1602–1604.
51. V. P. W. Böhm, C. W. K. Gstöttmayr, T. Weskamp and W. A. Herrmann, *Angew. Chem., Int. Ed*, 2001, **40**, 3387–3389.
52. D. Kremzow, G. Seidel, C. W. Lehmann and A. Fürstner, *Chem.–Eur. J.*, 2005, **11**, 1833–1853.
53. K. Matsubara, K. Ueno and Y. Shibata, *Organometallics*, 2006, **25**, 3422–3427.
54. K. Matsubara, K. Ueno, Y. Koga and K. Hara, *J. Org. Chem.*, 2007, **72**, 5069–5076.
55. K. Inamoto, J.-I. Kuroda, T. Sakamoto and K. Hiroya, *Synthesis*, 2007, 2853–2861.
56. Z. Xi, B. Liu and W. Chen, *J. Org. Chem.*, 2008, **73**, 3954–3957.

57. S. K. Schneider, C. F. Rentzsch, A. Krueger, H. G. Raubenheimer and W. A. Herrmann, *J. Mol. Catal. A: Chim.*, 2007, **265**, 50–58.
58. A. Leleu, Y. Fort and R. Schneider, *Adv. Synth. Catal.*, 2006, **348**, 1086–1092.
59. S. B. Blakey and D. W. C. MacMillan, *J. Am. Chem. Soc.*, 2003, **125**, 6046–6047.
60. T. Schaub, M. Backes and U. Radius, *J. Am. Chem. Soc.*, 2006, **128**, 15964–15965.
61. J. Liu and M. J. Robins, *Org. Lett.*, 2004, **6**, 3421–3423.
62. Z. X. Xi, X. M. Zhang, W. Z. Chen, S. Z. Fu and D. Q. Wang, *Organometallics*, 2007, **26**, 6636–6642.
63. Z. Xi, Y. Zhou and W. Chen, *J. Org. Chem.*, 2008, **73**, 8497–8501.
64. J. Louie, in *N-Heterocyclic Carbenes in Synthesis*, ed. S. P. Nolan, Wiley-VCH, Weinheim, 2006, pp. 163–182.
65. J. Louie, J. E. Gibby, M. V. Farnworth and T. N. Tekavec, *J. Am. Chem. Soc.*, 2002, **124**, 15188–15189.
66. (a) Y. Inoue, Y. Itoh, H. Kazama and H. Hashimoto, *Bull. Chem. Soc. Jpn.*, 1980, **53**, 3329–3333. (b) T. Tsuda, K. Maruta and Y. Kitaike, *J. Am. Chem. Soc.*, 1992, **114**, 1498–1499.
67. H. A. Duong, M. J. Cross and J. Louie, *J. Am. Chem. Soc.*, 2004, **126**, 11438–11439.
68. H. A. Duong, M. J. Cross and J. Louie, *Org. Lett.*, 2004, **6**, 4679–4681.
69. H. Duong and J. Louie, *Tetrahedron*, 2006, **62**, 7552–7559.
70. M. M. McCormick, H. A. Duong, G. Zuo and J. Louie, *J. Am. Chem. Soc.*, 2005, **127**, 5030–5031.
71. T. N. Tekavec and J. Louie, *J. Org. Chem.*, 2008, **73**, 2641–2648.
72. S. Ogoshi, H. Ikeda and H. Kurosawa, *Pure Appl. Chem.*, 2008, **80**, 1115–1125.
73. (a) J. A. Varela and C. Saá, *Chem. Rev.*, 2003, **103**, 3787–3801. (b) Y. Yamamoto, *Curr. Org. Chem.*, 2005, **9**, 503–519. (c) V. Gandon, C. Aubert and M. Malacria, *Curr. Org. Chem.*, 2005, **9**, 1699–1712. (d) P. R. Chopade and J. Louie, *Adv. Synth. Catal.*, 2006, **348**, 2307–2327.
74. W. Reppe, O. Schlichting, K. Klager and T. Toepel, *Justus Liebigs Ann. Chem.*, 1948, **560**, 1–92.
75. N. Saito, K. Shiotani, A. Kinbara and Y. Sato, *Chem. Commun.*, 2009, 4284–4286.
76. G. Zuo and J. Louie, *Angew. Chem., Int. Ed.*, 2004, **43**, 2277–2279.
77. G. Zuo and J. Louie, *J. Am. Chem. Soc.*, 2005, **127**, 5798–5799.
78. G. Zuo, K. Zhang and J. Louie, *Tetrahedron Lett.*, 2008, **49**, 6797–6799.
79. (a) *Multicomponent Reactions*, ed. J. Zhu and H. Bienaymé, Wiley-VCH, Weinheim, 2005; (b) D. J. Ramón and M. Yus, *Angew. Chem., Int. Ed.*, 2005, **44**, 1602–1634.
80. Y. Sato, R. Sawaki and M. Mori, *Organometallics*, 2001, **20**, 5510–5512.
81. R. Sawaki, Y. Sato and M. Mori, *Org. Lett.*, 2004, **6**, 1131–1133.
82. Y. Sato, Y. Hinata, R. Seki, Y. Oonishi and N. Saito, *Org. Lett.*, 2007, **9**, 5597–5599.

83. N. Saito, T. Katayama and Y. Sato, *Org. Lett.*, 2008, **10**, 3829–3832.
84. (a) S. S. Ng and T. F. Jamison, *J. Am. Chem. Soc.*, 2005, **127**, 7320–7321. (b) S.-S. Ng and T. F. Jamison, *Tetrahedron*, 2006, **62**, 11350–11359.
85. K. D. Schleicher and T. F. Jamison, *Org. Lett.*, 2007, **9**, 875–878.
86. C.-Y. Ho and T. F. Jamison, *Angew. Chem., Int. Ed.*, 2007, **46**, 782–785.
87. G. M. Mahandru, G. Liu and J. Montgomery, *J. Am. Chem. Soc.*, 2004, **126**, 3698–3699.
88. M. R. Chaulagain, G. M. Mahandru and J. Montgomery, *Tetrahedron*, 2006, **62**, 7560–7566.
89. K. Sa-ei and J. Montgomery, *Org. Lett.*, 2006, **8**, 4441–4443.
90. M. R. Chaulagain, G. J. Sormunen and J. Montgomery, *J. Am. Chem. Soc.*, 2007, **129**, 9568–9569.
91. M. Murakami, S. Ashida and T. Matsuda, *J. Am. Chem. Soc.*, 2005, **127**, 6932–6933.
92. (a) W. Li, H. Sun, M. Chen, Z. Wang, D. Hu, Q. Shen and Y. Zhang, *Organometallics*, 2005, **24**, 5925–5928. (b) B. E. Ketz, X. G. Ottenwaelder and R. M. Waymouth, *Chem. Commun.*, 2005, 5693–5695. (c) J. Campora, L. O. de la Tabla, P. Palma, E. Alvarez, F. Lahoz and K. Mereiter, *Organometallics*, 2006, **25**, 3314–3316. (d) S. Sujith, E. K. Noh, B. Y. Lee and J. W. Han, *J. Organomet. Chem.*, 2008, **693**, 2171–2176.
93. (a) H. M. Sun, Q. Shao, D. M. Hu, W. F. Li, Q. Shen and Y. Zhang, *Organometallics*, 2005, **24**, 331–334. (b) A. T. Normand, K. J. Hawkest, N. D. Clement, K. J. Cavell and B. F. Yates, *Organometallics*, 2007, **26**, 5352–5363. (c) W. F. Li, H. M. Sun, M. Z. Chen, Q. Shen and Y. Zhang, *J. Organomet. Chem.*, 2008, **693**, 2047–2051.
94. D. S. McGuinness, W. Mueller, P. Wasserscheid, K. J. Cavell, B. W. Skelton, A. H. White and U. Englert, *Organometallics*, 2002, **21**, 175–181.
95. S.-L. Liu, T.-Y. Hsieh, G.-H. Lee and S.-M. Peng, *Organometallics*, 1998, **17**, 993–995.
96. (a) D. S. McGuinness, K. J. Cavell and B. F. Yates, *Chem. Commun.*, 2001, 355–356. (b) D. S. McGuinness, K. J. Cavell, B. F. Yates, B. W. Skelton and A. H. White, *J. Am. Chem. Soc.*, 2001, **123**, 8317–8323. (c) M. A. Duin, N. D. Clement, K. J. Cavell and C. J. Elsevier, *Chem. Commun.*, 2003, 400–401.
97. N. Schneider, S. Bellemin-Laponnaz, H. Wadepohl and L. H. Gade, *Eur. J. Inorg. Chem.*, 2008, 5587–5598.
98. C. P. Newman, R. J. Deeth, G. J. Clarkson and J. P. Rourke, *Organometallics*, 2007, **26**, 6225–6233.
99. M. Iglesias, D. J. Beetstra, A. Stasch, P. N. Horton, M. B. Hursthouse, S. J. Coles, K. J. Cavell, A. Dervisi and I. A. Fallis, *Organometallics*, 2007, **26**, 4800–4809.
100. R. Lindner, C. Wagner and D. Steinborn, *J. Am. Chem. Soc.*, 2009, **131**, 8861–8874.
101. F. E. Hahn, M. C. Jahnke and T. Pape, *Organometallics*, 2006, **25**, 5927–5936.

102. For a ground-breaking report, see: I. E. Markó, S. Stérin, O. Buisine, G. Mignani, P. Branlard, B. Tinant and J.-P. Declerq, *Science*, 2002, **298**, 204–206.
103. I. G. Jung, J. Seo, S. I. Lee, S. Y. Choi and Y. K. Chung, *Organometallics*, 2006, **25**, 4240–4242.
104. (a) D. Brissy, M. Skander, P. Retailleau, G. Frison and A. Marinetti, *Organometallics*, 2009, **28**, 140–151. (b) D. Brissy, M. Skander, H. Jullien, P. Retailleau and A. Marinetti, *Org. Lett.*, 2009, **11**, 2137–2139.
105. V. Lillo, J. A. Mata, J. Ramirez, E. Peris and E. Fernandez, *Organometallics*, 2006, **25**, 5829–5831.
106. V. Lillo, J. A. Mata, A. M. Segarra, E. Peris and E. Fernandez, *Chem. Commun.*, 2007, 2184–2186.

CHAPTER 11

NHC–Copper, Silver and Gold Complexes in Catalysis

NICOLAS MARION*

Department of Chemistry, Massachusetts Institute of Technology, 77 Massachusetts Avenue, Cambridge, MA 02139, USA

11.1 Introduction

The Group 11 elements,[1] commonly called *coinage metals*, hold a particular position in chemistry and more generally in humankind. Historically, copper, silver and gold, which have all been known since prehistoric times, were used not only as currency but also in diverse decorative objects. In modern times, their acute properties as electric conductors were recognized early on and copper has notably been used extensively for wiring, while gold was selected for precision devices due to its high resistance to corrosion. In addition, silver has been indispensable to the development of photography, in the form of silver nitrate. In sharp contrast to their large-scale industrial use for their *metallic* properties, their organometallic chemistry as well as their use in catalysis remained relatively unexplored until recently. While copper catalysis can be traced back over a century, organic chemists discovered silver and gold catalysis only recently. Nevertheless, the breathtaking pace at which new complexes and novel reactivity are being explored with Group 11 metals is bringing copper, silver and gold at the forefront of organic chemistry.[2]

In this review, the applications in homogeneous catalysis of N-heterocyclic carbene-containing copper, silver and gold complexes will be presented. N-Heterocyclic carbenes (NHCs) are two-electron σ-donor ligands that share a

*Present address: Laboratorium für Organische Chemie, ETH Zürich, Hönggerberg, HCI, CH-8093 Zürich, Switzerland

RSC Catalysis Series No. 6
N-Heterocyclic Carbenes: From Laboratory Curiosities to Efficient Synthetic Tools
Edited by Silvia Díez-González
© Royal Society of Chemistry 2011
Published by the Royal Society of Chemistry, www.rsc.org

number of common features with the ubiquitous phosphane ligands.[3] They have developed tremendously in the last 15 years and are now considered as one of the most important classes of ligands in organometallic chemistry. The reader is invited to consult general reviews[4] as well as Chapters 1–5 in the present book for further details about NHCs. The chemistry of NHC–Group 11 complexes has already been reviewed in numerous places and from different angles.[5] This Chapter is therefore intended as a general guide providing an overview—and no detailed description—of the main achievements realized so far, *i.e.* early 2010, with NHC–Group 11 species in homogeneous catalysis.

11.2 NHC–Cu in Catalysis

Copper has a long history in organic synthesis, dating back to the Ullmann reaction,[6] but a rather short one with NHC ligands. The first NHC–Cu complex was reported in 1993 by the Arduengo group,[7] soon after their groundbreaking isolation of a free NHC.[8] Advances in NHC–copper chemistry were then steadily reported, including better synthetic routes to well-defined complexes,[9] and theoretical studies of the NHC–Cu bond that allowed for a better understanding of NHC–metal interactions.[10] Despite these achievements, the first application in catalysis of a NHC–Cu species only came in 2001 with Woodward's demonstration of the efficiency of such catalysts in conjugate addition.[11] Since then, the field has grown exponentially and a wide variety of catalytic applications has been explored. Previous reviews on this topic can be found in the literature.[12]

11.2.1 Conjugate Addition Reactions

Following Woodward's seminal results,[11] the 1,4-addition of organozinc reagents to enones has become a benchmark reaction for new NHC–Cu catalysts. Most importantly, relying mainly on the early work of Alexakis and Roland,[13] it rapidly evolved into one of the most successful enantioselective applications of chiral NHCs.[14] Several strategies, ranging from *in situ* deprotonation of an azolium salt to the use of well-defined complexes and including transmetalation from a NHC–silver species in the presence of a simple Cu source, can be employed for the formation of the chiral NHC–Cu species.[15] Comparative studies have shown that results are sensibly identical.[16] Organozinc addition is still an active field of research with new NHC architectures being frequently examined.[17,18]

Recent developments focused on the formation of quaternary stereocentres and the use of alternative nucleophilic reagents to organozincs. Mauduit and Alexakis notably reported the highly enantioselective addition of alkyl and aryl Grignard reagents to trisubstituted cyclic enones using **1**·HPF_6 (Equation (11.1)).[19,20]

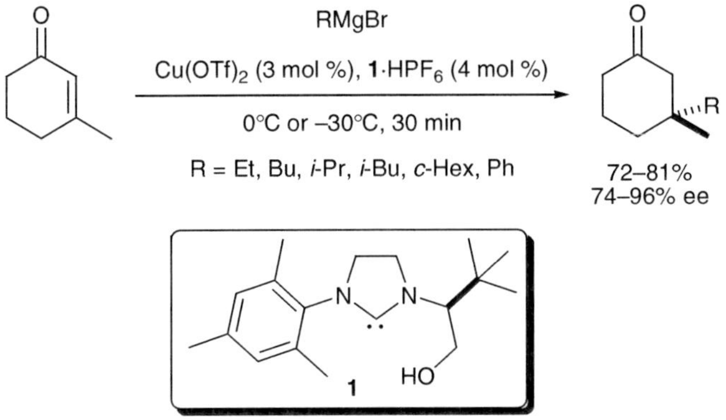

RMgBr
$Cu(OTf)_2$ (3 mol %), **1**·HPF_6 (4 mol %)
0°C or –30°C, 30 min
R = Et, Bu, *i*-Pr, *i*-Bu, *c*-Hex, Ph
72–81%
74–96% ee
1
HO

(11.1)

In the presence of doubly conjugated ketones, this catalytic system was found to promote 1,4-addition over 1,6-addition.[21] Subsequent work by Hoveyda revealed that organozinc and organoaluminium reagents were also amenable to the enantioselective formation of all-carbon quaternary stereogenic centres.[22] More recently, organosilanes, *i.e.* $RSiF_3$ or $RSi(OR')_3$, could also be used with NHC–Cu catalytic systems in conjugate addition to enones and allylic epoxides.[23]

Finally, even though generally labeled as hydrofunctionalization, the important work of Gunnoe should be mentioned here, in this section on conjugate addition.[24] Indeed, these extremely rare examples of anti-Markovnikov hydroamination, hydroalkoxylation and hydrothiolation of electron-deficient olefins (Equation (11.2)) have been rationalized mechanistically within the framework of conjugate additions.[25] Interestingly, reaching further than Michael-acceptor olefins, Gunnoe recently demonstrated the efficiency of this catalytic system with electron-deficient styrenes possessing a strong electron-withdrawing group in the *para* position of the aryl ring.[26]

EWG + R—H
[(**NHC**)Cu(NHPh)] or [(**NHC**)Cu(OEt)]
R = OR', NR'_2, SR'
EWG = ester, ketone, nitrile, *p*-NO_2-C_6H_4

(11.2)

11.2.2 Allylic Alkylation Reactions

Allylic alkylation is probably the only reaction that could dispute the honorific title of *benchmark test for new chiral NHCs* to conjugate addition. In fact, most chiral NHCs are tested in both reactions and, as a consequence, main advances in both topics have come from the same research groups. This chemistry was thoroughly reviewed by Mauduit and Douthwaite,[27] from Okamoto's first report[28] to the year 2007.

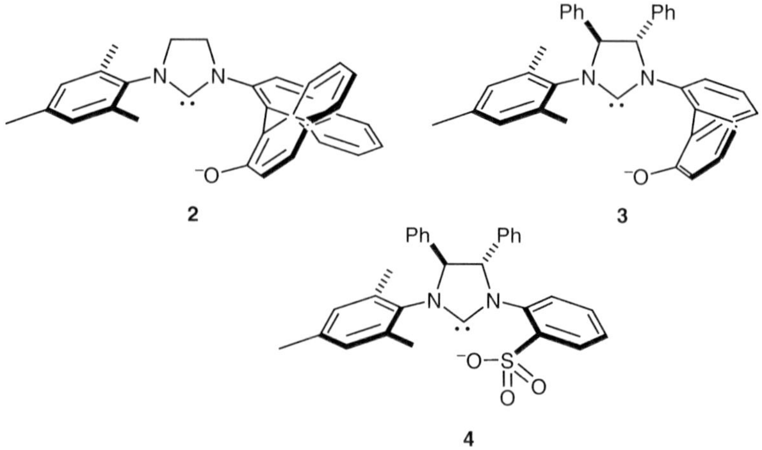

Figure 11.1 Structures of chiral NHCs **2**–**4**.

The Hoveyda group has arguably achieved the most important improvements in this field. The use of different families of axially chiral binaphthoxy- and biphenoxy-derived NHCs **2** and **3** (Figure 11.1), which proved efficient also in conjugate addition and olefin metathesis, allowed for excellent chiral induction with different combinations of allylic precursors and nucleophiles. This was notably showcased in the total synthesis of Baconipyrone C, where NHC **4** (Figure 11.1) was actually the most efficient ligand.[29] The most recent advances include the enantioselective synthesis of allylsilanes with organozinc reagents[30] and the use of vinylaluminium nucleophiles to form chiral 1,4-dienes.[31]

11.2.3 Reduction Reactions

The seminal work of Sadighi and Buchwald on the NHC–Cu-catalyzed 1,4-reduction of cyclic enones[32] triggered amazing developments in NHC–Cu-based reduction reactions. Today, this is one of the most investigated fields within NHC–Cu catalysis and thorough reviews have already been published covering this topic.[33]

Shortly after Sadighi and Buchwald's 2003 report, Nolan reported that [(IPr)CuCl] could be used in the hydrosilylation of simple ketones, leading to the silylated alcohols in high yields.[34] Subsequent work permitted the development of fine-tuned catalytic systems for different classes of carbonyl compounds. Hence, while [(IPr)CuCl] was found optimal for the 1,4-reduction of conjugated carbonyls[32] and 1,2-reduction of 'simple' ketones, [(ICy)CuCl] proved more efficient for hindered carbonyl substrates and [(SIMes)CuCl] turned out to be the best choice for heteroaromatic ketones (Scheme 11.1).[35] Other NHC architectures (*e.g.* six-membered diamino carbenes) and alternative

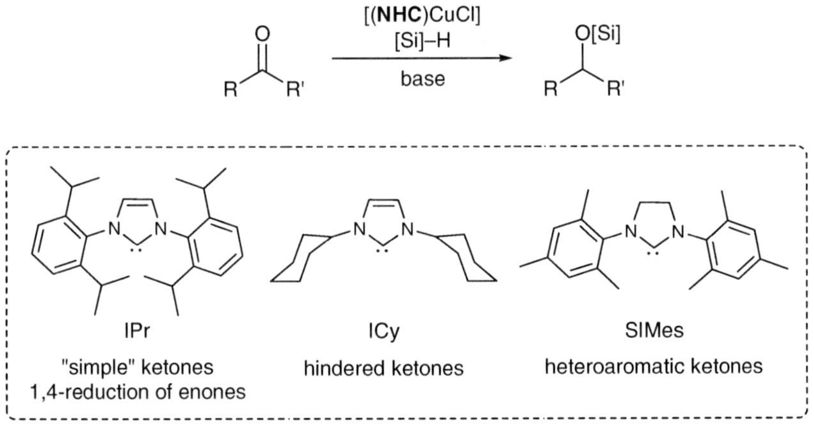

[(NHC)CuCl]
[Si]–H
base

Scheme 11.1 NHC–Cu-catalyzed reduction of carbonyl compounds.

copper sources (*e.g.* Cu^{II}) have also been investigated.[36] Some of these catalytic systems have already been applied to total synthesis. For instance, Poli and Madec recently reported the successful use of [(IPr)CuCl] for the 1,4-reduction of an α,β-unsaturated ethylester in a formal synthesis of kainic acid.[37]

Polymer science is also an interesting axis of development and the Buchmeiser group recently investigated the immobilization of NHC–Cu catalysts on amphiphilic block-copolymer and their utilization in different hydrosilylation reactions.[38]

Nolan and Díez-González introduced a new family of NHC–Cu complexes to the field of carbonyl hydrosilylation.[39] These bis-NHC cationic compounds of formulae $[(NHC)_2Cu]X$ were shown to be more efficient than their neutral mono-ligated counterparts under identical conditions and even allowed for the direct reduction of an ester into the corresponding protected alcohol (Equation (11.3)).

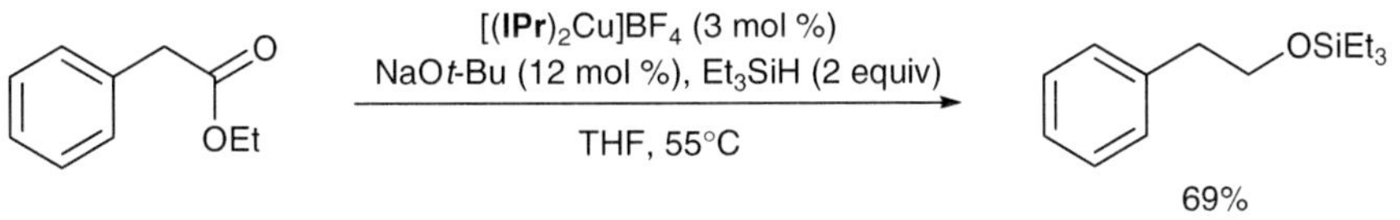

$[(IPr)_2Cu]BF_4$ (3 mol %), NaO*t*-Bu (12 mol %), Et_3SiH (2 equiv), THF, 55°C; 69%

(11.3)

More importantly, after full optimization, they allowed for decreasing the amount of hydrosilane required for the reduction and were compatible with lower reaction temperature.[39,40] Mechanistic studies using *in situ* ^1H NMR led the authors to propose decoordination of one NHC prior to entering the catalytic cycle and subsequent activation of the hydrosilane by the nucleophilic carbene released in solution.

Krause and co-workers recently examined the reduction of alkynes possessing a leaving group at the propargylic position. Such substrates underwent conjugate—or S_N2'—reduction and yielded allenic products. Hence, propargylic epoxides in the presence of SIMes or IBiox7, a copper salt, and polymethylhydrosiloxane (PMHS) furnished α-hydoxyallenes in good to high yields (Equation (11.4)).[41]

CuCl (3 mol %), **IBiox7**·HOTf (3 mol %)
NaO*t*-Bu (10 mol %), PHMS (2 equiv)
Toluene, 0°C

50–86%

IBiox7

(11.4)

This reactivity was extended to propargylic carbonates and represents a straightforward synthesis of di- and trisubstituted allenes.[42] In this case, IBiox12, the analogue of IBiox7 possessing a 12-membered ring in place of the 7-membered ring, proved more efficient than other NHC ligands.

In closing this section, it should be noted that this far there is no report of NHC–copper-mediated reduction of *unbiased* C–C multiple bonds,[43] a somewhat surprising feature considering the high efficiency of NHC-supported copper hydride species towards other π bonds.

11.2.4 Boration Reactions

Following advances made in reduction reactions (*vide supra*), hydroboration and diboration have been the subject of intense investigation with NHC–Cu catalysts. Early work by Sadighi revealed that [(ICy)Cu(O*t*-Bu)] efficiently catalyzed the 1,2-diboration of aldehydes.[44] Mechanistic studies permitted to rationalize a number of features of this reaction and notably ruled out a possible oxidative addition pathway to favour σ-activation of the diboron reagent by the copper centre.[45] [(ICy)Cu(O*t*-Bu)] was also used for the diastereoselective diboration—in fact, hydroboration after work-up—of sulfinyl aldimines.[46]

Lately, work in this area has focused on olefinic substrates. Sadighi first reported the stoichiometric insertion of an alkene into the Cu–B bond of [(IPr)Cu{B(pin)}].[47] A great deal of work was then performed by the groups of Pérez and Fernández on the β-boration of electron-deficient alkenes (*e.g.* conjugated enals).[48] In the case of α,β-unsaturated esters, the authors examined a number of monodentate chiral NHCs (Figure 11.2) which, despite high

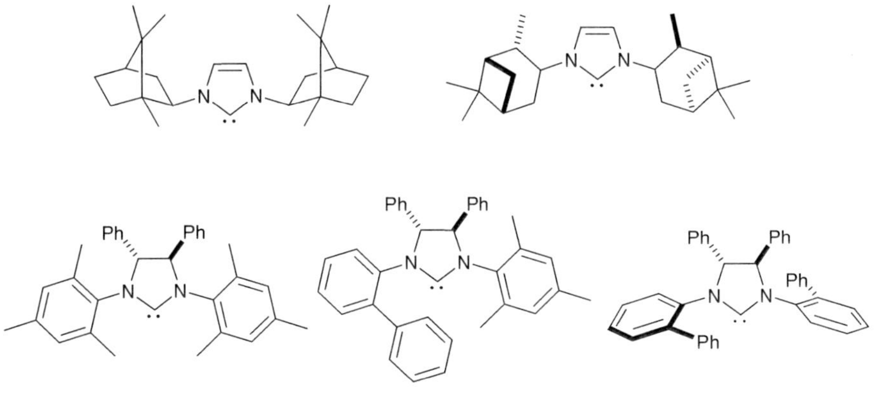

Figure 11.2 Structures of chiral NHCs used in β-boration of α,β-unsaturated esters.

conversions and excellent regioselectivities, led only to moderate *ee*'s (up to 73% *ee* but typically around 50% *ee*).[49]

The same groups, using [(NHC)CuCl] complexes, also reported the diboration of styrene and phenyl acetylene in good yields.[50] As in the case of the carbonyl diboration (*vide supra*), DFT studies for this transformation support σ-activation of the diboron reagent but also formation of an intermediate NHC–Cu–H, a species commonly thought to be involved in hydrosilylation reactions.[50,51] Very recently, Hoveyda developed a complementary method, using identical [(NHC)CuCl] complexes, to hydroborate β-substituted styrenyl compounds.[52] The optimized catalytic system, using $[B(pin)]_2$, was highly selective and represents an interesting alternative to typical hydroboration reactions. Preliminary results showed that this transformation can be rendered enantioselective by using chiral NHC **5** (Equation (11.5)).

Finally, further developments of this methodology led to the report of a double hydroboration of terminal alkynes, yielding vicinal diboronates. The main features of this reaction are the use of a diboron reagent, instead of a borane, and its high chemo-, regio- and enantioselectivity (Equation (11.5)).[53]

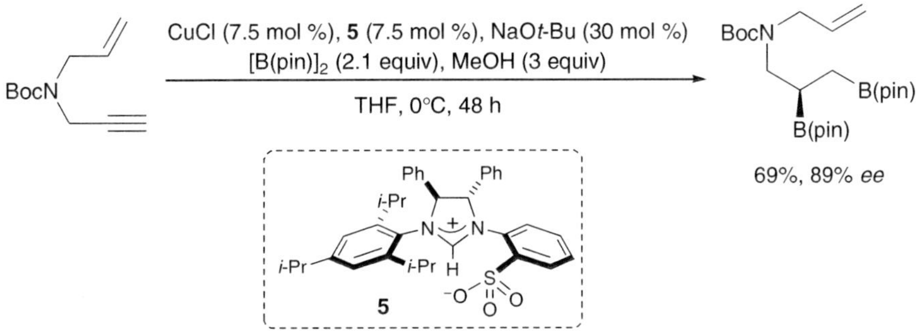

CuCl (7.5 mol %), **5** (7.5 mol %), NaO*t*-Bu (30 mol %), $[B(pin)]_2$ (2.1 equiv), MeOH (3 equiv); THF, 0°C, 48 h

69%, 89% *ee*

(11.5)

11.2.5 Cross-coupling Reactions

Despite the immense success of N-heterocyclic carbenes as supporting ligands in Pd-catalyzed cross-couplings,[54] they have been used only scarcely for the related Cu-mediated reactions. Hence, this far only three NHC–Cu complexes have been reported to successfully perform cross-coupling reactions. In 2004, Scholz showed that a biscarbene–copper system could achieve the mono-arylation of anilines with aryl chlorides.[55] Subsequent work by Biffis and co-workers made use of unusual trinuclear copper(I) complexes held together by triscarbene ligands.[56] Several member of this family were synthesized and displayed good activities in C–N, C–O, and C–C (Sonogashira-type) cross-couplings. Interestingly, comparative studies showed that, even though slightly less active, simple—and commercially available—[(IPr)CuCl] efficiently performed C–N coupling reactions.[56b] This observation should generate more investigations in a field where only poly-NHCs had proven useful.

Also, Vicic and co-workers developed an impressive NHC–Cu-mediated reaction allowing to couple a trifluoromethyl group with aryl iodides.[57] Even though the method was not catalytic in NHC–copper reagent, it represents a real breakthrough in Cu–CF_3 chemistry, a species that had remained elusive until this report and, more generally, in trifluoromethylation reactions. A variety of aryl iodides could be converted to their CF_3 analogues in the presence of [(NHC)Cu(CF_3)] **6** as shown in Scheme 11.2.

Of note, Vicic *et al.* also observed an unexpected reaction of these Cu–CF_3 complex in solution—two molecules of [(SIMes)Cu(CF_3)] being in equilibrium with [$(SIMes)_2$Cu][Cu$(CF_3)_2$]—that will probably render mechanistic investigations more difficult.[58]

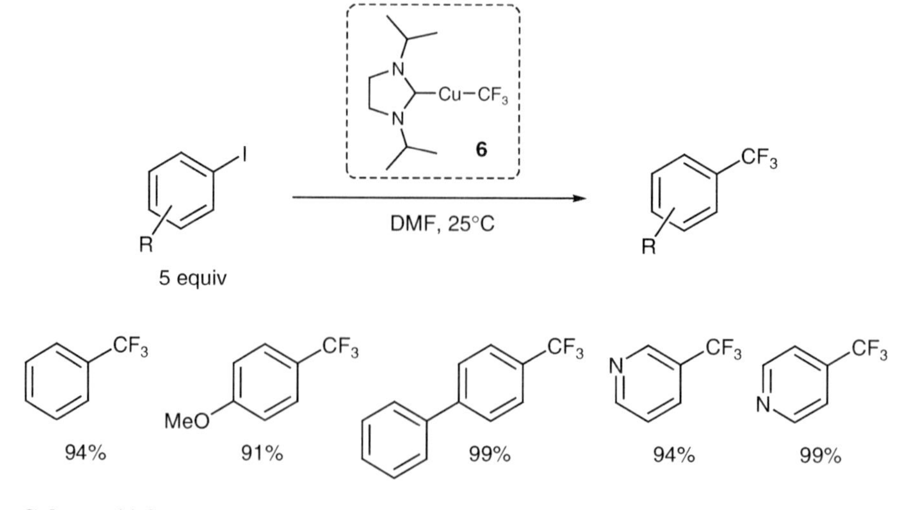

Scheme 11.2 NHC–Cu-mediated trifluoromethylating cross-coupling.

Finally, an oxidative coupling of 2-naphthols recently appeared using NHC–Cu species. While synthetically interesting—heterocoupling between electronically differentiated naphthols was performed—the nature of the NHC–Cu species (coordination number, copper oxidation state) remains uncertain.[59]

11.2.6 Miscellaneous Reactions

Carbene transfer reactions and related transformations—which have been reviewed thoroughly[5d,12]—could have been the subject of a section considering the number of studies published in that area. It is clearly an important field in which NHC–Cu catalysts have proven highly active, in part because, unlike phosphanes, the NHC ligand does not undergo carbene/nitrene transfer. Important work, mostly using [(NHC)CuX] systems, was achieved in diazo cyclopropanation of olefins,[60] diazo insertion into N–H, O–H, and C–H bonds,[61] and aziridination of olefins,[62] the latter methodology having proved its efficiency in total synthesis.[63] The most recent development in diazo compound decomposition in the presence of a NHC–Cu species came from Chang and co-workers with the report of an indene synthesis from α-aryldiazoesters and terminal alkynes.[64] This formal [3 + 2] cycloaddition (Equation (11.6)) required the use of [(IPr)CuCl] in conjugation with $AgSbF_6$ and a bulky non-coordinating counteranion to perform best. The corresponding NHC–Au complex could also be used but afforded indene products in much lower yields.

N_2 OMe O R + Ph; [(IPr)CuCl] (10 mol %), $AgSbF_6$/NaBARF (12 mol %), DCM, 25°C, 30 min → Ph OMe O R

R = H, 4-Br, 4-Me, 4-OMe, 2-Me

45–89%

(11.6)

Closely related to diazo insertions, the work of Lebel using TMS–N_2 in the presence of [(IPr)CuCl] for ketone methylenation should also be mentioned.[65]

Another recent important area of application for NHC–Cu species is the copper-catalyzed azide-alkyne cycloaddition. The main studies in this field were carried out by Nolan and Díez-González,[66] using both [(NHC)CuX] and $[(NHC)_2Cu]X$ complexes families, and were recently thoroughly reviewed by these authors.[12] Their latest achievement disclosed an interesting latent catalytic system.[67]

Finally, isolated reports on the ability of NHC–Cu complexes to catalyze diverse organic transformations can be found throughout the literature and notably include heterocycles synthesis,[68] atom-transfer radical cyclization leading to chloronaphthalenes,[69] and allylation reactions using [(IPr)CuF] and an allylsiloxane.[70] An increasing number of studies involving the use of CO_2 with NHC–Cu catalysts has also been reported lately, they encompass reduction of CO_2 to CO,[71] carboxylation of boronic acids (Equation (11.7)),[72] carbonylation of aminoalcohols,[73] and amino biscarbonylation of aryl iodides.[74]

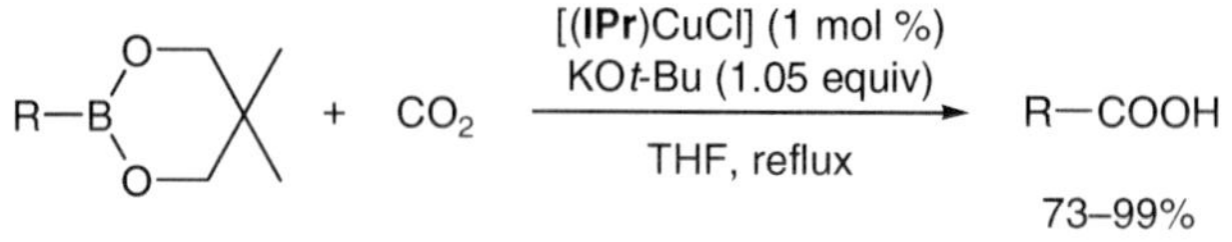

$$\text{R–B(pin-type dioxaborinane)} + CO_2 \xrightarrow[\text{THF, reflux}]{[(\textbf{IPr})\text{CuCl}]\ (1\ \text{mol}\ \%),\ \text{KO}t\text{-Bu}\ (1.05\ \text{equiv})} \text{R–COOH}\quad 73\text{–}99\% \tag{11.7}$$

11.3 NHC–Ag in Catalysis

NHC–Ag complexes have met with tremendous success in the last 10 years due to their extensive use as transmetalating agents.[75] This was made possible by the development of extremely efficient synthetic routes to these compounds, the most prominent example being the Lin synthesis which uses silver(I) oxide and azolium salts.[76] In sharp contrast, the applications of NHC–Ag complexes in homogeneous catalysis are surprisingly scarce, and barely amount to five at the beginning of 2010.[77]

The first report of a NHC–Ag-catalyzed reaction appeared only in 2005 when Peris and Fernandez disclosed the diboration of alkenes in the presence of a bis-NHC silver complex (Scheme 11.3).[78] The menthol-derived NHC ligand **7** showed only limited scope in this reaction but provided a proof-of-concept for NHC–Ag-based catalysis. Subsequent studies by the same groups examined chiral NHCs **8**–**10** (Figure 11.3) in this reaction, but enantiomeric excess remained very low (9% *ee* at best).[79]

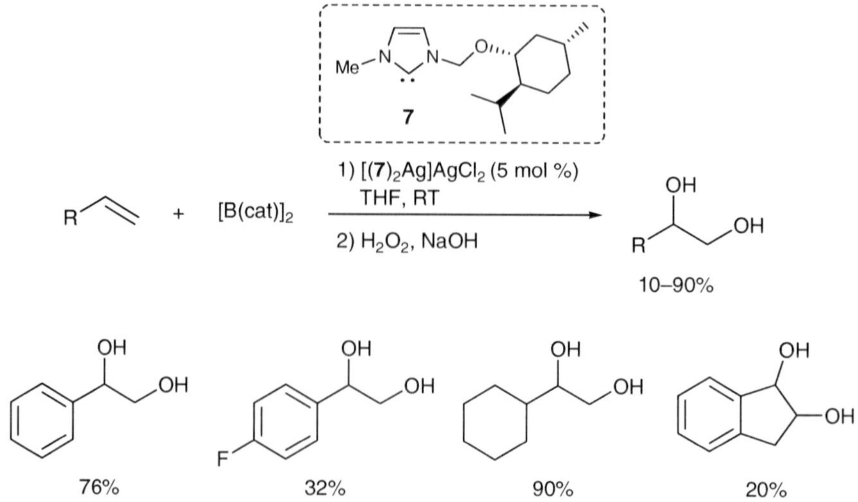

Me–N N–CH₂–O (menthyl) **7**

R–CH=CH₂ + $[B(cat)]_2$ → 1) $[(\mathbf{7})_2Ag]AgCl_2$ (5 mol %), THF, RT; 2) H_2O_2, NaOH → R–CH(OH)–CH₂OH 10–90%

76% | 32% (F) | 90% | 20%

Scheme 11.3 NHC–Ag-catalyzed diboration of alkenes.

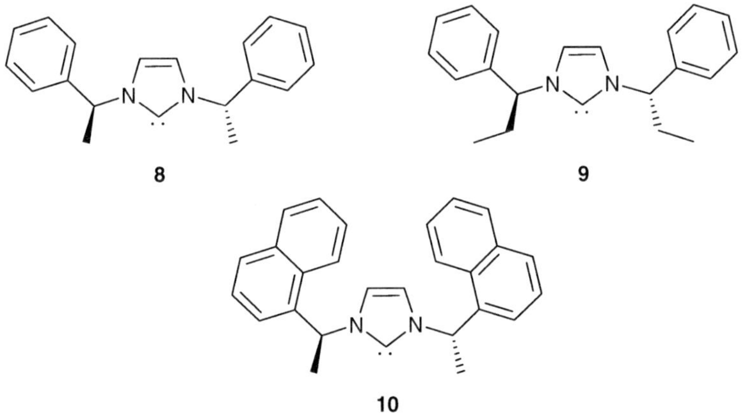

Figure 11.3 Structures of chiral NHCs used in diboration of alkenes.

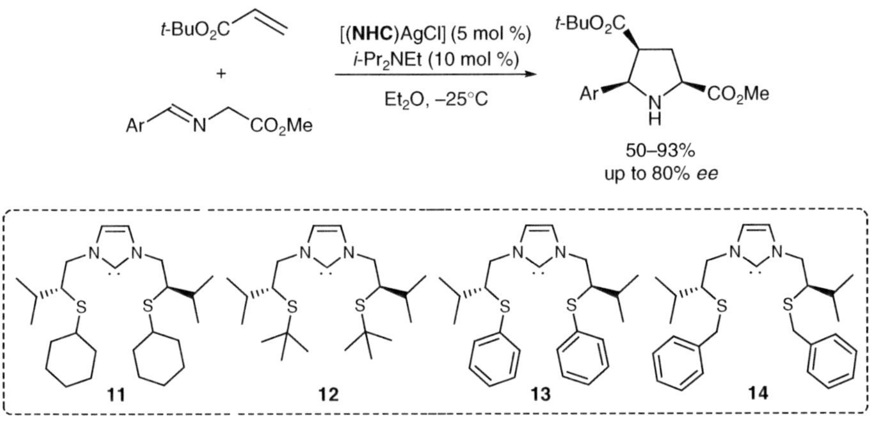

Scheme 11.4 NHC–Ag-catalyzed enantioselective cycloaddition of azomethines and acrylates.

More recently, a study by Lassaletta, Fernández and co-workers showed that enantioselective NHC–Ag catalysis was not an impossible goal. Hence, the use of polydentate thiolato-NHCs **11**–**14** afforded good enantioselectivities in the 1,3-dipolar cycloaddition of azomethine ylides with acrylates (Scheme 11.4).[80]

Ligand **11** systematically produced the highest *ee*'s (up to 80% *ee*) with a number of imino esters. The authors proposed that a tridentate SCS coordination of these C_2-symmetric ligands onto the silver center was key for obtaining the good chiral induction observed in the cycloaddition.

In a comparative study encompassing NHC complexes of all three coinage metals, Pérez and Echavarren reported that [(IPr)AgCl] and [(SIPr)AgCl] were able to mediate the cyclopropanation reaction between styrene and phenyldiazoacetate, albeit in lower yield than their copper and gold analogues.[81]

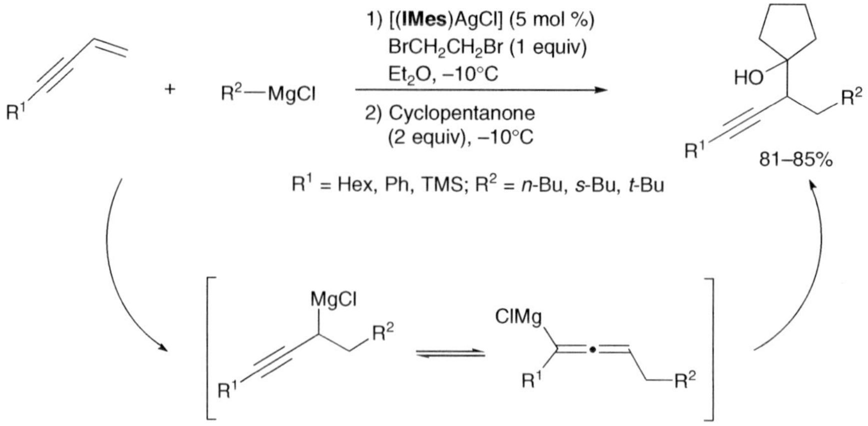

Scheme 11.5 NHC–Ag-catalyzed carbomagnesiation.

Preliminary results by Biffis and co-workers on the catalytic activity of a number of bis- and tris-NHC silver complexes showed that they could promote the Sonogashira coupling of phenylacetylene and 4-iodoacetophenone under rather harsh conditions.[82] The authors also observed noticeable amounts of halide homocoupling in some reactions. Somewhat related to cross-coupling, the carbomagnesiation of terminal alkenes was shown to proceed efficiently in the presence of [(IMes)AgCl] and 1,2-dibromoethane as oxidant (Scheme 11.5).[83] Remarkably, the use of conjugated enynes proved highly regioselective—despite the propargyl-allene equilibrium of the formed Grignard—provided an electrophile was used to trap the intermediate.

Finally, two reports describing the use of polymer-supported NHC–Ag compounds in three-component aldehyde–alkyne–amine reaction and Suzuki coupling have been published.[84]

11.4 NHC–Au in Catalysis

It is only in recent years, following what is commonly referred to as the *Gold Rush* in the scientific literature, that NHC–Au complexes have gained popularity and have seen their potential unfold. The first application in catalysis of a NHC–gold species appeared in 2003,[85] but the real surge in publication only started in 2006. Since then, there is hardly a week without a report on the activity of NHC–Au catalysts in organic synthesis is published. The field is dominated by two main topics of crucial importance in modern catalysis, namely cycloisomerization of polyunsaturated substrates and hydrofunctionalization of π-bonds. NHC–Au species have been the subject of several reviews both in organometallic chemistry[86] and in catalysis,[87] therefore, while presenting exhaustive referencing, only an overview of key transformations is provided here with an emphasis on very recent reports.

11.4.1 Enyne Cycloisomerization and Related Reactions

Interestingly, and rather surprisingly, NHC–Au catalysts have rarely been used for 'pure' enyne cycloisomerization (*i.e.* the cyclization/rearrangement of an enyne without other functionalities than the alkene and the alkyne being involved).[88,89] They have mainly found applications in cascade reactions taking advantage of the reactivity of the enyne–Au adduct after the initial enyne cycloisomerization step. In this context, 1,5- and 1,6-enynes are the most widely studied substrates and give rise to a variety of products upon different types of *trapping*.

This was notably exploited by Echavarren with the intermolecular trapping of the cyclopropyl auracarbene intermediate with olefins[90] and carbonyl derivatives,[91] leading respectively to **15** and **16**, and by Toste with the oxidation of this carbene moiety to give **17** (Scheme 11.6).[92] Similarly, Chung reported an interesting intramolecular bis-cyclopropanation in dienyne systems, leading to strained carbocycles.[93] Subsequent studies by the same group with 1,5-bisallenes allowed for the synthesis of compounds possessing the rather uncommon bicyclo[3.1.1]heptane framework.[94] In all these transformations, IPr and IMes are generally the NHCs of choice.

The most recent results in that area include trapping with 1,3-dicarbonyl compounds, electron-rich aromatics[95] and with alcohols.[96] In the latter case, Tomioka and co-workers examined the activity of a number of chiral NHCs **18–22** (Scheme 11.7). Even though chiral inductions were limited, this report provided the basis for future design of chiral NHCs in gold catalysis.

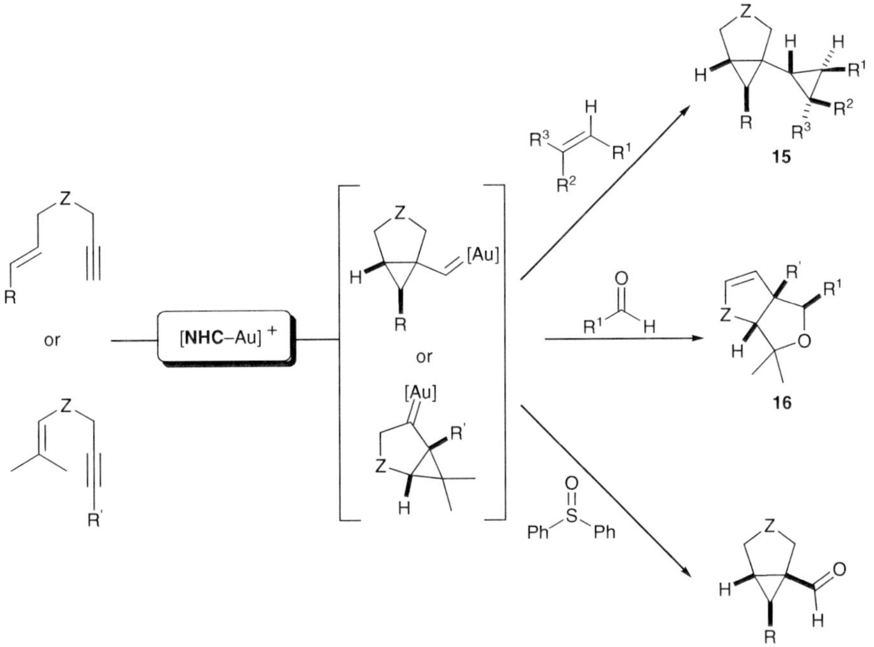

Scheme 11.6 NHC–Au-catalyzed cascade reactions initiated by enyne cycloisomerization.

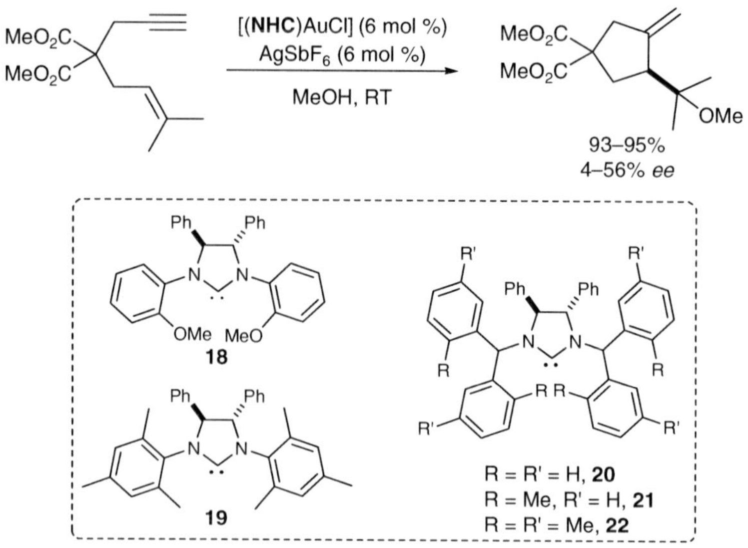

[(**NHC**)AuCl] (6 mol %)
$AgSbF_6$ (6 mol %)
MeOH, RT

93–95%
4–56% *ee*

R = R' = H, **20**
R = Me, R' = H, **21**
R = R' = Me, **22**

Scheme 11.7 Activity of chiral NHCs in enantioselective alkoxy-enyne cycloisomerization.

Another subfield of enyne cycloisomerization chemistry is best described as cycloaddition reactions. For instance, allenedienes were found to react smoothly in the presence of an IPr-based gold(I) cationic species to produce the formal [4 + 3] cycloadducts (Equation (11.8)).[97] This method is a valuable approach to fused seven-membered ring systems and was applied to a transannular reaction, allowing for the construction of a pentacyclic system.[98] Of note, Toste showed that, by replacing the NHC on gold by a triarylphosphite, an alternative [4 + 2] cycloaddition was observed from the same linear allenedienes.[99]

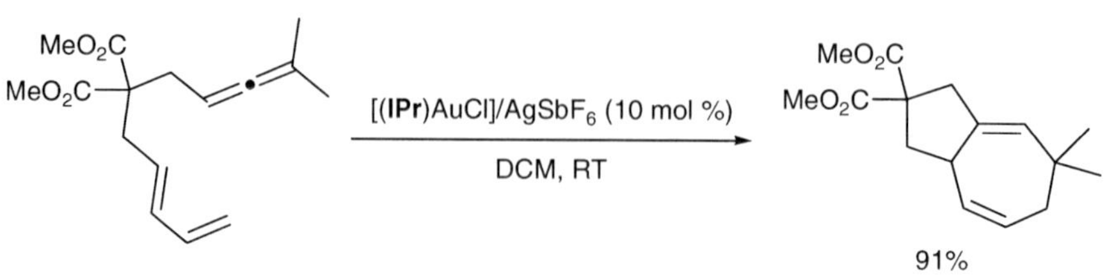

[(**IPr**)AuCl]/$AgSbF_6$ (10 mol %)
DCM, RT

91%

(11.8)

Even though not strictly speaking belonging to enyne cycloisomerization, an alkyne and an alkene can react intermolecularly with each other provided the alkyne is properly activated. Zhang and co-workers reported a number of such cycloadditions where a gold-mediated cascade was initiated by carbonyl addition to the alkyne moiety. The resulting 1,4- or 1,3-dipoles could then participate to formal [4 + 2] and [3 + 2] cycloadditions with an olefinic partner, giving access to

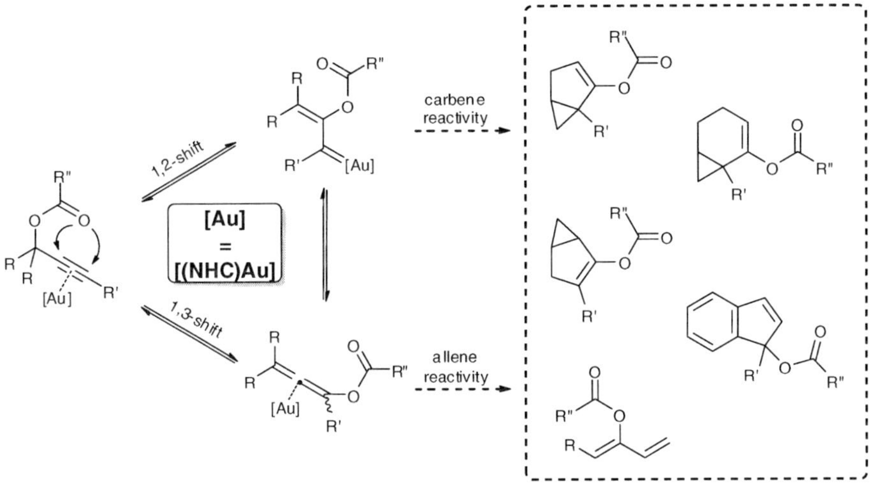

Scheme 11.8 Diverse reactivity of propargylic esters with NHC-Au catalysts.

interesting bicyclic scaffolds.[100] The IPr ligand systematically performed best in all these cycloadditions when compared to other NHCs or phosphines.

Often involved in cycloisomerization-type reactions, propargylic esters nevertheless have to be considered as a particular class of substrates.[101] This distinction stems from their unique ability to undergo 1,2- or 1,3-ester migration upon gold activation. NHC–Au catalysts have proven particularly efficient in this area and allowed accessing both types of reactivity of these systems (*i.e.* carbene-type after 1,2-shift and allene-type after 1,3-shift (Scheme 11.8). The use of NHC–Au complexes with propargylic esters led to efficient synthesis of cyclopropyl-fused bicycles,[102] conjugated dienes,[103] and indenes.[104] Detailed mechanistic questions related to the order of events in the cascade OAc-migration/olefin cyclopropanation and to product selectivity were also the subject of numerous experimental and theoretical studies with NHC–Au species.[102]

Propargylic tosylates were recently added to the family of viable migrating groups. González and co-workers were able to engage these substrates in different inter- and intramolecular cascade reactions, thereby expanding the scope of both NHC–Au catalysts and propargyl-activated alkynes.[105] Finally, it should be noted that *allylic* esters have also been investigated and shown to undergo [3,3] rearrangement.[106]

11.4.2 Hydrofunctionalization of π-Bonds

The addition of H–X bonds (X = C, N, O, F) across unsaturated C–C bonds has become a major field of research with gold catalysts,[107] and the use of NHCs as supporting ligands has proven beneficial in many instances.

In 2003, Herrmann reported the hydration of alkynes with a NHC–Au^I compound.[85,108] This was the first catalytic application involving NHCs in gold

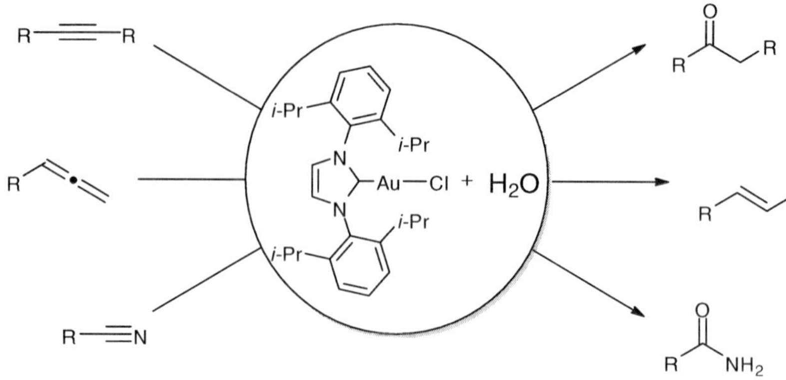

Scheme 11.9 Various hydration reactions catalyzed by [(IPr)AuCl].

chemistry. Following this seminal work, a recent study revealed the impressive efficiency of IPr in this reaction. Hence, a catalytic system comprising [(IPr)AuCl] and $AgSbF_6$ was found active at very low loadings (down to 10 ppm) with a variety of alkynes and under acid-free conditions.[109] Interestingly, preliminary mechanistic studies point towards two distinct pathways as a function of the solvent employed (methanol or dioxane). Far less common substrates than alkynes, allenes were shown to participate in NHC–Au-catalyzed hydration reactions, furnishing allylic alcohols in moderate to good yields and good regioselectivity.[110] The very good activity observed with the IPr ligand in hydration was then examined with organonitriles (Scheme 11.9). Remarkably, this reaction, leading to primary amides, had not precedent with gold catalyst prior to this study and represents the very first activation of the C≡N bond in gold catalysis.[111]

The same catalytic system also performed well in the hydroalkoxylation of allenes where the C–O bond was formed at the least hindered terminal carbon of the allene.[112] This report was later the subject of a DFT study by Maseras and Patton.[113] Interestingly, it was found that the addition of the alcohol likely occurred, in fact, at the most hindered carbon and was followed by a second hydroalkoxylation of the allylic ether and subsequent elimination, the whole process accounting for the observed regioselectivity. This second addition could nevertheless be impeded through appropriate choice of reaction conditions (*i.e.* DMF as solvent at 0 °C with 10 equivalents of the alcohol), leading to the formation of the most substituted ether.[114]

Hydroamination of π-bonds is one of the most straightforward methods for the construction of C–N bonds and, as such, has attracted a lot of attention. NHC–Au catalysts, in line with results obtained in hydration and hydroalkoxylation reactions (*vide supra*), proved highly efficient in this field and the inter- and intramolecular hydroamination of various alkenes,[115] allenes,[116] and alkynes[117] were reported with a number of NHC–Au complexes. Among these reports, Widenhoefer published an elegant bis-hydroamination of allenes, leading to fused imidazolidinones, where IPr proved much more selective than an *ortho*-biarylphosphine ligand.[118] Recently, Bertrand and co-workers disclosed an extremely efficient alkyne and allene hydroamination catalyzed by a CAAC–Au complex **23** (CAAC = cyclic alkyl(amino)carbene) (Equation (11.9)). These peculiar NHCs allowed for the challenging use of ammonia[119] and proved useful in a cascade hydroamination/hydroarylation leading to 1,2-dihydroquinolines.[120]

$$R{-}C{\equiv}C{-}R' + NH_3 \xrightarrow[\text{Toluene, 90°C; R, R}^1 = \text{H, Aryl, Alkyl}]{[(\mathbf{23})\text{AuCl}]/\text{KB}(C_6F_5)_4\ (1\text{–}5\ \text{mol \%})} R{-}C({=}NH){-}CH_2{-}R' \quad (11.9)$$

23

Even though less studied than the *O*- and *N*-nucleophiles, other important H–X bonds can participate in hydrofunctionalization catalyzed by gold. Hence, the intermolecular hydroarylation (*i.e.* C^{sp2}–H bond addition) of allenes with indoles was reported in the presence of [(IPr)AuCl].[121] Using the same catalyst, Che and co-workers investigated an interesting cascade terminated by an alkyne hydroarylation step.[122] It should also be noted that the intermolecular hydroarylation of alkynes with indoles was used by Echavarren in the synthesis of the tetracyclic core skeleton of lundurine molecules.[123] Unfortunately, the selectivity of the cyclization (*i.e.* 7-*exo vs.* 8-*endo*) was not satisfactory with an NHC ligand.

Finally, hydrofluorination of alkynes could also be performed with NHC–Au catalysts. Sadighi and co-workers reported the use of a chlorinated IPr ligand in **24** as efficient ligand in this challenging reaction (Equation (11.10)).[124] Subsequent studies by Miller improved the scope and the regioselectivity of the reaction by appending a directing group at the homopropargylic position.[125]

$$R{-}C{\equiv}C{-}R' \xrightarrow[\text{DCM, RT}]{\mathbf{24}/\text{AgBF}_4\ (2.5\ \text{mol \%});\ \text{PhNMe}_2{\cdot}\text{HOTf}\ (10\ \text{mol \%});\ \text{Et}_3\text{N}{\cdot}3\text{HF}\ (1.5\ \text{equiv})/\text{KHSO}_4\ (1\ \text{equiv})} (R)(H)C{=}C(F)(R') \quad 63\text{–}86\% \quad (11.10)$$

Ar = 2,6-(*i*-Pr)$_2$-C_6H_3

24

11.4.3 Miscellaneous Reactions

It would be vain to detail here—mainly for reasons of space—every application for which a NHC–Au catalytic system was used. Nevertheless, a number of cutting-edge discoveries, advancing organic chemistry beyond the 'simple' field of NHC–Au catalysis deserve to be mentioned.

An impressive synthesis of allenes has notably been reported by the group of Bertrand. It consists in the condensation of an enamine and a terminal alkyne mediated by a cationic gold catalyst bearing a CAAC ligand.[126] Recently, taking advantage of the excellent activity of CAACs in alkyne hydroamination (see Equation (11.9)), the same group developed a one-pot procedure where an internal alkyne reacted with a secondary amine furnishing the enamine, which in turn reacted with a terminal alkyne to yield the expected allene (Equation (11.11)).[127]

$$R^1{-}{\equiv}{-}R^2 + Bn_2NH + R^3{-}{\equiv} \xrightarrow{\mathbf{23}\ (5\ mol\ \%),\ 120°C} R^3CH{=}C{=}C(R^1)R^2 \qquad (11.11)$$

The use of diazo compounds has also attracted some attention. NHC–Au catalysts, and especially those bearing the IPr and IMes ligands, mediated the decomposition of diazo species, leading to organogold carbenes, which could be observed in the gas phase.[128] Once formed, the NHC–Au carbene could participate in a variety of reactions that include insertions into X–H bonds (X = C, N, O),[129] Buchner reactions,[130] and olefin cyclopropanation.[81]

Generation of gold carbene was also shown to proceed *via* internal redox processes involving homopropargylic sulfoxides (Scheme 11.10). The carbenoid can then be trapped by insertion into an aryl C–H bond[131] or alkyl 1,2-migration.[132]

Somewhat similar reactivity was also observed with oximes and nitrones[133] and, still related to internal redox processes, the Meyer–Schuster-type conversion of propargylic alcohols into conjugated enones was found to proceed well with NHC–Au catalysts.[134]

Finally, two fields of homogeneous catalysis in which gold has barely proven to be active are oxidation and reduction reactions. Nevertheless, preliminary studies seem to show that NHCs are well suited for these transformations. The group of Corma, in line with their investigation on NHC–Au-catalyzed cross-couplings,[135] demonstrated that olefin hydrogenation could be performed with NHC–Au compounds.[136] Remarkably, they recently extended their findings to asymmetric hydrogenation using proline-derived pincer NHC ligands on a gold(III) center.[137] Of note, the reductive diboration of olefins was also reported.[79] As for oxidation reactions catalyzed by NHC–Au complexes, it

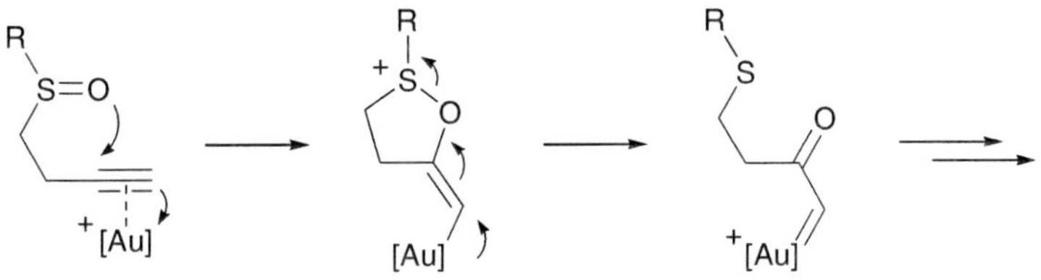

Scheme 11.10 Generation of gold carbenoid *via* internal redox of homopropargylic sulfoxides.

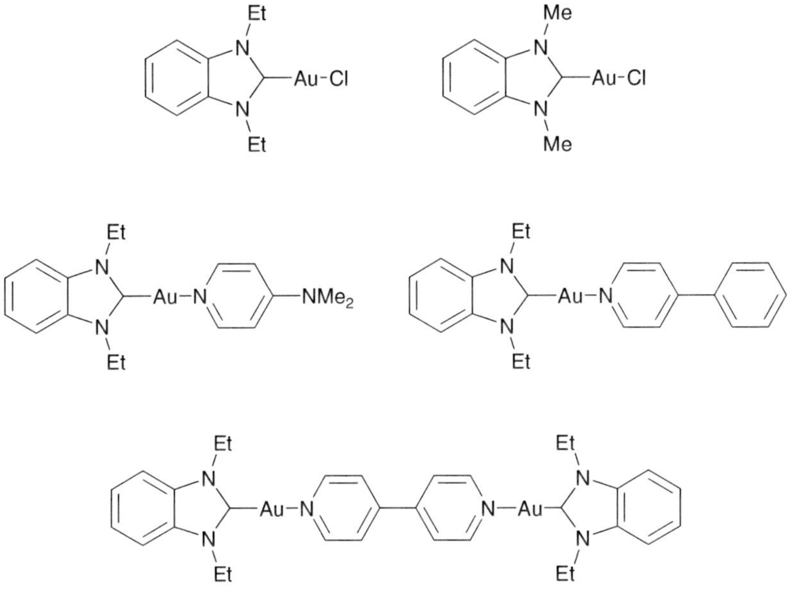

Figure 11.4 Structures of NHC–Au complexes active in oxidation reactions.

consists only for the moment in the oxidation of benzyl alcohol to benzaldehyde, the active catalysts are shown in Figure 11.4.[138]

11.5 Outlook

The use of NHC–Cu complexes in catalysis is still in its first decade but these catalysts have already reached a level of efficiency, notably in key C–C bond forming enantioselective reactions, which makes copper the leader of the Group 11 triad. In addition, NHC–Cu species have also shown impressive properties in other fields than catalysis (biochemistry, materials, *etc.*),[139] making these compounds all the more appealing for further studies.

Silver complexes containing NHC ligands seem to have interested—this far—more organometallic chemists than synthetic ones. In this area, everything remains to be tried and the expectations are immense.

As for gold, the development of its NHC chemistry is intimately linked to the amazing surge in reports on gold homogeneous catalysis in general. It should be noted that this is perhaps one of the few domains of catalysis where NHCs do not have to catch up with phosphanes, because both types of ligands are being developed concomitantly. In addition to catalysis, NHC–Au species have greatly contributed to gain a better understanding of NHC–metal interactions,[10] and are now center to the developing bonding model for gold carbenoids,[140] as well as the isolation of relevant catalytic intermediates.[141]

Overall, the NHC chemistry of Group 11 metals is clearly less investigated than with other transition metals, especially those of Groups 8, 9 and 10. Nevertheless, it is currently being developed at an extremely rapid pace and, given notably the low cost and toxicity of Group 11 metals, it makes no doubt that this field will see a lot more discoveries in the coming years.

Acknowledgement

Professor Gregory C. Fu is acknowledged for his support during the redaction of this work.

References and notes

1. Even though its electronic configuration would allow roentgenium (Rg) to be part of Group 11, its unstable nature (short-lived transactinide with a 3.6 s half-life) precludes such classification.
2. As a testimony to this recent and overwhelming interest, a special issue of *Chemical Reviews*, guest-edited by Bruce Lipshutz and Yoshinori Yamamoto, was dedicated to *Coinage Metals in Organic Synthesis*, see: *Chem. Rev.*, 2008, **108**, issue 8.
3. For reviews about the properties of NHCs, see: (a) P. de Frémont, N. Marion and S. P. Nolan, *Coord. Chem. Rev.*, 2009, **253**, 862–872; (b) H. Jacobsen, A. Correa, A. Poater, C. Costabile and L. Cavallo, *Coord. Chem. Rev.*, 2009, **253**, 687–703; (c) U. Radius and F. M. Bickelhaupt, *Coord. Chem. Rev.*, 2009, **253**, 678–686; (d) F. E. Hahn and M. C. Jahnke, *Angew. Chem., Int. Ed.*, 2008, **47**, 3122–3172; (e) S. Díez-González and S. P. Nolan, *Coord. Chem. Rev.*, 2007, **251**, 874–883.
4. Most recent general reviews about NHCs: (a) *Carbenes*, ed. A. J. Arduengo and G. Bertrand, *Chem. Rev.*, 2009, **109**, issue 8; (b) *N-Heterocyclic Carbenes in Transition Metal Catalysis*, ed. F. Glorius, *Top. Organomet. Chem., Vol. 21*, Springer-Verlag, Berlin/Heidelberg, 2007; (*c*) *N-Heterocyclic Carbenes in Synthesis*, ed. S. P. Nolan, Wiley-VCH, New York, 2006.

5. (a) K. M. Hindi, M. J. Panzner, C. A. Tessier, C. L. Cannon and W. J. Youngs, *Chem. Rev.*, 2009, **109**, 3859–3884; (b) S. Díez-González, N. Marion and S. P. Nolan, *Chem. Rev.*, 2009, **109**, 3612–3676; (c) J. C. Y. Lin, R. T. W. Huang, C. S. Lee, A. Bhattacharyya, W. S. Hwang and I. J. B. Lin, *Chem. Rev.*, 2009, **109**, 3561–3598; (d) P. J. Pérez and M. M. Díaz-Requejo, in *N-Heterocyclic Carbenes in Synthesis*, ed. S. P. Nolan, Wiley-VCH, New York, 2006, pp. 257–274.
6. J. Hassan, M. Sévignon, C. Gozzi, E. Schulz and M. Lemaire, *Chem. Rev.*, 2002, **102**, 1359–1470.
7. A. J. Arduengo III, H. V. R. Dias, J. C. Calabrese and F. Davidson, *Organometallics*, 1993, **12**, 3405–3409.
8. (a) A. J. Arduengo III, R. L. Harlow and M. Kline, *J. Am. Chem. Soc.*, 1991, **113**, 361–363. For the first isolation of a carbene, which was not a NHC, see: (b) A. Igau, H. Grutzmacher, A. Baceiredo and G. Bertrand, *J. Am. Chem. Soc.*, 1988, **110**, 6463–6466.
9. (a) H. G. Raubenheimer, S. Cronje, P. J. Olivier, J. G. Toerien and P. H. van Rooyen, *Angew. Chem., Int. Ed. Engl.*, 1994, **33**, 672–673; (b) H. G. Raubenheimer, S. Cronje and P. J. Olivier, *J. Chem. Soc., Dalton Trans.*, 1995, 313–316; (c) A. A. D. Tulloch, A. A. Danopoulos, S. Kleinhenz, M. E. Light, M. B. Hursthouse and G. Eastham, *Organometallics*, 2001, **20**, 2027–2031.
10. C. Boehme and G. Frenking, *Organometallics*, 1998, **17**, 5801–5809.
11. P. K. Fraser and S. Woodward, *Tetrahedron Lett.*, 2001, **42**, 2747–2749.
12. (a) S. Díez-González and S. P. Nolan, *Aldrichimica Acta*, 2008, **41**, 43–51; (b) S. Díez-González and S. P. Nolan, *Synlett*, 2007, 2158–2167.
13. For selected references, see: (a) F. Guillen, C. L. Winn and A. Alexakis, *Tetrahedron: Asymmetry*, 2001, **12**, 2083–2086; (b) J. Pytkowicz, S. Roland and P. Mangeney, *Tetrahedron: Asymmetry*, 2001, **12**, 2087–2089; (c) A. Alexakis, C. L. Winn, F. Guillen, J. Pytkowicz, S. Roland and P. Mangeney, *Adv. Synth. Catal.*, 2003, **345**, 345–348.
14. For a review on chiral NHCs, see: V. César, S. Bellemin-Laponnaz and L. H. Gade, *Chem. Soc. Rev.*, 2004, **33**, 619–636.
15. For selected references, see: (a) P. L. Arnold, M. Rodden, K. M. Davis, A. C. Scarisbrick, A. J. Blake and C. Wilson, *Chem. Commun.*, 2004, 1612–1613; (b) C. L. Winn, F. Guillen, J. Pytkowicz, S. Roland, P. Mangeney and A. Alexakis, *J. Organomet. Chem.*, 2005, **690**, 5672–5692; (c) H. Clavier, L. Contable, L. Toupec, J.-C. Guillemin and M. Mauduit, *J. Organomet. Chem.*, 2005, **690**, 5237–5254; (d) J. J. Van Veldhuizen, J. E. Campbell, R. E. Giudici and A. H. Hoveyda, *J. Am. Chem. Soc.*, 2005, **127**, 6877–6882.
16. K.-S. Lee, M. K. Brown, A. W. Hird and A. H. Hoveyda, *J. Am. Chem. Soc.*, 2006, **128**, 7182–7184.
17. A number of ligands belonging to the family of hydroxyalkyl and hydroxyaryl NHCs were synthesized and evaluated in the last five years, for a review see: J. Wencel, M. Mauduit, H. Hénon, S. Kehrli and A. Alexakis, *Aldrichimica Acta*, 2009, **42**, 43–50.

18. Selected examples: (a) T. Uchida and T. Katsuki, *Tetrahedron Lett.*, 2009, **50**, 4741–4743; (b) M. Okamoto, Y. Yamamoto and S. Sakaguchi, *Chem. Commun.*, 2009, 7363–7365.
19. D. Martin, S. Kehrli, M. d'Augustin, H. Clavier, M. Mauduit and A. Alexakis, *J. Am. Chem. Soc.*, 2006, **128**, 8416–8417.
20. For a related study, see: Y. Matsumoto, K.-i. Yamada and K. Tomioka, *J. Org. Chem.*, 2008, **73**, 4578–4581.
21. H. Hénon, M. Mauduit and A. Alexakis, *Angew. Chem., Int. Ed.*, 2008, **47**, 9122–9124.
22. For R_2Zn, see: (a) M. K. Brown, T. L. May, C. A. Baxter and A. H. Hoveyda, *Angew. Chem., Int. Ed.*, 2007, **46**, 1097–1100. For R_3Al, see: (b) T. L. May, M. K. Brown and A. H. Hoveyda, *Angew. Chem., Int. Ed.*, 2008, **47**, 7358–7362.
23. For enones, see: (a) K.-s. Lee and A. H. Hoveyda, *J. Org. Chem.*, 2009, **74**, 4455–4462. For allylic epoxides, see: (b) J. R. Herron, V. Russo, E. J. Valente and Z. T. Ball, *Chem.–Eur. J.*, 2009, **15**, 8713–8716.
24. (a) C. Munro-Leighton, E. D. Blue and T. B. Gunnoe, *J. Am. Chem. Soc.*, 2006, **128**, 1446–1447; (b) L. A. Goj, E. D. Blue, S. A. Delp, T. B. Gunnoe, T. R. Cundari, A. W. Pierpont, J. L. Petersen and P. D. Boyle, *Inorg. Chem.*, 2006, **45**, 9032–9045.
25. C. Munro-Leighton, S. A. Delp, E. D. Blue and T. B. Gunnoe, *Organometallics*, 2007, **26**, 1483–1493.
26. C. Munro-Leighton, S. A. Delp, N. M. Alsop, E. D. Blue and T. B. Gunnoe, *Chem. Commun.*, 2008, 111–113.
27. (a) R. E. Douthwaite, *Coord. Chem. Rev.*, 2007, **251**, 702–717; (b) M. Mauduit and H. Clavier, in *N-Heterocyclic Carbenes in Synthesis*, ed. S. P. Nolan, Wiley-VCH, New York, 2006, pp. 183–222; (c) See also ref. 14.
28. S. Tominaga, Y. Oi, T. Kato, D. K. An and S. Okamoto, *Tetrahedron Lett.*, 2004, **45**, 5585–5588.
29. D. G. Gillingham and A. H. Hoveyda, *Angew. Chem., Int. Ed.*, 2007, **46**, 3860–3864.
30. M. A. Kacprzynski, T. L. May, S. A. Kazane and A. H. Hoveyda, *Angew. Chem., Int. Ed.*, 2007, **46**, 4554–4558.
31. (a) Y. Lee, K. Akiyama, D. G. Gillingham, M. K. Brown and A. H. Hoveyda, *J. Am. Chem. Soc.*, 2008, **130**, 446–447; (b) K. Akiyama, F. Gao and A. H. Hoveyda, *Angew. Chem., Int. Ed.*, 2010, **49**, 419–423.
32. V. Jurkauskas, J. P. Sadighi and S. L. Buchwald, *Org. Lett.*, 2003, **5**, 2417–2420.
33. S. Díez-González and S. P. Nolan, *Acc. Chem. Res.*, 2008, **41**, 349–358. See also ref. 12.
34. H. Kaur, F. K. Zinn, E. D. Stevens and S. P. Nolan, *Organometallics*, 2004, **23**, 1157–1160.
35. S. Díez-González, H. Kaur, F. K. Zinn, E. D. Stevens and S. P. Nolan, *J. Org. Chem.*, 2005, **70**, 4784–4796.
36. For selected examples, see: (a) J. Yun, D. Kim and H. Yun, *Chem. Commun.*, 2005, 5181–5183; (b) B. Bantu, D. Wang, K. Wurst and M. R.

Buchmeiser, *Tetrahedron*, 2005, **61**, 12145–12152; (c) M. Nonnenmacher, D. Kunz and F. Rominger, *Organometallics*, 2008, **27**, 1561–1568; (d) J. Peng, L. Chen, Z. Xu, Y. Hu, J. Li, Y. Bai, H. Qiu and G. Lai, *Chin. J. Chem.*, 2009, **27**, 2121–2124.

37. M. B. T. Thuong, S. Sottocornola, G. Prestat, G. Broggini, D. Madec and G. Poli, *Synlett*, 2007, 1521–1524.
38. (a) M. J. Beier, W. Knolle, A. Prager-Duschke and M. R. Buchmeiser, *Macromol. Rapid Commun.*, 2008, **29**, 904–909; (b) G. M. Pawar, B. Bantu, J. Weckesser, S. Blechert, K. Wurst and M. R. Buchmeiser, *Dalton Trans.*, 2009, 9043–9051.
39. S. Díez-González, N. M. Scott and S. P. Nolan, *Organometallics*, 2006, **25**, 2355–2358.
40. S. Díez-González, E. D. Stevens, N. M. Scott, J. L. Petersen and S. P. Nolan, *Chem.–Eur. J.*, 2008, **14**, 158–168.
41. C. Deutsch, B. H. Lipshutz and N. Krause, *Angew. Chem., Int. Ed.*, 2007, **46**, 1650–1653.
42. C. Deutsch, B. H. Lipshutz and N. Krause, *Org. Lett.*, 2009, **11**, 5010–5012.
43. Sadighi reported the stoichiometric hydrocupration of 3-hexyne with a dimeric $[(NHC)CuH]_2$, see: N. P. Mankad, D. S. Laitar and J. P. Sadighi, *Organometallics*, 2004, **23**, 3369–3371.
44. D. S. Laitar, E. Y. Tsui and J. P. Sadighi, *J. Am. Chem. Soc.*, 2006, **128**, 11036–11037.
45. H. Zhao, L. Dang, T. B. Marder and Z. Lin, *J. Am. Chem. Soc.*, 2008, **130**, 5586–5594.
46. M. A. Beenen, C. An and J. A. Ellman, *J. Am. Chem. Soc.*, 2008, **130**, 6910–6911.
47. D. S. Laitar, E. Y. Tsui and J. P. Sadighi, *Organometallics*, 2006, **25**, 2405–2408.
48. A. Bonet, V. Lillo, J. Ramírez, M. M. Díaz-Requejo and E. Fernández, *Org. Biomol. Chem.*, 2009, **7**, 1533–1535.
49. V. Lillo, A. Prieto, A. Bonet, M. M. Díaz-Requejo, J. Ramírez, P. J. Pérez and E. Fernández, *Organometallics*, 2009, **28**, 659–662.
50. V. Lillo, M. R. Fructos, J. Ramírez, A. A. C. Braga, F. Maseras, M. M. Díaz-Requejo, P. J. Pérez and E. Fernández, *Chem.–Eur. J.*, 2007, **13**, 2614–2621.
51. L. Dang, H. Zhao, Z. Lin and T. B. Marder, *Organometallics*, 2008, **27**, 1178–1186.
52. Y. Lee and A. H. Hoveyda, *J. Am. Chem. Soc.*, 2009, **131**, 3160–3161.
53. Y. Lee, H. Jang and A. H. Hoveyda, *J. Am. Chem. Soc.*, 2009, **131**, 18234–18235.
54. For a thorough review, see: E. A. B. Kantchev, C. J. O'Brien and M. G. Organ, *Angew. Chem., Int. Ed.*, 2007, **46**, 2768–2813.
55. J. Haider, K. Kunz and U. Scholz, *Adv. Synth. Catal.*, 2004, **346**, 717–722.
56. (a) C. Tubaro, A. Biffis, E. Scattolin and M. Basato, *Tetrahedron*, 2008, **64**, 4187–4195; (b) A. Biffis, C. Tubaro, E. Scattolin, M. Basato, G. Papini, C. Santini, E. Alvarez and S. Conejero, *Dalton Trans.*, 2009, 7223–7229.

57. G. G. Dubinina, H. Furutachi and D. A. Vicic, *J. Am. Chem. Soc.*, 2008, **130**, 8600–8601.
58. G. G. Dubinina, J. Ogikubo and D. A. Vicic, *Organometallics*, 2008, **27**, 6233–6235.
59. A. Grandbois, M.-E. Mayer, M. Bédard, S. K. Collins and T. Michel, *Chem.–Eur. J.*, 2009, **15**, 9655–9659.
60. (a) M. R. Fructos, T. R. Belderrain, M. C. Nicasio, S. P. Nolan, H. Kaur, M. M. Díaz-Requejo and P. J. Pérez, *J. Am. Chem. Soc.*, 2004, **126**, 10846–10847; (b) R. E. Gawley and S. Narayan, *Chem. Commun.*, 2005, 5109–5111.
61. M. R. Fructos, P. de Frémont, S. P. Nolan, M. M. Díaz-Requejo and P. J. Pérez, *Organometallics*, 2006, **25**, 2237–2241.
62. (a) Q. Xu and D. H. Appella, *Org. Lett.*, 2008, **10**, 1497–1500; (b) S. Simonovic, A. C. Whitwood, W. Clegg, R. W. Harrington, M. B. Hursthouse, L. Male and R. E. Douthwaite, *Eur. J. Inorg. Chem.*, 2009, 1786–1795.
63. (a) B. M. Trost and G. Dong, *J. Am. Chem. Soc.*, 2006, **128**, 6054–6055; (b) R. Liu, S. R. Herron and S. A. Fleming, *J. Org. Chem.*, 2007, **72**, 5587–5591.
64. E. J. Park, S. H. Kim and S. Chang, *J. Am. Chem. Soc.*, 2008, **130**, 17268–17269.
65. (a) H. Lebel, M. Davi, S. Díez-González and S. P. Nolan, *J. Org. Chem.*, 2007, **72**, 144–149; (b) H. Lebel, C. Ladjel and L. Bréthous, *J. Am. Chem. Soc.*, 2007, **129**, 13321–13326; (c) H. Lebel and M. Parmentier, *Org. Lett.*, 2007, **9**, 3563–3566.
66. (a) S. Díez-González, A. Correa, L. Cavallo and S. P. Nolan, *Chem.–Eur. J.*, 2006, **12**, 7558–7564; (b) S. Díez-González and S. P. Nolan, *Angew. Chem., Int. Ed.*, 2008, **47**, 8881–8884.
67. S. Díez-González, E. D. Stevens and S. P. Nolan, *Chem. Commun.*, 2008, 4747–4749.
68. D. Benito-Garagorri, V. Bocokic and K. Kirchner, *Tetrahedron Lett.*, 2006, **47**, 8641–8644.
69. (a) J. A. Bull, M. G. Hutchings and P. Quayle, *Angew. Chem., Int. Ed.*, 2007, **46**, 1869–1872; (b) J. A. Bull, M. G. Hutchings, C. Luján and P. Quayle, *Tetrahedron Lett.*, 2008, **49**, 1352–1356.
70. (a) J. R. Herron and Z. T. Ball, *J. Am. Chem. Soc.*, 2008, **130**, 16486–16487; (b) V. Russo, J. R. Herron and Z. T. Ball, *Org. Lett.*, 2010, **12**, 220–223.
71. (a) D. S. Laitar, P. Müller and J. P. Sadighi, *J. Am. Chem. Soc.*, 2005, **127**, 17196–17197; (b) H. Zhao, Z. Lin and T. B. Marder, *J. Am. Chem. Soc.*, 2006, **128**, 15637–15643.
72. (a) T. Ohishi, M. Nishiura and Z. Hou, *Angew. Chem., Int. Ed.*, 2008, **47**, 5792–5795. For a DFT study of the catalytic cycle, see also: (b) L. Dang, Z. Lin and T. B. Marder, *Organometallics*, 2010, **29**, 917–927.
73. S. Zheng, F. Li, J. Liu and C. Xia, *Tetrahedron Lett.*, 2007, **48**, 5883–5886.

74. J. Liu, R. Zhang, S. Wang, W. Sun and C. Xia, *Org. Lett.*, 2009, **11**, 1321–1324.
75. I. J. B. Lin and C. S. Vasam, *Coord. Chem. Rev.*, 2007, **251**, 642–670.
76. H. M. J. Wang and I. J. B. Lin, *Organometallics*, 1998, **17**, 972–975.
77. Additionally, two reports are worth mentioning: (a) Stradiotto reported a single example of successful hydrosilylation of benzaldehyde using AgOTf and IPr, see: B. M. Wile and M. Stradiotto, *Chem. Commun.*, 2006, 4104–4106; (b) Echavarren described the unsuccessful use of $[(IMes)_2Ag][AgCl_2]$ in the carbostannylation of alkynes, see: S. Porcel and A. M. Echavarren, *Angew. Chem., Int. Ed.*, 2007, **46**, 2672–2676.
78. J. Ramírez, R. Corberán, M. Sanaú, E. Peris and E. Fernandez, *Chem. Commun.*, 2005, 3056–3058.
79. R. Corberán, J. Ramírez, M. Poyatos, E. Peris and E. Fernandez, *Tetrahedron: Asymmetry*, 2006, **17**, 1759–1762.
80. J. Iglesias-Sigüenza, A. Ros, E. Díez, A. Magriz, A. Vázquez, E. Álvarez, R. Fernández and J. M. Lassaletta, *Dalton Trans.*, 2009, 8485–8488.
81. A. Prieto, M. R. Fructos, M. M. Díaz-Requejo, P. J. Pérez, P. Pérez-Galán, N. Delpont and A. Echavarren, *Tetrahedron*, 2009, **65**, 1790–1793.
82. A. Biffis, G. C. Lobbia, G. Papini, M. Pellei, C. Santini, E. Scattolin and C. Tubaro, *J. Organomet. Chem.*, 2008, **693**, 3760–3766.
83. Y. Fujii, J. Terao and N. Kambe, *Chem. Commun.*, 2009, 1115–1117.
84. (a) P. Li, L. Wang, Y. Zhang and M. Wang, *Tetrahedron Lett.*, 2008, **49**, 6650–6654; (b) X. Zeng, T. Zhang, Y. Qin, Z. Wei and M. Luo, *Dalton Trans.*, 2009, 8341–8348.
85. S. K. Schneider, W. A. Herrmann and E. Herdtweck, *Z. Anorg. Allg. Chem.*, 2003, **629**, 2363–2370.
86. (a) H. G. Raubenheimer and S. Cronje, *Chem. Soc. Rev.*, 2008, **37**, 1998–2011; (b) I. J. B. Lin and C. S. Vasam, *Can. J. Chem.*, 2005, **83**, 812–825.
87. N. Marion and S. P. Nolan, *Chem. Soc. Rev.*, 2008, **37**, 1776–1782. See also ref. 5b, section 6.3.
88. For a recent comprehensive review, see: V. Michelet, P. Y. Toullec and J.-P. Genêt, *Angew. Chem., Int. Ed.*, 2008, **47**, 4268–4315.
89. L. Ricard and F. Gagosz, *Organometallics*, 2007, **26**, 4704–4707.
90. S. López, E. Herrero-Gómez, P. Pérez-Galán, C. Nieto-Oberhuber and A. M. Echavarren, *Angew. Chem., Int. Ed.*, 2006, **45**, 6029–6032.
91. A. Escribano-Cuesta, V. López-Carrillo, D. Janssen and A. M. Echavarren, *Chem.–Eur. J.*, 2009, **15**, 5646–5650.
92. C. A. Witham, P. Mauleón, N. D. Shapiro, B. D. Sherry and F. D. Toste, *J. Am. Chem. Soc.*, 2007, **129**, 5838–5839.
93. S. M. Kim, J. H. Park, S. Y. Choi and Y. K. Chung, *Angew. Chem., Int. Ed.*, 2007, **46**, 6172–6175.
94. S. M. Kim, J. H. Park, Y. K. Kang and Y. K. Chung, *Angew. Chem., Int. Ed.*, 2009, **48**, 4532–4535.
95. C. H. M. Amijs, V. López-Carrillo, M. Raducan, P. Pérez-Galán, C. Ferrer and A. M. Echavarren, *J. Org. Chem.*, 2008, **73**, 7721–7730.

96. Y. Matsumoto, K. B. Selim, H. Nakanishi, K.-i. Yamada, Y. Yamamoto and K. Tomioka, *Tetrahedron Lett.*, 2010, **51**, 404–406.
97. B. Trillo, F. López, S. Montserrat, G. Ujaque, L. Castedo, A. Lledós and J. L. Mascareñas, *Chem.–Eur. J.*, 2009, **15**, 3336–3339.
98. B. W. Gung, D. T. Craft, L. N. Bailey and K. Kirschbaum, *Chem.–Eur. J.*, 2010, **16**, 639–644.
99. P. Mauleón, R. M. Zeldin, A. Z. González and F. D. Toste, *J. Am. Chem. Soc.*, 2009, **131**, 6348–6349.
100. (a) G. Zhang, X. Huang, G. Li and L. Zhang, *J. Am. Chem. Soc.*, 2008, **130**, 1814–1815; (b) G. Li, X. Huang and L. Zhang, *J. Am. Chem. Soc.*, 2008, **130**, 6944–6945. For a related report, see also: (c) Y. Peng, M. Yu and L. Zhang, *Org. Lett.*, 2008, **10**, 5187–5190.
101. For reviews, see: (a) N. Marion and S. P. Nolan, *Angew. Chem., Int. Ed.*, 2007, **46**, 2750–2752; (b) J. Marco-Contelles and E. Soriano, *Chem.–Eur. J.*, 2007, **13**, 1350–1357.
102. (a) N. Marion, P. de Frémont, G. Lemière, E. D. Stevens, L. Fensterbank, M. Malacria and S. P. Nolan, *Chem. Commun.*, 2006, 2048–2050; (b) A. Correa, N. Marion, L. Fensterbank, M. Malacria, S. P. Nolan and L. Cavallo, *Angew. Chem., Int. Ed.*, 2008, **47**, 718–721; (c) N. Marion, G. Lemière, A. Correa, C. Costabile, R. S. Ramón, X. Moreau, P. de Frémont, R. Dahmane, A. Hours, D. Lesage, J.-C. Tabet, J.-P. Goddard, V. Gandon, L. Cavallo, L. Fensterbank, M. Malacria and S. P. Nolan, *Chem.–Eur. J.*, 2009, **15**, 3243–3260.
103. G. Li, G. Zhang and L. Zhang, *J. Am. Chem. Soc.*, 2008, **130**, 3740–3741.
104. N. Marion, S. Díez-González, P. de Frémont, A. R. Noble and S. P. Nolan, *Angew. Chem., Int. Ed.*, 2006, **45**, 3647–3650.
105. (a) S. Suárez-Pantiga, E. Rubio, C. Alvarez-Rúa and J. M. González, *Org. Lett.*, 2009, **11**, 13–16; (b) S. Suárez-Pantiga, D. Palomas, E. Rubio and J. M. González, *Angew. Chem., Int. Ed.*, 2009, **48**, 7857–7861.
106. (a) N. Marion, R. Gealageas and S. P. Nolan, *Org. Lett.*, 2007, **9**, 2653–2656; (b) P. de Frémont, N. Marion and S. P. Nolan, *J. Organomet. Chem.*, 2009, **694**, 551–560. For a DFT study, see: (c) C. Gourlaouen, N. Marion, S. P. Nolan and F. Maseras, *Org. Lett.*, 2009, **11**, 81–84.
107. For reviews, see: (a) J. Muzart, *Tetrahedron*, 2008, **64**, 5815–5849; (b) R. A. Widenhoefer and X. Han, *Eur. J. Org. Chem.*, 2006, 4555–4563.
108. NHC–Au^{III} complexes have also been shown to catalyze alkyne hydration, see: P. de Frémont, R. Singh, E. D. Stevens, J. L. Petersen and S. P. Nolan, *Organometallics*, 2007, **26**, 1376–1385.
109. N. Marion, R. S. Ramón and S. P. Nolan, *J. Am. Chem. Soc.*, 2009, **131**, 448–449.
110. Z. Zhang, S. D. Lee, A. S. Fisher and R. A. Widenhoefer, *Tetrahedron*, 2009, **65**, 1794–1798.
111. (a) R. S. Ramón, N. Marion and S. P. Nolan, *Chem.–Eur. J.*, 2009, **15**, 8695–8697. For a related study, see also: (b) R. S. Ramón, J. Bosson, S. Díez-González, N. Marion and S. P. Nolan, *J. Org. Chem*, 2010, **75**, 1197–1202.

112. Z. Zhang and R. A. Widenhoefer, *Org. Lett.*, 2008, **10**, 2079–2081.
113. R. S. Paton and F. Maseras, *Org. Lett.*, 2009, **11**, 2237–2240.
114. M. S. Hadfield and A.-L. Lee, *Org. Lett.*, 2010, **12**, 484–487.
115. (a) C. F. Bender and R. A. Widenhoefer, *Org. Lett.*, 2006, **8**, 5303–5305; (b) A. Corma, C. González-Arellano, M. Iglesias, M. T. Navarro and F. Sanchéz, *Chem. Commun.*, 2008, 6218–6220.
116. (a) R. E. Kinder, Z. Zhang and R. A. Widenhoefer, *Org. Lett.*, 2008, **10**, 3157–3159; (b) A. Saito, T. Konishi and Y. Hanzawa, *Org. Lett.*, 2010, **12**, 372–374.
117. (a) S. Chessa, N. J. Clayden, M. Bochmann and J. A. Wright, *Chem. Commun.*, 2009, 797–799; (b) K. Wilckens, M. Uhlemann and C. Czekelius, *Chem.–Eur. J.*, 2009, **15**, 13323–13326.
118. H. Li and R. A. Widenhoefer, *Org. Lett.*, 2009, **11**, 2671–2674.
119. V. Lavallo, G. D. Frey, S. Kousar, B. Donnadieu, M. Soleilhavoup and G. Bertrand, *Angew. Chem., Int. Ed.*, 2008, **47**, 5224–5228.
120. (a) X. Zeng, G. D. Frey, R. Kinjo, B. Donnadieu and G. Bertrand, *J. Am. Chem. Soc.*, 2009, **131**, 8690–8696. For mechanistic insights, see: (b) X. Zeng, R. Kinjo, B. Donnadieu and G. Bertrand, *Angew. Chem., Int. Ed.*, 2010, **49**, 942–945.
121. K. L. Toups, G. T. Liu and R. A. Widenhoefer, *J. Organomet. Chem.*, 2009, **694**, 571–575.
122. X.-Y. Liu, P. Ding, J.-S. Huang and C.-M. Che, *Org. Lett.*, 2007, **9**, 2645–2648.
123. C. Ferrer, A. Escribano-Cuesta and A. M. Echavarren, *Tetrahedron*, 2009, **65**, 9015–9020.
124. J. A. Akana, K. X. Bhattacharyya, P. Müller and J. P. Sadighi, *J. Am. Chem. Soc.*, 2007, **129**, 7736–7737.
125. B. C. Gorske, C. T. Mbofana and S. J. Miller, *Org. Lett.*, 2009, **11**, 4318–4321.
126. V. Lavallo, G. D. Frey, S. Kousar, B. Donnadieu and G. Bertrand, *Proc. Natl. Acad. Sci. U. S. A.*, 2007, **104**, 13569–13573.
127. X. Zeng, G. D. Frey, S. Kousar and G. Bertrand, *Chem.–Eur. J.*, 2009, **15**, 3056–3060.
128. (a) A. Fedorov and P. Chen, *Organometallics*, 2009, **28**, 1278–1281; (b) A. Fedorov, M.-E. Moret and P. Chen, *J. Am. Chem. Soc.*, 2008, **130**, 8880–8881.
129. (a) M. R. Fructos, M. M. Díaz-Requejo and P. J. Pérez, *Chem. Commun.*, 2009, 5153–5155; (b) P. de Frémont, E. D. Stevens, M. R. Fructos, M. M. Díaz-Requejo, P. J. Pérez and S. P. Nolan, *Chem. Commun.*, 2006, 2045–2047.
130. M. R. Fructos, T. R. Belderrain, P. de Frémont, N. M. Scott, S. P. Nolan, M. M. Díaz-Requejo and P. J. Pérez, *Angew. Chem., Int. Ed.*, 2005, **44**, 5284–5288.
131. N. D. Shapiro and F. D. Toste, *J. Am. Chem. Soc.*, 2007, **129**, 4160–4161.
132. G. Li and L. Zhang, *Angew. Chem., Int. Ed.*, 2007, **46**, 5156–5159.
133. H.-S. Yeom, Y. Lee, J.-E. Lee and S. Shin, *Org. Biomol. Chem.*, 2009, **7**, 4744–4752.

134. (a) R. S. Ramón, N. Marion and S. P. Nolan, *Tetrahedron*, 2009, **65**, 1767–1773. For a related study using propargylic esters, see: (b) N. Marion, P. Carlqvist, R. Gealageas, P. de Frémont, F. Maseras and S. P. Nolan, *Chem.–Eur. J.*, 2007, **13**, 6437–6451.
135. A. Corma, C. González-Arellano, M. Iglesias, S. Pérez-Ferreras and F. Sánchez, *Synlett*, 2007, 1771–1774.
136. A. Corma, E. Gutiérrez-Puebla, M. Iglesias, A. Monge, S. Pérez-Ferreras and F. Sánchez, *Adv. Synth. Catal.*, 2006, **348**, 1899–1907.
137. M. Boronat, A. Corma, C. González-Arellano, M. Iglesias and F. Sánchez, *Organometallics*, 2010, **29**, 134–141.
138. J. Y. Z. Chiou, S. C. Luo, W. C. You, A. Bhattacharyya, C. S. Vasam, C. H. Huang and I. J. B. Lin, *Eur. J. Inorg. Chem.*, 2009, 1950–1959.
139. (a) X. Hu, I. Castro-Rodriguez, K. Olsen and K. Meyer, *Organometallics*, 2004, **23**, 755–764; (b) D. Nemcsok, K. Wichmann and G. Frenking, *Organometallics*, 2004, **23**, 3640–3646; (c) A. Kausamo, H. M. Tuononen, K. E. Krahulic and R. Roesler, *Inorg. Chem.*, 2008, **47**, 1145–1154; (d) M.-L. Teyssot, A.-S. Jarrousse, A. Chevry, A. De Haze, C. Beaudoin, M. Manin, S. P. Nolan, S. Díez-González, L. Morel and A. Gautier, *Chem.–Eur. J.*, 2009, **15**, 314–318.
140. D. Benitez, N. D. Shapiro, E. Tkatchouk, Y. Wang, W. A. Goddard III and F. D. Toste, *Nat. Chem.*, 2009, **1**, 482–486.
141. (a) D. Zuccaccia, L. Belpassi, F. Tarantelli and A. Macchioni, *J. Am. Chem. Soc.*, 2009, **131**, 3170–3171; (b) T. J. Brown, M. G. Dickens and R. A. Widenhoefer, *J. Am. Chem. Soc.*, 2009, **131**, 6350–6351; (c) S. Flügge, A. Anoop, R. Goddard, W. Thiel and A. Fürstner, *Chem.–Eur. J.*, 2009, **15**, 8558–8565; (d) A. S. K. Hasmi, A. M. Schuster and F. Rominger, *Angew. Chem., Int. Ed.*, 2009, **48**, 8247–8249.

CHAPTER 12

Oxidation Reactions with NHC–Metal Complexes

SUSANNE M. PODHAJSKY AND
MATTHEW S. SIGMAN*

Department of Chemistry, University of Utah, 315 South 1400 East,
Salt Lake City, Utah 84112, USA

12.1 Introduction

Over the course of the past decade, N-heterocyclic carbenes (NHCs) have been established as an important class of catalytic reagents, both as nucleophilic catalysts themselves[1,2] and as ligands on metal catalysts.[3–7] As ligands, they have been compared extensively to phosphines,[7] since both are monodentate ligands, strong sigma donors, and tunable both sterically and electronically. However, both free NHCs and their metal complexes are significantly more oxidatively stable than phosphines. Thus, NHC ligands have been used in a steadily expanding number of oxidative reactions,[8] where ligands have traditionally been limited to oxidatively stable N and O ligands, both to tune the reactivity of the metal by changing its electronic and steric environment, and to provide a chiral environment around the metal in order to develop enantioselective reactions.

NHC catalysts have been reviewed extensively recently, in the context of oxidative reactions[8,9] and otherwise.[3,5–7,10] Therefore, this Chapter will provide an overview of key systems as well as some of the most interesting recent developments.

RSC Catalysis Series No. 6
N-Heterocyclic Carbenes: From Laboratory Curiosities to Efficient Synthetic Tools
Edited by Silvia Díez-González
© Royal Society of Chemistry 2011
Published by the Royal Society of Chemistry, www.rsc.org

12.2 O_2 Activation by NHC–Metal Complexes

Several catalytic oxidative reactions have been developed using O_2 as the terminal oxidant, most notably using Pd. In these reactions, the substrate is typically oxidized by Pd, which is in turn oxidized by O_2, without transfer of oxygen to the substrate.[11] The mechanistic details of the oxidation of Pd complexes by O_2 are therefore of great interest. At the centre of this debate is the question whether a Pd^{II} hydride (formed upon product release in most oxidative Pd-catalyzed reactions) reacts directly with O_2 (Scheme 12.1, **1**→**2**), or the Pd^{II} hydride undergoes reductive elimination to form Pd^0 followed by re-oxidation (Scheme 12.1, **3**→**4**→**2**).

In initial studies, Pd^{II} hydride **1** was found to react with O_2 to give Pd hydroperoxide **2**. Similarly, Pd^0 carbene complex **3** was found to react rapidly with O_2 to form complex **4**, which could be converted to **2** by reaction with acid.[12,13] However, the exact pathway for the reaction of Pd hydride **1** remained unclear, since it could undergo reductive elimination to form Pd^0 followed by re-oxidation.

In addition to this, several pathways can be envisioned for a Pd hydride reacting directly with O_2. The pathway lowest in energy according to DFT calculations was a hydrogen abstraction pathway (Scheme 12.2a).[14] In this mechanistic scenario, a hydrogen atom would be abstracted by O_2, forming a Pd^I adduct of protonated superoxide **5**. This would rearrange to form the Pd^{II} hydroperoxide **2**, which was the experimentally observed product. While this was formally the lowest energy pathway, the reductive elimination pathway (Scheme 12.2b) was calculated to be only 4.8 kcal mol^{-1} higher, and thus could not be excluded based on calculations alone. Kinetic studies were thus performed, since the two mechanisms outlined above should be distinguishable, for example by kinetic isotope effects (KIEs) and O_2 dependence. When these studies and others were performed, a zero order dependence on O_2 was observed as well as a small deuterium KIE of 1.3 ± 1. A significantly larger KIE

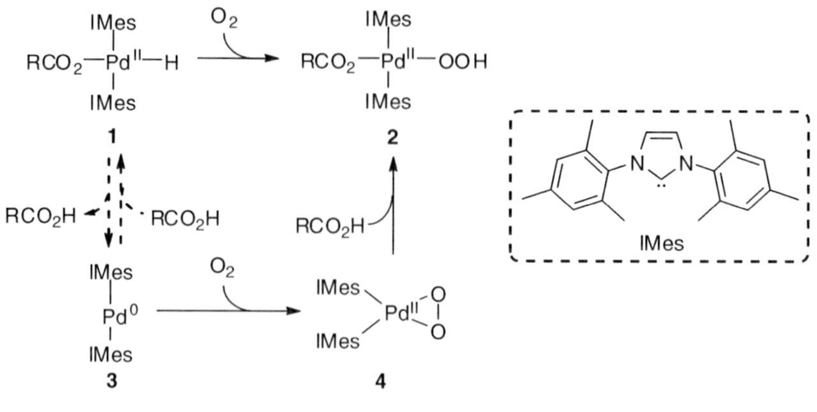

Scheme 12.1 Pd reaction with O_2 *via* two possible pathways (**1**→**2** or **3**→**4**→**2**).

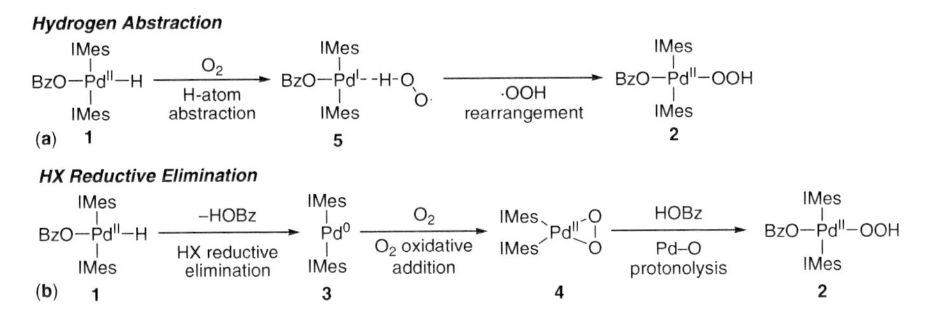

Scheme 12.2 Pd hydride reaction with O_2 *via* two possible pathways.

for O_2 reacting with a Pd hydride had been previously measured by Goldberg and co-workers in a reaction proposed to proceed *via* hydrogen abstraction;[15] thus the KIE measured in this system contradicted this pathway. Even more significantly, since O_2 was involved in the first step of the hydrogen abstraction pathway, a first-order dependence should be observed. In the reductive elimination pathway, however, O_2 was involved in the second step, thus, if reductive elimination (the first step) was rate determining, a zero order dependence on O_2 would be observed. Based on this evidence, it was concluded that the traditionally proposed reductive elimination pathway was indeed operative, at least for the dicarbene complex studied.[16] In addition to these results, Stahl and co-workers also found that the addition of acid accelerated the conversion of Pd hydride to Pd hydroperoxide. This seemed counter-intuitive at first glance, since acid was released *via* reductive elimination. However, when acid was added to the Pd hydride, the ^{1}H NMR shift of the Pd–H changed, indicating an adduct of acid and Pd hydride in solution. Based on this, it was proposed that the acid was catalyzing the reductive elimination step through stabilizing the anionic charge build-up on the dissociating benzoate. Unfortunately, general conclusions about aerobic Pd-catalyzed reactions cannot be drawn from these studies, since the ligand on the palladium centre changes the barriers to different paths in varying degree, allowing different mechanisms to operate depending on the environment of the metal.[17]

The interactions of O_2 with NHC complexes of other metals have been studied far less extensively than those of Pd, likely due to the fact that aerobic oxidative reactions with other metals are not as prevalent. However, the reaction of different NHC–Ni complexes with O_2 were studied by Dible and Sigman, and it was found that their reactivity towards O_2 varied with the substituents on the metal.[18] While [(IPr)Ni(C_3H_5)Cl], [(SIPr)Ni(C_3H_5)Cl] and [(IMes)Ni(C_3H_5)Cl] reacted rapidly with O_2 at room temperature to form (NHC)nickel(II)-bis-μ-hydroxochloro dimers, [(ItBu)Ni(C_3H_5)Cl] and [(IAd)-Ni(C_3H_5)Cl] were stable to O_2 even when refluxed in benzene (Scheme 12.3a). Through extensive studies, it was concluded that an open coordination site on Ni was required for coordination of O_2. Thus, NHCs exhibiting free rotation about the Ni–C bond appeared to reveal an axial coordination site during the

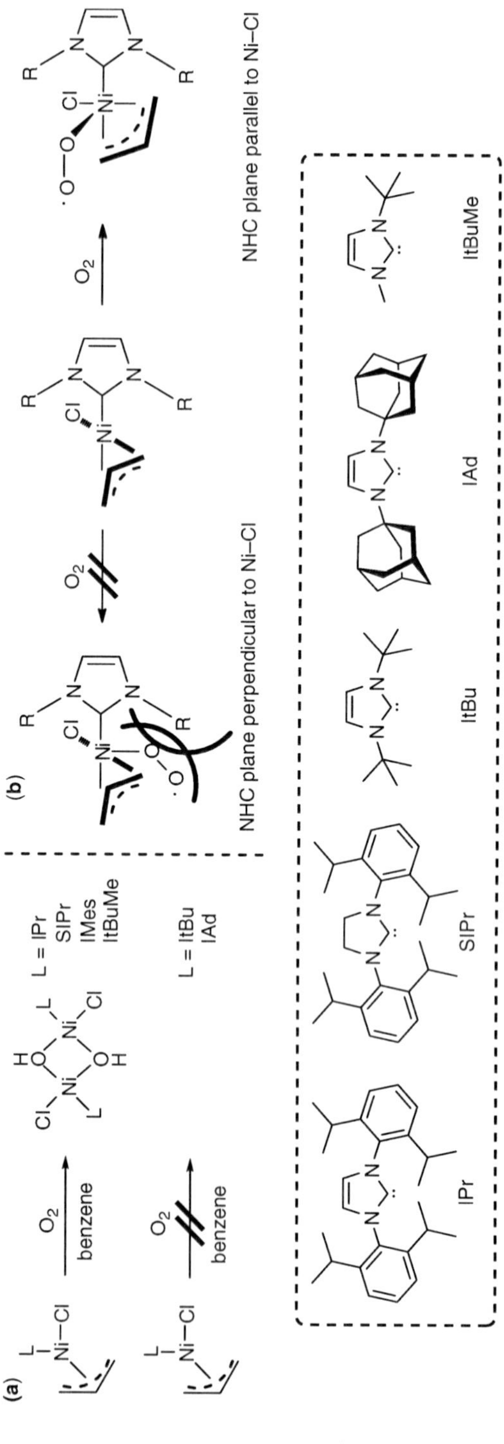

Scheme 12.3 (a) Reactivity of different Ni carbene complexes toward O_2. (b) Reactivity of different conformers toward O_2.

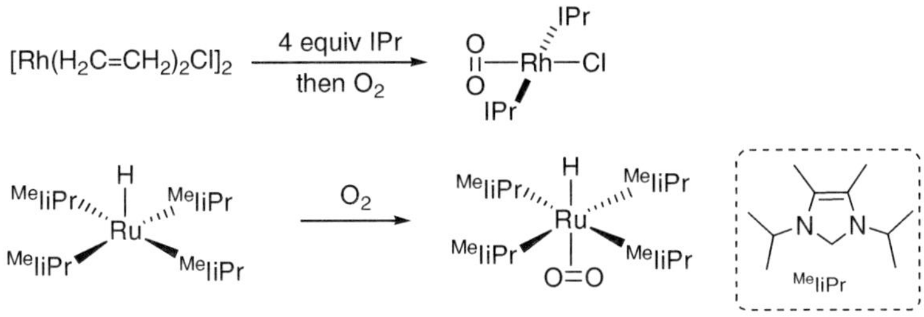

Scheme 12.4 Formation of singlet-oxygen complexes.

rotation through the square plane of Ni, whereas larger NHCs with hindered rotation inhibited O_2 coordination by blocking the axial sites (Scheme 12.3b). To confirm this hypothesis, ItBuMe was synthesized, which showed hindered rotation, but contained an open coordination site and was thus oxidized by O_2.

Interestingly, NHC complexes of Ru and Rh could be converted to their O_2 adducts, best described as metal-singlet oxygen complexes (Scheme 12.4).[19,20] In both cases, unusually short O–O bond lengths were observed in the crystal structures, which led to closer examination of the complexes by other spectroscopical techniques as well as computational methods. It was then concluded that the complexes could be regarded as singlet O_2 adducts, where the metal had not been oxidized by O_2. Moreover, in the case of the Ru complex reported by Häller *et al.*,[20] the coordinated O_2 could be removed by a simple freeze–pump–thaw technique to regenerate the starting complex.

12.3 Alcohol Oxidation

NHC–palladium complexes have been well developed as catalysts for alcohol oxidations. They were used initially by Sigman and co-workers in 2003 in the oxidative kinetic resolution of alcohols, replacing (–)-sparteine as a ligand on Pd.[21] It was shown that using (–)-sparteine as base along with an achiral ligand was sufficient to obtain satisfactory k_{rel} values (Scheme 12.5a). Using chiral NHC ligands, matched/mismatched diastereomeric interactions were observed. However, a successful catalyst using a chiral NHC ligand with an achiral base was not developed until 2007, when Shi and co-workers published a system using axially chiral 1,1′-bi(2-naphthylamine) (BINAM)-derived carbene complex **6** (Scheme 12.5b).[22]

In addition to the oxidative kinetic resolution of alcohols, general alcohol oxidation catalysts were also developed using NHC–Pd complexes.[23] In previous Pd-catalyzed aerobic alcohol oxidation systems using nitrogen-based ligands, it had been found that on the one hand, excess ligand was necessary to prevent catalyst decomposition, but on the other hand, a ligand had to dissociate to allow alcohol binding and/or β-hydride elimination, thus slowing catalysis. It was thought that a NHC–Pd catalyst could overcome this issue, since carbenes are monodentate ligands that are known to stabilize

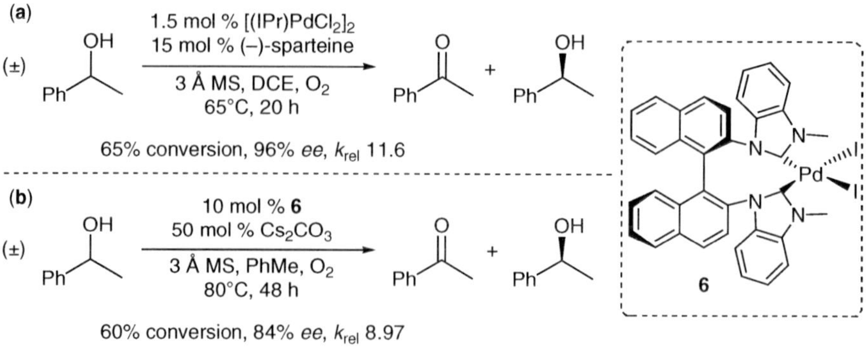

(a)

(±) Ph–CH(OH)–CH$_3$ → 1.5 mol % [(IPr)PdCl$_2$]$_2$, 15 mol % (–)-sparteine; 3 Å MS, DCE, O_2, 65°C, 20 h → Ph–C(O)–CH$_3$ + Ph–CH(OH)–CH$_3$

65% conversion, 96% *ee*, k_{rel} 11.6

(b)

(±) Ph–CH(OH)–CH$_3$ → 10 mol % **6**, 50 mol % Cs_2CO_3; 3 Å MS, PhMe, O_2, 80°C, 48 h → Ph–C(O)–CH$_3$ + Ph–CH(OH)–CH$_3$

60% conversion, 84% *ee*, k_{rel} 8.97

6

Scheme 12.5 Oxidative kinetic resolution of *sec*-phenethyl alcohol using carbenes.

palladium(II) as well as palladium(0),[7] thus eliminating the need for excess ligand in the reaction. In addition, it was known that exogenous base was typically needed for alcohol; however, when acetate was used as a counter-ion on the palladium complex, it could act as a base itself.

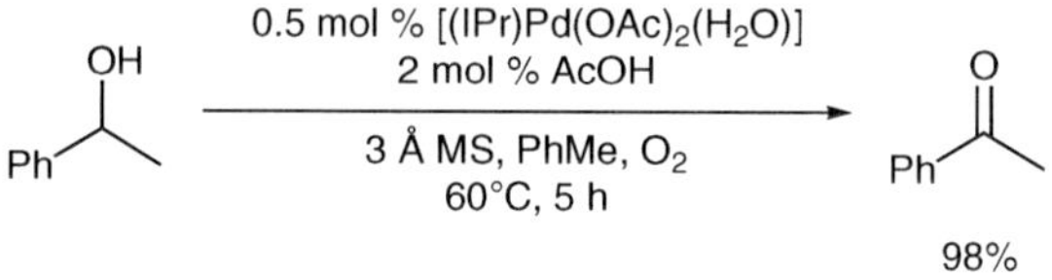

Ph–CH(OH)–CH$_3$ → 0.5 mol % [(IPr)Pd(OAc)$_2$(H_2O)], 2 mol % AcOH; 3 Å MS, PhMe, O_2, 60°C, 5 h → Ph–C(O)–CH$_3$ 98% (12.1)

Thus, [(IPr)Pd(OAc)$_2$] was designed as a general alcohol oxidation catalyst (Equation (12.1)). The proposed mechanism for this system is shown in Scheme 12.6. [(IPr)Pd(OAc)$_2$] crystallized as its hydrate **7**, and thus the water would be initially replaced by the alcohol forming **8**. The coordinated alcohol may then be deprotonated intramolecularly by the acetate counter-ion to generate Pd alkoxide **9**, which would undergo β-hydride elimination to form Pd hydride **10**. Reductive elimination of AcOH would lead to [(IPr)Pd^0] **11**, which would be oxidized by O_2 forming Pd peroxide **12**, and subsequently protonated to yield **13**. Surprisingly, alcohol oxidation using 0.5 mol% [(IPr)Pd(OAc)$_2$] gave inconsistent results. This problem was addressed by the addition of acetic acid to the reaction, which was proposed to influence three steps in the catalytic cycle: (i) Pd alkoxide re-protonation (**9**→**8**); (ii) the equilibrium between a Pd hydride and Pd^0 (**11**→**10**); and (iii) protonation of the Pd peroxide (**12**→**13**). This modulation served to decelerate the alcohol oxidation sequence of the catalytic cycle, while accelerating the oxidation of Pd^0, and thus establish a robust catalytic system.

The role(s) of acid in this system was further elucidated using mechanistic studies.[24] When the rate dependence of *sec*-phenethyl alcohol oxidation on AcOH concentration was determined, a rate increase at low concentrations of acid (up to 0.62 mol%) was observed followed by first-order inhibition at higher concentrations (up to 15 mol%) (Figure 12.1a).

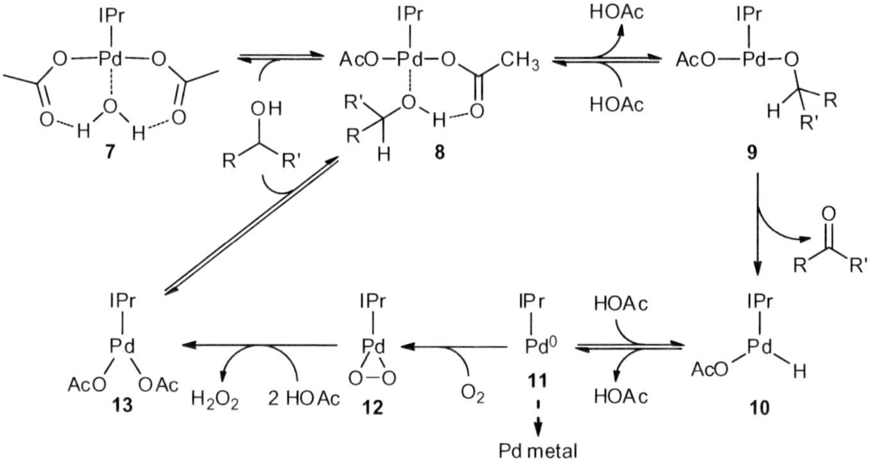

Scheme 12.6 Mechanism of $[(IPr)Pd(OAc)_2]$-catalyzed aerobic oxidation of alcohols.

This pointed towards a change in rate-determining steps from low to high acid concentration. Additionally, the reaction was determined to be first order in *sec*-phenethyl alcohol at 2 and 3 mol% AcOH, with an uncommonly large KIE of 5.5 ± 1 for α-D *sec*-phenethyl alcohol. This indicated that the alcohol oxidation was rate-determining under these conditions, and, more specifically, β-hydride elimination was the rate-determining step. As can also be seen clearly from Figure 12.1b, the rate of alcohol oxidation was significantly lower at 3 than at 2 mol% AcOH, consistent with inhibition by AcOH. The reason for this was proposed to be re-protonation of the Pd alkoxide, which could slow the overall reaction. The KIE was significantly larger than those measured for other Pd-catalyzed aerobic alcohol oxidation using bidentate nitrogen ligands. This was attributed to a late transition state in β-hydride elimination, with almost complete Pd–H bond formation (Scheme 12.7a). This, in turn, could be rationalized with the open coordination site on Pd, which was due to the monodentate NHC ligand. In the case of bidentate ligands, an associative mechanism *via* a 5-coordinate transition state was proposed, leading to an earlier transition state and lower KIEs (Scheme 12.7b).

Using no added AcOH, the same parameters were evaluated with different outcomes. The reaction order in *sec*-phenethyl alcohol was determined to be 0.55, with a KIE of 1.7 ± 2. This partial order indicated two competing rate-determining steps, with only one of them involving *sec*-phenethyl alcohol. The KIE could be interpreted in the same way, in that the rate of β-hydride elimination was then competitive with another step in the mechanism. Additionally, it was observed that catalyst decomposition was occurring over time at low concentrations of AcOH (see Figure 12.1b, 0 and 0.75 mol% AcOH). From this, it could be concluded that the other rate-limiting step is in the catalyst regeneration part of the catalytic cycle. To determine whether oxygenation of Pd^0 or protonation of the Pd peroxide was rate-determining, the rate

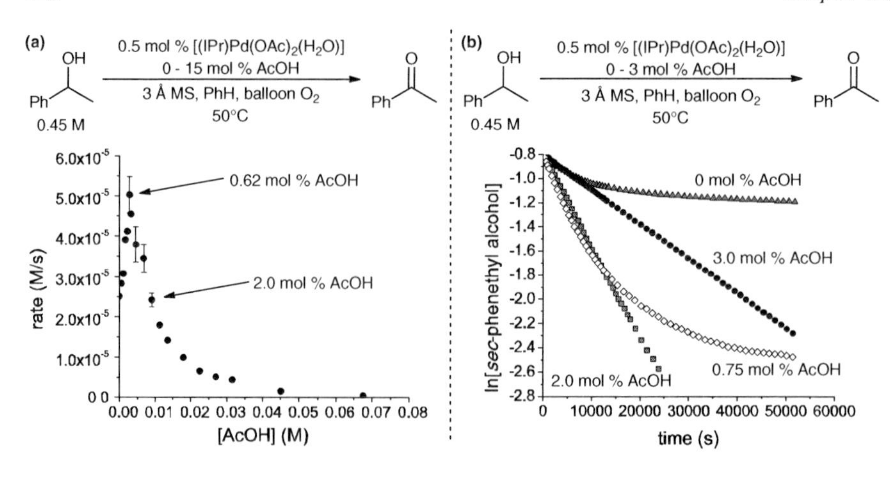

Figure 12.1 (a) Dependence of alcohol oxidation on [AcOH]. (b) ln[*sec*-phenethyl alcohol] over time at different [AcOH].

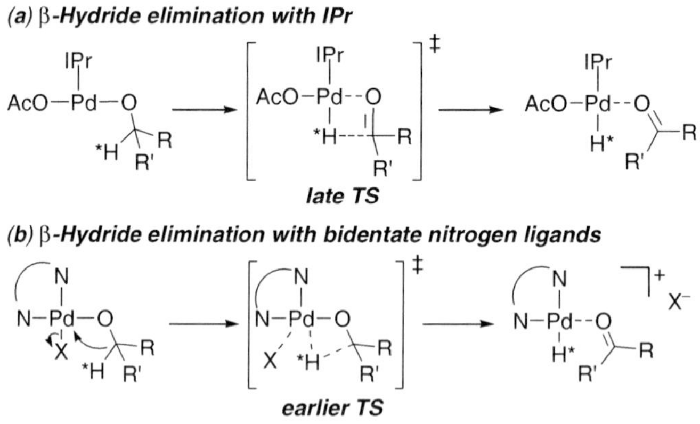

(a) β-Hydride elimination with IPr

late TS

(b) β-Hydride elimination with bidentate nitrogen ligands

earlier TS

Scheme 12.7 Proposed transition states for β-hydride elimination with IPr (a) and bidentate nitrogen ligands (b).

dependence on oxygen pressure was investigated using mixtures of O_2 and N_2 at balloon pressure. These experiments showed zero order dependencies in O_2, although a minimum amount of O_2 was required to avoid catalyst decomposition. Therefore, it was concluded that protonation of the Pd peroxide or Pd hydroperoxide was the competitive rate-determining step under these conditions. Interestingly, the same overall observation was made in Stahl's stoichiometric studies (see Scheme 12.2); however, different conclusions were drawn in each case as to which step specifically was acid-catalyzed.

The knowledge garnered about different alcohol oxidation systems was further employed to develop hydro-functionalizations of olefins, wherein the hydride incorporated into product stemmed from the oxidation of the alcohol solvent. Using this concept, a hydroarylation of styrenes was developed using $[(SIPr)PdCl_2]_2$, which successfully catalyzed both the oxidation of *i*-PrOH and the olefin functionalization (Equation (12.2)).[25]

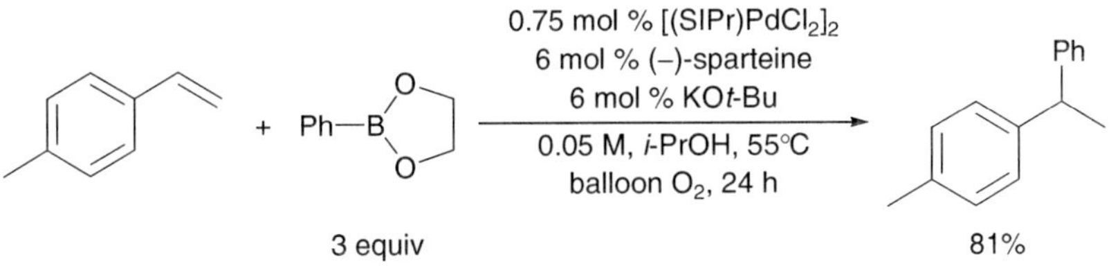

0.75 mol % [(SIPr)PdCl2]2
6 mol % (–)-sparteine
6 mol % KOt-Bu
0.05 M, i-PrOH, 55°C
balloon O2, 24 h
Ph—B
O
O
Ph
3 equiv
81%

(12.2)

Mechanistically, the reaction was proposed to be initiated by an alcohol oxidation generating Pd hydride **15** (Scheme 12.8). The styrene (or diene) substrate would coordinate to **15**, followed by insertion into the Pd–H bond to give Pd alkyl **17**, which might be stabilized as a Pd π-benzyl (or π-allyl) species **18**. Transmetalation would then occur to give **19**, which would undergo reductive elimination to form the hydroarylation product along with Pd^0, which may be then re-oxidized by O_2.

In a different approach, Lebel and co-workers designed a multicatalytic process combining the $[(IPr)Pd(OAc)_2]$-catalyzed alcohol oxidation and their $[RhCl(PPh_3)_3]$-catalyzed carbonyl methylenation.[26] They were able to perform both reactions sequentially in one pot, effectively converting alcohols into terminal alkenes (Equation (12.3)). To further highlight this approach, they performed a three-reaction cascade comprised of the two former reactions, which resulted in the formation of a diene, followed by an alkene metathesis in

(CH3)2CHOH
(CH3)2CO
IPr
Pd
Cl
Cl
14
15
16
17
18
19
20
H2O2
O2
+
2 HCl
MCl
R'M
Pd0
R
R'
H

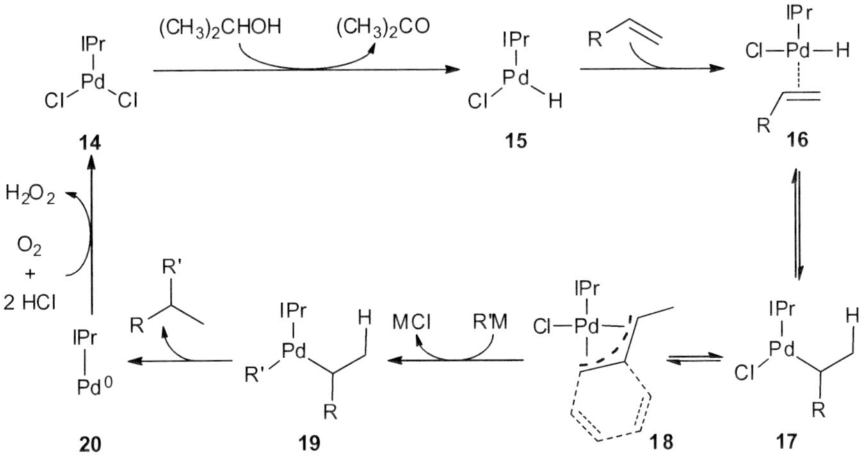

Scheme 12.8 Pd-catalyzed hydroarylation of styrenes and dienes.

the same pot (Equation (12.4)).

1) 2.5 mol % [(IPr)Pd(OAc)$_2$(H$_2$O)]
5 mol % Bu$_4$NOAc, 3 Å MS
PhMe, O$_2$, 60°C
2) 2.5 mol % [RhCl(PPh$_3$)$_3$]
1.12 equiv PPh$_3$, 10 equiv *i*-PrOH
1.6 equiv TMSCHN$_2$
1,4-dioxane, 50°C

84% (12.3)

1) 2.6 mol % [(IPr)Pd(OAc)$_2$(H$_2$O)]
5.2 mol % Bu$_4$NOAc, 3 Å MS
PhMe, O$_2$, 60°C
2) 2.6 mol % [RhCl(PPh$_3$)$_3$]
1.1 equiv PPh$_3$, 10 equiv *i*-PrOH
1.6 equiv TMSCHN$_2$
1,4-dioxane, 50°C
3) 1 equiv AlCl$_3$
4) 10 mol % [(IMes)PCy$_3$Cl$_2$Ru=CHPh]
1,4-dioxane, 50°C

70% (12.4)

Both Ir and Ru NHC complexes were successfully used for "transfer-hydrogenation" reactions, wherein an alcohol or amine is oxidized, followed by reaction of the oxidized species and hydrogenation to give the final product.[27–33] Both Crabtree[27] and Peris[28] reported the use of Ir carbene complexes as catalysts in reactions of this type, such as the *N*-alkylation by alcohols (Equations (12.5) and (12.6)).

Crabtree and co-workers:

Ph⁀OH (1 equiv) + H$_2$N—Ph (1 equiv) → [1 mol % **21**, 0.5 equiv NaHCO$_3$; 4 ÅMS, PhMe, 110°C] → PhCH$_2$NHPh 98%

21 (PF$_6^-$) (12.5)

Peris and co-workers:

Ph⁀OH (5 equiv) + H$_2$N—Ph (1 equiv) → [5 mol % **22**, 15 mol % AgOTf; neat, 110°C] → PhCH$_2$NHPh >95%

22 (12.6)

The reaction was proposed to proceed *via* initial oxidation of the alcohol by the Ir catalyst, with concomitant formation of an Ir hydride. The resulting aldehyde then would react with the amine to form an imine, which would be subsequently reduced to the final amine product by the Ir hydride.[34] Although these systems will not be discussed further here, it should be noted that NHC–Ir complexes were also used to catalyze the Oppenauer-type oxidation of alcohols.[35–37]

During a study aimed at the development of a Ru catalyst for the amine alkylation reaction described above, Madsen and co-workers observed exclusive formation of amides instead of amines when using NHC–Ru complexes.[32] Interestingly, it was found that in addition to the NHC, a phosphine ligand was necessary for the reaction to proceed. The authors proposed a shift in mechanism, wherein the aldehyde formed upon oxidation of the alcohol substrate would not dissociate from the Ru catalyst. Instead, it would be attacked by the amine to form a Ru-coordinated hemiaminal, which may be then dehydrogenated to give the amide product (Equation (12.7), COD = 1,5-cyclooctadiene, Cyp = cyclopentyl). Based on this finding, Hong and co-workers also developed a system using nitrogen ligands instead of phosphines under otherwise similar conditions.[38]

Ph–CH₂CH₂–OH (1 equiv) + H_2N–CH₂–Ph (1 equiv) → [2 mol % $[Ru(COD)Cl_2]$, 2 mol % **23**, 8 mol % KO*t*-Bu, 2 mol % $PCyp_3 \cdot HBF_4$; PhMe, 110°C] → Ph–CH₂–C(=O)–NH–CH₂–Ph (93%)

23: *i*-Pr–N(imidazolium)N–*i*-Pr, Cl^-

(12.7)

Using another Ru carbene catalyst, Williams and co-workers were able to develop a tandem process involving alcohol oxidation followed by a Wittig reaction and hydrogenation of the resulting alkene.[33] Interestingly, C–H bond activation of the carbene ligand was observed in this reaction (Scheme 12.9).

In stoichiometric studies, it was shown that **25** could be converted into **24** by reacting it with an alkene, and the reverse reaction could be performed using H_2 or *i*-PrOH (Scheme 12.9c). Based on these results, it was proposed that in the catalytic reaction, the alcohol substrate would initially be oxidized by **24**, forming an aldehyde along with dihydride **25**. The aldehyde would then undergo a Wittig reaction, and the resulting alkene would be hydrogenated by **25**.

12.4 Alkene Oxidation

One of the best known oxidative transformations of alkenes is the conversion of terminal alkenes to methyl ketones, known as the Wacker oxidation. While this reaction is used industrially on extremely large scale, its application in laboratory settings (known as the Tsuji–Wacker oxidation) has been problematic.[39–41] This is due to selectivity problems with several substrate classes as well as the need for co-oxidants, typically copper salts. The use of copper has prevented the use of ligands on Pd to modulate catalysis, which could

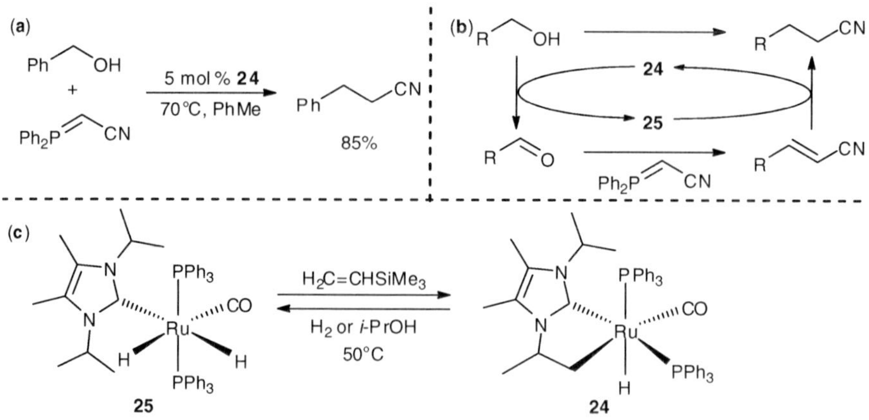

Scheme 12.9 (a) Tandem alcohol oxidation–Wittig reaction–hydrogenation. (b) Proposed catalytic cycle. (c) Stoichiometric study showing the interconversion between **24** and **25**.

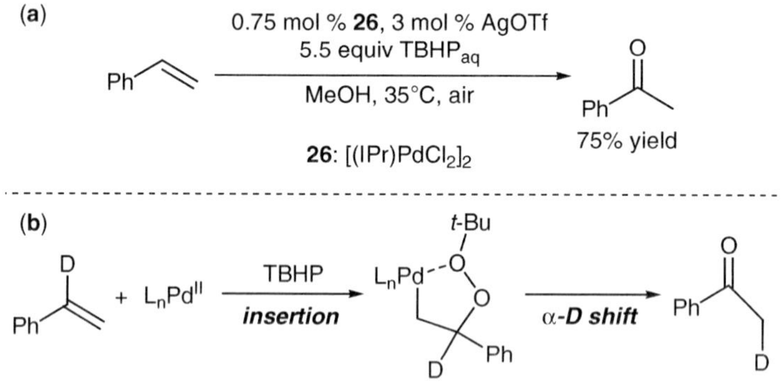

Scheme 12.10 (a) Pd carbene-catalyzed Tsuji–Wacker oxidation of styrenes. (b) Proposed mechanism based on labelling experiments.

potentially help overcome some of the issues associated with this reaction. One of the first ligand-modulated Tsuji–Wacker reactions[42,43] was developed by Cornell and Sigman using a Pd carbene catalyst with *t*-butyl hydrogen peroxide (TBHP) as the sole oxidant (Scheme 12.10).[44]

In order to elucidate the mechanism of this transformation, several labelling experiments were carried out. TBHP was used as an aqueous solution, and H_2O is known to be the source of oxygen incorporated into product in traditional Wacker chemistry.[39] To determine whether H_2O or TBHP was the source of oxygen in the present system, $^{18}OH_2$ was used along with unlabelled TBHP, revealing TBHP to be the origin of oxygen. Additionally, deuterium labelling experiments were carried out using α-deutero styrene, which showed conservation of deuterium in the product. Based on this and previous work by Mimoun,[45] the mechanism shown in Scheme 12.10b was proposed, which was initiated by alkene coordination to Pd and TBHP insertion, followed by an α-D shift to form the product.

In addition to the Tsuji–Wacker reaction itself, Wacker cyclizations as well as the related oxidative aminations were explored using Pd carbene complexes. Muñiz reported a Wacker cyclization of allyl phenols in 2004 using $[Pd(CF_3CO_2)_2]$ and IMes under basic conditions (Equation (12.8), DMAP = 4-(dimethylamino)pyridine).[46]

5 mol % $[Pd(CF_3CO_2)_2]$
6 mol % IMes
20 mol % DMAP
2 equiv Na_2CO_3
4 Å MS, PhMe
O_2, 80°C
OH
O
91%
(12.8)

A general mechanism for this reaction is shown in Scheme 12.11 (L_n = IMes, Nu = O). It was proposed to proceed *via* initial nucleopalladation of the alkene to give Pd alkyl **28**. This intermediate would undergo β-hydride elimination, forming the product along with a Pd hydride **29**, which would give Pd^0 **30** upon reductive elimination of HX. Finally, **30** may be re-oxidized by O_2 to regenerate the active catalyst **27**.

Palladium-catalyzed oxidative amination reactions are proposed to occur by the same basic mechanism as Wacker cyclizations (Scheme 12.11, Nu = NR). Stahl and co-workers developed a catalytic system for oxidative amination using pyridine as a ligand,[47,48] but found some key challenges in this system, similar to the alcohol oxidation developed by Sigman and co-workers: (i) pyridine, a kinetically labile ligand, could dissociate under the reaction

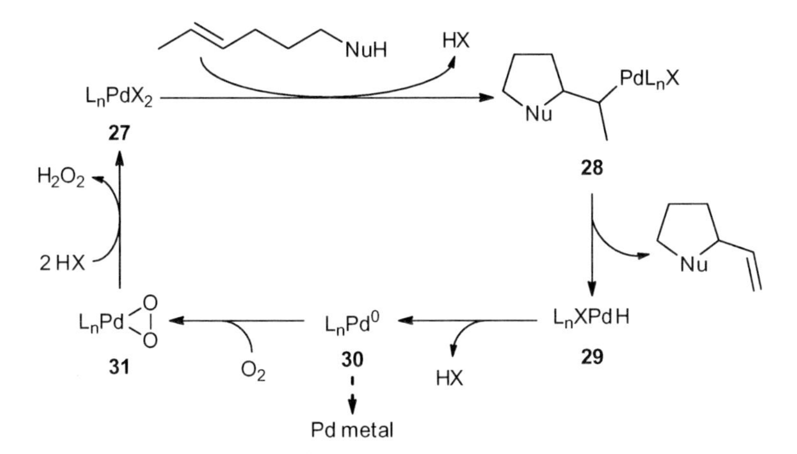

Scheme 12.11 Mechanism of Pd-catalyzed aerobic Wacker cyclization (Nu = O)/ oxidative amination (Nu = NTs) of alkenes.

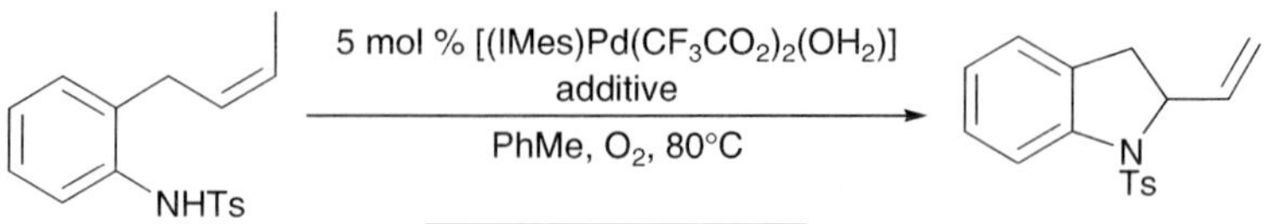

5 mol % [(IMes)Pd(CF_3CO_2)$_2$(OH_2)]
additive
PhMe, O_2, 80°C

NHTs → N Ts

additive	yield
-	88%
1 equiv $NaHCO_3$	75%
10 mol % AcOH	94%

Scheme 12.12 Effect of additives on [(IMes)Pd(CF_3CO_2)$_2$(OH_2)]-catalyzed oxidative amination.

conditions, and thus lead to catalyst instability at low ligand concentration (4 equivalents of pyridine per Pd were required); and (ii) chelating ligands, which might circumvent this problem and also be useful for asymmetric catalysis, significantly reduced catalytic activity. Based on these observations, Stahl and co-workers began to investigate NHC ligands for their reactions.

IMes was found to be a competent ligand under conditions close to those developed for pyridine complexes.[48,49] Interestingly, it was found that basic additives in general had a detrimental effect on the reactions, whereas acidic additives gave improved yields (Scheme 12.12). This was in contrast to Muñiz's Wacker cyclization, where base was needed to achieve robust catalysis (*vide supra*). While these effects were not investigated in detail, it was speculated that acetic acid might help stabilize the catalyst and/or promote the oxidation of Pd, analogous to the findings in Stahl's Pd oxidation studies and Sigman's alcohol oxidation (*vide supra*).

Stahl and co-workers then sought to develop chiral NHC ligands in order to achieve asymmetric catalysis. Toward this end, axially chiral seven-membered NHCs were developed and tested in their racemic forms initially, giving comparable results to IMes (Equation (12.9)).[49]

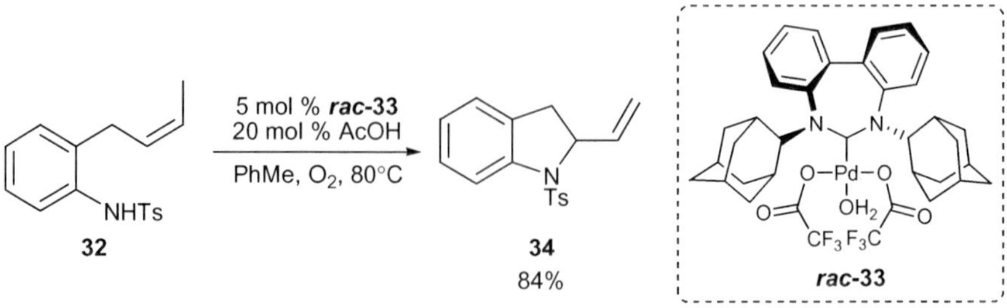

5 mol % ***rac*-33**
20 mol % AcOH
PhMe, O_2, 80°C

NHTs **32** → N Ts **34** 84%

O–Pd–O, OH_2, CF_3 F_3C ***rac*-33**

(12.9)

Having demonstrated that these were competent catalysts, methods were developed to resolve the ligands. These were then tested for asymmetric catalysis and optimized;[50] however, they met with limited success, the highest *ee* obtained being 63% (Equation (12.10)).

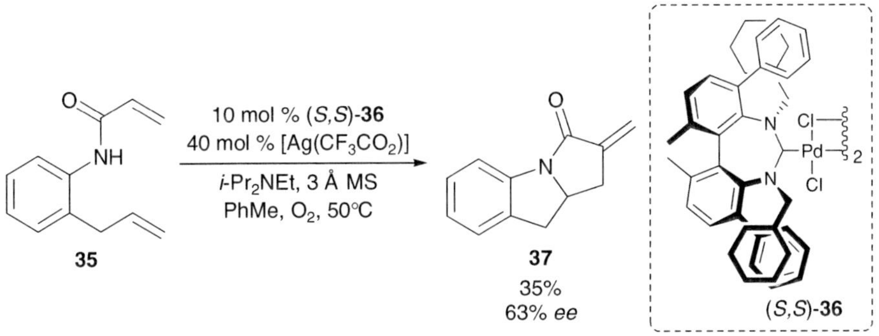

10 mol % (*S,S*)-**36**
40 mol % $[Ag(CF_3CO_2)]$
i-Pr_2NEt, 3 Å MS
PhMe, O_2, 50°C

35 → **37** 35% 63% *ee*

(*S,S*)-**36**

(12.10)

In addition to substrate **35**, substrate **32** was tested for asymmetric catalysis as well, but gave only racemic products. Finally, five-membered NHC ligands with chiral *N*-substituents were synthesized; however, no asymmetric induction was observed.

Using a different ligand manifold, specifically CNC pincer ligands, Peris and co-workers found that Ru complex **38** catalyzed both the transfer hydrogenation of ketones and the oxidative cleavage of alkenes using $NaIO_4$, show-casing again the good oxidative stability of NHC complexes (Equation (12.11)).[51]

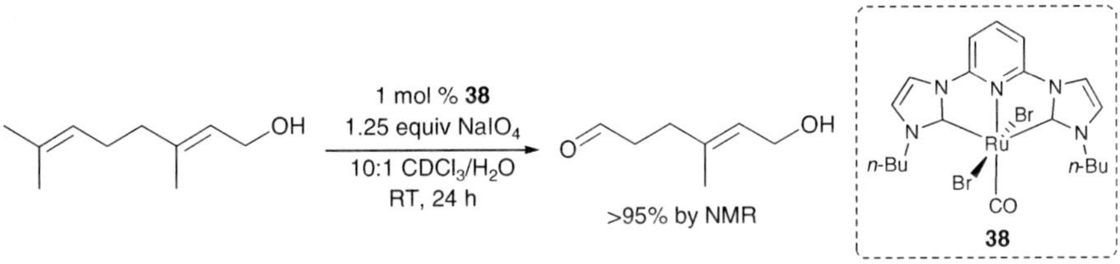

1 mol % **38**
1.25 equiv $NaIO_4$
10:1 $CDCl_3/H_2O$
RT, 24 h

>95% by NMR

38

(12.11)

Several different alkene substrates were tested, and it was found that electron-rich substrates generally reacted faster and more efficiently than slightly electron-poor ones (such as aryl-substituted alkenes). However, the selectivity for oxidative cleavage over epoxidation or diol formation was excellent in all cases. Remarkably, when geraniol was used as a substrate, clean oxidation of only one double bond was observed (Equation (12.11)), which the authors attributed to a combination of both steric and electronic parameters.

12.5 Alkane and Arene Oxidation

Both Pd and Pt are known to catalyze the oxidation of methane to methanol *via* C–H activation. This transformation poses several daunting challenges.[52] First, the product of the initial oxidation, methanol, is significantly more reactive than methane, and thus overoxidation is a problem. This was solved by performing the reaction in highly acidic media, such as H_2SO_4[53] or a mixture of $K_2S_2O_8$, trifluoroacetic acid anhydride (TFAA) and trifluoroacetic acid (TFA).[54] This led to the formation of methyl esters, which are significantly more electron poor than methane and are thus not oxidized further by the Lewis-acidic metal. However, these conditions required highly acid stable and oxidatively stable catalysts, which preclude the use of the majority of widely used ligands. Strassner and co-workers published several reports on the use of NHC ligands on Pd[54–57] and Pt[58,59] to affect this transformation in an attempt to develop new and more modular catalysts, which could then be systematically optimized. In their initial report, it was found that Pd bis-carbene complex **39** was a competent catalyst, giving a TON of 24 (Table 12.1, entry 1).[54]

Following their initial communication, Strassner and co-workers studied the influence of several parameters on the reaction. When different counter-ions were compared, no significant difference was found between chloride, bromide, and TFA complexes, while the corresponding iodide complex was unreactive (Table 12.1, entries 2 and 3).[55] It was hypothesized that the halides might be displaced by TFA counter-ions to form the active catalyst, which would explain the similar results obtained. DFT calculations were performed calculating the Pd–X bond strengths; however, the difference in bond strength between the Pd–Br and Pd–I bonds was not large enough to justify the sudden drop in activity.[55] Next, the effect of the alkane linker was examined by synthesizing ligands with 1-, 2- and 3-carbon linkers (Table 12.1, entries 1, 4, and 5).[56] Interestingly, it was found that the catalyst containing the ethylene linker **42** was the most active (both for methane oxidation and the Mizoroki–Heck reaction), with the methylene-linked bis-carbene **39** giving intermediate results and the propylene-linked bis-carbene **43** being the least effective.[56] Finally, to further modify the ligands,

Table 12.1

CH_4 (30 bar) → $F_3CCO_2CH_3$ (1 mol % [Pd}, 1 equiv $K_2S_2O_8$, 4:3 TFA:TFAA, 90°C, 17 h)

entry	[Pd]	TON^a
1	**39**	24.0
2	**40**	24.0
3	**41**	-
4	**42**	33.0
5	**43**	16.8
6	**44**	19.0
7^b	**45**	41.0

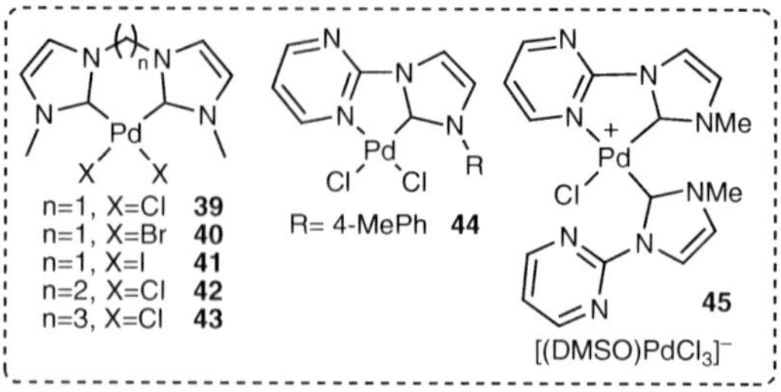

n=1, X=Cl **39**
n=1, X=Br **40**
n=1, X=I **41**
n=2, X=Cl **42**
n=3, X=Cl **43**

R= 4-MePh **44**

45 $[(DMSO)PdCl_3]^-$

^{a} Measured by GC. ^{b} 0.5 mol % catalyst were employed to keep the amount of Pd constant.

pyrimidine-functionalized NHCs were synthesized.[57] Pt–bipyrimidine complexes had been previously used as methane oxidation catalysts by Periana and co-workers,[53] and thus it was thought that a mixed system might be beneficial. A series of complexes with the general structure of **44** was synthesized with differing R groups. Interestingly, when the methyl-substituted complex **45** was synthesized, its structure was found to be drastically different from the rest of the series. Instead of a bidentate CN-ligand and two chloride anions, the Pd was coordinated to one chloride and two NHCs, with one binding bidentate and the other monodentate through the carbene, with only a weak second Pd–N interaction observed in the crystal structure. The counter-ion was an unusual $[(DMSO)PdCl_3]^-$ species. Interestingly, this complex was found to be the most active catalyst, giving a TON of 41, with all other pyrimidine-NHC catalysts in the range of 10–20.

It should be noted that Pt–biscarbene complexes were also prepared.[58,59] However, while **40** was an active methane oxidation catalyst, its Pt analogue decomposed in TFA. Aryl-substituted NHCs, however, were found to be substantially more stable to acid, and could be used for the conversion of methane to methanol, albeit in extremely low turnover (Equation (12.12)).

$$CH_4 \ (30\ bar) \xrightarrow[4:3\ TFA:TFAA,\ 90^\circ C,\ 17\ h]{1\ mol\ \%\ \mathbf{46},\ 1\ equiv\ K_2S_2O_8} F_3CCO_2CH_3 \ (TON: 4.1) \quad (12.12)$$

R = 4-MeOPh **46**

In addition to C–H bond activation in methane, NHC–Pd complexes were used as catalysts for aryl C–H activation. C–H bond activation of substrates containing a directing group using Pd^{II}/Pd^{IV} catalysis was developed by Sanford and co-workers.[60] However, these catalytic systems did not employ ligands, which would be potentially beneficial in order to achieve enhanced activity, broader scope, and asymmetric catalysis by tuning the catalyst. The set of potential ligands is limited in similar ways as in the case of methane oxidation, in that the reaction requires strongly oxidizing conditions. Again, NHCs were chosen as a modular ligand scaffold, and an initial study of their use in the bromination of benzo[*h*]quinoline was reported.[61]

$$\xrightarrow[MeCN,\ 100^\circ C,\ 48\ h]{5\ mol\ \%\ \mathbf{47},\ 1.25\ equiv\ NBS} \quad 74\% \quad (12.13)$$

It was shown in stoichiometric studies (Scheme 12.13) as well as a catalytic reaction (Equation (12.13)) that NHC–Pd complex **47** was a capable catalyst

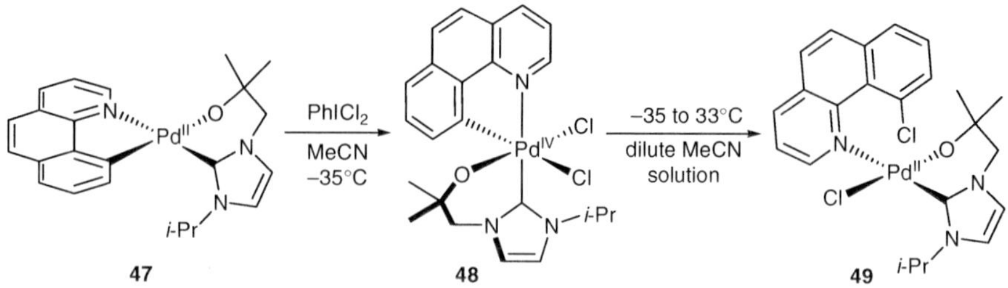

Scheme 12.13 Stoichiometric studies on a NHC–Pd^{IV} complex.

for the halogenation of arenes. It should be noted that complex **48** could be isolated and characterized, and underwent reductive elimination to form **49** only upon warming in solution. Complex **47** was found to be a slower bromination catalyst than $[Pd(OAc)_2]$, which had been previously used for the same reaction. This was attributed to slower cyclometalation with a more electron-rich Pd, which is proposed to be the rate-limiting step in this type of reaction.[62]

12.6 Conclusion

In summary, the use of N-heterocyclic carbene ligands in oxidation reactions is an exciting and growing area of research, as the publication of a number of reports over just the last 2 years shows. Although the number of mature systems and mechanistically understood oxidations using NHC catalysts is still relatively small, an expansion of the field is ongoing. As NHCs are being established as a highly versatile and robust class of ligands, more and more types of oxidative reactions are being explored using NHC complexes. Unfortunately, asymmetric oxidations using chiral NHCs are still sparse; however, as more diverse types of NHCs are developed, the development of enantioselective reactions should become more facile in the future. Thus, it should be an exciting field to observe since there are still many areas of oxidation chemistry that can be explored further.

References

1. D. Enders, O. Niemeier and A. Henseler, *Chem. Rev.*, 2007, **107**, 5606–5655.
2. N. Marion, S. Díez-González and S. P. Nolan, *Angew. Chem., Int. Ed.*, 2007, **46**, 2988–3000.
3. S. P. Nolan, (ed.), *N-Heterocyclic Carbenes in Synthesis*, Wiley-VCH, New York, 2006.
4. F. Glorius, (ed.), *N-Heterocyclic Carbenes in Transition Metal Catalysis*, Springer, Berlin, 2007.

5. P. de Frémont, N. Marion and S. P. Nolan, *Coord. Chem. Rev.*, 2009, **253**, 862–892.
6. S. Díez-González, N. Marion and S. P. Nolan, *Chem. Rev.*, 2009, **109**, 3612–3676.
7. W. A. Herrmann, *Angew. Chem., Int. Ed*, 2002, **41**, 1290–1309.
8. M. J. Schultz and M. S. Sigman, in *N-Heterocyclic Carbenes in Synthesis*, ed. S. P. Nolan, Wiley-VCH, New York, 2006, pp. 103–118.
9. M. M. Rogers and S. S. Stahl, *Top. Organomet. Chem.*, 2007, **21**, 21–46.
10. T. Strassner, *Top. Organomet. Chem.*, 2007, **22**, 125–148.
11. S. S. Stahl, *Science*, 2005, **309**, 1824–1826.
12. M. M. Konnick, I. A. Guzei and S. S. Stahl, *J. Am. Chem. Soc.*, 2004, **126**, 10212–10213.
13. M. M. Konnick, B. A. Gandhi, I. A. Guzei and S. S. Stahl, *Angew. Chem., Int. Ed.*, 2006, **45**, 2904–2907.
14. J. Muzart, *Chem.–Asian J.*, 2006, **1**, 508–515.
15. M. C. Denney, N. A. Smythe, K. L. Cetto, R. A. Kemp and K. I. Goldberg, *J. Am. Chem. Soc.*, 2006, **128**, 2508–2509.
16. M. M. Konnick and S. S. Stahl, *J. Am. Chem. Soc.*, 2008, **130**, 5753–5762.
17. J. M. Keith, W. A. Goddard III and J. Oxgaard, *J. Am. Chem. Soc.*, 2007, **129**, 10361–10369.
18. B. R. Dible and M. S. Sigman, *Inorg. Chem.*, 2006, **45**, 8430–8441.
19. J. M. Praetorius, D. P. Allen, R. Wang, J. D. Webb, F. Grein, P. Kennepohl and C. M. Crudden, *J. Am. Chem. Soc.*, 2008, **130**, 3724–3725.
20. L. J. L. Häller, E. Mas-Marzá, A. Moreno, J. P. Lowe, S. A. Macgregor, M. F. Mahon, P. S. Pregosin and M. K. Whittlesey, *J. Am. Chem. Soc.*, 2009, **131**, 9618–9619.
21. D. R. Jensen and M. S. Sigman, *Org. Lett.*, 2003, **5**, 63–65.
22. T. Chen, J.-J. Jiang, Q. Xu and M. Shi, *Org. Lett.*, 2007, **9**, 865–868.
23. D. R. Jensen, M. J. Schultz, J. A. Mueller and M. S. Sigman, *Angew. Chem., Int. Ed.*, 2003, **42**, 3810–3813.
24. J. A. Mueller, C. P. Goller and M. S. Sigman, *J. Am. Chem. Soc.*, 2004, **126**, 9724–9734.
25. Y. Iwai, K. M. Gligorich and M. S. Sigman, *Angew. Chem., Int. Ed.*, 2008, **47**, 3219–3222.
26. H. Lebel and V. Paquet, *J. Am. Chem. Soc.*, 2004, **126**, 11152–11153.
27. D. Gnanamgari, E. L. O. Sauer, N. D. Schley, C. Butler, C. D. Incarvito and R. H. Crabtree, *Organometallics*, 2009, **28**, 321–325.
28. A. Prades, R. Corberán, M. Poyatos and E. Peris, *Chem.–Eur. J.*, 2008, **14**, 11474–11479.
29. A. Pontes da Costa, M. Viciano, M. Sanaú, S. Merino, J. Tejeda, E. Peris and B. Royo, *Organometallics*, 2008, **27**, 1305–1309.
30. A. Prades, M. Viciano, M. Sanaú and E. Peris, *Organometallics*, 2008, **27**, 4254–4259.
31. M. Viciano, M. Sanaú and E. Peris, *Organometallics*, 2007, **26**, 6050–6054.
32. L. U. Nordstrøm, H. Vogt and R. Madsen, *J. Am. Chem. Soc.*, 2008, **130**, 17672–17673.

33. S. Burling, B. M. Paine, D. Nama, V. S. Brown, M. F. Mahon, T. J. Prior, P. S. Pregosin, M. K. Whittlesey and J. M. J. Williams, *J. Am. Chem. Soc.*, 2007, **129**, 1987–1995.
34. D. Balcells, A. Nova, E. Clot, D. Gnanamgari, R. H. Crabtree and O. Eisenstein, *Organometallics*, 2008, **27**, 2529–2535.
35. F. Hanasaka, K.-i. Fujita and R. Yamaguchi, *Organometallics*, 2004, **23**, 1490–1492.
36. F. Hanasaka, K.-i. Fujita and R. Yamaguchi, *Organometallics*, 2005, **24**, 3422–3433.
37. F. Hanasaka, K.-i. Fujita and R. Yamaguchi, *Organometallics*, 2006, **25**, 4643–4647.
38. S. C. Ghosh, S. Muthaiah, Y. Zhang, X. Xu and S. H. Hong, *Adv. Synth. Catal.*, 2009, **351**, 2643–2649.
39. J. M. Takacs and X.-t. Jiang, *Curr. Org. Chem.*, 2003, **7**, 369–396.
40. J. Tsuji, *Synthesis*, 1984, 369–383.
41. C. N. Cornell and M. S. Sigman, *Inorg. Chem.*, 2007, **46**, 1903–1909.
42. G.-J. ten Brink, I. W. C. E. Arends, G. Papadogianakis and R. A. Sheldon, *Chem. Commun.*, 1998, 2359–2360.
43. T. Nishimura, N. Kakiuchi, T. Onoue, K. Ohe and S. Uemura, *J. Chem. Soc., Perkin Trans. 1*, 2000, 1915–1918.
44. C. N. Cornell and M. S. Sigman, *J. Am. Chem. Soc.*, 2005, **127**, 2796–2797.
45. H. Mimoun, R. Charpentier, A. Mitschler, J. Fischer and R. Weiss, *J. Am. Chem. Soc.*, 1980, **102**, 1047–1054.
46. K. Muñiz, *Adv. Synth. Catal.*, 2004, **346**, 1425–1428.
47. S. R. Fix, J. L. Brice and S. S. Stahl, *Angew. Chem., Int. Ed.*, 2002, **41**, 164–166.
48. V. Kotov, C. C. Scarborough and S. S. Stahl, *Inorg. Chem.*, 2007, **46**, 1910–1923.
49. M. M. Rogers, J. E. Wendlandt, I. A. Guzei and S. S. Stahl, *Org. Lett.*, 2006, **8**, 2257–2260.
50. C. C. Scarborough, A. Bergant, G. T. Sazama, I. A. Guzei, L. C. Spencer and S. S. Stahl, *Tetrahedron*, 2009, **65**, 5084–5092.
51. M. Poyatos, J. A. Mata, E. Falomir, R. H. Crabtree and E. Peris, *Organometallics*, 2003, **22**, 1110–1114.
52. B. L. Conley, W. J. Tenn III, K. J. H. Young, S. K. Ganesh, S. K. Meier, V. R. Ziatdinov, O. Mironov, J. Oxgaard, J. Gonzales, W. A. Goddard III and R. A. Periana, *J. Mol. Catal. A: Chem.*, 2006, **251**, 8–23.
53. R. A. Periana, D. J. Taube, S. Gamble, H. Taube, T. Satoh and H. Fujii, *Science*, 1998, **280**, 560–564.
54. M. Muehlhofer, T. Strassner and W. A. Herrmann, *Angew. Chem., Int. Ed. Engl.*, 2002, **41**, 1745–1747.
55. T. Strassner, M. Muehlhofer, A. Zeller, E. Herdtweck and W. A. Herrmann, *J. Organomet. Chem.*, 2004, **689**, 1418–1424.
56. S. Ahrens, A. Zeller, M. Taige and T. Strassner, *Organometallics*, 2006, **25**, 5409–5415.

57. D. Meyer, M. A. Taige, A. Zeller, K. Hohlfeld, S. Ahrens and T. Strassner, *Organometallics*, 2009, **28**, 2142–2149.
58. S. Ahrens, E. Herdtweck, S. Goutal and T. Strassner, *Eur. J. Inorg. Chem.*, 2006, 1268–1274.
59. S. Ahrens and T. Strassner, *Inorg. Chim. Acta*, 2006, **359**, 4789–4796.
60. A. R. Dick and M. S. Sanford, *Tetrahedron*, 2006, **62**, 2439–2463.
61. P. L. Arnold, M. S. Sanford and S. M. Pearson, *J. Am. Chem. Soc.*, 2009, **131**, 13912–13913.
62. L. V. Desai, K. J. Stowers and M. S. Sanford, *J. Am. Chem. Soc.*, 2008, **130**, 13285–13293.

CHAPTER 13

Reduction Reactions with NHC-bearing Complexes

BEKIR ÇETINKAYA

Ege University, Department of Chemistry, 35100, Bornova, Izmir, Turkey

13.1 Introduction

The aim of this Chapter is to examine the application of well-defined N-heterocyclic carbene (NHC) complexes as well as the systems prepared *in situ* which involve free NHCs or the precursor salt for the reduction of unsaturated organic molecules such as alkynes, alkenes and carbonyl compounds. The most active complexes for such reductions contain electron-rich, late transition metals in low oxidation states. Herein, reductions useful for organic synthesis will be classified into four types according to reductants used: (i) hydrogenations, (ii) transfer hydrogenation, (iii) hydrosilylation and (iv) hydroboration. For examples of reduction reactions with systems containing non-classical NHC ligands, the reader is referred to Chapter 5.

13.2 Hydrogenation

Hydrogenation refers to the addition of hydrogen to unsaturated molecules with ideal atom economy, whereby both hydrogen atoms are incorporated in the reduction product. Homogeneously catalyzed hydrogenations play an important role in a number of synthetic routes in the chemical and pharmaceutical industry, as well as in the preparation of fine chemicals such as synthetic fragrances, odours and agrochemicals. A particular advantage of homogeneous catalysts lies in their high potential to direct the reaction in a chemoselective, regioselective and stereoselective manner.

RSC Catalysis Series No. 6
N-Heterocyclic Carbenes: From Laboratory Curiosities to Efficient Synthetic Tools
Edited by Silvia Díez-González
© Royal Society of Chemistry 2011
Published by the Royal Society of Chemistry, www.rsc.org

13.2.1 Hydrogenation of Alkenes and Carbonyl Compounds

The earliest homogeneous hydrogenation catalyst was Wilkinson catalyst $[Rh(PPh_3)_3Cl]$ **1** which was active at 1 atm of H_2 pressure at room temperature for monosubstituted and *cis*-disubstituted alkenes such as cyclohexene.[1] The second, $[Rh(COD)(PPh_3)_2]PF_6$ **2**, and third, $[Ir(COD)(PCy_3)(pyr)]PF_6$ **3**, generation catalysts were developed by Osborn–Schrock[2] and Crabtree,[3] respectively.

Nolan and co-workers first reinvestigated Crabtree's original catalyst in 2001 replacing PCy_3 in **3** with SIMes.[4] Although, the new cationic complex [(SIMes) Ir(COD)(pyr)]PF_6 **4** was less efficient than **3** at room temperature, but it exhibited higher activity at 50 °C for the hydrogenation of simple olefins such as cyclohexene. Additionally, **4** was thermally more stable than $[Ir(COD)(PCy_3)(pyr)]PF_6$.

Buriak and co-workers chose a different modification strategy to Crabtree's catalysts and they replaced the pyridine with various alkyl phosphines.[5] Optimization studies revealed that complex $[(IMes)Ir(COD)(PBu_3)]BARF$ **5** (BARF = tetrakis[3,5-bis(triflouromethylphenyl)borate]) was the most stable and displayed similar activity to **3**. Furthermore, **5** was active for highly hindered tri- and tetrasubstituted alkenes at 1 atm pressure of H_2 in the presence of air.

Comparison of $[(IMes)RuCl(H)(PCy_3)(CO)]$ **6** with the parent $[RuCl(H)(PCy_3)_2(CO)]$ in the hydrogenation of 1-alkenes at 4 atm of H_2 showed that **6** was less active at room temperature, whereas a comparable activity was achieved at 100 °C.[6] This decreased activity was attributed to the enhanced congestion around the ruthenium centre.

A study by Fogg and co-workers compared the activities of the carbonyl complexes $[RuCl(H)(CO)(PCy_3)_2]$/$[(IMes)RuCl(H)(CO)(PCy_3)]$ with the dihydrogen complexes, $[RuCl(H)(H_2)(PCy_3)_2]$/$[(IMes)RuCl(H)(H_2)(PCy_3)]$. The former were found to be more efficient than the latter for the hydrogenation of various alkenes. This difference was attributed to the higher stability of the carbonyl complexes.[7] In related work, Whittlesey and co-workers found that $[(IMes)_2Ru(CO)H(\mu^6\text{-}BH_4)]$ **7** could be used as a catalyst for the hydrogenation of aryl methyl ketones under 10 atm of H_2.[8]

Water-soluble NHC–Ru complex **8** was tested in the biphasic hydrogenation of styrene to ethylbenzene and found to be moderately active (Figure 13.1).[9] Related complex **9** exhibited a higher activity in biphasic systems and it was also applied to carbonyl compounds.[10] Albrecht and co-workers reported the catalytic activity of several η^5-Cp and η^6-arene bearing ruthenium complexes bearing donor-functionalized imidazol-2-ylidene.[11] Whereas a thioether or a second NHC site deactivated the metal centre (**10**, **11**), the highest activity (99% conversion) was obtained with the neutral complex **12**, with a carboxylate donor group (Figure 13.1).

Nolan and Lebel reported $[(IMes)RhCl(PPh_3)_3]$ **13**, an analogue of Wilkinson catalyst, and tested it for the hydrogenation of cyclohexene.[12] Crudden and co-workers prepared more analogues with different phosphines.[13] The highest activity for the hydrogenation of β-methylstyrene was obtained with $P(p\text{-}FC_6H_4)_3$. Furthermore, the additive CuCl, known to be a phosphine sponge, enhanced the catalytic activity.

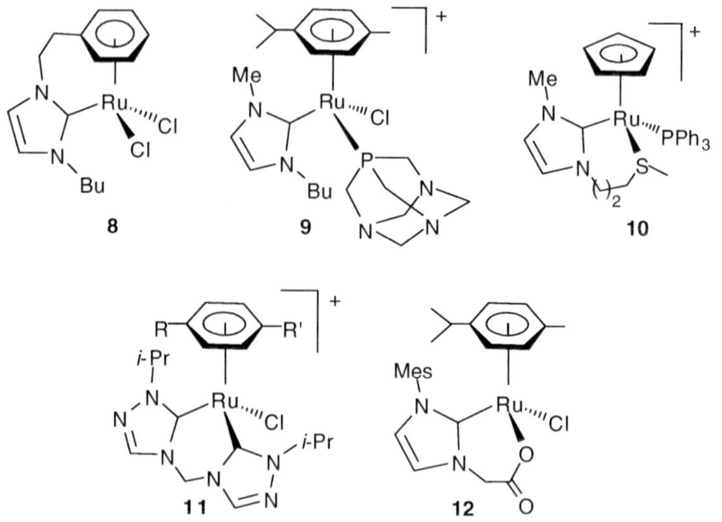

Figure 13.1 [(NHC)Ru] catalysts for hydrogenation reactions.

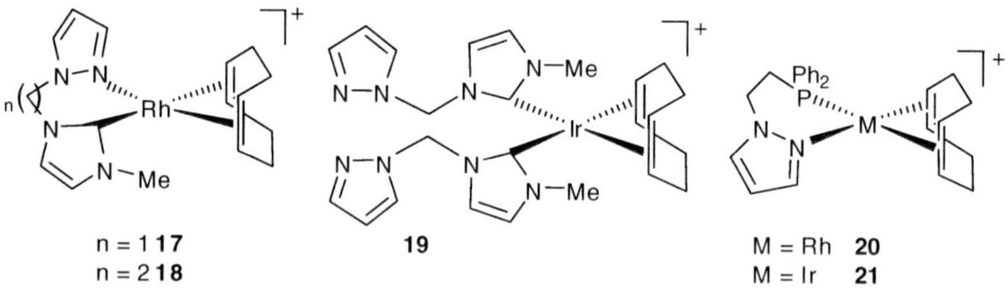

Figure 13.2 [(NHC)Rh] and [(NHC)Ir] catalysts for hydrogenation reactions.

Herrmann *et al.* reported the hydrogenation activity of various NHC–Rh complexes [(ICy)Rh(COD)Cl] **14**, [$(IMe)_2$Rh(COD)]Cl **15**, and [(IBn)RhBr$(CO)_2$] **16**.[14] ICy-containing **14** produced the best results, whereas the carbonyl complex **16** was inert in the hydrogenation of cyclohexene. In these reactions, PPh_3 was used as additive to stabilized low coordinated catalytic species.

Rh^I and Ir^I complexes containing NHC–pyrazolyl chelate ligands were tested for hydrogenation of styrene using 1 mol% of complexes **17–21**.[15] Complex **19**, with a non-chelating bis-carbene, showed no detectible activity. However, **20** and **21** were active under the same reaction conditions (Figure 13.2). This result clearly indicates the importance of the donor functionality, regardless of the size of the chelating ring.

James and co-workers studied five-coordinated dihydrido bis-carbene NHC–Rh species [$(IPr)_2$RhCl$(H)_2$] **22**, [$(IMes)_2$RhCl$(H)_2$] **23** and [(IMes)(IPr)RhCl$(H)_2$] **24** for the hydrogenation of cyclooctene and 1-octene under ambient conditions.[16]

Conversion of cyclooctene did not exceed 15% and the formation of colloidal metal was noticeable. In the case of 1-octene, the major reaction was isomerization to *cis*- and *trans*-2-octene.

In contrast with the well-developed [(NHC)Pd]-catalyzed cross-coupling reactions, reduction reactions involving NHC–Pd catalysts have only seen limited developments so far. This may be due to their propensity to decomposition through reductive elimination, under hydrogenation conditions.[17] Indeed, $[(ItBu)_2Pd]$ was reported to decompose in the presence of H_2 to form palladium black and 1,3-di-*tert*-butylimidazolidine.[18] However, very recently, Nolan and co-workers reported that the NHC–phosphine mixed complex, $[(SIPr)Pd(PCy_3)]$ **25**, was assumed to undergo an oxidative–addition reaction with a hydrogen molecule to give a *trans*-dihydride *trans*-$[(SIPr)Pd(PCy_3)(H)_2]$ **26**, a very efficient hydrogenation catalyst for a variety of alkenes under ambient conditions (93–99% conversion).[19] This catalytic system was also active for sterically hindered substrates such as tri- and tetra substituted alkenes.

13.2.2 Asymmetric Catalysis

The application of NHC ligands in this reaction has naturally been extended to asymmetric processes. In general, the chiral induction of NHCs was low, probably due to the rapid internal rotation of the chiral substituents around the C–N axis. Also, metalation procedures involving deprotonation of a chiral azolium precursor with strong bases may cause loss of chirality in the desired complex. In contrast, the silver transmetalation route generally yields the optically active target complex.[20] Since the first reports by Herrmann *et al.*, and shortly after by Enders *et al.* in 1996–1997 on the synthesis and applications of chiral NHC–Rh complexes in the hydrosilylation of carbonyl compounds,[21] increased efforts were made in this area.[22]

Encouraged by the good results obtained with the phosphine-oxazoline ligands developed by Pfaltz (**27**),[23] several research groups synthesized Ir^I complexes with NHC–oxazoline-chelating ligands and used them as reduction catalysts. Asymmetric hydrogenations with excellent *ee* values (>99%) were first reported by Burgess and co-workers in 2001 (Figure 13.3). After several modifications, they applied it to very complex substrates, giving products with up to 99% *ee* in 98% yield.[24] In 2003, Chung and co-workers synthesized complex **29**, which contained a chiral ferrocenylmethyl group as a substituent on a nitrogen atom. They applied **29** to the hydrogenation of dimethyl itaconate at 50 °C and at 10 atm of H_2. After 12 h, **29** gave 44% yield with 18% *ee*. Surprisingly, a very similar complex, containing a PPh_2 unit instead of SPh did not show any activity.[25]

Herrmann *et al.* studied a number of monodentate NHC ligands bearing chiral groups on the backbone derived from 2,2′-bipiperidine or partially reduced bis-isoquinoline. The NHC complexes of Rh^I and Ir^I were used for the hydrogenation of methyl 2-acetamidoacrylate and all complexes exhibited

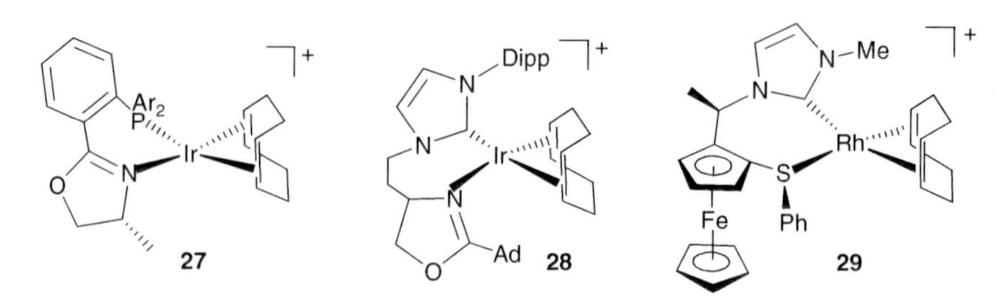

Figure 13.3 Chiral [(NHC)M] catalysts for hydrogenation reactions (Dipp = 2,6-diisopropylphenyl).

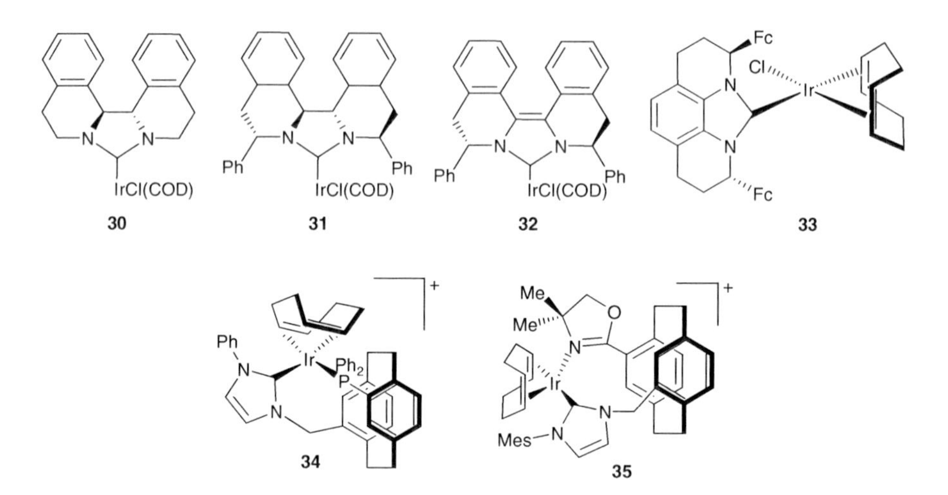

Figure 13.4 Iridium-based catalysts for asymmetric hydrogenation reactions.

significant catalytic activity over 16 h at room temperature and 30 bar of H_2. However, the optical induction was poor, with the best catalyst **30** giving only 22% *ee* (Figure 13.4).[26] The incorporation of substituents into the 3,3′-positions of a partially reduced bis-isoquinoline system was investigated in order to obtain a C_2-symmetric rigid chiral architecture. The corresponding complexes were applied to the hydrogenation of methyl 2-acetamidoacrylate, but only iridium complexes **31** and **32** displayed moderate enantioselectivity, 60% and 67%, respectively. The difference between 3,3′-disubstituted and unsubstituted NHCs were attributed to the strict rigidity.[27]

Complex **33**, containing a partially reduced 1,10-phenantroline-based monodentate NHC ligand led to 97% yield and 81% *ee* in the hydrogenation of methyl acetamidoacrylate (Figure 13.4).[28] A bidentate NHC bearing a PPh_2 unit with a chiral pseudo-*o*-[2.2]paracyclophane linker (**34**) was found to be a selective catalysts for non-functionalized trisubstituted alkenes, although functionalized alkenes were found to be sensitive to the H_2 pressure.[29] Also, the

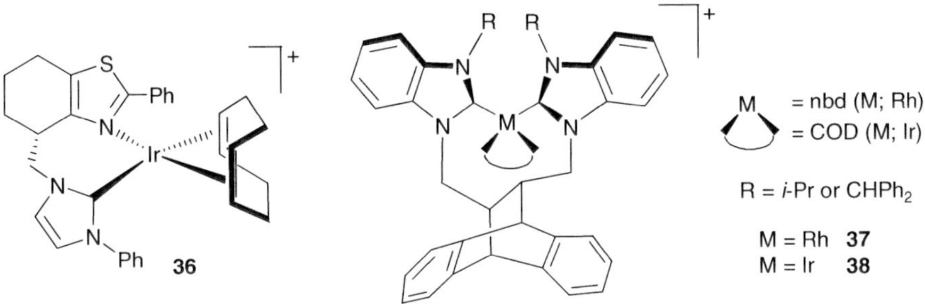

Figure 13.5 Further catalysts for hydrogenation reactions (nbd = norbornadiene).

planar chirality of pseudo-*o*-[2.2]paracyclophane was combined with a chiral oxazoline and imidazol-2-ylidene and the resulting complex **35** achieved a modest enantioselectivity in the hydrogenation of dimethyl itaconate (46% *ee*).[30] It is worth noting that the combination of two chiral groups in one ligand does not necessarily lead to better selectivity.

Another interesting chiral ligand was prepared by incorporation of a partially reduced benzothiazole scaffold to an NHC ligand.[31] This ligand coordinated to Ir^{I} to give catalytically active cationic complex **36**, which efficiently catalyzed the reduction of a number of trisubstituted functional alkenes with high *ee* values (>90%) (Figure 13.5). Finally, C_2-symmetric chelating bis-NHC ligands based on a *trans*-ethanoanthrecene backbone and their monometallic Rh^{I} and Ir^{I} complexes were applied to the hydrogenation of *trans*-methylstilbene and methyl 2-acetamidoacrylate.[32] *trans*-Methylstilbene is difficult to hydrogenate and iridium catalyst **38** was not effective. However, the observed conversions with **37** were greater than 98% even though this was accompanied by the formation of rhodium black. There was no optical induction and mercury addition inhibited the hydrogenation reaction, indicating that M^{0} species formed under catalytic conditions may be the active catalysts. Therefore, even chelating bis-NHC ligands seem susceptible to reductive elimination under hydrogenation conditions.

13.3 Transfer Hydrogenation

Transfer hydrogenation (TH) is the addition of hydrogen to an unsaturated molecule by a reagent other than H_2. Typically a sacrificial reagent (hydrogen donor) such as isopropanol, together with a strong base and a Ru, Rh or Ir catalyst are used (Scheme 13.1). TH generates few by-products, avoids hazardous reagents, and generally uses readily available and benign starting materials.

TH is preferred for large-scale industrial uses with the objective of developing a greener process by reducing toxicity, waste production and energy use.[33] However, TH is more difficult for imines than for aldehydes and ketones since imine nitrogen lone pair coordination to a metal centre is believed to inhibit

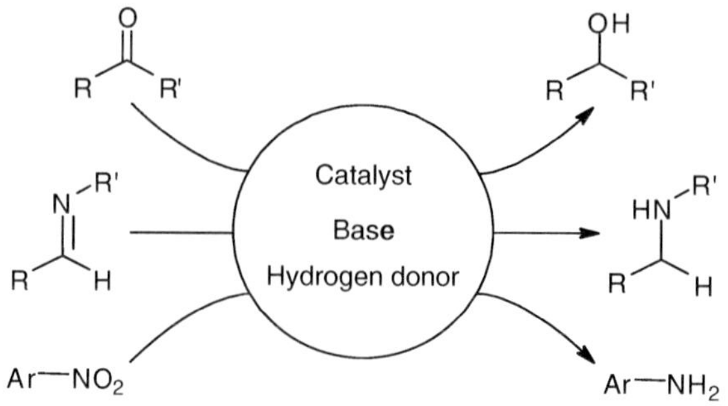

Scheme 13.1 Transfer hydrogenation reactions.

isomerization to the π-bound form that is needed for catalysis. Alkene TH are also limited by lower polarity of the C=C bond.

13.3.1 Carbonyl and Imine Reductions

TH of ketones by ruthenium-based catalysts represents the vast majority of protocols, following the seminal studies of the groups of Noyori and Zhang.[34] For TH of carbonyl compounds by *i*-PrOH several mechanisms have been proposed. The inner-sphere TH mechanism involves four steps: (i) coordination of the substrate to a coordinatively unsaturated metal-hydride species, (ii) formation of an alkoxy-metal intermediate *via* hydrogen migration from metal to the carbonyl carbon, (iii) exchange of alkoxy group with the alcohol acting as hydrogen donor (here *i*-PrOH), and (iv) β-elimination (here of isopropoxy group as acetone).[35]

In 2001, the Grubbs group reported the use of their second generation ruthenium complex [(SIMes)RuCl$_2$(=CHPh)(PCy$_3$)] **39** in tandem metathesis/TH reactions leading to the formation of a series of unsaturated heterocyclic alcohols in up to 56% overall yield.[36] This strategy was notably used for the preparation of (*R*)-(−)-muscone, a natural product with a pleasant fragrance.

N-Functionalized pincer bis-carbene ruthenium complex **40**, in the presence of KO*i*-Pr or KO*t*-Bu, catalyzed the reduction of C=O and C=N functionalities by TH from isopropanol at 82 °C (Figure 13.6).[37] Crabtree, Peris and co-workers used a related pincer ruthenium complex **41** under similar conditions to convert aryl and alkyl ketones to the corresponding alcohols in quantitative yields.[38] A 1,2,4-triazol-5-ylidene-bearing complex **42** was applied to the TH of various arylalkyl, cyclic and dialkyl ketones at 82 °C in NaOH/*i*-PrOH.[39] The substrates were converted almost quantitatively to the corresponding alcohols within a few minutes, with TOF values of $>50\,000\ h^{-1}$. This high activity was attributed to the combination of bidentate carbene and the 2-aminomethylpyridine ligand. Chiu and Lee reported that *fac*-[Ru$_2$(μ-Cl)$_3$(PCP)$_2$]Cl **43** was an efficient TH catalyst for a variety of ketones at 82 °C using isopropanol as hydrogen donor.[40]

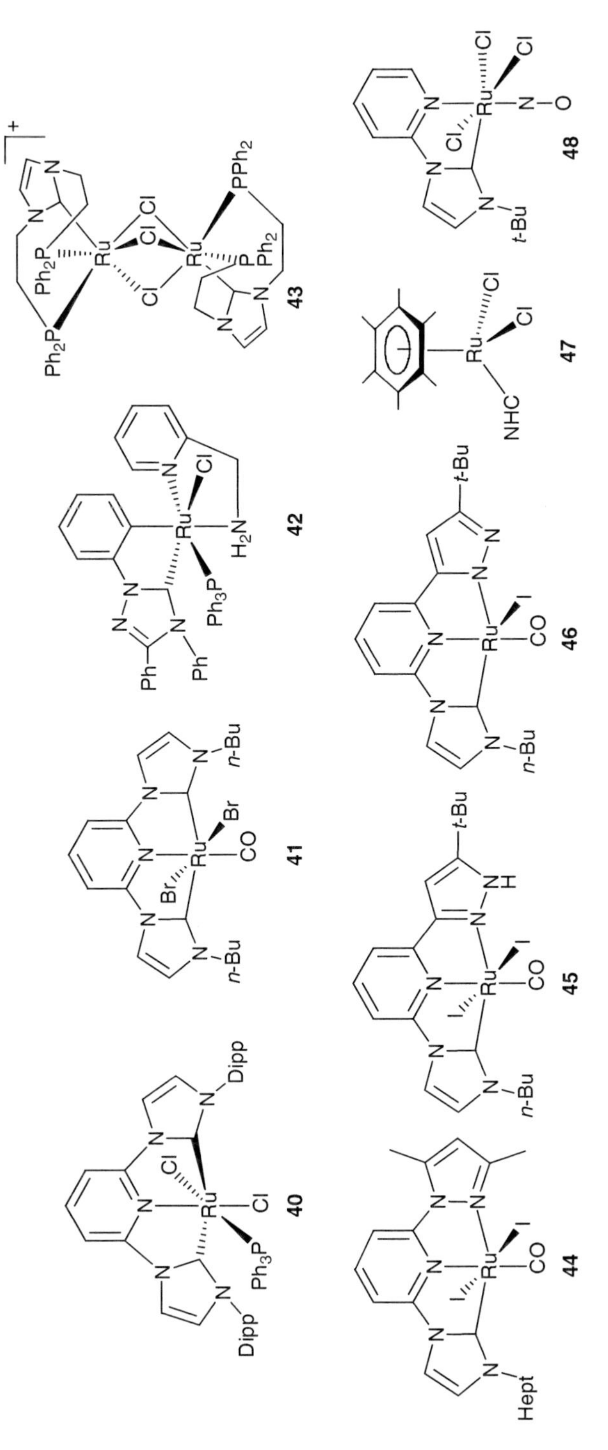

Figure 13.6 [(NHC)Ru] catalysts for TH.

18-Electron and 16-electron ruthenium complexes bearing pyridyl-supported pyrazolyl–NHC ligands were also investigated for TH of ketones at 82 °C, using KO*i*-Pr/*i*-PrOH (Figure 13.6).[41] Under the same conditions complexes **45** and **46** showed almost the same activity, whereas complex **44** was less efficient. This similarity was attributed to the transformation of **45** into **46** in the presence of the base. A number of η^6-arene ruthenium NHC complexes **47** were investigated for TH of arylalkyl ketones. In the presence of silver triflate (1 equiv) in isopropanol at 82 °C, good to excellent yields (62–99%) were obtained.[42] Ruthenium(II)/nitrosyl complex **48** bearing a pyridine-functionalized–NHC ligand was only moderately efficient with sterically hindered and electronically rich substrates.[43]

Cationic complex **49**, related to Crabtree's catalyst, was one of the first NHC complexes used for TH of ketones, alkenes and nitro compounds in refluxing isopropanol (Figure 13.7).[44] Related neutral complexes **50** were also very active in the reduction of acetophenone in KOH/*i*-PrOH.[45] Complexes of Rh^I and Ir^I with benzimidazol-2-ylidenes **51** provided 99% yield after 90 min in the TH of acetophenone and cyclohexanone in KOH/*i*-PrOH.[46] Such activity was attributed to the hemilabile–OMe group. Also, pyridinylimidazol-2-ylidene Ir^I complex **52** was applied to the reduction of benzophenone and *p*-nitrobromobenzene.[47]

Quinoline-functionalized NHC complex **53** was tested with various ketones in refluxing KOH/*i*-PrOH.[48] Substrates with electron-withdrawing substituents were more reactive which was taken as indicative of the hydridic nature of the reducing species.

Complex **54** displayed a high activity in the reduction of cyclohexanone with a TOF value of 6000 h^{-1} (Figure 13.7).[49] Tetrahydropyridoimidazolin-2-ylidene-containing **55** was also evaluated for TH of acetophenone and cyclohexanone in *i*-PrOH/KOH.[50] Cationic triazol-5-ylidene complexes **56** were found highly active for the TH of C=O and C=C functions as well as for reductive amination of aldehydes.[51]

Bis-carbene complexes of Rh^{III} **57** were applied to the TH of alkenes and arylalkyl, cyclic and dialkyl ketones at 82 °C in NaOH/*i*-PrOH (Figure 13.8).[52] Whereas TH of alkenes failed, ketones were converted almost quantitatively to the corresponding alcohols within 2 h, with TON > 19 000. Neopentyl-substituted NHC complexes **58** were also very efficient for reductions of carbonyl groups in refluxing K_2CO_3/*i*-PrOH.[53] For CNC-pincer-containing **59**, the rate of TH was found to be faster for aromatic ketones than for aliphatic ketones.[54]

Chelating pyrimidine–NHC complex **60** was active in TH of C=O and C=N bonds in KOH/*i*-PrOH at 82 °C,[55] whereas Cp*–NHC-containing **61** was applied to TH of ketones in KOH/*i*-PrOH at 82 °C (Figure 13.8).[56] Also, compound **62** with AgOTf (1:3) led to quantitative conversions over 15 min. Interestingly, this TH reaction proceeded in the absence of base, of importance for the reduction of base sensitive aldehydes and ketones such as phenyl benzyl ketone.[57] Ir^{III} complexes **63** with a bis-chelating *N*-alkenyl NHC were applied to the TH of ketones and imines in *i*-PrOH/KOH at 82 °C. Quantitative conversions were obtained and the authors proposed a pre-activation step to generate a vacant site by hydrogenation of the coordinating alkene, to afford an alkyl fragment.[58]

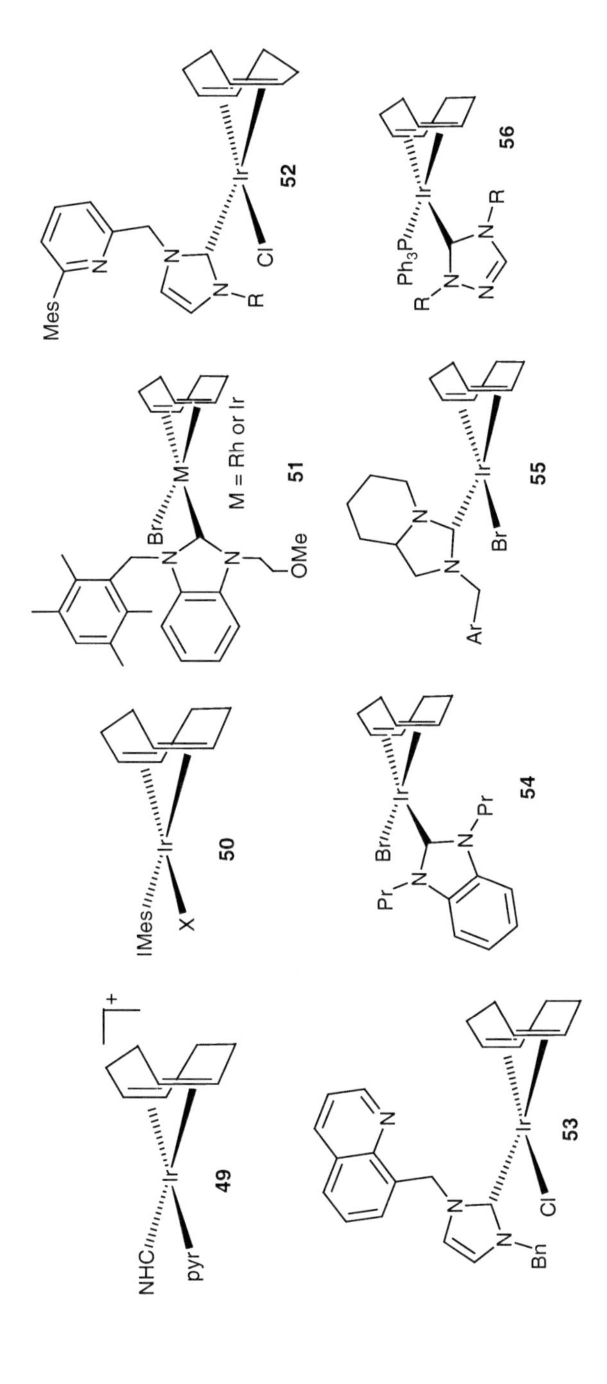

Figure 13.7 Rhodium(I) and iridium(I) catalysts for TH.

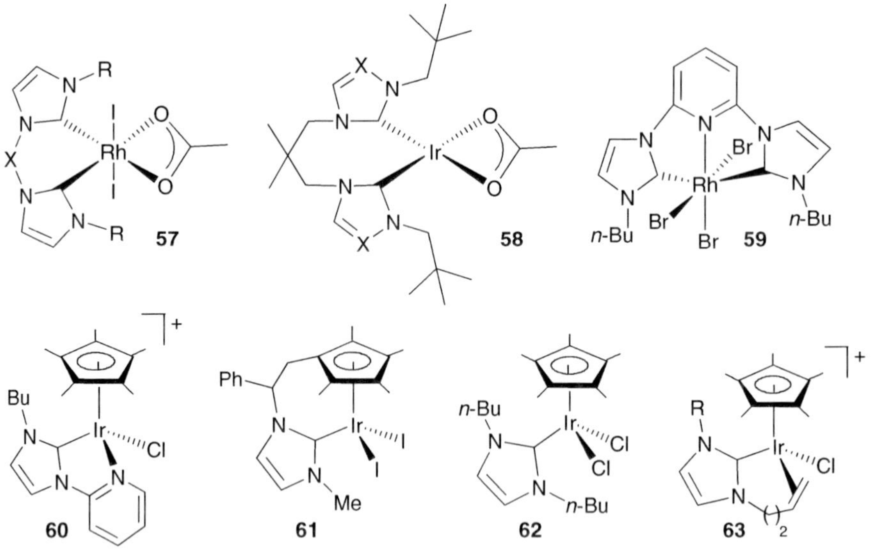

Figure 13.8 Rhodium(III) and iridium(III) catalysts for TH.

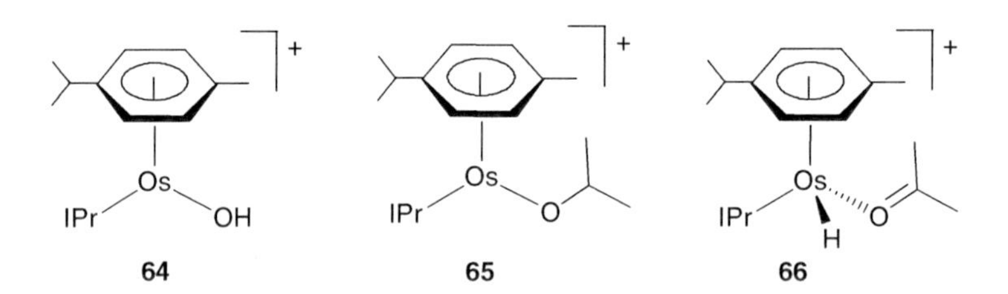

Figure 13.9 [(NHC)Os] complexes in TH.

A NHC–Os complex **64** was reported active in TH of various ketones in refluxing *i*-PrOH (Figure 13.9).[59] It is worth noting that no base was required for these reactions. Furthermore, the authors observed that in isopropanol at $-30\,^{\circ}C$, the hydroxo complex converted into **65**, an active monohydrido catalyst and subsequently led to **66**, which was stable enough to isolate. Thus, this study threw some light into the mechanism of TH of ketones by *i*-PrOH.[35]

For TH of various imines, an *in situ* formed IMes–Ni^0 catalyst was studied in dioxane, and among the bases tested, sodium pentane-3-alkoxide gave the best results (Equation (13.1)).[60]

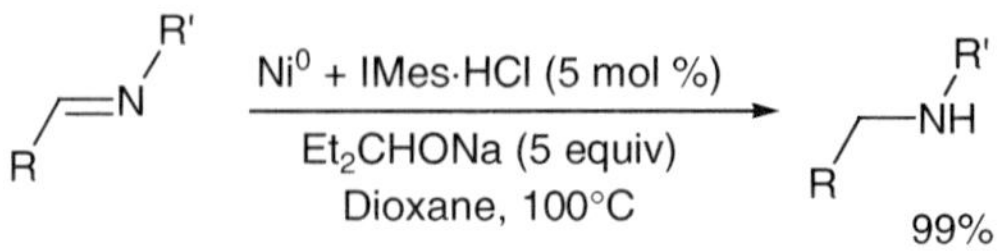

$$\mathrm{RCH{=}NR'} \xrightarrow[\text{Et}_2\text{CHONa (5 equiv), Dioxane, 100°C}]{\text{Ni}^0\text{ + IMes·HCl (5 mol \%)}} \mathrm{RCH_2NHR'}\ \ (99\%) \qquad (13.1)$$

Finally, a Pd^0 complex **67** was applied to TH of alkynes (Equation (13.2)). Internal substrates were selectively hydrogenated with excellent yields and over-reduction to alkane was almost negligible.[61]

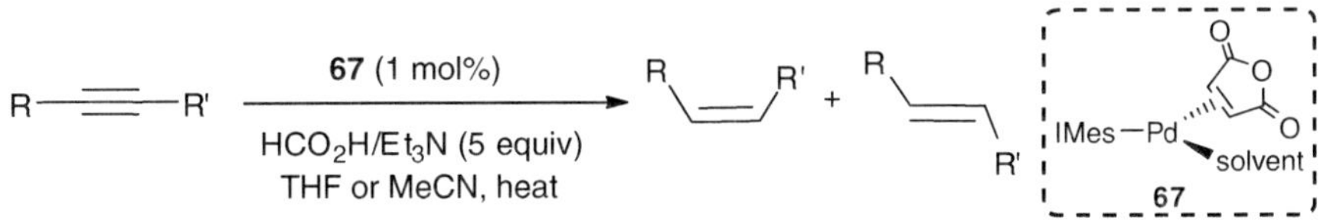

$$R{-}C{\equiv}C{-}R' \xrightarrow[\text{THF or MeCN, heat}]{\textbf{67}\text{ (1 mol\%)},\ HCO_2H/Et_3N\text{ (5 equiv)}} \text{(Z)-}RCH{=}CHR' + \text{(E)-}RCH{=}CHR' \quad (13.2)$$

13.3.2 Asymmetric Transfer Hydrogenation

Asymmetric reduction of prochiral ketones using TH is an effective and mild route for the formation of optically active secondary alcohols.[62] Generally, NHC-based systems display good catalytic activity but the obtained *ee* values are low to moderate.

Chiral ferrocenyl-functionalized NHC–rhodium(I) and iridium(I) complexes were studied by Chung and co-workers in asymmetric transfer hydrogenation (ATH) of ketones in refluxing KO*t*-Bu/*i*-PrOH.[63] With the exception of 4′-chloroacetophenone, the yields were quantitative, but the enantioselectivities only moderate even if **68** led to higher *ee* values (>53%) than its rhodium analogue (Figure 13.10). Iridium catalysts derived from (1*R*,2*R*)-*trans*-diaminocyclohexane hybrid NHC **69** were used *in situ* for TH of acetophenone in isopropanol at 82 °C. Yields ranged from 68% to 96% for 1–5 h heating; however, *ee* values and did not exceed 34%.[64] The chiral carbene rhodium complex **70** was applied to ATH of acetophenone to give (*S*)-1-phenylethanol

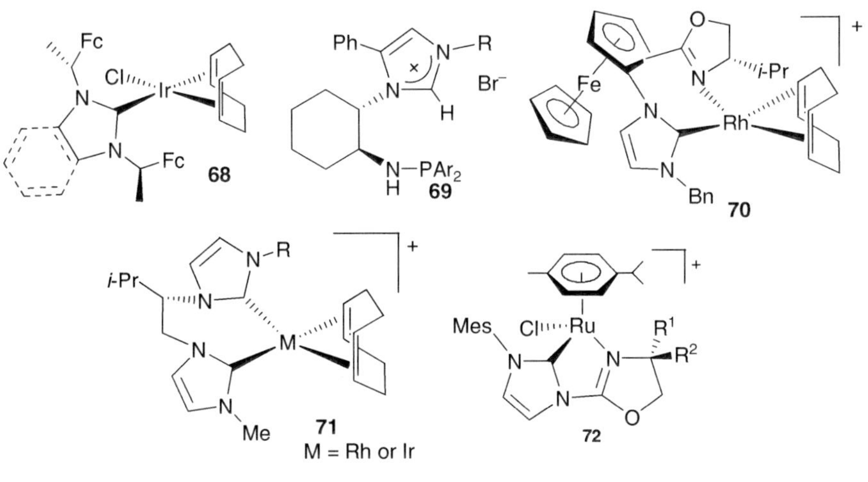

Figure 13.10 NHC-based catalysts for ATH.

with 88% yield and 60% *ee*. The bulky groups in the ketone showed a positive effect on the enantioselectivity, but lowered the reaction rate.[65]

Rh^I and Ir^I NHC derived from amino acids such as **71** were found to be promising catalysts.[66] Finally, Ru^{II} complexes prepared from chiral oxazolines containing imidazol-2-ylidenes were used for ATH of acetophenone and alkyl ketones in refluxing KOH/*i*-PrOH. However, **72** only exhibited a moderate activity and no chiral induction.[67]

13.3.3 Borrowing Hydrogen Methodology

The borrowing hydrogen approach (also called auto-transfer process) is another version of TH and refers to a recently developed method for the formation of C–C and C–N bonds in the presence of ruthenium or iridium catalysts.[68] Initially, the catalysts, [M], remove H_2 from a primary and secondary alcohol to give carbonyl compounds which then undergo addition reactions and "borrowed" hydrogen is consumed in the final stage (Scheme 13.2). Thus, in this mechanism both dehydrogenation and hydrogenation occurs and as a consequence there is no net oxidation or reduction.

A number of *N*-alkyl substituted NHC–Ru-hydride complexes were studied in this context. Ruthenecycle **73** was readily hydrogenated to **74** displaying the

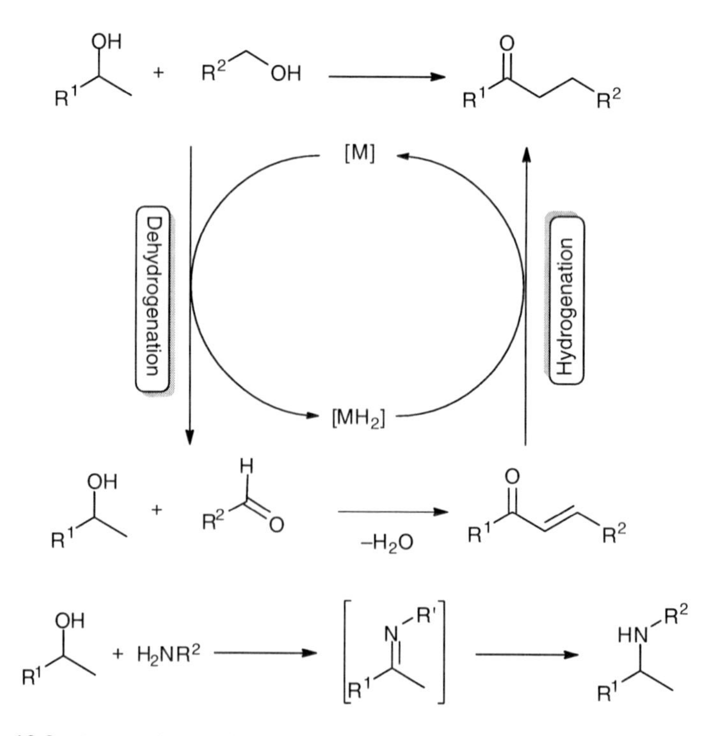

Scheme 13.2 Borrowing-hydrogen reactions.

highest activity and allowing the reaction temperature to be lowered and the reaction time to be dramatically shortened (Equation (13.3)).[69]

$$\mathbf{73} \xrightarrow{H_2,\ 50^{\circ}C} \mathbf{74} \qquad (13.3)$$

[(NHC)Ru(*p*-cymene)] complexes (NHC = imidazol-2-ylidene, imidazolin-4-ylidene or pyrazolin-2-ylidene) were also used as catalysts for the β-alkylation of secondary alcohols in the presence of KOH, in refluxing toluene.[70] Notably, abnormal NHC complexes were better performing than "normal" [(IMe)Ru-Cl_2(*p*-cymene)] **75**. For further details the reader is referred to Chapter 5.

Binuclear complexes with bridging trimethyltriazoldiylidene ligands were also used in TH and β-alkylation reactions. Thus, complexes **76** and **77** were applied for TH of ketones and imines in *i*-PrOH/KOH (Figure 13.11).[71] The reductions of ketones and *N*-benzylidene anilines were complete within 1 h. Related **78** and **79** ruthenium complexes were investigated as catalysts in the β-alkylation of secondary alcohols with primary alcohols. Such catalytic systems were very selective and more active than the previously reported catalysts.[72]

Finally, a new method was developed for the amidation of amines with primary alcohols in the presence of *in situ* formed [(NHC)Ru(PR_3)] complexes.[73] According to the authors, the reaction proceeded *via* an aldehyde formed by the dehydrogenation of the substrate, which would stay coordinated to the ruthenium centre. Subsequent attack of the amine would give a coordinated hemiaminal and β-hydride elimination from this intermediate would result in the *N*-alkyl amide.

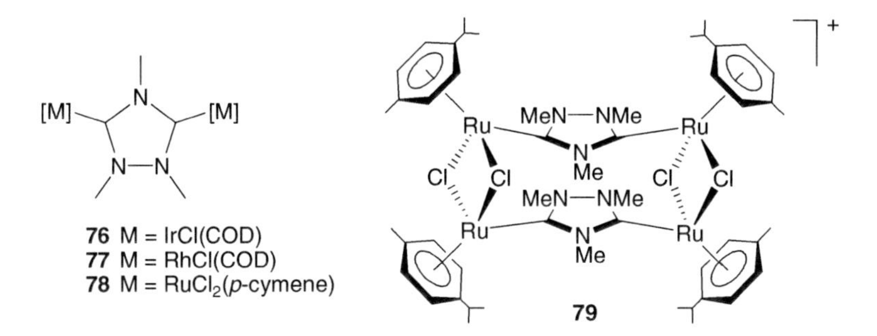

Figure 13.11 NHC-based catalysts for borrowing-hydrogen processes.

13.4 Hydrosilylation

The Si–H bond is polar, and reactive, and its addition across multiple bonds, such as C=C, C≡C and C=O is an essential reaction in organo silicon chemistry.[74] Generally trichloro- or dichlorosilanes are employed for this reaction since the reactivity of trialkylsilanes is significantly lower.

13.4.1 Hydrosilylation of Alkenes

Internal alkenes are difficult to hydrosilylate with NHC–M catalysts. In the case of terminal alkenes, the silyl group is usually added to the terminal carbon, although α-isomers and dehydrogenative products might be also formed (Equation (13.4)). Most of these catalysts operate according to the Chalk–Harrod mechanism (CHM)[75] or one of its variations.[76] However, for $PhSiH_3$ an alternative mechanism, involving a silylene intermediate, was proposed.[77]

$$\text{R} \diagup\!\!\!\!= \; + \; \text{R}'_3\text{SiH} \xrightarrow{[\text{M}]} \text{R}\diagup\!\diagdown\text{SiR}'_3 \; + \; \text{R}\diagup\!\!\overset{\text{SiR}'_3}{|}\!\!\diagdown \; + \; \text{R}\diagup\!\!=\!\!\diagdown\text{SiR}'_3 \; + \; \text{R}\diagup\!\diagdown \qquad (13.4)$$

Markó and co-workers reported the activity of monomeric and moisture and air-stable NHC–Pt^0 such as **80** (Figure 13.12). Complex **80** was applied to the hydrosilylation of a variety of terminal alkenes and only the anti-Markovnikov adducts were obtained, isomerization by-products representing less than 2% of the reaction products.[78] This catalyst tolerated functional groups such as ethers, carbonyl groups and esters, but internal alkenes were inert under similar conditions.

The *in situ* generated **81** was the most efficient NHC–Pt^0 complex for the hydrosilylation of styrene with triethylsilane (Figure 13.12).[79] A high yield (99%) and selectivity (82.5:17.2:0.3, β/α/dehydrogenative respectively) towards the desired product were the most attractive aspects of the catalyst.

A NHC–Rh catalyst, [(1,3-dimesityltetrahydropyrimidin-2-ylidene)Rh (CF_3COO)(COD)], was also studied for the hydrosilylation of alkenes, alkynes

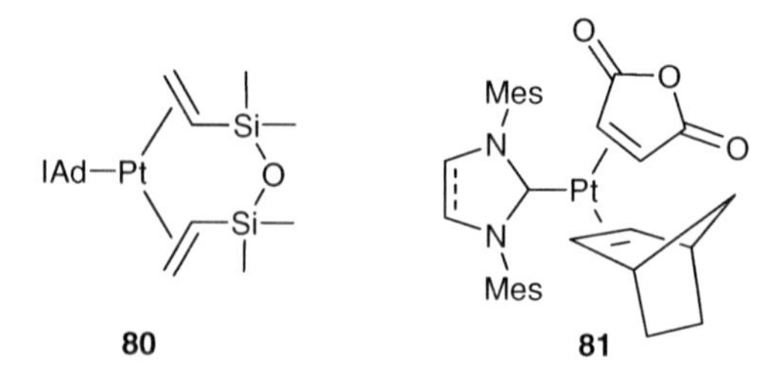

Figure 13.12 [(NHC)Pt] catalysts for alkene hydrosilylation.

and carbonyl compounds and up to 100% conversion and 1 000 TON were observed.[80]

13.4.2 Hydrosilylation of Alkynes

Alkynes are hydrosilylated more readily than alkenes, offering a simple and direct means of producing vinylsilanes.[74] However, a mixture of three isomeric vinylsilanes including the branched α-isomer as well as the β-(*E*)- and β-(*Z*)-vinylsilanes may be obtained (Equation (13.5)).

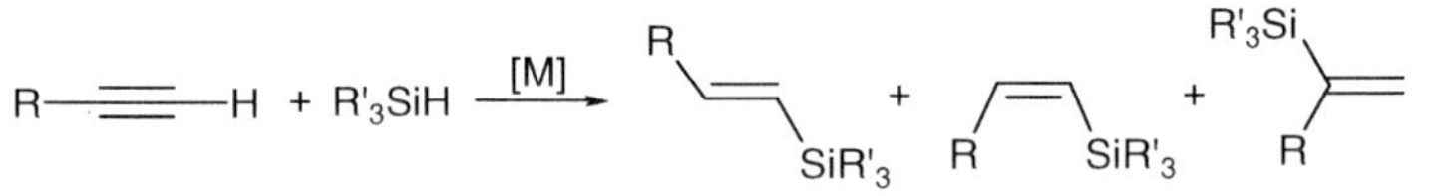

(13.5)

Several factors including substrate, solvent, temperature and catalyst are known to influence the regio- and stereoselectivity of the alkyne hydrosilylation reaction. For instance, β-(*E*) stereoisomers are best obtained using Pt-based complexes, whilst the selective formation of β-(*Z*)-vinylsilanes is generally achieved using Rh- or Ru-based catalysts. α-Vinylsilanes are privileged by substrates with a coordinating function so the Si–H delivery to the alkyne is intramolecular.[81] In the absence of such a directing group, ruthenium derivatives can also be used.[82]

Markó and co-workers extended their hydrosilylation reactions of alkenes to terminal alkynes. Initially, complexes **80**-type were screened in the hydrosilylation of 1-octyne by bis(trimethylsilyloxy)methylsilane as model reaction and enhanced selectivity was observed when bulky aryl substituted NHCs were employed (IPr and SIPr).[83]

A variety of NHC–Pt^{II} complexes **82** were studied by Hor and co-workers, although, the activity and selectivities were considerably lower than for NHC–Pt^{0} complexes (Figure 13.13).[84] Complex **82** was highly selective for the formation of the α-isomer in the hydrosilylation of phenylacetylene with bis(trimethylsilyloxy)methylsilane. Bis-carbene derivative **83** selectively afforded the β-(*Z*)-isomer (25:41:15:19, β-(*E*)/β-(*Z*)/α/dehydrogenative), whereas **84** behaved similarly to **82** but less efficiently.

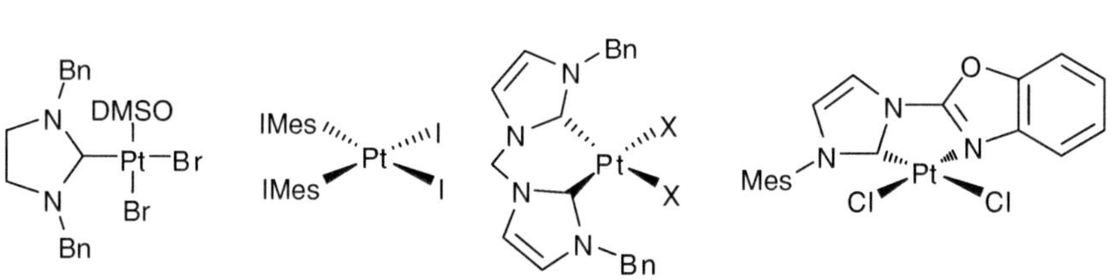

82 **83** **84** **85**

Figure 13.13 [(NHC)Pt] catalysts for alkyne hydrosilylation.

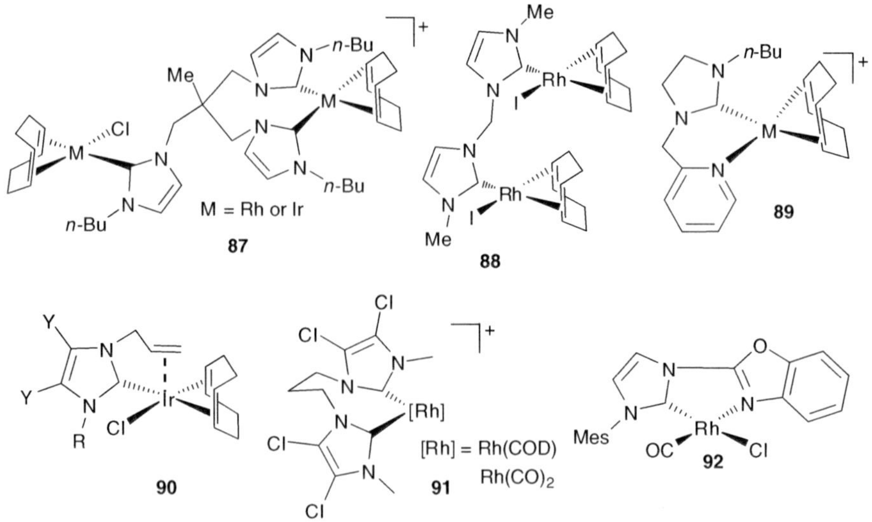

Figure 13.14 Rhodium(I) and iridium(I) catalysts for alkyne hydrosilylation.

Poyatos *et al.* investigated the activity of the N–C-chelating oxazole–NHC-containing Pt^{II} complex **85** with various substrates.[85] From phenylacetylene, the β-(*E*)-isomer and the α-isomer were the only two reaction products (60:40). Styrene hydrosilylation led to the formation of the linear product with 85% selectivity (70% conv.).

A series of mixed PCy_3/NHC Ru complexes, such as [(IMes)-$RuCl_2$(=CHPh)(PCy_3)] **86**, favored the formation of the *Z*-isomer, while the α-isomer was only formed in minor amounts.[86]

Peris and co-workers reported dimetallic complexes **87**, where the neutral metal fragment was proposed to be the active site in the hydrosilylation of phenylacetylene or 1-hexyne (Figure 13.14).[87] These catalysts produced a mixture of β-(*E*), β-(*Z*) and α-isomers, although the β-(*Z*) isomers were major (up to 5:93:2, β-(*E*)/β-(*Z*)/α) in all cases. Related **88** was also found active. In contrast with other cationic rhodium complexes, **89** produced β-(*E*)-vinylsilanes as the major product.[88]

Alkenyl-functionalized NHC–Ir^{I} complexes **90** showed a high selectivity for the β-(*Z*)-isomers (up to 0:100:0, β-(*E*)/β-(*Z*)/α), with no α-isomers or dehydrogenative silylation processes observed (Figure 13.14).[89] Similar Rh^{I} complexes with an amino-alkyl instead of alkenyl NHC displayed high regio- and stereoselectivities (up to 2:98:0, β-(*E*)/β-(*Z*)/α) in the hydrosilylation of 1-hexyne with $PhMe_2SiH$.[90] Viciano *et al.* reported the hydrosilylation of 1-hexyne or phenylacetylene by Me_2PhSiH in the presence of rhodium catalyst **91**.[91] Notably, the use of $HSi(OEt)_3$ as a hydride source only afforded the dehydrogenative silylation products. The activity of **92** was tested in the hydrosilylation of phenylacetylene with triethylsilane, and complete conversion was obtained with a 50:20:30 ratio for β-(*E*)/β-(*Z*)/α-isomer, respectively.[85]

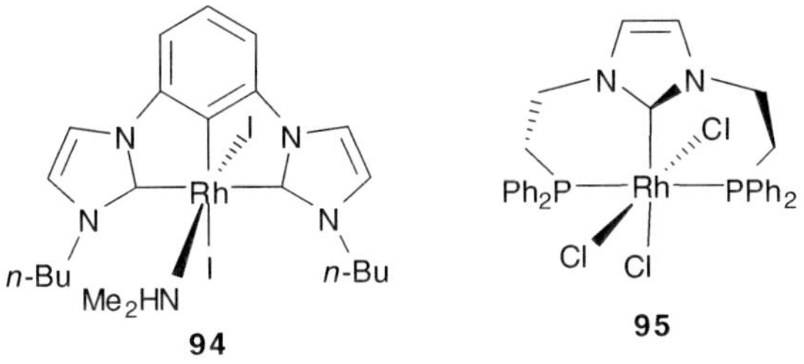

Figure 13.15 [(NHC)Rh^{III}] catalysts for alkyne hydrosilylation.

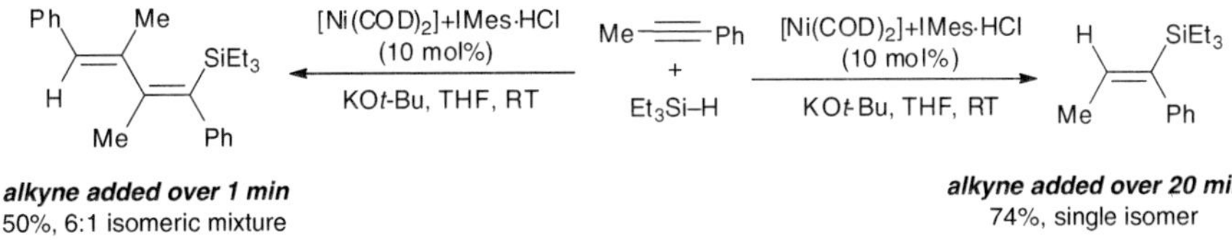

Scheme 13.3 [(NHC)Ni]-catalyzed hydrosilylation of alkynes.

Additionally, the first NHC–Rh-catalyzed hydrosilylation/cyclization of enynes was achieved using Rh^{I} catalyst **93** by Chung and co-workers (Equation (13.6)).[92]

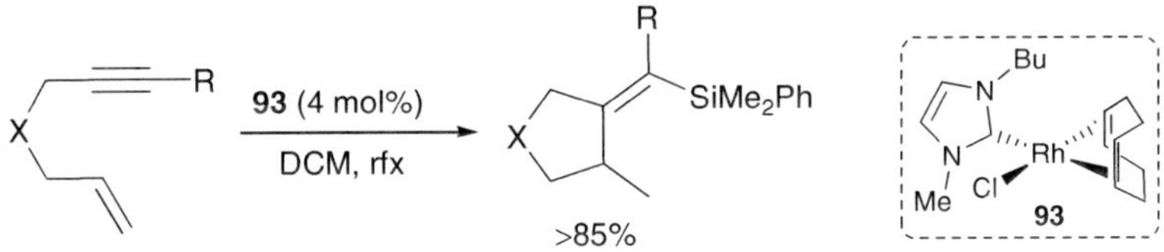

(13.6)

CCC–NHC–Rh^{III} pincer complex **94** generated β-(*Z*)-isomers as major products,[93] whereas **95** afforded the β-(*E*)-isomer as the major product (Figure 13.15). Intriguingly, this β-(*E*)/β-(*Z*) selectivity improved significantly with a prolonged reaction time, indicating that after consumption of the starting phenylacetylene, a relatively slow isomerization pathway converted the β-(*Z*)-alkenylsilane into its β-(*E*)-isomer.[94]

Finally, NHC–Ni^{0} *in situ* generated species was also applied to the hydrosilylation of terminal alkynes with silanes. Of note, the alkyne had to be added slowly to the reaction mixture in order to obtain the expected hydrosilylated product. Otherwise, a 2:1 adduct was isolated as the major product instead (Scheme 13.3).[95]

13.4.3 Hydrosilylation of Carbonyl Compounds

The hydrosilylation of aldehydes and ketones is of interest since it provides the corresponding alcohols protected as silyl ethers and the hydroxy function can subsequently be regenerated under mild conditions. Although the mechanism of carbonyl hydrosilylation is not as well-established as for olefin hydrosilylation, a few variations have been discussed by Ojima,[96] Zhang and Chan[97] and, more recently, Gade.[98]

Hydrosilylation reactions catalyzed by achiral NHC–Rh^I complexes were first reported by Lappert and Maskell in 1984, who found **96** and **97** effective for ketones and alkynes (Figure 13.16).[99] In recent studies by Özdemir and co-workers, different NHC–Rh^I complexes were applied to the hydrosilylation of acetophenone derivatives with triethylsilane and bimetallic complex **98** was particularly efficient.[100] Also, Rh^I complexes **99**, bearing a hydrophilic tetra-ethylene glycol and/or hydrophobic long-chain alkyl-functionalized NHC, were active catalyst for the hydrosilylation of ketones with Ph_2SiH_2.[101]

The hydrosilylation of cyclohexanone by Ph_2SiH_2 with the related dendimer containing catalyst **100** gave the corresponding alcohol in moderate to good yields (Figure 13.16). Notably, the yields increased with higher dendimer generations.[102] Complex **101**, with a NHC bearing 2,3,4,5-tetraphenylphenyl substituents, and higher dendritic analogues were active in the hydrosilylation of α,β-unsaturated ketones with Ph_2SiH_2 to give 1,4-adducts predominantly.[103] In contrast, 1,2-adducts were obtained as the major product with an IMes ligand.

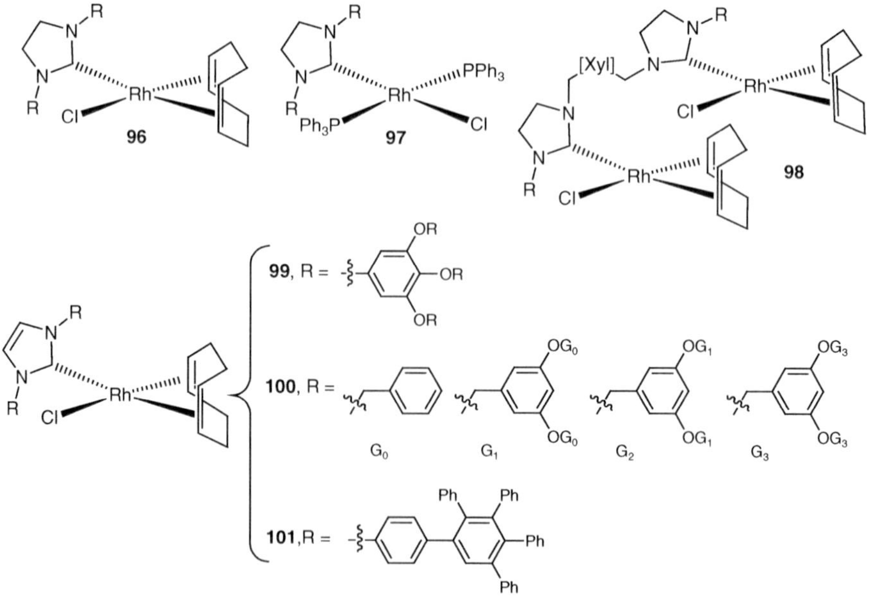

Figure 13.16 [(NHC)Rh^I] catalysts for ketone hydrosilylation.

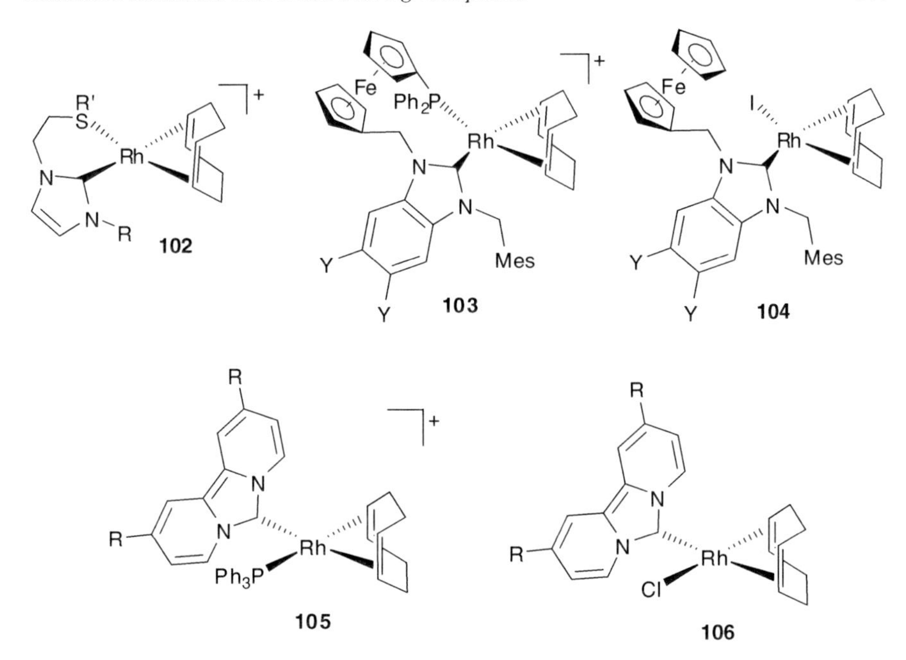

Figure 13.17 Further [(NHC)Rh^I] catalysts.

NHC–Rh^I complexes **102** with chelating carbene-thioether were active catalysts in the hydrosilylation of acetophenone and its derivatives with Ph_2SiH_2 (Figure 13.17).[104] Significantly, other hydrosilanes such as Et_3SiH, $MeEt_2SiH$ or Cl_3SiH were unreactive under identical conditions. Benzimidazol-2-ylidene–Rh^I complexes **103** and **104** were good catalysts for the hydrosilylation of acetophenone by Ph_2SiH_2 (Figure 13.17).[105] Furthermore, bidentate NHC ligands bearing a phosphanyl group displayed a much higher activity and selectivity than monodentate NHC ligands.

Cationic Rh^I complex **105** with a phosphine and a dipyrido-anullated NHC ligand was also active catalyst in the hydrosilylation of ketones by Ph_2SiH_2.[106] Of note, the conversion of acetophenone was almost complete in 10 min, while the neutral analogue **106** required 20 h under identical conditions.

In 2007, Tsuji and co-workers showed NHC–Rh^{III} complexes **107** with dendrimer substituents were more active than Rh^I analogues[102] in the hydrosilylation of 2-cyclohexenone (Figure 13.18).[107] Furthermore, **107** gave the 1,4-adduct as major product, whereas Rh^I complexes **100** led preferentially to the 1,2-adduct instead. Alternatively, related complex **108** was fairly effective for the hydrosilylation of ketones with diphenylsilane to give the corresponding alcohols in moderate to good yields (67–93%).[108]

In 2006, Díez-González *et al.* synthesized air- and moisture-stable cationic copper complexes bearing two NHC ligands. These complexes were found highly active toward the hydrosilylation of different ketones, aldehydes and esters with Et_3SiH, and generally, starting substrates were transformed into the

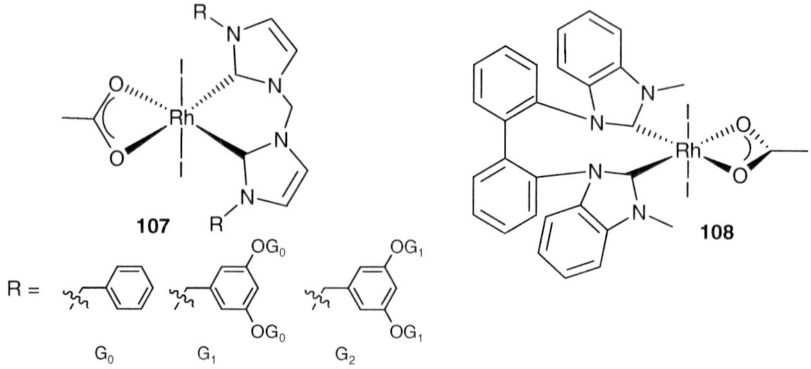

Figure 13.18 [(NHC)Rh^{III}] catalysts for ketone hydrosilylation.

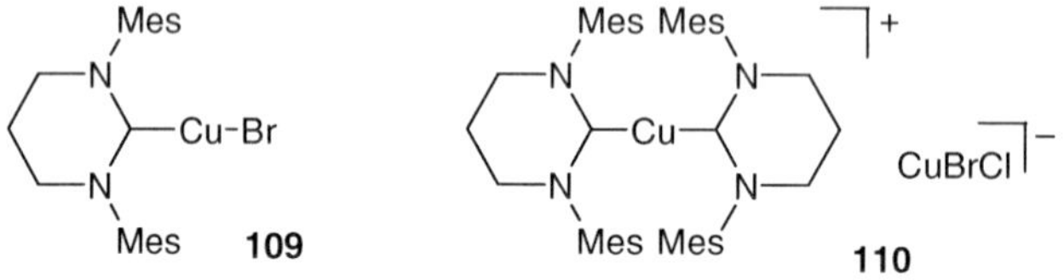

Figure 13.19 [(NHC)Cu] catalysts for ketone hydrosilylation.

desired silyl ethers in greater than 90% yield.[109] An important counter-ion effect was observed for these complexes with BF_4^- as counter-ion were consistently superior to their PF_6^- analogues.

In situ generated NHC–Cu (3 mol%) complexes were also used for hydrosilylation of various ketones with triethylsilane. IPr·BF_4 and IAd·HBF_4 salts showed good activities (99% conversion) after 2 h. However, there was only 3% conversion after 24 h when using the IBuMe·HBF_4 salt for the hydrosilylation of cyclohexanone.[110] The catalytic hydrosilylation of highly hindered and functionalized ketones in the *in situ* generated catalyst system has also been studied by Díez-González *et al.* The authors notably observed that the isolated NHC–Cu complexes displayed better activities than the *in situ* generated systems.[111]

Bantu *et al.* synthesized neutral **109** and cationic **110** copper complexes bearing 1,3-dimesityl-3,4,5,6-tetrahydropyrimidin-2-ylidene ligands and used them as catalysts for the hydrosilylation of carbonyl compounds by triethylsilane (Figure 13.19).[112] Up to 100 000 and 50 000 TONs were observed for bimetallic complex **109** and neutral **110**, respectively.

Finally, Yun and co-workers reported that Cu^{II} salts in combination with a NHC ligand, catalyzed the hydrosilylation of carbonyl compounds. A good catalytic activity was observed when using pre-catalyst [(IMes)Cu(OAc)$_2$] **111** in the presence of various hydrosilanes.[113]

13.4.4 Asymmetric Hydrosilylation

Herrmann and co-workers reported the first use of chiral NHC complexes in asymmetric hydrosilylations in 1996. The hydrosilylation of acetophenone in the presence of catalyst **111** at $-20\,^{\circ}C$ gave more than 90% conversion into the desired silyl ether and afforded optical inductions of up to 30% *ee* (Figure 13.20).[21a] Subsequently, Enders and co-workers reported the preparation of chiral [(triazolylidene)Rh(COD)] complexes **112**, which incorporated the C_I-symmetrical NHC ligand for the asymmetric hydrosilylation of methyl ketones. After hydrolysis, the desired alcohols were obtained in 40–90% yield and up to 44% *ee*.[21b] Recently, Ros *et al.* reported the evaluation of triazol-5-ylidene–rhodium(I) complexes **113** and **114** in the asymmetric hydrosilylation of acetophenone with diphenylsilane.[114] For these complexes the selectivity was slightly dependent on the nature of the NHC and strongly dependent on the solvent and reaction temperature. The most active catalyst, cationic **114**, catalyzed the reaction to give the product in 88% yield with 62% *ee*. In 2005, Yuan *et al.* reported a series of rhodium complexes **115** bearing a chiral NHC with oxazolinyl ferrocenyl substituents.[115] At very low catalyst loading (0.001 mol%), these complexes showed a high catalytic activity in the hydrosilylation of acetophenone (92–99% yield); however, none of the complexes showed a significant enantioselectivity. Related **70** improved the enantioselectivity, although it remained low to moderate (30–53% *ee*).[116] It was suggested that this may be because of the more rigid backbone of the ligand, which would hinder its rotation with respect to the substituted cyclopentadienyl ring of the ferrocene.

Gade, Bellemin-Laponnaz and co-workers reported the use of rhodium complexes with chelating NHC–oxazoline ligands in the asymmetric hydrosilylation of substituted ketones.[98,117] The best results were obtained at $-60\,^{\circ}C$ giving a 92% yield and 90% *ee* for the hydrosilylation of acetophenone by diphenylsilane in the presence of a slight excess of $AgBF_4$ to generate the corresponding cationic species. When the reaction was performed without a silver salt the yield (53%) and enantioselectivity (13%) were lower. Subsequently, the same authors extended their studies to complexes **117** in which the two heterocycles of the oxazoline–NHC ligand were connected by a methylene linker. However, catalyst **117** displayed relatively poor activity and selectivity in comparison with the ligands with directly connected oxazoline–NHCs.[118]

Very good enantioselectivities were observed with monodentate NHC complexes **118** (Figure 13.20), and their cationic derivatives with an isoquinoline ligand, when they were applied to the hydrosilylation of acetophenone by diphenylsilane.[119] In 2005, Chianese and Crabtree reported complex **119** with a (*S*)-BINAM-derived NHC and its application to the asymmetric hydrosilylation of acetophenone by diphenylsilane.[120] **119** was found very active in this reaction (>98% conversion) but only low enantioselectivities were achieved. Slightly better selectivities were obtained with an iridium analogue.

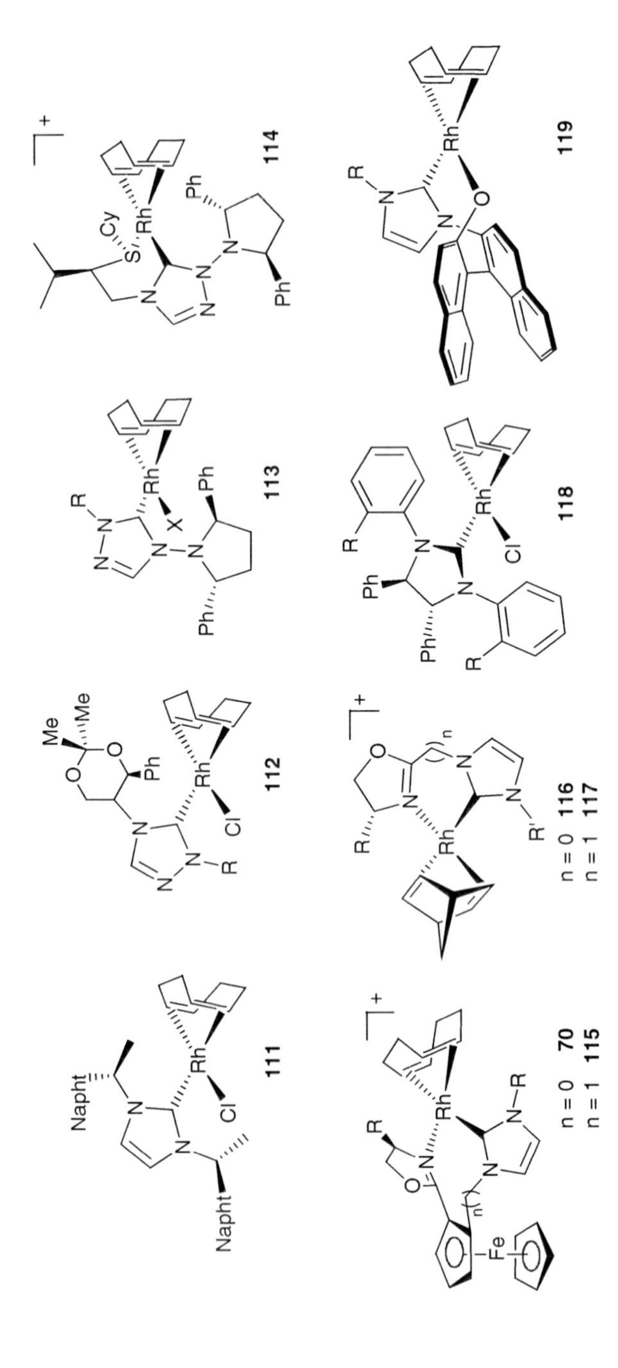

Figure 13.20 [(NHC)Rh^{I}] catalysts for asymmetric hydrosilylation of ketones.

BINAM-derived NHCs (BINAM = 1,1′-bi(2-naphthylamine)) were also present in Rh^{III} complex **120**, in the presence of which various aryl alkyl ketones and dialkyl ketones were reduced to the desired alcohols in high yields (82–96%) and enantioselectivities (67–98% *ee*) (Scheme 13.4).[121] Related **121** was investigated by Shi and co-workers and in toluene good enantioselectivities (up to 98% *ee*) were obtained after 48 h.[122] Shi also developed complex **122** with H_8-BINAM as a chiral skeleton and applied it, along with **120** to the enantioselective hydrosilylation of various 3-oxo-3-arylpropionic acid esters to yield the corresponding 3-hydroxy-3-arylpropionic acid esters in good to high yields and greater than 80% *ee* in most cases.[123]

Finally, in 2005, Song *et al.* reported the use of chiral bis-paracyclophane NHC ligand **123** in the ruthenium-catalyzed asymmetric hydrosilylation of various ketones with Ph_2SiH_2.[124] Under optimized conditions, the desired alcohols were obtained in 89–96% conversion and up to 93% *ee*. (Figure 13.21)

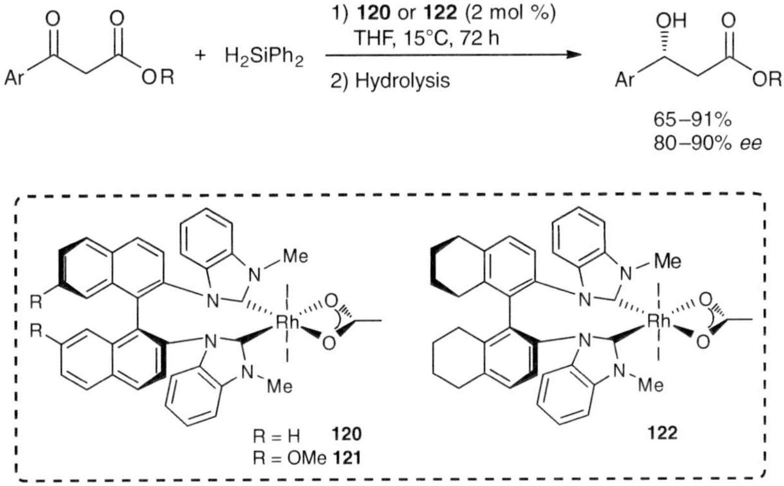

Scheme 13.4 [(NHC)Rh^{III}] catalysts for asymmetric hydrosilylation.

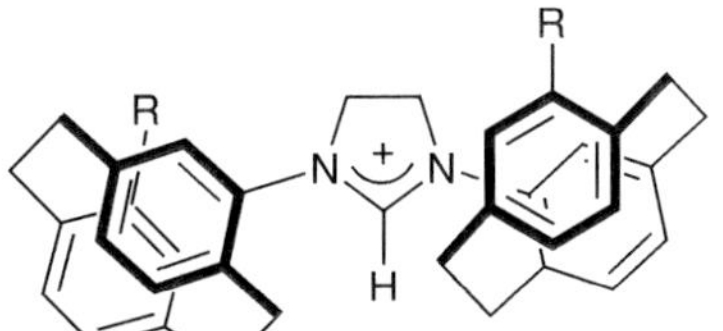

Figure 13.21 Bis-paracyclophane NHC ligand for asymmetric hydrosilylation.

13.5 Hydroboration

Hydroboration, the addition of B–H bonds to carbon–carbon multiple bonds, is an attractive route to organoboranes,[125] which can be converted into a variety of functional groups such as alcohols upon alkaline hydrogen peroxide treatment to give the corresponding anti-Markovnikov products. Metal catalysts allow for these hydroborations to be carried out under milder reaction conditions, improved or even altered reaction selectivity and more importantly, enantioselectively.[126]

In contrast to hydrosilylations, to date only a limited number of NHC complexes has been employed to catalyze hydroboration reactions and neutral and cationic complexes of Rh and Cu have proven to be the most effective.

Nolan, Lebel and co-workers first reported [(IMes)RhCl(PPh_3)$_2$] **5** and [(SIMes)Ir(COD)(pyr)]PF_6 **1** and tested them in the hydroboration of 1-hexene by catecholborane at room temperature (Equation (13.7)), a reaction which normally requires elevated temperatures (80–100 °C). The observed selectivities were comparable to Wilkinson's and Crabtree's catalysts, respectively, although longer reaction times (24 instead of 1 h) were required.[12]

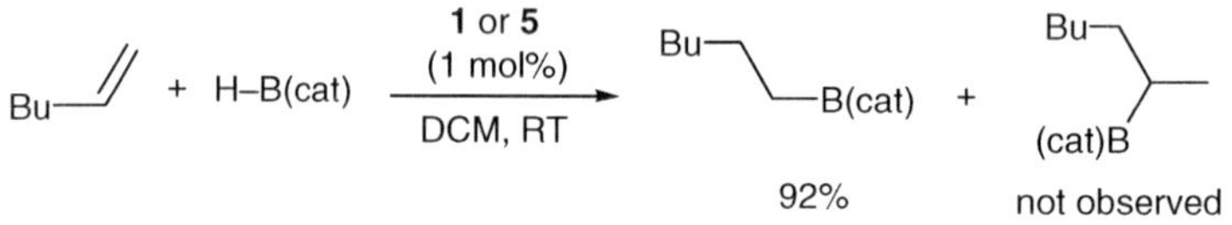

Bu + H–B(cat) —[**1** or **5** (1 mol%), DCM, RT]→ Bu–CH₂CH₂–B(cat) (92%) + branched (cat)B product (not observed) (13.7)

Fernandez and co-workers reported that with [(NHC)Pt] catalysts **124** instead, branched alcohols were the major products (Scheme 13.5).[127] Gratifyingly, hydrogenated derivatives, common by-products in this reaction, were not observed for most entries. Alkynes were also used as substrates with **124**. Although good activities were obtained, the selectivity was only moderate and styrene derivatives were formed in all cases.

Hydroboration of terminal alkynes catalyzed by NHC–Rh complexes **125** was found to be influenced by the electronic nature of the substituents on the NHC backbone, π-withdrawing groups affording even lower yields than

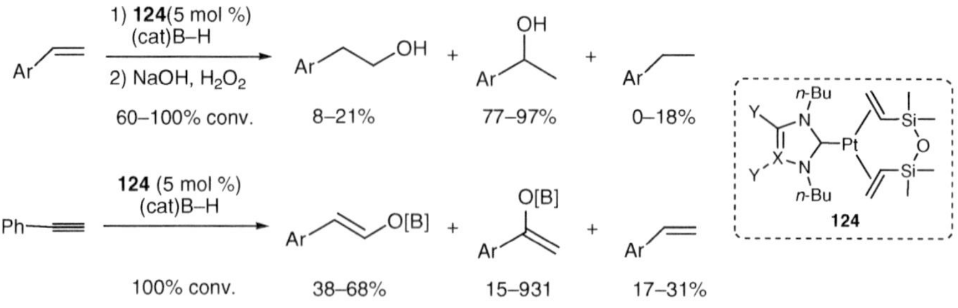

Ar–CH=CH₂ —[1) **124**(5 mol %), (cat)B–H; 2) NaOH, H_2O_2; 60–100% conv.]→ Ar–CH₂CH₂–OH (8–21%) + Ar–CH(OH)–CH₃ (77–97%) + Ar–CH₂CH₃ (0–18%)

Ph–C≡CH —[**124** (5 mol %), (cat)B–H; 100% conv.]→ Ar–CH=CH–O[B] (38–68%) + Ar–C(O[B])=CH₂ (15–931) + Ar–CH=CH₂ (17–31%)

124: n-Bu, Y, X, N, Pt, Si, O

Scheme 13.5 [(NHC)Pt]-catalyzed hydroboration of alkenes and alkynes.

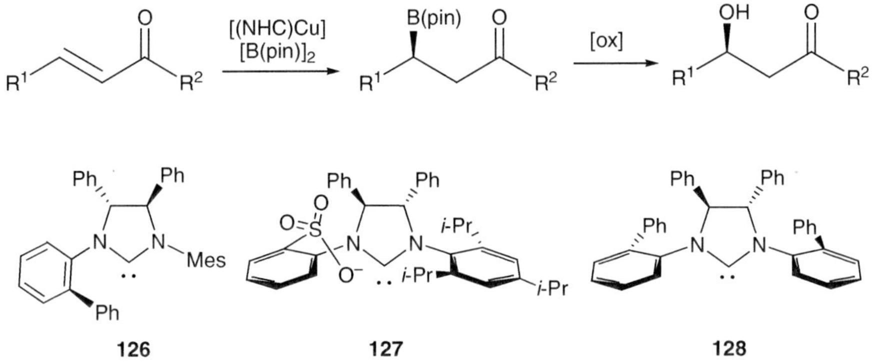

Scheme 13.6 [(NHC)Cu]-catalyzed asymmetric hydroborations.

π-withdrawing ones.[128] In all cases the *E*-alkene was obtained as the major product, but the selectivities remained modest (Equation (13.8)).

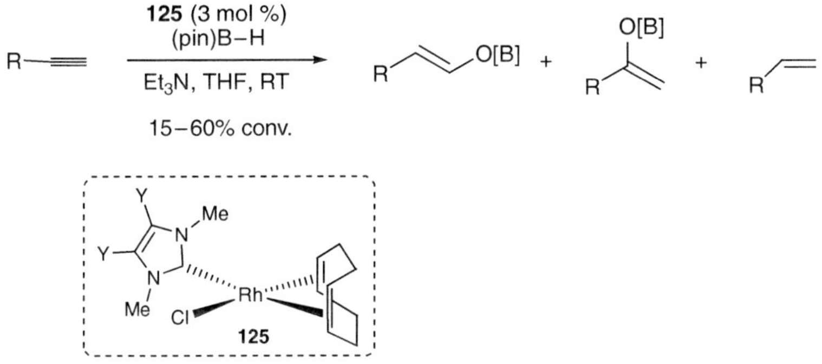

(13.8)

13.5.1 Asymmetric Hydroborations

Asymmetric hydroboration of internal olefins was developed by Pérez, Fernández and co-workers,[129] with $[B(pin)]_2$ as boration reagent. Ligands **126** provided the best chiral induction (up to 59% *ee*) (Scheme 13.6). Improved results were obtained by Hoveyda and co-workers[130] using **127** and **128** with acyclic and cyclic internal alkenes.

13.6 Conclusion

NHC ligands have led to great advances in metal catalysis thanks to their high stability and reactivity of the resulting species. However, the reaction conditions for reduction processes need to be carefully optimized to avoid irreversible cleavage of NHC from metal. For these transformations, sterically hindered and donor functional NHC ligands have played an important role.

But asymmetric applications remain arguably the area to further develop since with notable exceptions, the level of chiral induction for these reactions remains disappointing. Hopefully the known studies will lead to new strategies for the development of more reactive and selective catalysts in the reduction of multiple bonds.

References

1. J. A. Osborn, F. H. Jardine, J. F. Young and G. Wilkinson, *J. Chem. Soc. A.*, 1966, 1711–1732.
2. R. R. Schrock and J. A. Osborn, *J. Am. Chem. Soc.*, 1976, **98**, 4450–4455.
3. (a) R. H. Crabtree, *Acc. Chem. Res.*, 1979, **12**, 331–337; (b) R. H. Crabtree, H. Felkin and G. E. Morris, *J. Organomet. Chem.*, 1977, **141**, 205–215.
4. H. M. Lee, T. Jiang, E. D. Stevens and S. P. Nolan, *Organometallics*, 2001, **20**, 1255–1258.
5. (a) L. D. Vaquez-Serrano, B. T. Owens and J. M. Buriak, *Inorg. Chim. Acta.*, 2006, **359**, 2786–2797; (b) L. D. Vaquez-Serrano, B. T. Owens and J. M. Buriak, *Chem. Commun.*, 2002, 2518–2519.
6. (a) H. M. Lee, D. C. Smith, Jr., Z. He, E. D. Stevens, C. S. Yi and S. P. Nolan, *Organometallics*, 2001, **20**, 794–797. For a related study, see: (b) U. L. Dharmasena, H. M. Foucault, E. N. Dos Santos, D. E. Fogg and S. P. Nolan, *Organometallics*, 2005, **24**, 1056–1058.
7. N. J. Beach, J. M. Blacquiere, S. D. Drouin and D. E. Fogg, *Organometallics*, 2009, **28**, 441–447.
8. V. L. Chantler, S. L. Chatwin, R. F. R. Jazzar, M. F. Mahon, O. Saker and M. K. Whittlesey, *Dalton Trans.*, 2008, 2603–2614.
9. T. J. Geldbach, G. Laurenczy, R. Scopelliti and P. J. Dyson, *Organometallics*, 2006, **25**, 733–742.
10. P. Csabai and F. Joo, *Organometallics*, 2004, **23**, 5640–5643.
11. C. Gandolfi, M. Heckenroth, A. Neels, G. Laurenczy and M. Albrecht, *Organometallics*, 2009, **28**, 5112–5121.
12. G. A. Grasa, Z. Moore, K. L. Martin, S. P. Nolan, V. Paquet and H. Lebel, *J. Organomet. Chem.*, 2002, **658**, 126–131.
13. D. P. Allen, C. M. Crudden, L. A. Calhoun, R. Wang and A. Decken, *J. Organomet. Chem.*, 2005, **690**, 5736–5746.
14. W. A. Hermann, G. D. Frey, E. Herdtweck and M. Steinbeck, *Adv. Synth. Catal.*, 2007, **349**, 1677–1691.
15. B. A. Messerle, M. J. Page and P. Turner, *Dalton Trans.*, 2006, 3927–3933.
16. X.-Y. Yu, H. Sun, B. O. Patrick and B. R. James, *Eur. J. Inorg. Chem.*, 2009, 1752–1758.
17. D. S. McGuinness and K. J. Cavell, *Organometallics*, 1999, **18**, 1596–1605.
18. P. L. Arnold, F. G. N. Cloke, T. Geldbach and P. B. Hitchcock, *Organometallics*, 1999, **18**, 3228–3233.

19. V. Jurčík, S. P. Nolan and C. S. J. Cazin, *Chem.–Eur. J.*, 2009, **15**, 2509–2511.
20. S. C. Zinner, W. A. Herrmann and F. E. Kühn, *J. Organomet. Chem.*, 2008, **693**, 1543–1546.
21. (a) W. A. Hermann, L. J. Groosen, C. Köcher and G. R. J. Artus, *Angew. Chem., Int. Ed. Engl.*, 1996, **35**, 2805–2807; (b) D. Enders, H. Gielen and K. Breuer, *Tetrahedron: Asymmetry*, 1997, **8**, 3571–3574.
22. V. César, S. Bellemin-Laponnaz and L. H. Gade, *Chem. Soc. Rev.*, 2004, **33**, 619–636.
23. A. Lightfoot, P. Schnider and A. Pfaltz, *Angew. Chem., Int. Ed.*, 1998, **37**, 2897–2899.
24. (a) X. Cui, Y. Fan, M. B. Hall and K. Burgess, *Chem.–Eur. J.*, 2005, **11**, 6859–6868; (b) J. Zhou, J. W. Ogle, Y. Fan, V. Banphavichit(Bee), Y. Zu and K. Burgess, *Chem.–Eur. J.*, 2007, **13**, 7162–7170; (c) J. Zhou and K. Burgess, *Angew. Chem., Int. Ed.*, 2007, **46**, 1129–1131; (d) Y. Zhu and K. Burgess, *J. Am. Chem. Soc.*, 2008, **130**, 8894–8895; (e) Y. Zhu and K. Burgess, *Adv. Synth. Catal.*, 2008, **350**, 979–983; (f) J. Zhao and K. Burgess, *Org. Lett.*, 2009, **11**, 2053–2056.
25. H. Seo, H.-J. Park, B. Y. Kim, J. H. Lee, S. U. Son and Y. K. Chung, *Organometallics*, 2003, **22**, 618–620.
26. W. A. Herrmann, D. Baskakov, E. Herdtweck, S. D. Hoffmann, T. Bunlaksananusorn, F. Rampf and L. Rodefeld, *Organometallics*, 2006, **25**, 2449–2456.
27. D. Baskakov, W. A. Herrmann, E. Herdtweck and S. D. Hoffmann, *Organometallics*, 2007, **26**, 626–632.
28. C. Metallinos and X. Du, *Organometallics*, 2009, **28**, 1233–1242.
29. T. Focken, G. Raabe and C. Bolm, *Tetrahedron: Asymmetry*, 2004, **15**, 1693–1706.
30. C. Bolm, T. Focken and G. Raabe, *Tetrahedron: Asymmetry*, 2003, **14**, 1733–1746.
31. K. Källström and P. G. Andersson, *Tetrahedron Lett.*, 2006, **47**, 7477–7480.
32. M. S. Jeletic, M. T. Jan, I. Ghiviriga, K. A. Abboud and A. S. Veige, *Dalton Trans.*, 2009, 2764–2776.
33. (a) H.-U. Blasler, C. Malan, B. Pugin, F. Spindler, H. Steiner and M. Studer, *Adv. Synth. Catal.*, 2003, **345**, 103–151; (b) R. Noyori and S. Hashiguchi, *Acc. Chem. Res.*, 1997, **30**, 97–102.
34. (a) K. Matsamura, S. Hashiguchi, T. Ikariya and R. Noyori, *J. Am. Chem. Soc.*, 1997, **119**, 8738–8739; (b) Y. Jiang, Q. Jiang and X. Zhang, *J. Am. Chem. Soc.*, 1998, **120**, 3817–3818.
35. J. S. M. Samec, J.-E. Bäckvall, P. G. Anderson and P. Brandt, *Chem. Soc. Rev.*, 2006, **35**, 237–248.
36. J. Louie, C. W. Bielawski and R. H. Grubbs, *J. Am. Chem. Soc.*, 2001, **123**, 11312–11313.
37. A. A. Danopoulos, S. Winston and W. B. Motherwell, *Chem. Commun.*, 2002, 1376–1377.

38. M. Poyatos, J. A. Mata, E. Falomir, R. H. Crabtree and E. Peris, *Organometallics*, 2003, **22**, 1110–1114.
39. W. Baratta, J. Schütz, E. Herdtweck, W. A. Herrmann and P. Rigo, *J. Organomet. Chem.*, 2005, **690**, 5570–5575.
40. P. L. Chiu and H. M. Lee, *Organometallics*, 2005, **24**, 1692–1702.
41. F. Zeng and Z. Yu, *Organometallics*, 2008, **27**, 6025–6028.
42. (a) I. Özdemir, S. Yaşar and B. Çetinkaya, *Transition Met. Chem.*, 2005, **30**, 831–835; (b) M. Yiğit, B. Yiğit, Ý. Özdemir, E. Çetinkaya and B. Çetinkaya, *Appl. Organomet. Chem.*, 2006, **20**, 322–327.
43. Y. Cheng, J.-F. Sun, H.-L. Yang, H.-J. Xu, Y.-Z. Li, X.-T. Chen and Z.-L. Xue, *Organometallics*, 2009, **28**, 819–823.
44. A. C. Hillier, H. M. Lee, E. D. Stevens and S. P. Nolan, *Organometallics*, 2001, **20**, 4246–4252.
45. I. Kownacki, M. Kubicki, K. Szubert and B. Marciniec, *J. Organomet. Chem.*, 2008, **693**, 321–328.
46. H. Türkmen, T. Pape, F. E. Hahn and B. Çetinkaya, *Eur. J. Inorg. Chem.*, 2008, **5418–5423**.
47. C.-Y. Wang, C.-F. Fu, Y.-H. Liu, S.-M. Peng and S.-T. Liu, *Inorg. Chem.*, 2007, **46**, 5779–5786.
48. J.-F. Sun, F. Chen, B. A. Dougan, H.-J. Xu, Y. Cheng, Y.-Z. Li, X.-T. Chen and Z.-L. Xue, *J. Organomet. Chem.*, 2009, **694**, 2096–2105.
49. F. E. Hahn, C. Holtgrewe, T. Pape, M. Martin, E. Sola and L. A. Oro, *Organometallics*, 2005, **24**, 2203–2209.
50. H. Türkmen, T. Pape, F. E. Hahn and B. Çetinkaya, *Organometallics*, 2008, **27**, 571–575.
51. D. Gnanamgari, A. Moores, E. Rajaseelan and R. H. Crabtree, *Organometallics*, 2007, **26**, 1226–1230.
52. M. Albrect, R. H. Crabtree, J. Mata and E. Peris, *Chem. Commun.*, 2002, 32–33.
53. (a) J. R. Miecznikowski and R. H. Crabtree, *Organometallics*, 2004, **23**, 629–631. See also: (b) M. Albrecht, J. R. Miecznikowski, A. Samuel, J. W. Faller and R. H. Crabtree, *Organometallics*, 2002, **21**, 3596–3604.
54. M. Poyatos, E. Mas-Marzá, J. A. Mata, M. Sanaú and E. Peris, *Eur. J. Inorg. Chem.*, 2003, 1215–1221.
55. D. Gnanamgari, E. L. O. Sauer, N. D. Schley, C. Butler, C. D. Incarvito and R. H. Crabtree, *Organometallics*, 2009, **28**, 321–325.
56. A. Pontes da Costa, M. Viciano, M. Sanaú, S. Merino, J. Tejeda, E. Peris and B. Royo, *Organometallics*, 2008, **27**, 1305–1309.
57. R. Corberán and E. Peris, *Organometallics*, 2008, **27**, 1954–1958.
58. R. Corberán, M. Sanaú and E. Peris, *Organometallics*, 2007, **26**, 3492–3498.
59. R. Castarlenas, M. A. Esteruelas and E. Oñate, *Organometallics*, 2008, **27**, 3240–3247.
60. S. Kuhl, R. Schneider and Y. Fort, *Organometallics*, 2003, **22**, 4184–4186.

61. P. Hauwert, G. Maestri, J. W. Sprengers, M. Catellani and C. J. Elsevier, *Angew. Chem., Int. Ed.*, 2008, **47**, 3223–3226.
62. S. Gladiali and E. Alberico, *Chem. Soc. Rev.*, 2006, **35**, 226–236.
63. H. Seo, B. Y. Kim, J. H. Lee, H.-J. Park, S. U. Son and Y. K. Chung, *Organometallics*, 2003, **22**, 4783–4791.
64. R. Hodgson and R. E. Douthwaite, *J. Organomet. Chem.*, 2005, **690**, 5822–5831.
65. R. Jiang, X. Sun, W. He, H. Chen and Y. Kuang, *Appl. Organomet. Chem.*, 2009, **23**, 179–182.
66. U. Nagel and C. Diez, *Eur. J. Inorg. Chem.*, 2009, **1**, 1248–1255.
67. M. Poyatos, A. M. François, S. Bellemine-Laponnaz, E. Peris and L. H. Gade, *J. Organomet. Chem.*, 2006, **691**, 2713–2720.
68. (a) G. Guillena, D. J. Ramón and M. Yus, *Angew. Chem., Int. Ed.*, 2007, **46**, 2358–2364; (b) M. H. S. A. Hamid, P. A. Slatford and J. M. J. Williams, *Adv. Synth. Catal.*, 2007, **349**, 1555–1375; (c) T. D. Nixon, M. K. Whittlesey and J. M. J. Williams, *Dalton Trans.*, 2009, 753–762.
69. (a) R. F. R. Jazzar, S. A. Macgregor, M. F. Mahon, S. P. Richards and M. K. Whittlesey, *J. Am. Chem. Soc.*, 2002, **124**, 4944–4945; (b) S. Burling, M. K. Whittlesey and J. M. J. Williams, *Adv. Synth. Catal.*, 2005, **347**, 591–594; (c) S. Burling, B. M. Paine, D. Nama, V. S. Brown, M. F. Mahon, T. J. Prior, P. S. Pregosin, M. K. Whittlesey and J. M. J. Williams, *J. Am. Chem. Soc.*, 2007, **129**, 1987–1995.
70. A. Prades, M. Viciano, M. Sanaú and E. Peris, *Organometallics*, 2008, **27**, 4254–4259.
71. E. Mas-Marzá, J. A. Mata and E. Peris, *Angew. Chem., Int. Ed.*, 2007, **46**, 3729–3731.
72. (a) R. Martínez, D. J. Ramón and M. Yus, *Tetrahedron*, 2006, **62**, 8982–8987; (b) R. Martínez, D. J. Ramón and M. Yus, *Tetrahedron*, 2006, **62**, 8988–9001; (c) M. Viciano, M. Sanaú and E. Peris, *Organometallics*, 2007, **26**, 6050–6054.
73. L. U. Nordstrøm, H. Vogt and R. Madsen, *J. Am. Chem. Soc.*, 2008, **130**, 17672–17673.
74. B. Marciniec, (ed.), *Comprehensive Handbook on Hydrosilylation*, Elsevier, Oxford, 1992.
75. A. J. Chalk and J. F. Harrod, *J. Am. Chem. Soc.*, 1965, **87**, 16–21.
76. I. Ojima, T. Fuchikami and M. Yatabe, *J. Organomet. Chem.*, 1984, **260**, 335–346.
77. H. Brunner, *Angew. Chem., Int. Ed.*, 2004, **43**, 2749–2750.
78. (a) I. E. Markó, S. Stérin, O. Buisine, G. Berthon, G. Michaud, B. Tinant and J.-P. Declercq, *Adv. Synth. Catal.*, 2004, **346**, 1429–1434; (b) G. Berthon-Gelloz, O. Buisine, J.-F. Briere, G. Michaud, S. Sterin, G. Mignani, B. Tinant, J.-P. Declercq, D. Chapon and I. E. Markó, *J. Organomet. Chem.*, 2005, **690**, 6156–6168; (c) O. Buisine, G. Berthon-Gelloz, J.-F. Briere, S. Sterin, G. Mignani, P. Branlard, B. Tinant, J.-P. Declercq and I. E. Markó, *Chem. Commun*, 2005, 3856–3858.

79. J. W. Sprengers, M. J. Mars, M. A. Duin, K. J. Cavell and C. J. Elsevier, *J. Organomet. Chem.*, 2003, **679**, 149–152.
80. N. Imlinger, K. Wurst and M. R. Buchmeiser, *Monatsh. Chem.*, 2005, **136**, 47–57.
81. C. S. Arico and L. R. Cox, *Org. Biomol. Chem.*, 2004, **2**, 2558–2562.
82. B. M. Trost and Z. T. Ball, *J. Am. Chem. Soc.*, 2001, **123**, 12726–12727.
83. G. D. Bo, G. Berthon-Gelloz, B. Tinant and I. E. Markó, *Organometallics*, 2006, **25**, 1881–1890.
84. J. J. Hu, F. Li and T. S. A. Hor, *Organometallics*, 2009, **28**, 1212–1220.
85. M. Poyatos, A. Maisse-François, S. Bellemin-Laponnaz and L. H. Gade, *Organometallics*, 2006, **25**, 2634–2641.
86. S. V. Maifeld, M. N. Tran and D. Lee, *Tetrahedron Lett.*, 2005, **46**, 105–108.
87. (a) M. Poyatos, M. Sanaú and E. Peris, *Inorg. Chem.*, 2003, **42**, 2572–2576; (b) E. Mas-Marzá, M. Poyatos, M. Sanaú and E. Peris, *Inorg. Chem.*, 2004, **43**, 2213–2219.
88. E. Mas-Marzá, M. Sanaú and E. Peris, *Inorg. Chem.*, 2005, **44**, 9961–9967.
89. A. Zanardi, E. Peris and J. A. Mata, *New. J. Chem.*, 2008, **32**, 120–126.
90. M. V. Jiménez, J. J. Pérez-Torrente, M. I. Bortolomé, V. Gierz, F. J. Lahoz and L. A. Oro, *Organometallics*, 2008, **27**, 224–234.
91. M. Viciano, E. Mas-Marzá, M. Sanaú and E. Peris, *Organometallics*, 2006, **25**, 3063–3069.
92. K. H. Park, S. Y. Kim, S. U. Son and Y. K. Chung, *Eur. J. Org. Chem.*, 2003, 4341–4345.
93. G. T. S. Andavan, E. B. Bauer, C. S. Letko, T. K. Hollis and F. S. Tham, *J. Organomet. Chem.*, 2005, **690**, 5938–5947.
94. J. Y. Zeng, M.-H. Hsieh and H.-M. Lee, *J. Organomet. Chem.*, 2005, **690**, 5662–5671.
95. M. R. Chaulagain, G. M. Mahandru and J. Montgomery, *Tetrahedron*, 2006, **62**, 7560–7566.
96. I. Ojima, M. Nihonyanagi, T. Kogure, M. Kumagai, S. Horiuchi and K. Nakatsugawa, *J. Organomet. Chem.*, 1975, **94**, 449–461.
97. G. Z. Zheng and T. H. Chan, *Organometallics*, 1995, **14**, 70–79.
98. N. Schneider, M. Finger, C. Haferkemper, S. Bellemin-Laponnaz, P. Hofmann and L. H. Gade, *Angew. Chem., Int. Ed.*, 2009, **48**, 1609–1613.
99. M. F. Lappert and R. K. Maskell, *J. Organomet. Chem.*, 1984, **264**, 217–228.
100. (a) M. Yiğit, I. Özdemir, B. Çetinkaya and E. Çetinkaya, *J. Mol. Catal. A: Chem.*, 2005, **241**, 88–92; (b) I. Özdemir, M. Yiğit, B. Yiğit, B. Çetinkaya and E. Çetinkaya, *J. Coord. Chem.*, 2007, **60**, 2377–2384; (c) I. Özdemir, S. Demir, O. Şahin, O. Büyükgüngör and B. Çetinkaya, *Appl. Organomet. Chem.*, 2008, **22**, 59–66.
101. H. Ohta, T. Fujihara and Y. Tsuji, *Dalton Trans.*, 2008, 379–385.

102. T. Fujihara, Y. Obora, M. Tokunaga, H. Sato and Y. Tsuji, *Chem. Commun.*, 2005, 4526–4528.
103. For an example of dendritic NHC–Ir catalyst, see: (a) H. Sato, T. Fujihara, Y. Obora, M. Tokunaga, J. Kiyosu and Y. Tsuji, *Chem. Commun.*, 2007, 269–271; (b) A. R. Chianese, A. Mo and D. Data, *Organometallics*, 2009, **28**, 465–472.
104. J. Wolf, A. Labande, J.-C. Daran and R. Poli, *Eur. J. Inorg. Chem.*, 2007, 5069–5079.
105. (a) S. Gülcemal, A. Labande, J.-C. Daran, B. Çetinkaya and R. Poli, *Eur. J. Inorg. Chem.*, 2009, 1806–1815; See also: (b) J. Wolf, A. Labande, J.-C. Daran and R. Poli, *Eur. J. Inorg. Chem.*, 2007, 5069–5079.
106. M. Nonnenmacher, D. Kunz and F. Rominger, *Organometallics*, 2008, **27**, 1561–1568.
107. T. Fujihara, Y. Obora, M. Tokunaga and Y. Tsuji, *Dalton Trans.*, 2007, 1567–1569.
108. T. Chen, X.-G. Liu and M. Shi, *Tetrahedron*, 2007, **63**, 4874–4880.
109. (a) S. Díez-González, N. M. Scott and S. P. Nolan, *Organometallics*, 2006, **25**, 2355–2358; (b) S. Díez-González, E. D. Stevens, N. M. Scott, J. L. Petersen and S. P. Nolan, *Chem.–Eur. J.*, 2008, **14**, 158–168.
110. H. Kaur, F. K. Zinn, E. D. Stevens and S. P. Nolan, *Organometallics*, 2004, **23**, 1157–1160.
111. S. Díez-González, H. Kaur, F. K. Zinn, E. D. Stevens and S. P. Nolan, *J. Org. Chem.*, 2005, **70**, 4784–4796.
112. B. Bantu, D. Wang, K. Wurst and M. R. Buchmeiser, *Tetrahedron*, 2005, **61**, 12145–12152.
113. J. Yun, D. Kim and H. Yun, *Chem. Commun.*, 2005, 5181–5183.
114. A. Ros, M. Alcarazo, J. Iglesias-Sigüenza, E. Díez, E. Álvarez, R. Fernández and J. M. Lassaletta, *Organometallics*, 2008, **27**, 4555–4564.
115. Y. Yuan, G. Raabe and C. Bolm, *J. Organomet. Chem.*, 2005, **690**, 5747–5752.
116. Y. Kuang, X. Sun, H. Chen, P. Liu and R. Jiang, *Catal. Commun.*, 2009, **10**, 1493–1496.
117. (a) L. H. Gade, V. César and S. Bellemin-Laponnaz, *Angew. Chem., Int. Ed.*, 2004, **43**, 1014–1017; (b) V. César, S. Bellemin-Laponnaz, H. Wadepohl and L. H. Gade, *Chem.–Eur. J.*, 2005, **11**, 2862–2873.
118. N. Schneider, M. Kruck, S. Bellemin-Laponnaz, H. Wadepohl and L. H. Gade, *Eur. J. Inorg. Chem.*, 2009, 493–500.
119. J. W. Faller and P. P. Fontaine, *Organometallics*, 2006, **25**, 5887–5893.
120. A. R. Chianese and R. H. Crabtree, *Organometallics*, 2005, **24**, 4432–4436.
121. W.-L. Duan, M. Shi and G.-B. Rong, *Chem. Commun.*, 2003, 2 916–2917.
122. L.-J. Liu, F. Wang and M. Shi, *Organometallics*, 2009, **28**, 4416–4420.
123. Q. Xu, X. Gu, S. Liu, Q. Dou and M. Shi, *J. Org. Chem.*, 2007, **72**, 2240–2242.

124. C. Song, C. Ma, Y. Ma, W. Feng, S. Ma, Q. Chai and M. B. Andrus, *Tetrahedron Lett.*, 2005, **46**, 3241–3244.
125. H. C. Brown and S. K Gupta, *J. Am. Chem. Soc.*, 1975, **97**, 5249–5255.
126. (a) A.-M. Carroll, T. P. O'Sullivan and P. J. Guiry, *Adv. Synth. Catal.*, 2005, **347**, 609–631; (b) C. M. Crudden and D. Edwards, *Eur. J. Org. Chem.*, 2003, 4695–4712; (c) K. Burgess and M. J. Ohlmeyer, *Chem. Rev.*, 1991, **91**, 1179–1191.
127. V. Lillo, J. A. Mata, A. M. Segarra, E. Peris and E. Fernandez, *Chem. Commun.*, 2007, 2184–2186.
128. D. M. Khramov, E. L. Rosen, J. A. V. Er, P. D. Vu, V. M. Lynch and C. W. Bielawski, *Tetrahedron*, 2008, **64**, 6853–6862.
129. V. Lillo, A. Prieto, A. Bonet, M. Mar Díaz-Requejo, J. Ramírez, P. J. Pérez and E. Fernández, *Organometallics*, 2009, **28**, 659–662.
130. Y. Lee and A. H. Hoveyda, *J. Am. Chem. Soc.*, 2009, **131**, 3160–3161.

CHAPTER 14

N-Heterocyclic Carbenes as Organic Catalysts

PEI-CHEN CHIANG* AND JEFFREY W. BODE*

Department of Chemistry, University of Pennsylvania, 231 S. 34th St, Philadelphia, PA 19104, USA

14.1 Introduction

The idea that azolium salts could serve as nucleophilic catalysts dates back to the pioneering work of Ugai, who in 1943 demonstrated that thiamine (vitamin B1) isolated from natural sources catalyzed the benzoin condensation of benzaldehyde.[1] The elegant mechanistic work of Breslow established the currently understood mechanism for thiamine's catalytic reaction, and provided the first suggestion of the resonance structure represented by the N-heterocyclic carbene (NHC) (Scheme 14.1).[2]

Despite these detailed studies and the suggestion that the N-heterocyclic carbene might be involved in this chemistry, researchers did not generally adopt the NHC nomenclature when discussing thiazolium-catalyzed reactions. This stems in large part from the belief, prior to Arduengo's demonstration that related N-heterocyclic carbenes can be isolated as stable compounds,[3] that if such carbenes were formed they would immediately undergo nucleophilic reactions.[4] Furthermore, in most of these reactions the NHCs were not used as the catalysts themselves. More commonly, the combination of a thiazolium, triazolium or imidazolium salt and a base were employed to generate the reactive ylide or carbene.

There is still room for healthy debate about the true nature of the catalyst in the benzoin reaction and the many exciting and novel processes catalyzed by azolium salts discussed in this Chapter. Whatever the active species, these

RSC Catalysis Series No. 6
N-Heterocyclic Carbenes: From Laboratory Curiosities to Efficient Synthetic Tools
Edited by Silvia Díez-González
© Royal Society of Chemistry 2011
Published by the Royal Society of Chemistry, www.rsc.org

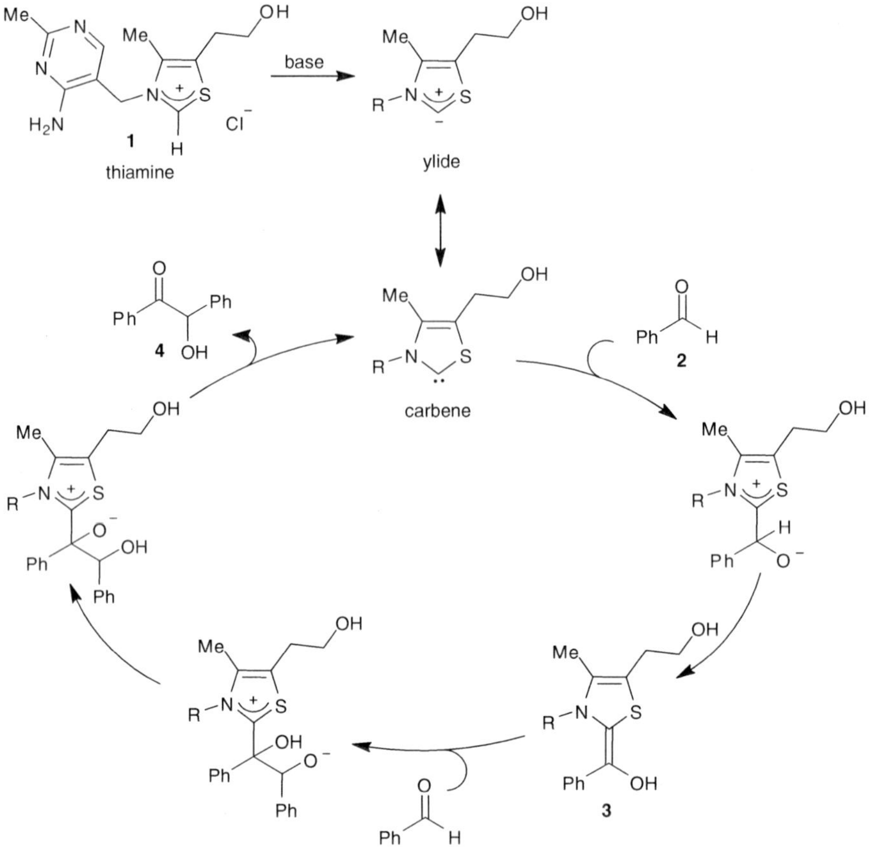

Scheme 14.1 Mechanism of benzoin condensations as proposed by Breslow in 1958.

reactions have become collectively known as 'N-heterocyclic carbene-catalyzed reactions'. We therefore adopt this nomenclature for the purpose of providing an update account of advances in this field. We focus on new transformations catalyzed by NHCs and attempt to provide a breakdown of the many new reactions reported in this field in the past 6 years.

14.2 Benzoin and Stetter Reactions

Benzoin, and the related Stetter reactions proceed *via* the catalytic generation of the Breslow intermediate (**3** in Scheme 14.1), which serves as an acyl anion equivalent. In the benzoin reaction, this intermediate adds to an aldehyde, while in the Stetter reaction it adds to an activated alkene in a conjugate addition fashion. Both of these reactions have been extensively reviewed and will not be the subject of this Chapter.[5]

It is worth noting, however, that the rise of enantioselective benzoin and Stetter reactions were one of the forerunners of the new generation of NHC-catalyzed processes that have been reported since 2004. These studies, particularly the work of Knight and Leeper on chiral triazolium salts for benzoin and Stetter reactions,[6] formed the basis for the design and synthesis of the now widely used chiral NHCs (Scheme 14.2). These designs were elegantly extended by Enders and by Rovis[7] for the development of catalytic enantioselective benzoin condensations,[8] intramolecular Stetter reactions,[9] and the wide variety of enantioselective NHC-catalyzed transformations detailed below.

14.3 NHC-catalyzed Transesterification Reactions

The first reaction designed specifically as an N-heterocyclic carbene catalyzed process was the transesterification of esters in the presence of IMes **13**. This process, reported simultaneously by Hendrick and Waymouth[10] and by Nolan,[11] allowed the facile transesterification of esters with an added alcohol (Scheme 14.3). A number of researchers have also developed chiral NHCs for kinetic resolutions of alcohols using these processes.[12]

An interesting aspect of this reaction was reported by Movassaghi, who found that the preparation of amides from esters, expected to be possible under similar conditions, failed except for the special case of 1,2-amino alcohols.[13] Similar procedures were also found to promote *O*- to *C*-carboxyl transfer reactions of oxazolyl carbonates (Scheme 14.4).[14]

14.4 Catalytic Generation of Activated Carboxylates

The explosion in novel reactions catalyzed by N-heterocyclic carbenes began with the 2004 disclosure of the first internal redox reactions of α-functionalized aldehydes by N-heterocyclic carbenes.[15] Bode reported the thiazolium-catalyzed conversion of α,β-epoxyaldehydes to β-hydroxyesters (Scheme 14.5).[16] This reaction produced esters without the need for stoichiometric coupling reagents, and provided the *anti*-propionate esters in good yield and diastereoselectivity. Mechanistic studies supported the catalytic generation of an acyl thiazolium as the key acylating reagent. Almost simultaneously, Rovis reported a similar process from α-halo aldehydes.[17]

Since these initial reports, the NHC-catalyzed generation of activated carboxylates from α-functionalized aldehydes has been extended to a remarkably wide range of substrate classes. Bode first reported the generation of activated carboxylates from α,β-unsaturated aldehydes as part of his work on the catalytic generation of homoenolates (*vide infra*).[18] These studies were further explored to develop improved catalysts and render this a general process.[19] This work, which identified the uniquely high reactivity of *N*-mesityl substituted triazolium pre-catalysts as superior to all other catalyst classes for NHC-catalyzed redox reactions of α,β-unsaturated aldehydes, was crucial to the development of the remarkable annulation reactions described below. Zeitler reported analogous

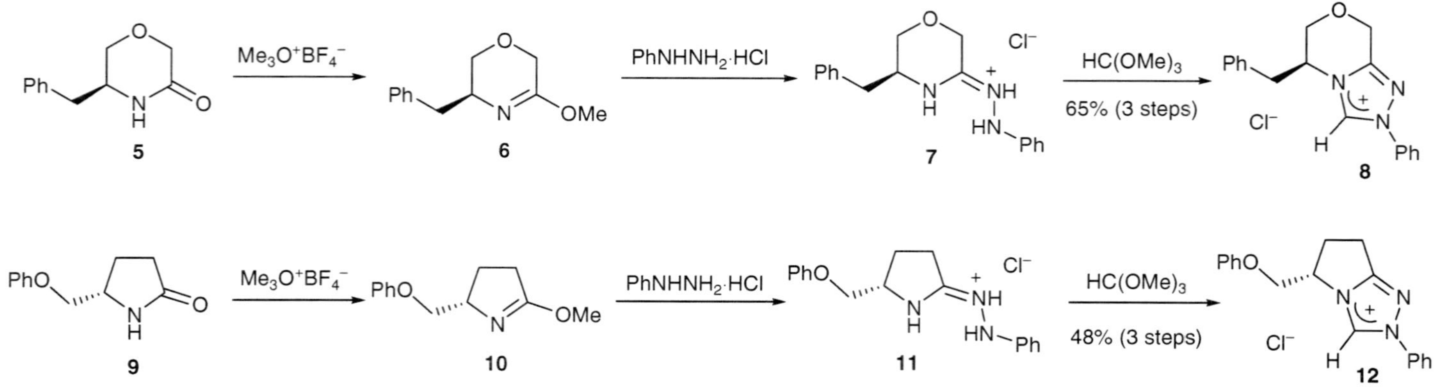

Scheme 14.2 Knight and Leeper's chiral triazolium syntheses.

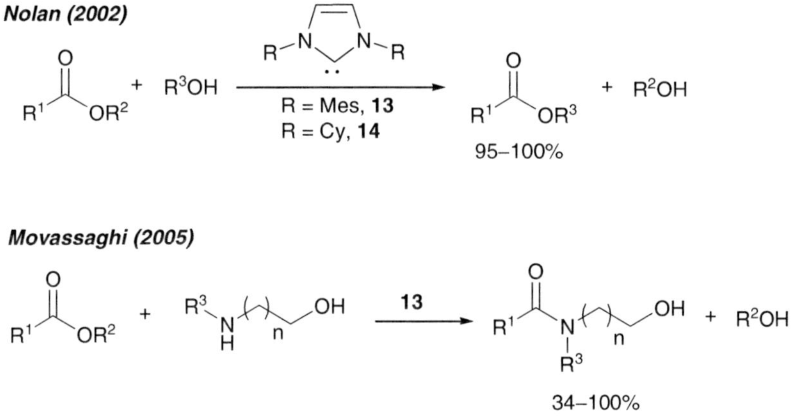

Nolan (2002)

$R^1C(O)OR^2$ + R^3OH → $R^1C(O)OR^3$ + R^2OH

R = Mes, **13**
R = Cy, **14**

95–100%

Movassaghi (2005)

$R^1C(O)OR^2$ + $R^3NH(CH_2)_nCH_2CH_2OH$ —**13**→ $R^1C(O)N(R^3)(CH_2)_nCH_2CH_2OH$ + R^2OH

34–100%

Scheme 14.3 NHC-catalyzed transesterification.

work using ynals as substrates,[20] and Rovis disclosed an exciting catalytic, asymmetric variant, leading to the formation of enantiomerically enriched α-chloroaldehydes *via* enolate protonation.[21] A vinologus version of this reactions led to (*Z*)-fluoroalkenes, which can be used as dipeptide isosteres.[22]

The NHC-catalyzed internal redox reactions also found use for ring opening and ring expansion processes. Bode reported the opening of enantioenriched formylcyclopropanes to give the acyclic esters in excellent yield,[23] and You applied this chemistry to ring expansions (Scheme 14.6).[24] Alcaide[25] and You[26] independently reported similar processes for the opening of formyl β-lactams. Recently, Gravel reported an attractive NHC-catalyzed ring expansion of formyl lactones.[27]

An intriguing feature of the acyl azoliums generated catalytically during these reactions is their reluctance to undergo reactions with amines, affording amides. Both Rovis[28] and Bode[29] reported the simple workaround of adding co-catalysts, such as HOAt (1-hydroxy-7-azabenzotriazole) or imidazole, to the reactions, leading to amides *via* tandem catalysis (Scheme 14.7). A limitation of this method was the propensity of the starting amines and aldehydes to generate imines. Bode disclosed that α′-hydroxyenones were surrogates for α,β-unsaturated aldehydes that circumvented this problem and led to saturated amides in high yield under truly catalytic conditions.[30]

These NHC-catalyzed redox acylations are already finding a number of synthetic applications. Rovis exploited the NHC-catalyzed ring opening of α,β-epoxyaldehyde to generate acyl azides that undergo Curtius rearrangement, leading to 1,2-aminoalcohols.[31] Vosberg elegantly applied this procedure to the synthesis of (+)-davanone; other approaches to the *anti*-propionate were unsuccessful (Scheme 14.8).[32] Forsyth used this NHC-catalyzed epoxide opening-generation of activated carboxylates as a fragment coupling reaction in the synthesis of largazole.[33] Remarkably, they were able to affect selective amide formation on unprotected amino acid **25**.

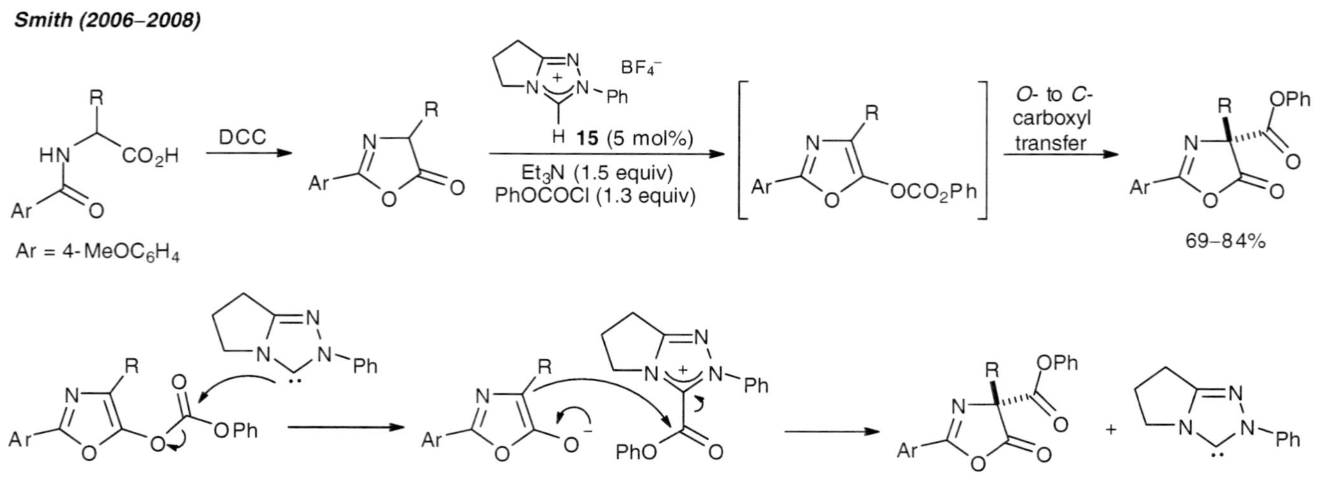

Smith (2006–2008)

R
HN
CO_2H
O
Ar
DCC
N
R
Ar
O
O
BF_4^-
N
N
N
Ph
H **15** (5 mol%)
Et_3N (1.5 equiv)
PhOCOCl (1.3 equiv)
N
R
Ar
O
OCO_2Ph
O- to *C*-
carboxyl
transfer
N
R
OPh
O
Ar
O
O
69–84%

Ar = 4-MeOC_6H_4

N
R
Ar
O
O
O
OPh
N
N
N
Ph
N
R
Ar
O
O^-
N
N
N
Ph
PhO
O
N
R
OPh
O
Ar
O
O
+
N
N
N
Ph

Scheme 14.4 NHC-promoted *O*- to *C*-carboxyl transfer (DCC = *N*,*N*′-dicyclohexylcarbodiimide).

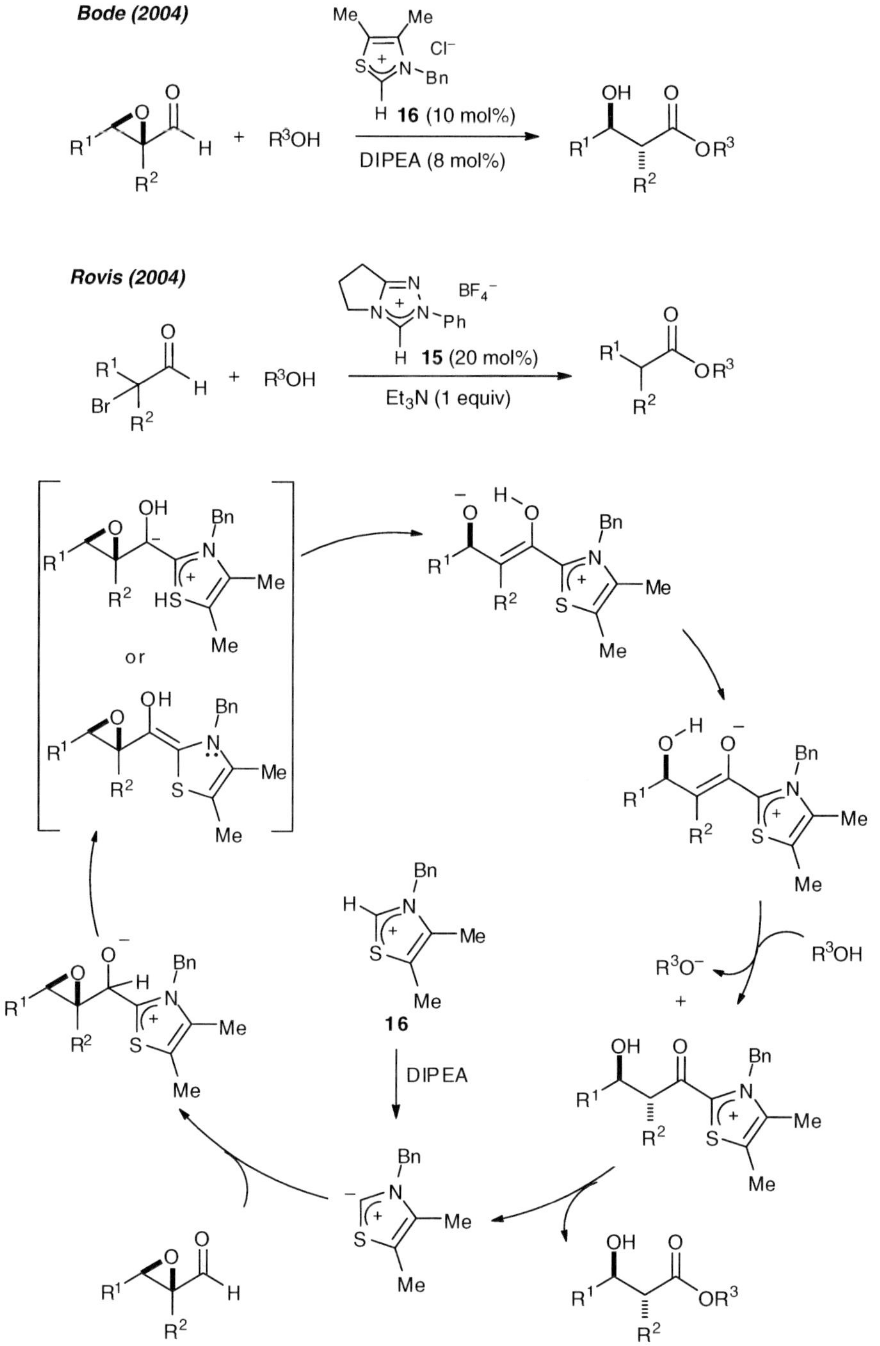

Bode (2004)

Rovis (2004)

Scheme 14.5 Catalytic generation of activated carboxylates (DIPEA = N,N-diisopropylethylamine).

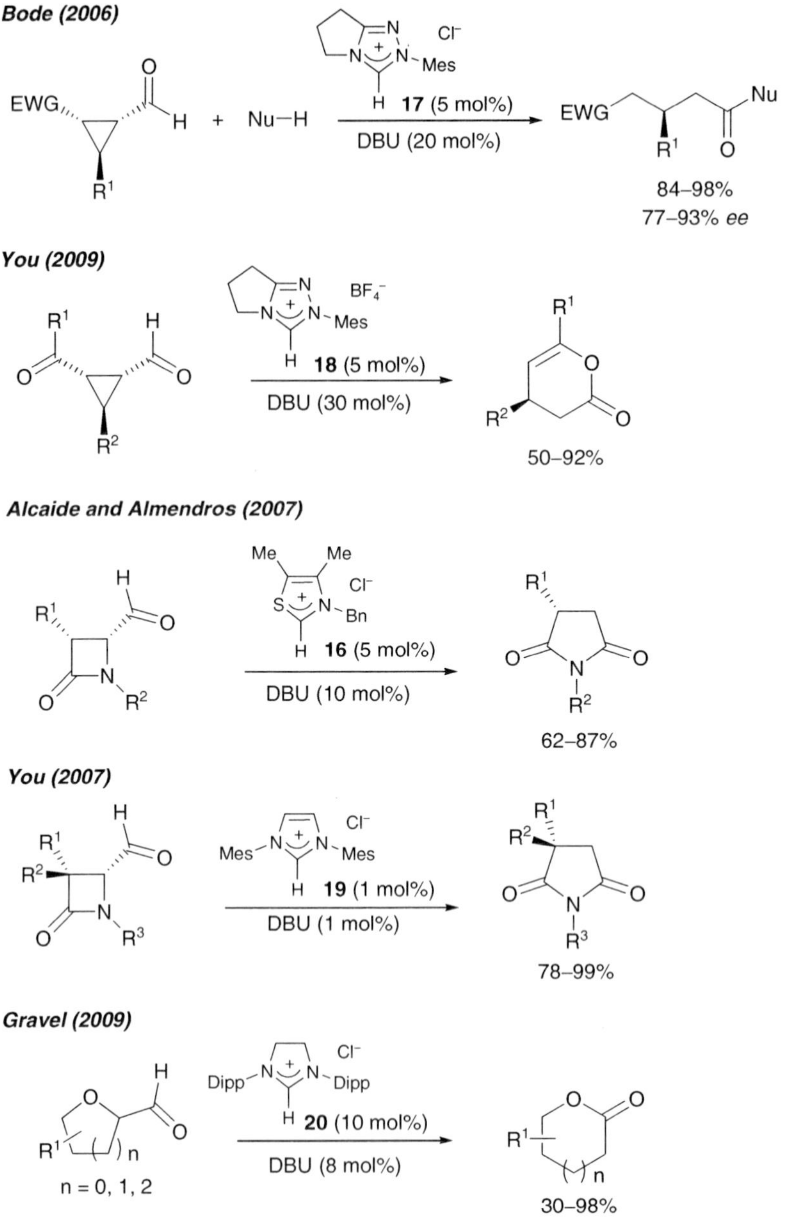

Bode (2006)

EWG, O, H, R[1] + Nu–H → **17** (5 mol%), DBU (20 mol%) → EWG, R[1], O, Nu

Cl^-, Mes, H

84–98%
77–93% *ee*

You (2009)

R^1, H, O, O, R^2 → **18** (5 mol%), DBU (30 mol%) → R^1, O, R^2, O

BF_4^-, Mes, H

50–92%

Alcaide and Almendros (2007)

H, R^1, O, O, N, R^2 → **16** (5 mol%), DBU (10 mol%) → R^1, O, N, O, R^2

Me, Me, Cl^-, S, N, Bn, H

62–87%

You (2007)

H, R^1, R^2, O, O, N, R^3 → **19** (1 mol%), DBU (1 mol%) → R^1, R^2, O, N, O, R^3

Cl^-, Mes, N, N, Mes, H

78–99%

Gravel (2009)

H, O, O, R^1, n; n = 0, 1, 2 → **20** (10 mol%), DBU (8 mol%) → O, O, R^1, n

Cl^-, Dipp, N, N, Dipp, H

30–98%

Scheme 14.6 NHC-catalyzed internal redox reactions of α-substituted aldehydes (Mes = mesityl; DBU = 1,8-diazabicyclo[5.4.0]undec-7-ene; Dipp = 2,6-diisopropylphenyl).

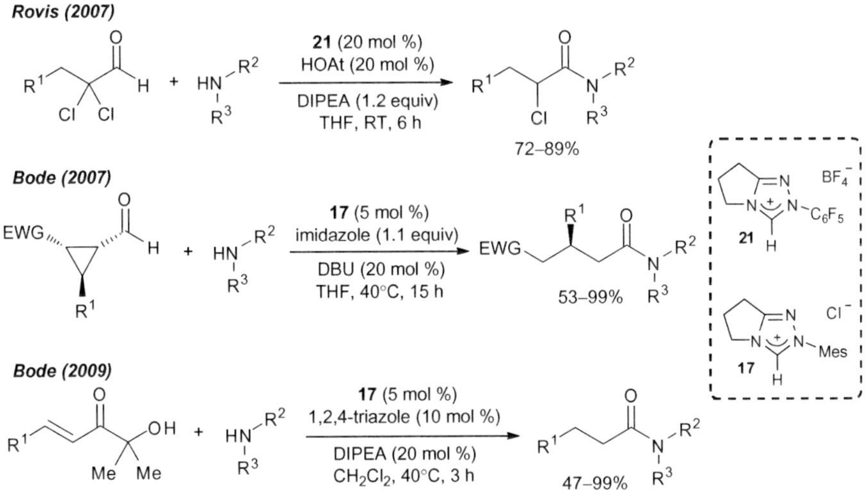

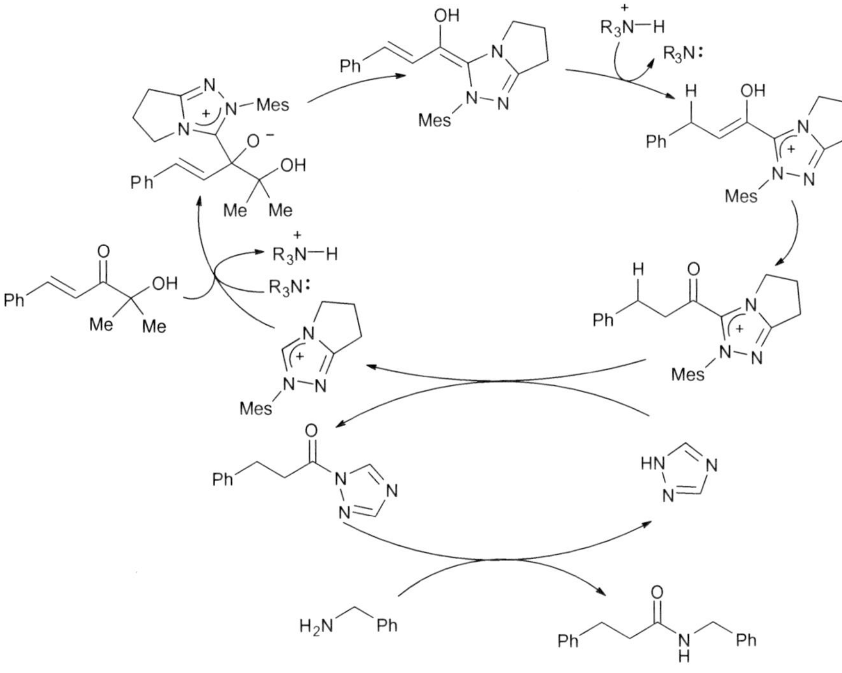

Scheme 14.7 Tandem catalytic cycle of for NHC-promoted amidation reactions.

14.5 NHC-catalyzed Oxidative Esterification

Unlike the NHC-catalyzed redox esterifications described above, which were first reported in 2004, chemists have long recognized the potential of azolium

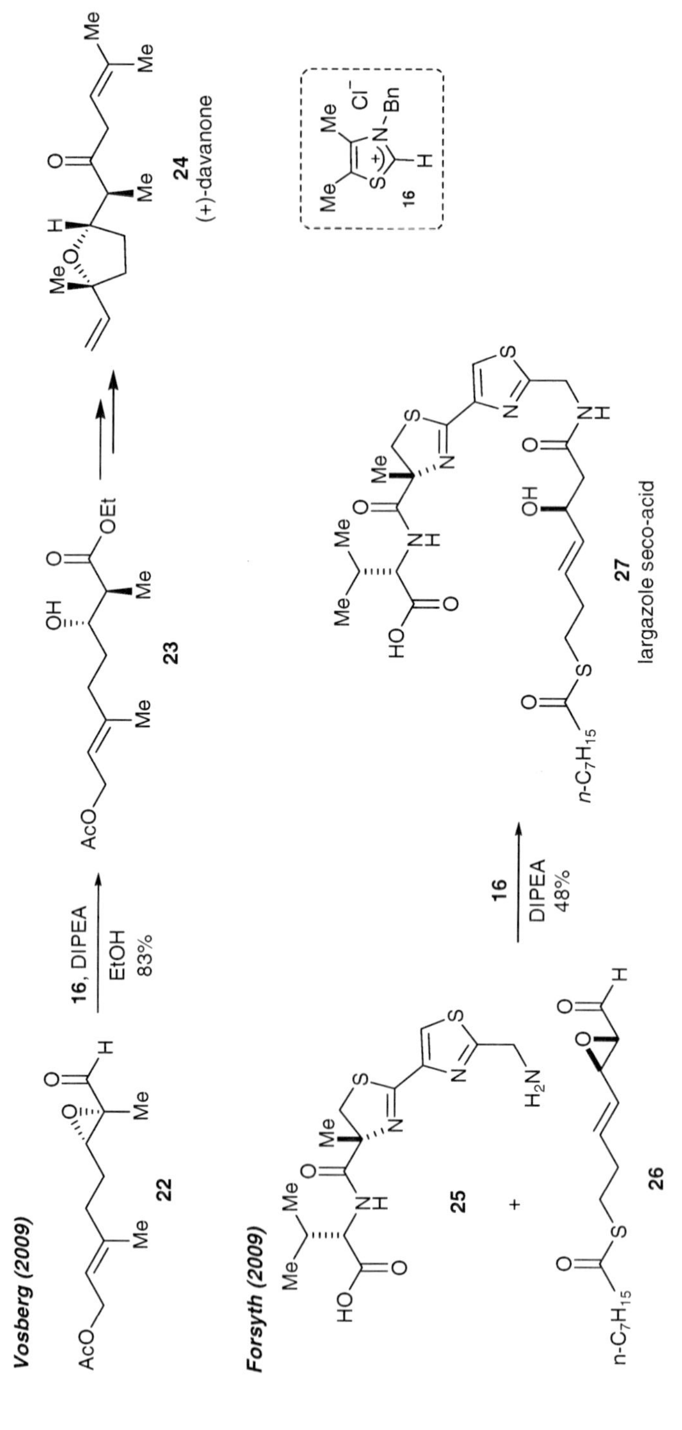

Vosberg (2009)

22 → **16**, DIPEA / EtOH / 83% → **23** → **24** (+)-davanone

Forsyth (2009)

25 + **26** → **16** / DIPEA / 48% → **27** largazole seco-acid

Scheme 14.8 Total syntheses of davanone and largazole using NHC-promoted redox reactions.

salts such as thiamine to promote esterification reactions in the presence of an external oxidant. These reactions are inspired by the biological process whereby an activated aldehyde or Breslow intermediate is oxidized by reaction with lipoic acid.[34] Shinkai reported a biomimetic system consisting of thiamine, flavin, and cationic micelles.[35] In developing biomimetic chemistry, Diederich reported an electrochemical variant using catalytic amount of both the thiamine and flavin to produce esters from aldehydes.[36]

The resurgence of interest in NHC as catalysts in the last 6 years led several chemists to reinvestigate external oxidants for the azolium-catalyzed esterification of aldehydes. Using cyanide, which shares many similarities to azolium salts in its reactions with aldehyde, Corey had shown in 1968 that the combination of an aldehyde and sodium cyanide in the presence of MnO_2 and an alcohol led to esters.[37] Scheidt executed this work with triazolium carbene as a promoter, leading to similar results (Scheme 14.9).[38] Other oxidants including TEMPO radicals (2,2,6,6-tetramethylpiperidine 1-oxyl)[39] and azobenzene[40] were also found suitable for NHC-catalyzed oxidative esterifications.

Alternatively, the oxidation may occur by coupling to a suitable reactant. In 1997, Miyashita reported the formation of oxidative aroylation with nitrobenzene (Scheme 14.10).[41] With the same oxidative reagent, nitrobenzene, benzoin reactions had already been developed in 1982 using cyanide or thiamine.[42] The combination of an aldehyde and nitrosobenzene, in the presence of an N-heterocyclic carbene, led to the formation of *N*-phenyl hydroxamic acids in good yields.[43]

Perhaps the most interesting case of NHC-catalyzed oxidative esterification reported to date is the opening of aziridines reported by Chen (Scheme 14.11).[44] During the attempted addition of *N*-tosyl aziridines by nucleophilic addition of the Breslow intermediate, the authors observed the formation of amino ester **30**. Their mechanistic postulate involved a pathway of *O*-alkylation to give **31**, which would undergo rapid oxidation by molecular oxygen to provide the observed ester.

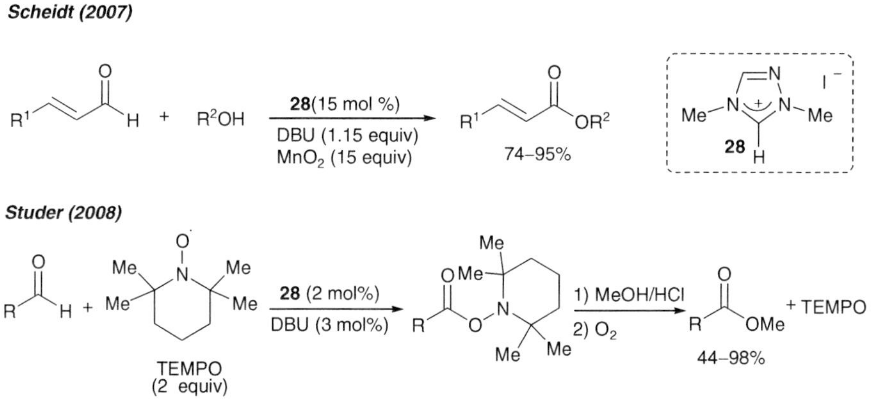

Scheidt (2007)

Studer (2008)

Scheme 14.9 Oxidative esterification with MnO_2 and TEMPO.

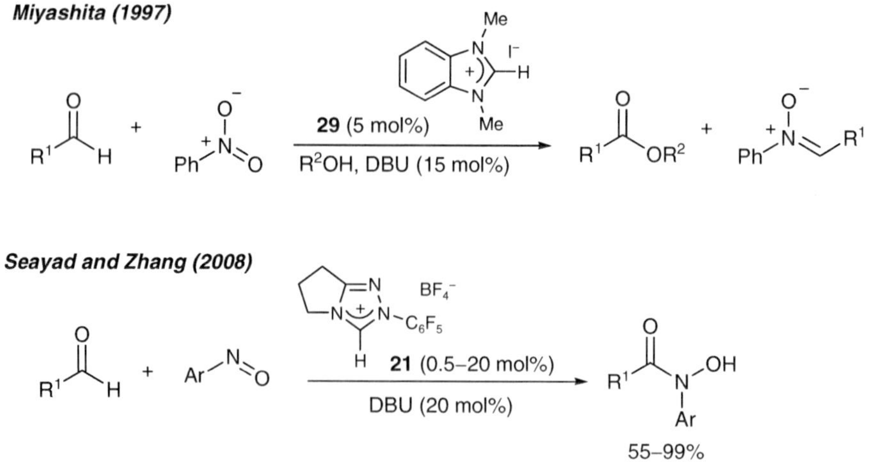

Miyashita (1997)

29 (5 mol%)

R^2OH, DBU (15 mol%)

Seayad and Zhang (2008)

21 (0.5–20 mol%)

DBU (20 mol%)

55–99%

Scheme 14.10 Oxidative esterification with nitrobenzene and nitrosobenzene.

14.6 NHC-catalyzed Reactions of α,β-Unsaturated Aldehydes

The power of NHC catalysts lies in the ability of these heterocycles to promote the transient generation of reactive species, such as acyl anion equivalents or activated carboxylates. Using the mechanistic postulates for these processes, it is possible to predict that the combination of an NHC catalyst and an α,β-unsaturated aldehyde could lead to the generation of a wide variety of catalytically generated reactive intermediates (Scheme 14.12). Over the past 5 years, the rapid developments of new catalysts and reaction conditions have made possible the selective generation of each of these classes of reactive species, including the synthetically powerful homoenolate and ester enolate equivalents.

14.6.1 NHC-catalyzed Generation of Homoenolates

The discovery, in 2004, that NHCs can induce the catalytic generation of homoenolate equivalents opened an entirely new field in NHC-promoted, stereoselective C–C bond forming reactions. In nearly simultaneous reports, Bode[18] and Glorius[45] disclosed the generation of reactive homoenolate equivalents from α,β-unsaturated aldehydes using readily available IMes·HCl **19** as the pre-catalyst. The resulting catalytically generated homoenolates added to aldehydes and α-trifluoromethylketones to give γ-butyrolactones in good yields and moderate diastereoselectivities. The reactions proceeded under simple, mild reactions conditions and tolerate water, air, and many functional groups. This reactivity is best explained by the catalytic cycle shown in Scheme 14.13.

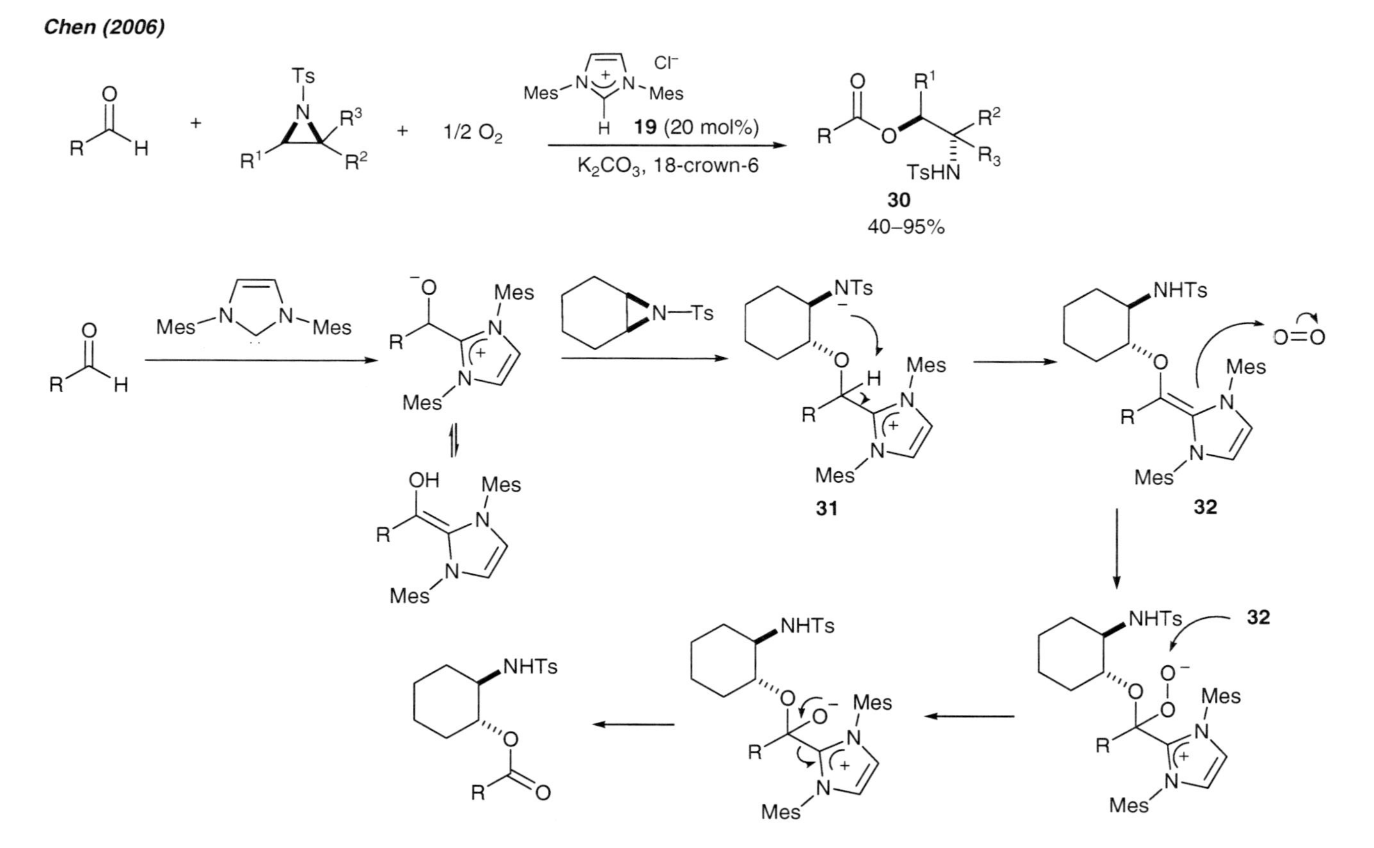

Scheme 14.11 NHC-catalyzed oxidative, ring-opening reactions of *N*-tosyl aziridines with aldehydes.

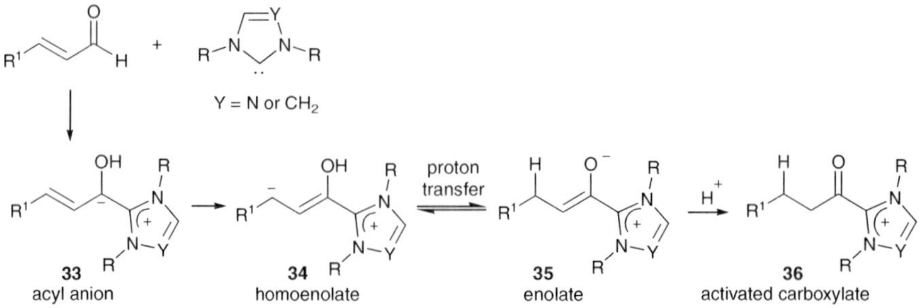

Scheme 14.12 Reactive intermediates generated from α,β-unsaturated aldehydes.

Since this initial disclosure, numerous groups have reported further applications of the catalytically generated homoenolates. Bode[18] found that imidazolium-derived NHC-carbenes promoted the addition of enals to *N*-sulfonylimines, leading to γ-lactams in good yield and diastereoselectivity (Scheme 14.14). Importantly, this was the first report of a general method for the addition of homoenolates, regardless of their method of generation, to imines and opened a new pathway to the synthesis of γ-lactams and related compounds. Scheidt extended these findings by reporting that other imine derivatives, including azomethine ylides[46] and certain nitrones[47] could serve as electrophiles. The latter could be prepared with good enantioselectivity using a chiral triazolium catalyst. Ketone-derived imines are notoriously difficult to prepare and unstable to most nucleophilic reaction conditions. Bode presented a solution to this in NHC-catalyzed γ-lactam formations by employing saccharine-derived imines.[48] These exceptionally reactive substrates required only 0.5 mol% catalyst loadings for most substrates.

Trapping of the catalytically generated homoenolates with electrophilic nitrogen reagents is also possible, as first demonstrated by Zhang and Ying in an intriguing report on the trapping of catalytically generated homoenolates with nitrosobenzene followed by a spontaneous Bamberger rearrangement to afford *N*-*p*-methoxyphenyl-β-amino esters (Scheme 14.15).[49] Recently, Zhong disclosed a variant of this reaction leading to seven-membered 4-azalactones under similar conditions.[50] In a related fashion, Scheidt reported the trapping of homoenolates with electronically tuned azo-compounds to give similar amidation products.[51]

14.6.2 NHC-catalyzed Cyclopentene and Cyclopentane Formations

From the first reports by Glorius and Bode on the NHC-catalyzed generation of homoenolate equivalents, it was recognized that conjugate additions of these species to unsaturated carbonyls could lead to cyclopentanones (Scheme 14.16). This was first achieved by Nair,[52] who made the surprising finding that imidazolium-catalyzed additions of α,β-unsaturated aldehydes to chalcones led to

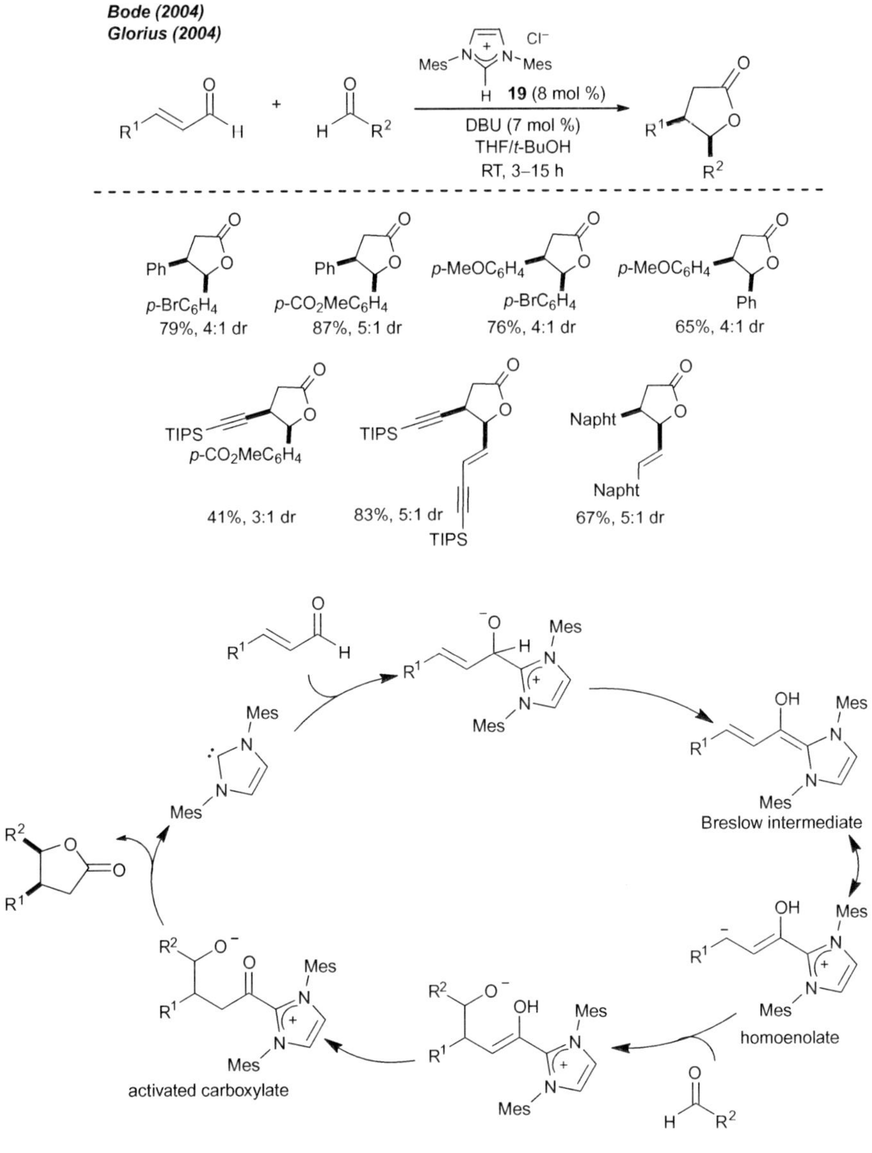

Scheme 14.13 NHC-promoted γ-lactone forming annulations *via* catalytically generated homoenolate equivalents (TIPS = triisopropylsilyl).

trans-disubstituted cyclopentenes, rather than the expected cyclopentanones, in good yield and as single stereoisomers. Key to this process was the spontaneous decarboxylation of β-lactone **41** to give the observed cyclopentene products.

Shortly thereafter, the Bode group reported the remarkable finding that a similar cyclopentene formation catalyzed by a chiral triazolium salt led to the formation of *cis*-disubstituted cyclopentenes in good yield and with

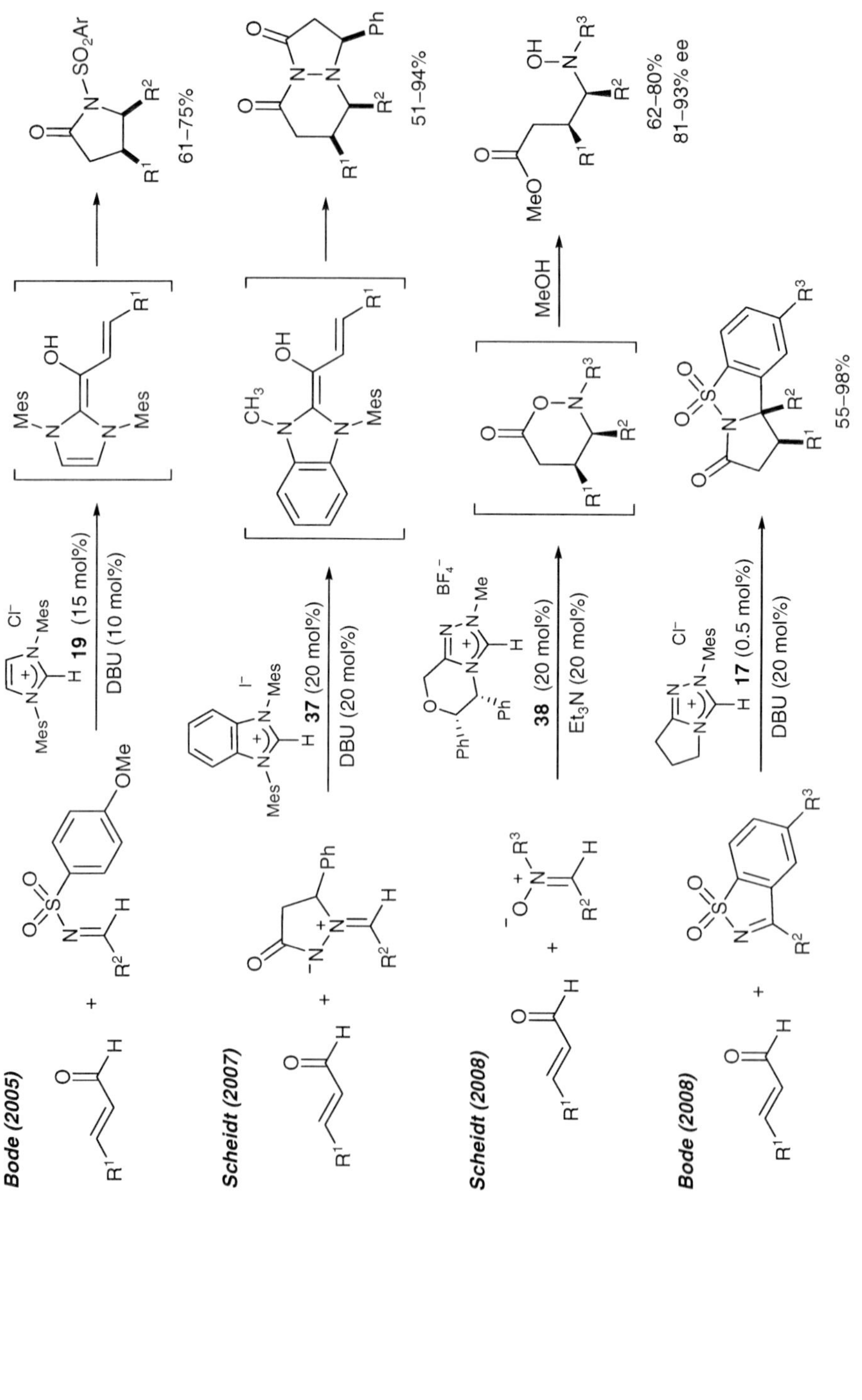

Scheme 14.14 γ-Lactam formation reactions *via* homoenolate intermediates.

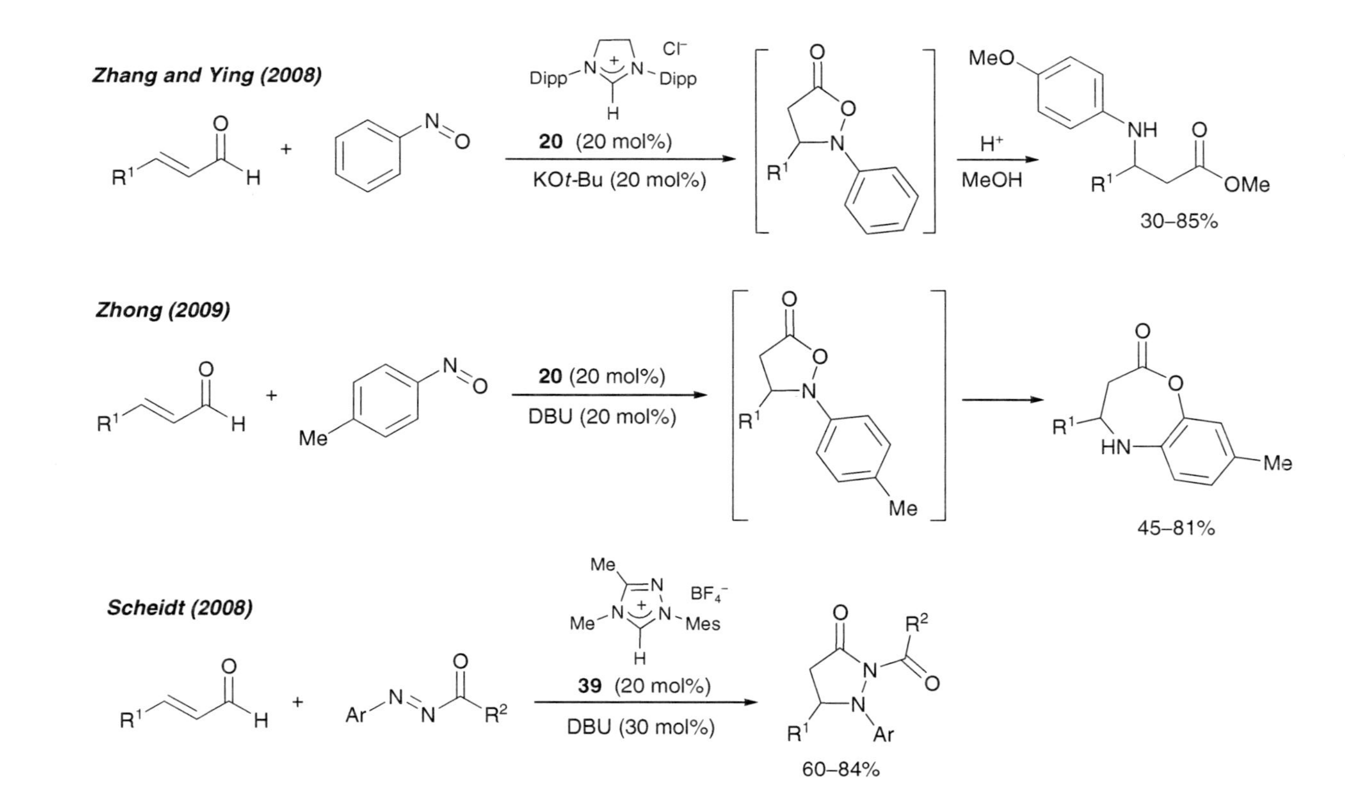

Scheme 14.15 NHC-promoted reactions of enals and electrophilic nitrogen reagents.

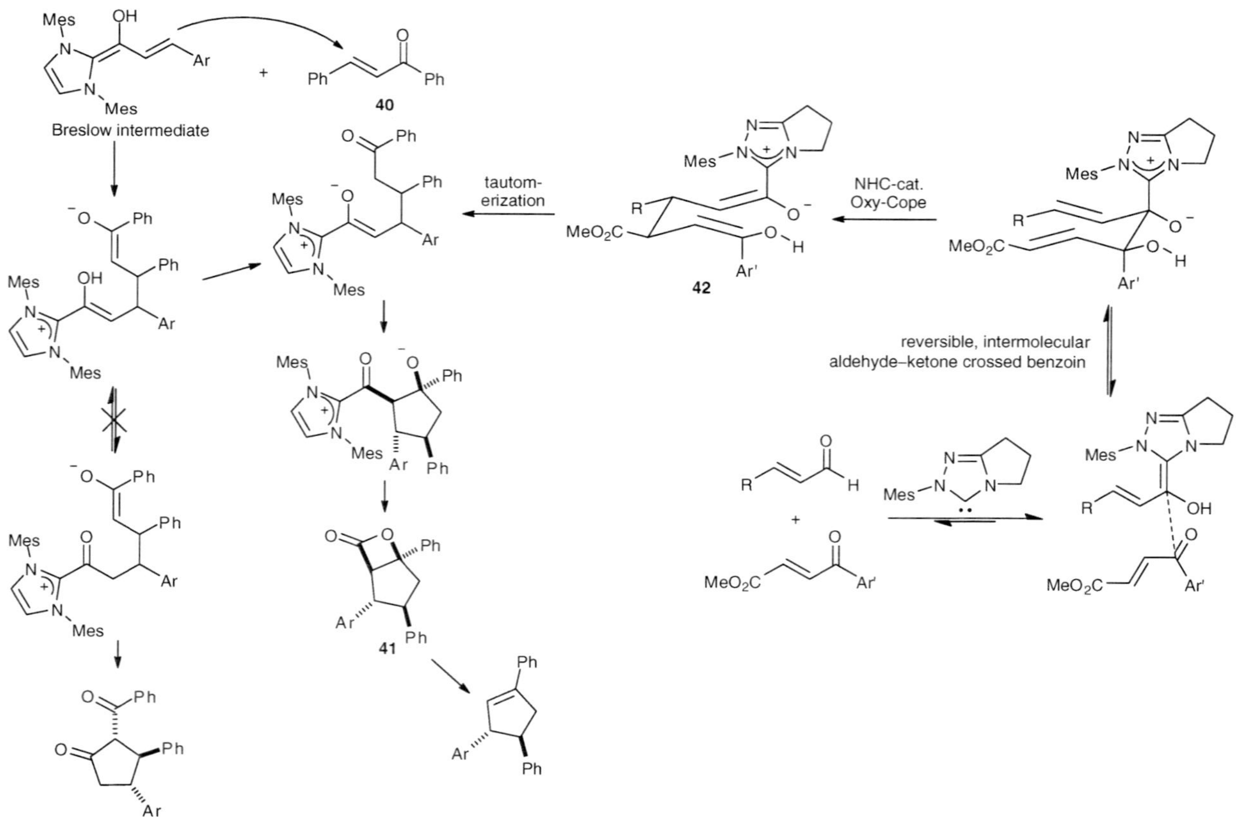

Scheme 14.16 Postulated mechanisms for NHC-promoted cyclopentene formation.

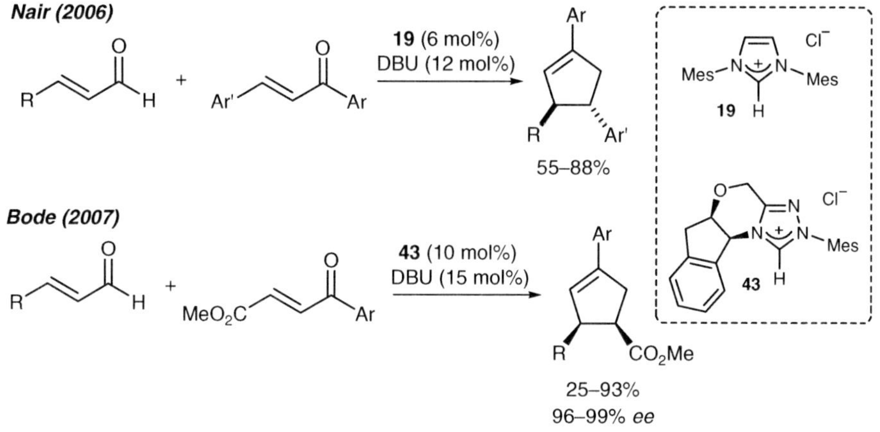

Scheme 14.17 NHC-promoted cyclopentene forming annulations.

outstanding enantioselectivity (Scheme 14.17).[53] The discrepancy between the stereochemical outcomes in comparison to Nair's work was rationalized by an alternative mechanism promoted by the chiral triazolium salt. In Bode's proposal, the key C–C bond forming even occurred not *via* direct homoenolate addition to the unsaturated enone, but rather by a benzoin–oxy-Cope pathway in which an initially formed benzoin adduct would undergo a fast, stereoselective oxy-Cope rearrangement leading to intermediate **42**. This proposal was supported by both control reactions and stereochemical investigations.

Several groups built upon these results by developing methods that prevent decarboxylation and produce highly functionalized cyclopentanes. Bode reported the highly enantioselective synthesis of bicyclic cyclopentyl fused β-lactams[54] and lactones,[55] including an intriguing case of stereodivergency arising from the use of a chiral imidazolium, rather than chiral triazolium-derived carbene (Scheme 14.18). Nair intercepted the β-lactone with MeOH to give cyclopentyl esters.[56] Scheidt reported an intramolecular version of this reaction to afford enantioenriched cyclopentenes and cyclopentane.[57] Interestingly, Nair found that the use of cyclopentanone-derived dibenzylidines afford spirocyclic cyclopentanones *via* the Claisen condensation pathway.[58]

14.6.3 NHC-catalyzed Generation of Enolates from Enals

Consideration of the reaction pathways shown in Scheme 14.12 for the NHC-catalyzed generation of activated carboxylates from enals reveals the intermediacy of an NHC-bound ester enol or enolate equivalent that could be trapped by a suitable electrophile. This was first achieved by Bode, who in 2006 reported highly enantioselective inverse electron-demand Diels–Alder reactions of the catalytically generated enolate equivalents and α,β-unsaturated *N*-sulfonyl imines (Scheme 14.19).[59] At the time, this was the first report of a highly

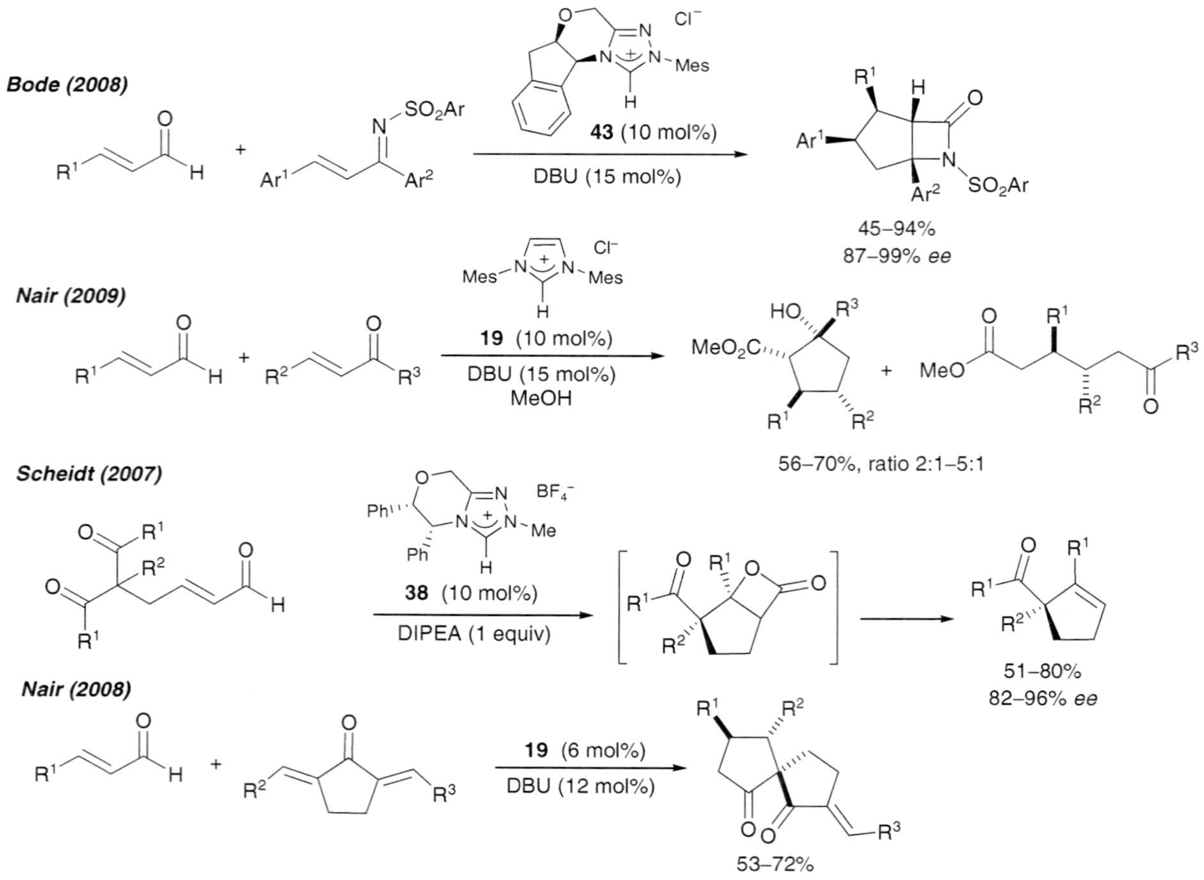

Bode (2008)

43 (10 mol%), DBU (15 mol%)

45–94%
87–99% *ee*

Nair (2009)

19 (10 mol%), DBU (15 mol%), MeOH

56–70%, ratio 2:1–5:1

Scheidt (2007)

38 (10 mol%), DIPEA (1 equiv)

51–80%
82–96% *ee*

Nair (2008)

19 (6 mol%), DBU (12 mol%)

53–72%

Scheme 14.18 Other cyclopentane-forming annulations.

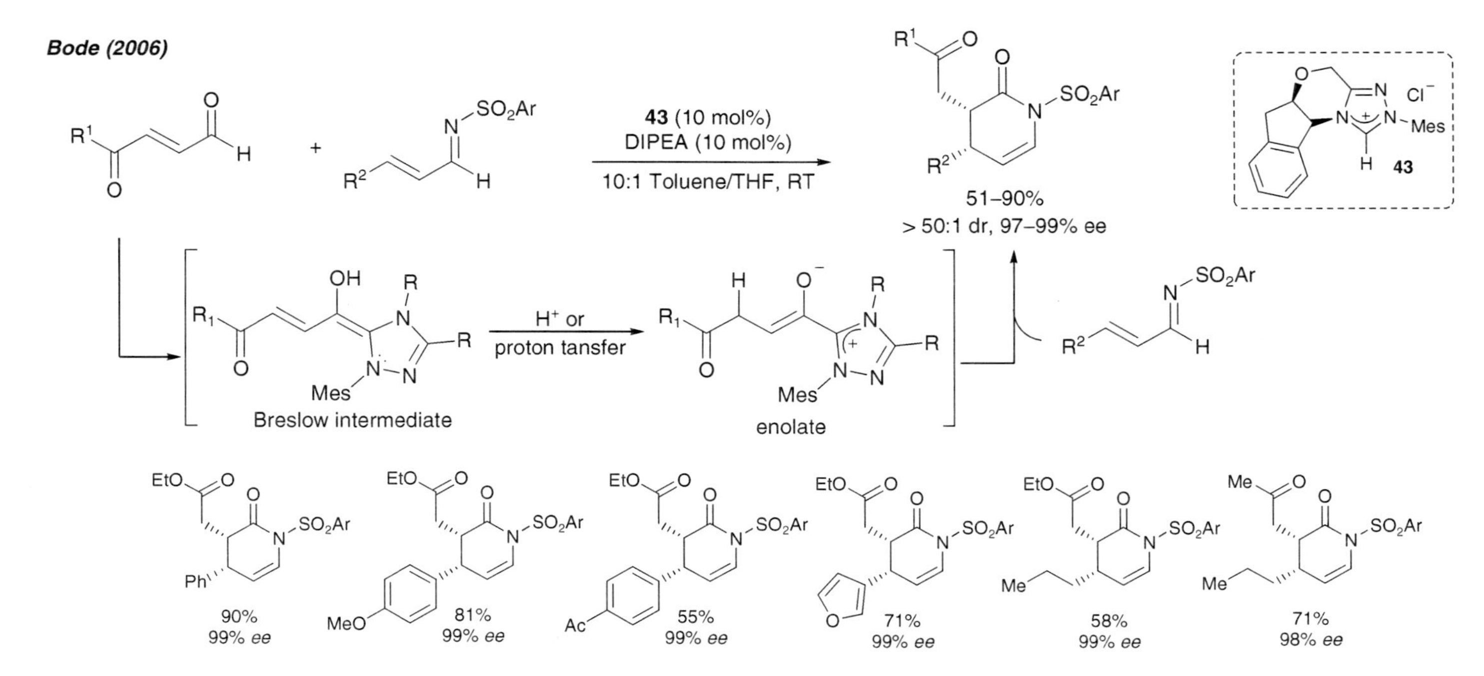

Scheme 14.19 Chiral NHC-promoted inverse electron demand Diels-Alder reactions of catalytically generated enolate equivalents.

enantioselective NHC-catalyzed cross-coupling reaction. This process afforded dihydropyridinone products in good yields and with outstanding levels of enantio- and diastereoselectivity for a broad range of substrates. It also paved the way for a new generation of chiral NHC-catalyzed reactions proceeding *via* this enolate intermediate.

14.6.4 α-Hydroxyenones as Enal Surrogates in NHC-catalyzed Reactions

The scope of the NHC-catalyzed reactions is generally excellent, particularly when applied to the cyclopentene-forming annulations, making this chemistry ideal for diversity oriented synthesis. Unfortunately, it can be quite difficult to prepare the starting α,β-unsaturated aldehydes, especially when they contain basic functionalities that interfere with the common methods for their synthesis. In seeking to address this, Bode[30,60] reported that readily prepared α′-hydroxyenones[61] serve as surrogates for most of the NHC-promoted annulations reaction, albeit only with the less hindered achiral triazolium catalysts (Scheme 14.20). Although this procedure currently affords only racemic products, the improved access to the starting materials makes possible the synthesis of libraries of attractive compounds from a single starting material, simply by changing the electrophile.

14.7 Enantioselective Annulations with NHC-bound Ester Enolate Equivalents

Following the demonstration that NHC-bound enolates **35** (Scheme 14.12) could be catalytically generated and were reactive as nucleophiles and dipoles, a number of impressive annulation methods emerged based on this principle. Bode reported that enolates generated from racemic, α-chloroaldehydes underwent inverse electron-demand oxo-diene Diels–Alder reactions to give dihydropyranones in excellent yields and with outstanding enantioselectivities (Scheme 14.21).[62] Particularly notable the use of very low catalyst loadings, a first for enantioselective organocatalysis at that time. The only limitation of this method was the need to freshly prepare the α-chloroaldehydes, later addressed by developing biphasic reaction conditions for using the corresponding bisulfite salts.[63] These results also make possible formal enantioselective acetate additions by using the commercially available bisulfite adduct of α-chloroacetaldehyde. Building on the NHC-catalyzed generation of enolates, Scheidt reported an intramolecular variant affording identical products using the same *N*-mesityl substituted, aminoindanol derived catalyst.[64]

Ketenes also serve as starting materials to NHC-bound enolate equivalents. Ye reported that enantiomerically enriched dihydropyranones bearing quaternary chiral centers could be generated from aryl-substituted ketenes (Scheme 14.22).[65] Likewise, these substrates underwent diastereoselective protonation, presumably *via* the intermediacy of acyl triazoliums, to give chiral α-arylesters with good

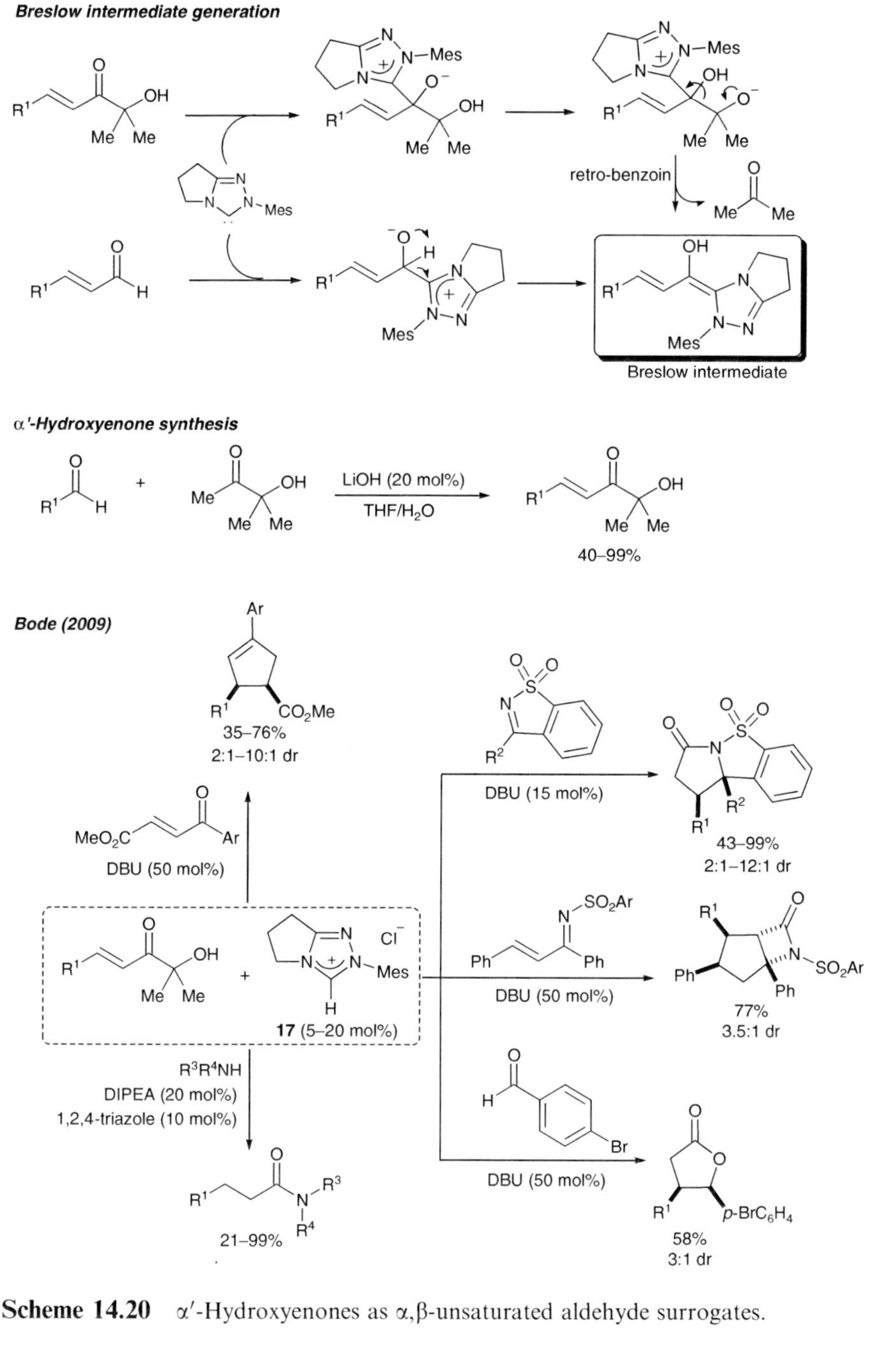

Breslow intermediate generation

retro-benzoin

Breslow intermediate

α′-Hydroxyenone synthesis

LiOH (20 mol%)

THF/H_2O

40–99%

Bode (2009)

35–76%

2:1–10:1 dr

DBU (50 mol%)

17 (5–20 mol%)

DBU (15 mol%)

43–99%

2:1–12:1 dr

DBU (50 mol%)

77%

3.5:1 dr

DBU (50 mol%)

58%

3:1 dr

R^3R^4NH

DIPEA (20 mol%)

1,2,4-triazole (10 mol%)

21–99%

Scheme 14.20 α′-Hydroxyenones as α,β-unsaturated aldehyde surrogates.

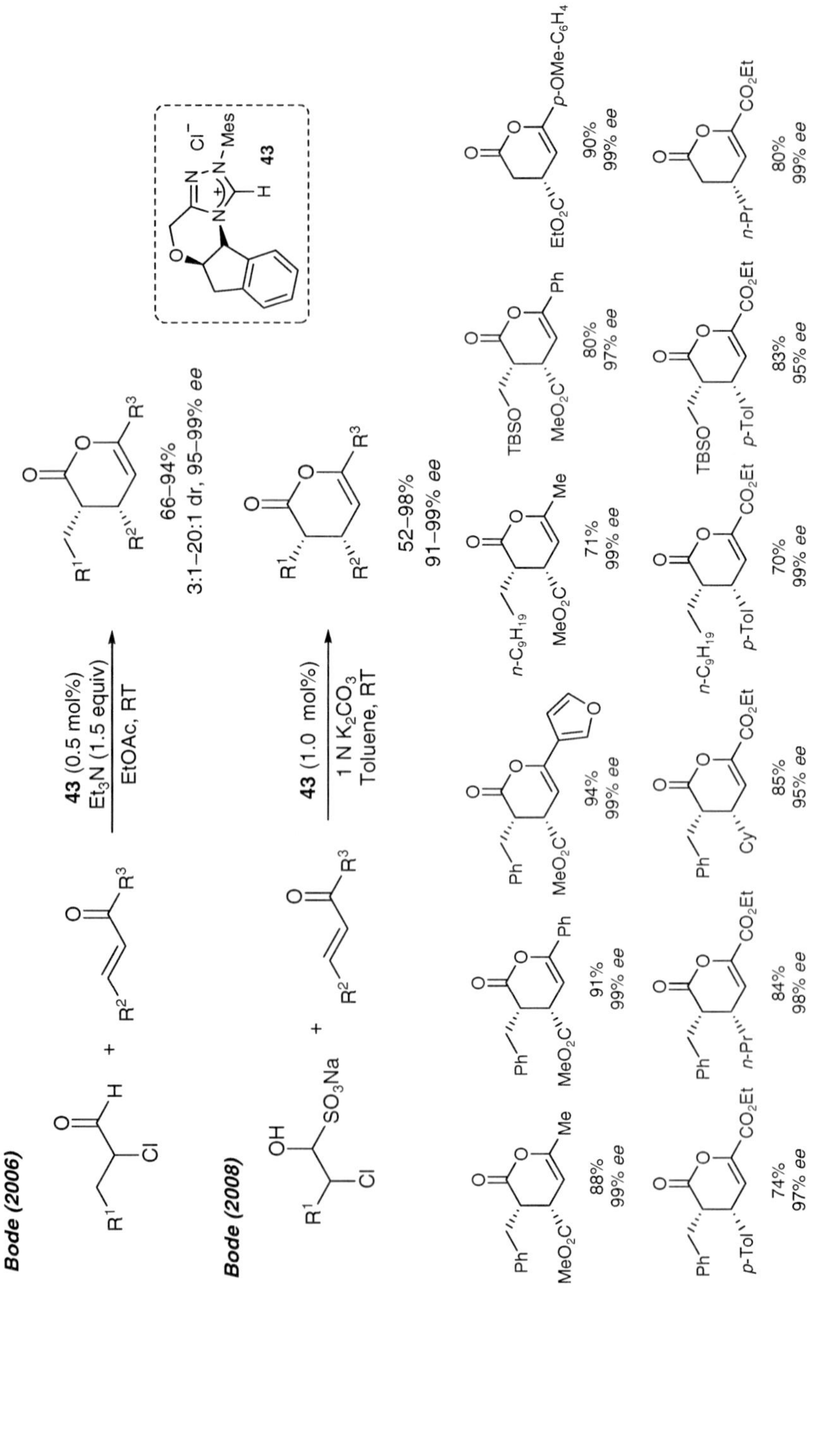

Bode (2006)

43 (0.5 mol%)
Et_3N (1.5 equiv)
EtOAc, RT

66–94%
3:1–20:1 dr, 95–99% *ee*

Bode (2008)

43 (1.0 mol%)
1 N K_2CO_3
Toluene, RT

52–98%
91–99% *ee*

43

88% 99% *ee*; 91% 99% *ee*; 94% 99% *ee*; 71% 99% *ee*; 80% 97% *ee*; 90% 99% *ee*

74% 97% *ee*; 84% 98% *ee*; 85% 95% *ee*; 70% 99% *ee*; 83% 95% *ee*; 80% 99% *ee*

Scheme 14.21 Enantioselective, NHC-catalyzed oxo-diene Diels-Alder reactions.

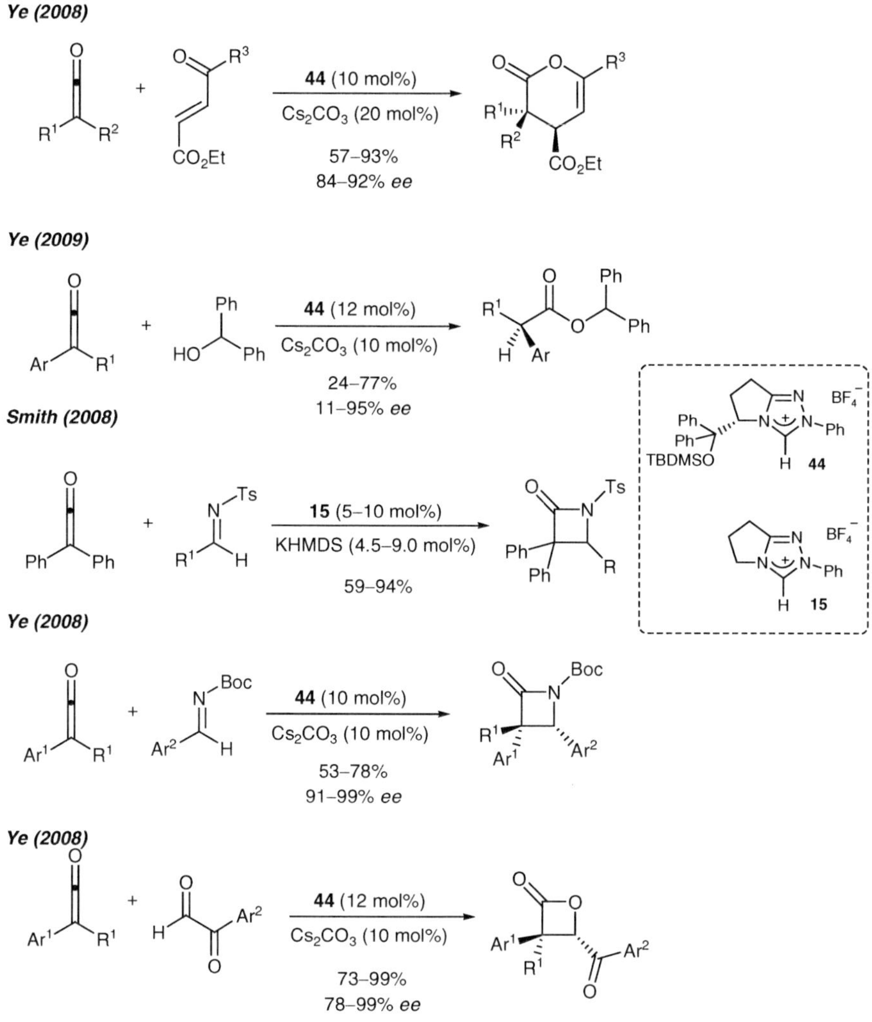

Ye (2008)

44 (10 mol%), Cs_2CO_3 (20 mol%); 57–93%, 84–92% *ee*

Ye (2009)

44 (12 mol%), Cs_2CO_3 (10 mol%); 24–77%, 11–95% *ee*

Smith (2008)

15 (5–10 mol%), KHMDS (4.5–9.0 mol%); 59–94%

Ye (2008)

44 (10 mol%), Cs_2CO_3 (10 mol%); 53–78%, 91–99% *ee*

Ye (2008)

44 (12 mol%), Cs_2CO_3 (10 mol%); 73–99%, 78–99% *ee*

Scheme 14.22 Catalytic generation and reactions of NHC-enolates from ketenes (HMDS = hexamethyldisilazide; TBDMS = *t*-butyldimethylsilyl).

enantioselectivity.[66] Both Ye and Smith applied this system to β-lactam and β-lactone formations *via* Staudinger-like reactions with imines or ketones.[67]

14.8 1,2 Additions Catalyzed by N-Heterocyclic Carbenes

Beyond their role as catalysts for the benzoin reaction, which is formally a 1,2-addition of an acyl anion equivalent to an aldehyde, NHCs were also employed as catalysts for a number of interesting 1,2 addition processes.

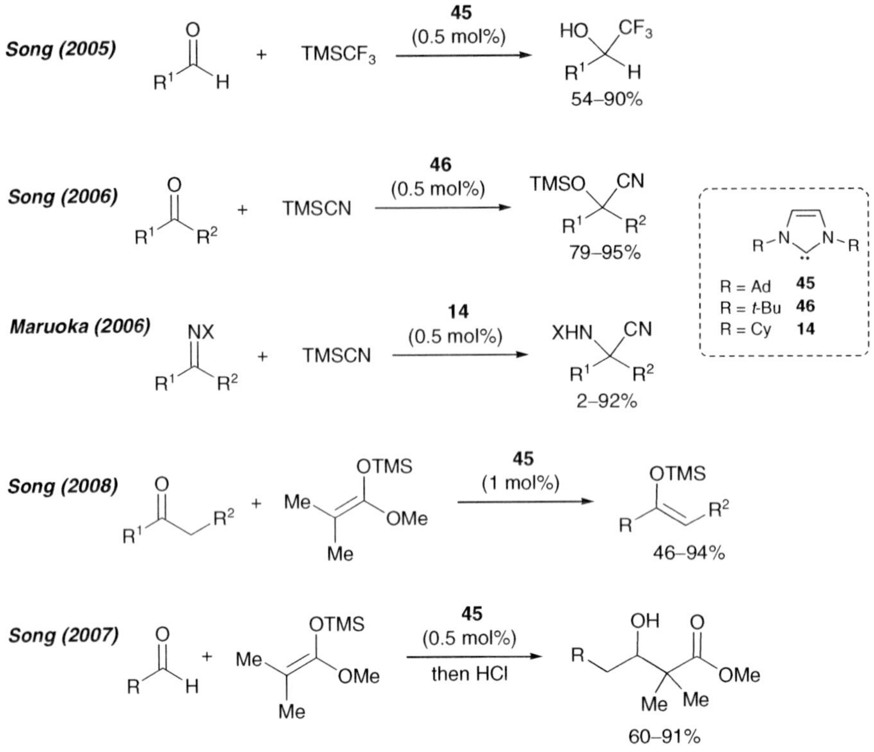

Scheme 14.23 NHC-catalyzed 1,2-additions.

Song *et al.* have exploited the Lewis base properties of pre-generated N-heterocyclic carbenes to activate silylated nucleophiles. In 2005, these researchers reported the NHC-catalyzed addition of $TMSCF_3$ to alcohols using only 0.5 mol% of the preformed carbene as a catalyst (Scheme 14.23).[68] This protocol was also applicable to the addition of other silylated nucleophiles including TMSCN additions to aldehydes[69] and ketones,[70] and imines.[71]

The ability of NHC to activate silylated nucleophiles led to further investigations and useful procedures. Song reported the formation of silyl enol ethers by an NHC-catalyzed silyl exchange reaction that transferred a silyl group from a silyl ketene acetal to a ketone.[72] When aldehydes, rather than ketones, were used as substrates the NHC catalysts promoted a Mukaiyma aldol reaction to give β-hydroxy esters and ketones in good yields.[73]

14.9 Alkylations Catalyzed by NHCs

To date, there have been no reports of *C*-alkylations of the homoenolates or enolates catalytically generated with NHCs. Fischer and Fu, however, reported an interesting example of an NHC-catalyzed cyclization that proceeds *via* the

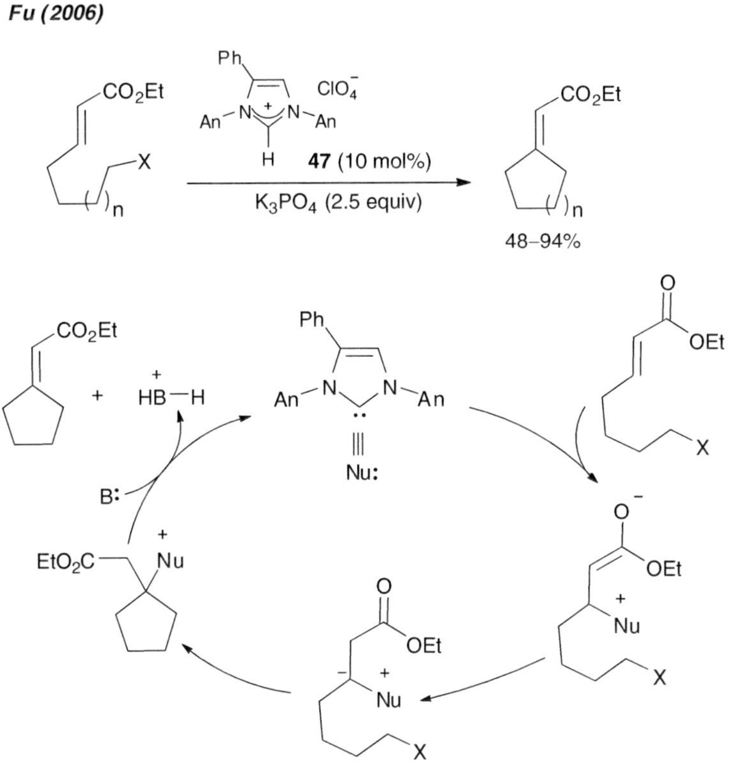

Fu (2006)

Scheme 14.24 NHC-catalyzed cyclizations of electron-deficient alkenes (An = *p*-anisyl).

catalytic generation of a vinyl anion equivalent (Scheme 14.24).[74] This reaction is akin to a Morita–Baylis–Hillman reaction in which a homoenolate, rather than an enolate, was generated. Ye also applied a similar strategy for aza-Morita–Baylis–Hillman reactions of cyclic enones with imines.[75]

14.10 NHC–CO_2 Adducts

Based on the extremely good nucleophilicity of NHCs, the direct addition of imidazolium derived carbenes to CO_2 cleanly afforded the corresponding carboxylates.[76] In 2004, Louie reported a more convenient route to prepare CO_2 adducts by deprotonating the imidazolium salts with potassium *tert*-butoxide under an atmosphere of CO_2 (Scheme 14.25).[77] Recently, Louie also disclosed the decarboxylation factors of imidazolium carboxylates in a systematic investigation,[78] showing that the decarboxylating ability was related to the steric bulk of the *N*-substituent. Thermal gravimetric analysis established that the larger the substituents of the NHC–CO_2 adducts, the more prone it was to decarboxylation.

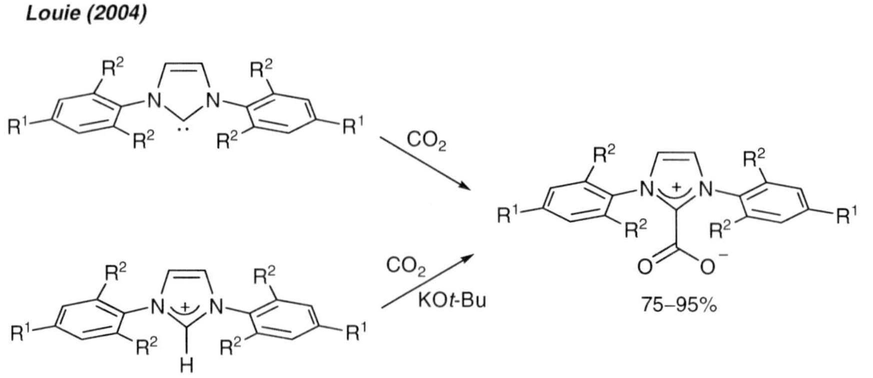

Louie (2004)

CO_2

CO_2, KO*t*-Bu

75–95%

Scheme 14.25 NHC-CO_2 adducts formations.

The coupling of CO_2 with epoxides was applied to imidazolium-based ionic liquids. Using NHC–CO_2 adducts as pre-catalysts was effective for efficient CO_2 fixation reactions. Lu tracked the reaction progress by *in situ* IR monitoring with various epoxides.[79] In 2009, Ikariya and Tommassi extended the substrate scopes widely under milder conditions (Scheme 14.26).[80,81] Ying discovered the NHC–induced hydrosilylation of CO_2 to methanol *via* the NHC–CO_2 adducts.[82]

14.11 NHC-promoted Polymerizations

Some of the earliest work on the use of imidazolium-derived NHCs as catalysts focused on their ability to serve as initiators for polymerization reaction. Pioneers of this area, Waymouth and Hedrick extensively investigated the synthesis of polyesters under the influence of an NHC catalyst. Their results have been summarized in a number of excellent accounts.[83] In 2002, Hedrick first investigated catalytic living polymerization using NHCs for cyclic esters (Scheme 14.27).[84] Then, Waymouth and Hedrick studied polymerization of lactides and ε-caprolactones and explored the chemical and mechanical properties of the resulting polymers.[85] Also, the use of multifunctional alcohol initiators, such as ethylene glycol or pentaerythritol, produced star polymers with good conversion and low PDI (polydispersity index).[86]

In parallel, NHCs were also used as initiators for the ring-opening polymerization of epoxides. This could be executed for the synthesis of polyethers of ethylene oxides. The living zwitterionic intermediate could further react with ε-caprolactone to generate di-block co-polymers in one pot.[87]

14.12 Choice of Azolium Pre-catalysts

One of the most common questions concerning reactions catalyzed by N-heterocyclic carbenes is which azolium pre-catalyst to use. One consideration is the

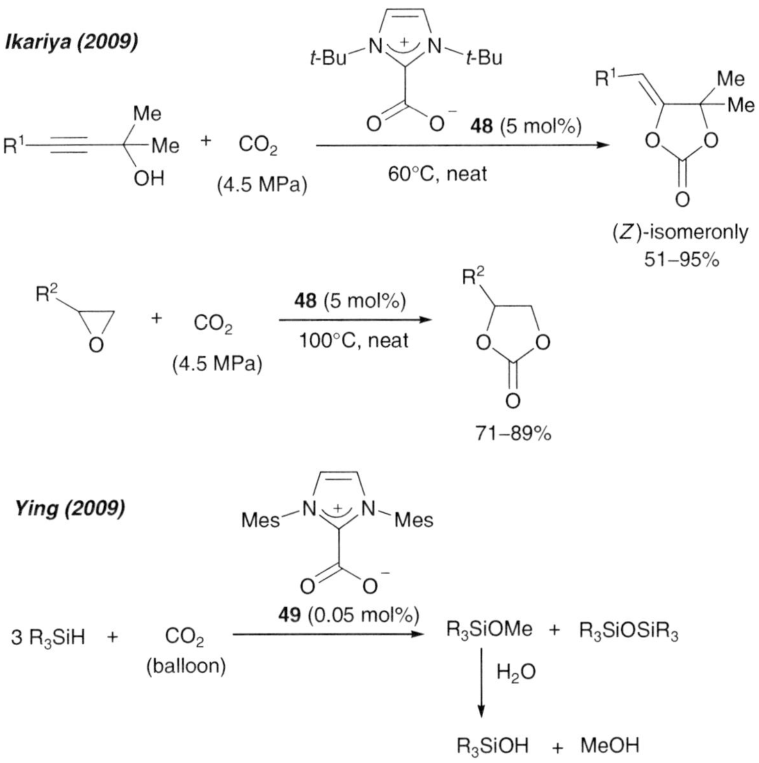

Ikariya (2009)

t-Bu–N(+)N–t-Bu, O, O⁻ **48** (5 mol%)

R1, Me, Me, OH + CO_2 (4.5 MPa) → 60°C, neat → R1, Me, Me, O, O, O

(Z)-isomeronly
51–95%

R2, O + CO_2 (4.5 MPa) → **48** (5 mol%), 100°C, neat → R2, O, O, O

71–89%

Ying (2009)

Mes–N(+)N–Mes, O, O⁻

3 R_3SiH + CO_2 (balloon) → **49** (0.05 mol%) → R_3SiOMe + $R_3SiOSiR_3$

H_2O

R_3SiOH + MeOH

Scheme 14.26 NHC-promoted CO_2 fixation reactions.

choice of imidazolium or triazolium-derived catalysts and an equally important aspect is the *N*-substituent.

For nearly the entire NHC-promoted C–C bond forming reactions, and most of the redox reactions as well, catalysts bearing an *N*-mesityl substituent are both more reactive and selective. This was first noted in the redox esterification of α,β-unsaturated aldehydes, where other *N*-substituents were either inferior or completely inert.[19] In general, NHC-catalyzed C–C bond forming reactions from α,β-unsaturated aldehydes or α-chloroaldehydes required the *N*-mesityl substitution. This is not the case for C–C bond forming reactions that use preformed ketenes as the starting materials, suggesting that the *N*-substituent is important for the redox or protonation processes responsible for the generation of the reactive intermediate. For benzoin and Stetter reactions, which proceed *via* the catalytic generation of an acyl anion equivalent, the *N*-mesityl substituent is not critical and other catalysts types are viable.[6–9] In many cases, the pentafluorophenyl azoliums developed by Rovis are the most active and selective. Interestingly, from the few results available, the *N*-mesityl substituted triazoliums are also excellent catalysts for intramolecular benzoin and Stetter reactions, suggesting that the increased steric demands of these catalysts are not detrimental to reactivity.

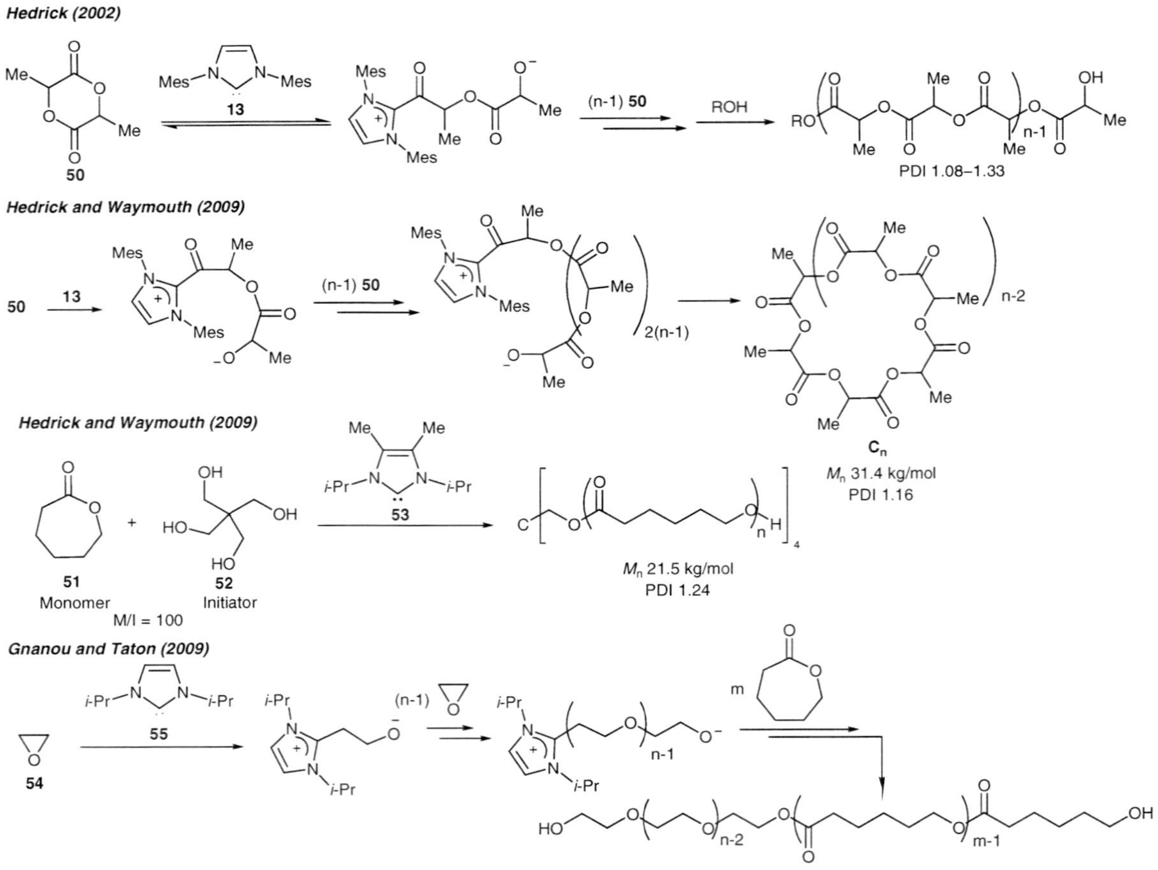

Hedrick (2002)

50 + 13 ⇌ ... → (n-1) 50 → ROH → PDI 1.08–1.33

Hedrick and Waymouth (2009)

50 → 13 → ... → (n-1) 50 → 2(n-1) → C_n (M_n 31.4 kg/mol, PDI 1.16)

Hedrick and Waymouth (2009)

51 Monomer + 52 Initiator, M/I = 100 → 53 → M_n 21.5 kg/mol, PDI 1.24

Gnanou and Taton (2009)

54 → 55 → (n-1) → m (ε-caprolactone) → n-2, m-1

Scheme 14.27 NHC-catalyzed polymerizations (M_n = number-average molar mass).

The choice of imidazolium *versus* triazolium pre-catalyst is subtler. Triazolium-derived NHCs are nearly always preferred with the exception of certain processes proceeding *via* catalytically generated homoenolate equivalents. For example, the γ-lactone forming annulations of α,β-unsaturated aldehydes and aromatic aldehydes give extremely poor conversion with triazolium-derived pre-catalysts but proceed in excellent yield with IMes · HCl **19**.[19] When more reactive electrophiles, such as α-trifluoromethylketone[88] or saccharine-derived imines[48] are employed, other catalyst classes including *N*-mesityl substituted triazoliums and thiazoliums again become viable pre-catalysts. This can be attributed to the increased electron donating ability of the imidazolium pre-catalysts, rending their corresponding homoenolate equivalents **34** (see Scheme 14.12) more nucleophilic. In contrast, the additional electronegative atom in the triazolium pre-catalysts decreases the nucleophilicity of this intermediate, leading eventually to other reaction manifolds such as protonation.

In order to separate structural effects from the electronic differences of these two catalyst classes, Bode synthesized chiral imidazolium salt **57** (Scheme 14.28).[89] This allowed direct comparison of imidazolium *versus* triazolium pre-catalysts across a number of different reaction manifolds including those involving the catalytic generation of homoenolate equivalents, ester enolate equivalents, and acyl anions.[90] These studies conclusively demonstrated that imidazolium-derived catalysts are superior for homoenolate reactions with less reactive electrophiles, while the triazolium-derived pre-catalysts are preferred for all other reactions. Interestingly, from the currently published body of the work, it does not appear to be any effects from the counterion of the azolium pre-catalysts in the presence of bases.

The electronic factors of these two catalyst types also play an important role in the leaving group ability of the NHC catalyst, an important step in all reactions described above and key to catalyst turnover. *N*-Mesityl substituted

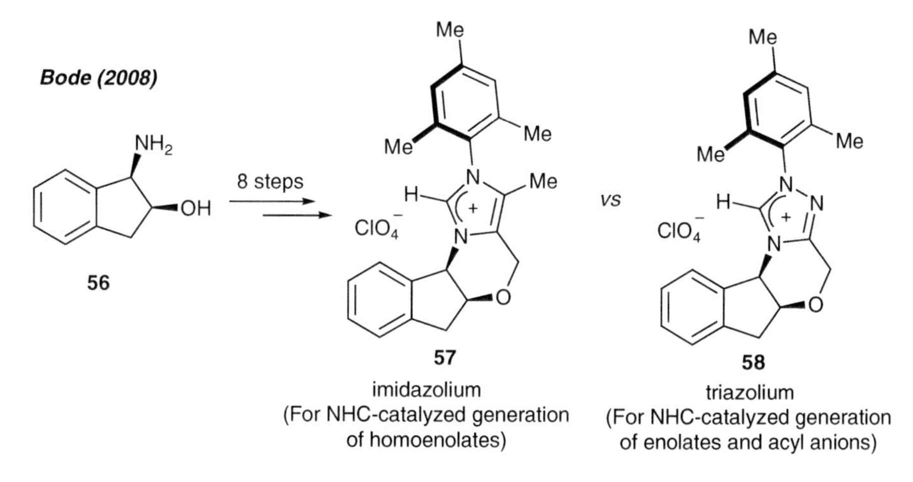

Scheme 14.28 *N*-Mesityl substituted chiral imidazolium and triazolium salts.

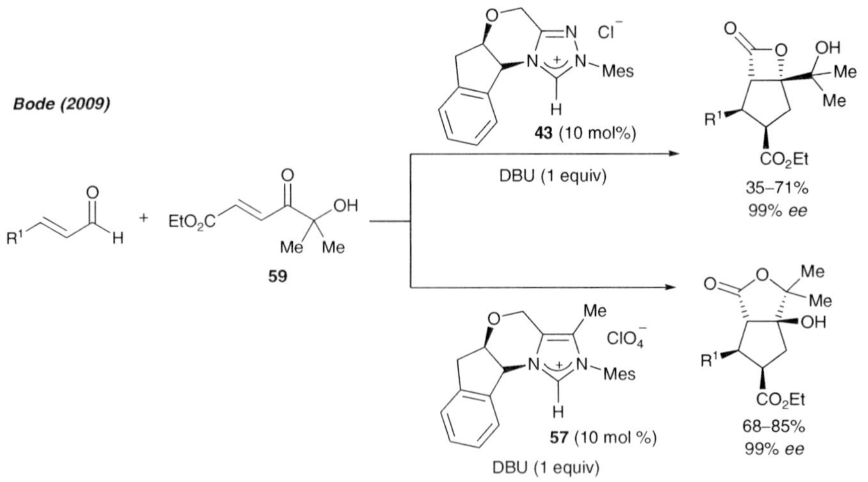

Scheme 14.29 Stereodivergency of chiral imidazolium- and triazolium-promoted cyclopentane-forming annulations.

imidazoliums and triazoliums both promote the protonation of Breslow intermediates derived from enals at the β-position to give catalytically generated activated carboxylates. The acyl-imidazolium derivatives, however, are less reactive in the acylation step, leading to slow catalyst turnover. This is dramatically illustrated by the stereodivergency of otherwise identical cyclopentane-forming annulation (Scheme 14.29).[55] The authors postulated that the stereodivergency arose from the poorer leaving group ability of the imidazolium-derived carbene, resulting in a change in the rate determining step of the cascade and leading to the formation of the more thermodynamically favored γ-lactone product.

14.13 Conclusions and Outlook

Although azolium catalysts have been known and studied for the past 50 years or more, it is only in the last 5 years that chemists have realized their long hidden potential as one of the most powerful and dramatic promoters of complex organic transformation. Unlike most other catalyst classes, NHCs unlock unexpected reactivity and engender complex molecular organization, reactions once thought to be the sole domain of transition metal catalysts. The sensitivity of these processes to subtle structural and electronic effects, such as the pivotal role of bulky aromatic substituents or the delicate effects of imidazolium *versus* triazolium precursors, promise many more refinements and discoveries to come.

For this field to continue to develop at this breakneck pace, several existing roadblocks must be overcome. First, the synthesis of the catalysts remains challenging. Although the amino alcohol-derived triazolium salts affect a remarkably broad range of transformations with exceptional enantiocontrol, it is difficult to rapidly modify the NHC pre-catalysts to improve those reactions that are less selective. Second, many of the newer discoveries have yet to be fully exploited. For example, it appears that NHC catalysts can promote long-sought enantioselective sigmatropic rearrangements, but the implementation of this potentially powerful finding remains to be demonstrated. Finally, major gaps remain in the mechanistic understanding of these reactions and the elegant dance of proton transfer and redox reactions that characterize the novel cascades that have become the hallmark of NHC-catalyzed processes.

References

1. T. Ugai, S. Tanaka and S. Dokawa, *J. Pharm. Soc. Jpn.*, 1943, **63**, 296–300.
2. R. Breslow, *J. Am. Chem. Soc.*, 1958, **80**, 3719–3726.
3. A. J. Arduengo III, R. L. Harlow and M. Kline, *J. Am. Chem. Soc.*, 1991, **113**, 361–363.
4. For a review on Wanzlick carbenes, see: T. Weskamp, V. P. W. Böhm and W. A. Herrmann, *J. Organomet. Chem.*, 2000, **600**, 12–22.
5. For general reviews including the benzoin and Stetter reactions, see: (a) D. Enders and T. Balensiefer, *Acc. Chem. Res.*, 2004, **37**, 534–541; (b) D. Enders, O. Niemeier and A. Henseler, *Chem. Rev.*, 2007, **107**, 5606–5655; (c) N. Marion, S. Díez-González and S. P. Nolan, *Angew. Chem., Int. Ed.*, 2007, **46**, 2988–3000; (d) T. Rovis, *Chem. Lett.*, 2008, **37**, 2–7.
6. (a) R. L. Knight and F. J. Leeper, *Tetrahedron Lett.*, 1997, **38**, 3611–3614; (b) R. L. Knight and F. J. Leeper, *J. Chem. Soc., Perkin Trans. 1*, 1998, 1891–1893.
7. M. S. Kerr, J. Read de Alaniz and T. Rovis, *J. Org. Chem.*, 2005, **70**, 5725–5728.
8. (a) D. Enders, K. Breuer and J. H. Teles, *Helv. Chim. Acta*, 1996, **79**, 1217–1221; (b) D. Enders and U. Kallfass, *Angew. Chem., Int. Ed.*, 2002, **41**, 1743–1745; (c) D. Enders, O. Niemeier and T. Balensiefer, *Angew. Chem., Int. Ed.*, 2006, **45**, 1463–1467; (d) D. Enders and J. Han, *Tetrahedron: Asymmetry*, 2008, **19**, 1367–1371; (e) Y. Ma, S. Wei, J. Wu, F. Yang, B. Liu, J. Lan, S. Yang and J. You, *Adv. Synth. Catal.*, 2008, **350**, 2645–2651.
9. (a) D. Enders, K. Breuer, J. Runsink and J. H. Teles, *Helv. Chim. Acta*, 1996, **79**, 1899–1902; (b) M. S. Kerr, J. Read de Alaniz and T. Rovis, *J. Am. Chem. Soc.*, 2002, **124**, 10298–10299; (c) M. S. Kerr and T. Rovis, *J. Am. Chem. Soc.*, 2004, **126**, 8876–8877; (d) Q. Liu and T. Rovis, *J. Am. Chem. Soc.*, 2006, **128**, 2552–2553; (e) Q. Liu, S. Perreault and T. Rovis, *J. Am. Chem. Soc.*, 2008, **130**, 14066–14067; (f) S. C. Cullen and T. Rovis, *Org. Lett.*, 2008, **10**, 3141–3144; (g) D. Enders and J. Han, *Synthesis*, 2008, 3864–3868; (h) D. Enders, J. Han and A. Henseler, *Chem. Commun.*, 2008, 3989–3991; (i) Q. Liu and T. Rovis, *Org. Lett.*, 2009, **11**, 2856–2859.
10. G. W. Nyce, J. A. Lamboy, E. F. Connor, R. M. Waymouth and J. L. Hedrick, *Org. Lett.*, 2002, **4**, 3587–3590.

11. G. A. Grasa, R. M. Kissling and S. P. Nolan, *Org. Lett.*, 2002, **4**, 3583–3586.
12. (a) T. Kano, K. Sasaki and K. Maruoka, *Org. Lett.*, 2005, **7**, 1347–1349; (b) Y. Suzuki, K. Muramatsu, K. Yamauchi, Y. Morie and M. Sato, *Tetrahedron*, 2006, **62**, 302–310.
13. M. Movassaghi and M. A. Schmidt, *Org. Lett.*, 2005, **7**, 2453–2456.
14. (a) J. E. Thomson, K. Rix and A. D. Smith, *Org. Lett.*, 2006, **8**, 3785–3788; (b) C. D. Campbell, N. Duguet, K. A. Gallagher, J. E. Thomson, A. G. Lindsay, A. C. O'Donoghue and A. D. Smith, *Chem. Commun.*, 2008, 3528–3530; (c) J. E. Thomson, C. D. Campbell, C. Concellon, N. Duguet, K. Rix, A. M. Z. Slawin and A. D. Smith, *J. Org. Chem.*, 2008, **73**, 2784–2791; (d) J. E. Thomson, A. F. Kyle, K. A. Gallagher, P. Lenden, C. Concellón, L. C. Morrill, A. J. Miller, C. Joannesse, A. M. Z. Slawin and A. D. Smith, *Synthesis*, 2008, 2805–2818.
15. For a review of the rapid developments in this area, see: K. Zeitler, *Angew. Chem., Int. Ed.*, 2005, **44**, 7506–7510.
16. K. Y. Chow and J. W. Bode, *J. Am. Chem. Soc.*, 2004, **126**, 8126–8127.
17. N. T. Reynolds, J. Read de Alaniz and T. Rovis, *J. Am. Chem. Soc.*, 2004, **126**, 9518–9519.
18. (a) S. S. Sohn, E. L. Rosen and J. W. Bode, *J. Am. Chem. Soc.*, 2004, **126**, 14370–14371; (b) M. He and J. W. Bode, *Org. Lett.*, 2005, **7**, 3131–3134.
19. S. S. Sohn and J. W. Bode, *Org. Lett.*, 2005, **7**, 3873–3876.
20. K. Zeitler, *Org. Lett.*, 2006, **8**, 637–640.
21. N. T. Reynolds and T. Rovis, *J. Am. Chem. Soc.*, 2005, **127**, 16406–16407.
22. Y. Yamaki, A. Shigenaga, K. Tomita, T. Narumi, N. Fujii and A. Otaka, *J. Org. Chem.*, 2009, **74**, 3272–3277.
23. S. S. Sohn and J. W. Bode, *Angew. Chem., Int. Ed.*, 2006, **45**, 6021–6024.
24. G.-Q. Li, L.-X. Dai and S.-L. You, *Org. Lett.*, 2009, **11**, 1623–1625.
25. B. Alcaide, P. Almendros, G. Cabrero and M. P. Ruiz, *Chem. Commun.*, 2007, 4788–4790.
26. G.-Q. Li, Y. Li, L.-X. Dai and S.-L. You, *Org. Lett.*, 2007, **9**, 3519–3521.
27. L. Wang, K. Thai and M. Gravel, *Org. Lett.*, 2009, **11**, 891–893.
28. H. U. Vora and T. Rovis, *J. Am. Chem. Soc.*, 2007, **129**, 13796–13797.
29. J. W. Bode and S. S. Sohn, *J. Am. Chem. Soc.*, 2007, **129**, 13798–13799.
30. P.-C. Chiang, Y. Kim and J. W. Bode, *Chem. Commun.*, 2009, 4566–4568.
31. H. U. Vora, J. R. Moncecchi, O. Epstein and T. Rovis, *J. Org. Chem.*, 2008, **73**, 9727–9731.
32. K. C. Morrison, J. P. Litz, K. P. Scherpelz, P. D. Dossa and D. A. Vosburg, *Org. Lett.*, 2009, **11**, 2217–2218.
33. B. Wang and C. J. Forsyth, *Synthesis*, 2009, 2873–2880.
34. (a) L. J. Reed and B. G. Debusk, *J. Biol. Chem.*, 1952, **199**, 881–888; (b) I. C. Gunsalus, *J. Cell Physiol. Suppl.*, 1953, **41**, 113–136.
35. S. Shinkai, T. Yamashita, Y. Kusano and O. Manabe, *J. Org. Chem.*, 1980, **45**, 4947–4952.
36. S.-w. Tam, L. Jimenez and F. Diederich, *J. Am. Chem. Soc.*, 1992, **114**, 1503–1505.

37. E. J. Corey, N. W. Gilman and B. E. Ganem, *J. Am. Chem. Soc.*, 1968, **90**, 5616–5617.
38. (a) B. E. Maki, A. Chan, E. M. Phillips and K. A. Scheidt, *Org. Lett.*, 2007, **9**, 371–374; (b) B. E. Maki, A. Chan, E. M. Phillips and K. A. Scheidt, *Tetrahedron*, 2009, **65**, 3102–3109.
39. J. Guin, S. De Sarkar, S. Grimme and A. Studer, *Angew. Chem., Int. Ed.*, 2008, **47**, 8727–8730.
40. C. Noonan, L. Baragwanath and S. J. Connon, *Tetrahedron Lett.*, 2008, **49**, 4003–4006.
41. A. Miyashita, Y. Suzuki, I. Nagasaki, C. Ishiguro, K. Iwamoto and T. Higashino, *Chem. Pharm. Bull.*, 1997, **45**, 1254–1258.
42. J. Castells, F. Pujol, H. Llitjos and M. Morenomanas, *Tetrahedron*, 1982, **38**, 337–346.
43. F. T. Wong, P. K. Patra, J. Seayad, Y. Zhang and J. Y. Ying, *Org. Lett.*, 2008, **10**, 2333–2336.
44. Y.-K. Liu, R. Li, L. Yue, B.-J. Li, Y.-C. Chen, Y. Wu and L.-S. Ding, *Org. Lett.*, 2006, **8**, 1521–1524.
45. C. Burstein and F. Glorius, *Angew. Chem., Int. Ed.*, 2004, **43**, 6205–6208.
46. A. Chan and K. A. Scheidt, *J. Am. Chem. Soc.*, 2007, **129**, 5334–5335.
47. E. M. Phillips, T. E. Reynolds and K. A. Scheidt, *J. Am. Chem. Soc.*, 2008, **130**, 2416–2417.
48. M. Rommel, T. Fukuzumi and J. W. Bode, *J. Am. Chem. Soc.*, 2008, **130**, 17266–17267.
49. J. Seayad, P. K. Patra, Y. Zhang and J. Y. Ying, *Org. Lett.*, 2008, **10**, 953–956.
50. L. Yang, B. Tan, F. Wang and G. Zhong, *J. Org. Chem.*, 2009, **74**, 1744–1746.
51. A. Chan and K. A. Scheidt, *J. Am. Chem. Soc.*, 2008, **130**, 2740–2741.
52. V. Nair, S. Vellalath, M. Poonoth and E. Suresh, *J. Am. Chem. Soc.*, 2006, **128**, 8736–8737.
53. P.-C. Chiang, J. Kaeobamrung and J. W. Bode, *J. Am. Chem. Soc.*, 2007, **129**, 3520–3521.
54. M. He and J. W. Bode, *J. Am. Chem. Soc.*, 2008, **130**, 418–419.
55. J. Kaeobamrung and J. W. Bode, *Org. Lett.*, 2009, **11**, 677–680.
56. V. Nair, B. P. Babu, S. Vellalath, V. Varghese, A. E. Raveendran and E. Suresh, *Org. Lett.*, 2009, **11**, 2507–2510.
57. M. Wadamoto, E. M. Phillips, T. E. Reynolds and K. A. Scheidt, *J. Am. Chem. Soc.*, 2007, **129**, 10098–10099.
58. V. Nair, B. P. Babu, S. Vellalath and E. Suresh, *Chem. Commun.*, 2008, 747–749.
59. M. He, J. R. Struble and J. W. Bode, *J. Am. Chem. Soc.*, 2006, **128**, 8418–8420.
60. P.-C. Chiang, M. Rommel and J. W. Bode, *J. Am. Chem. Soc.*, 2009, **131**, 8714–8718.
61. (a) C. Palomo, M. Oiarbide, J. M. García, A. González and E. Arceo, *J. Am. Chem. Soc.*, 2003, **125**, 13942–13943; (b) M. Reiter, H. Turner, R. Mills-Webb and V. Gouverneur, *J. Org. Chem.*, 2005, **70**, 8478–8485.

62. M. He, G. J. Uc and J. W. Bode, *J. Am. Chem. Soc.*, 2006, **128**, 15088–15089.
63. M. He, B. J. Beahm and J. W. Bode, *Org. Lett.*, 2008, **10**, 3817–3820.
64. E. M. Phillips, M. Wadamoto, A. Chan and K. A. Scheidt, *Angew. Chem., Int. Ed.*, 2007, **46**, 3107–3110.
65. Y.-R. Zhang, H. Lv, D. Zhou and S. Ye, *Chem.–Eur. J.*, 2008, **14**, 8473–8476.
66. X.-N. Wang, H. Lv, X.-L. Huang and S. Ye, *Org. Biomol. Chem.*, 2009, **7**, 346–350.
67. (a) N. Duguet, C. D. Campbell, A. M. Z. Slawin and A. D. Smith, *Org. Biomol. Chem.*, 2008, **6**, 1108–1113; (b) Y.-R. Zhang, L. He, X. Wu, P.-L. Shao and S. Ye, *Org. Lett.*, 2008, **10**, 277–280; (c) L. He, H. Lv, Y.-R. Zhang and S. Ye, *J. Org. Chem.*, 2008, **73**, 8101–8103.
68. J. J. Song, Z. Tan, J. T. Reeves, F. Gallou, N. K. Yee and C. H. Senanayake, *Org. Lett.*, 2005, **7**, 2193–2196.
69. (a) Y. Suzuki, M. D. A. Bakar, K. Muramatsu and M. Sato, *Tetrahedron*, 2006, **62**, 4227–4231; (b) Y. Fukuda, Y. Maeda, S. Ishii, K. Kondo and T. Aoyama, *Synthesis*, 2006, 589–590.
70. J. J. Song, F. Gallou, J. T. Reeves, Z. Tan, N. K. Yee and C. H. Senanayake, *J. Org. Chem.*, 2006, **71**, 1273–1276.
71. T. Kano, K. Sasaki, T. Konishi, H. Mii and K. Maruoka, *Tetrahedron Lett.*, 2006, **47**, 4615–4618.
72. J. J. Song, Z. Tan, J. T. Reeves, D. R. Fandrick, N. K. Yee and C. H. Senanayake, *Org. Lett.*, 2008, **10**, 877–880.
73. J. J. Song, Z. Tan, J. T. Reeves, N. K. Yee and C. H. Senanayake, *Org. Lett.*, 2007, **9**, 1013–1016.
74. C. Fischer, S. W. Smith, D. A. Powell and G. C. Fu, *J. Am. Chem. Soc.*, 2006, **128**, 1472–1473.
75. (a) L. He, T.-Y. Jian and S. Ye, *J. Org. Chem.*, 2007, **72**, 7466–7468; (b) L. He, Y.-R. Zhang, X.-L. Huang and S. Ye, *Synthesis*, 2008, 2825–2829.
76. N. Kuhn, M. Steimann and G. Weyers, *Z. Naturforsch., Teil B*, 1999, **54**, 427–433.
77. H. A. Duong, T. N. Tekavec, A. M. Arif and J. Louie, *Chem. Commun.*, 2004, 112–113.
78. B. R. Van Ausdall, J. L. Glass, K. M. Wiggins, A. M. Aarif and J. Louie, *J. Org. Chem.*, 2009, **74**, 7935–7942.
79. H. Zhou, W.-Z. Zhang, C.-H. Liu, J.-P. Qu and X.-B. Lu, *J. Org. Chem.*, 2008, **73**, 8039–8044.
80. Y. Kayaki, M. Yamamoto and T. Ikariya, *Angew. Chem., Int. Ed.*, 2009, **48**, 4194–4197.
81. I. Tommasi and F. Sorrentino, *Tetrahedron Lett.*, 2009, **50**, 104–107.
82. S. N. Riduan, Y. Zhang and Jackie Y. Ying, *Angew. Chem., Int. Ed.*, 2009, **48**, 3322–3325.
83. N. E. Kamber, W. Jeong, R. M. Waymouth, R. C. Pratt, B. G. G. Lohmeijer and J. L. Hedrick, *Chem. Rev.*, 2007, **107**, 5813–5840.

84. E. F. Connor, G. W. Nyce, M. Myers, A. Mock and J. L. Hedrick, *J. Am. Chem. Soc.*, 2002, **124**, 914–915.
85. (a) S. Csihony, D. A. Culkin, A. C. Sentman, A. P. Dove, R. M. Waymouth and J. L. Hedrick, *J. Am. Chem. Soc.*, 2005, **127**, 9079–9084; (b) O. Coulembier, A. P. Dove, R. C. Pratt, A. C. Sentman, D. A. Culkin, L. Mespouille, P. Dubois, R. M. Waymouth and J. L. Hedrick, *Angew. Chem., Int. Ed.*, 2005, **44**, 4964–4968; (c) O. Coulembier, B. G. G. Lohmeijer, A. P. Dove, R. C. Pratt, L. Mespouille, D. A. Culkin, S. J. Benight, P. Dubois, R. M. Waymouth and J. L. Hedrick, *Macromolecules*, 2006, **39**, 5617–5628; (d) W. Jeong, E. J. Shin, D. A. Culkin, J. L. Hedrick and R. M. Waymouth, *J. Am. Chem. Soc.*, 2009, **131**, 4884–4891.
86. N. E. Kamber, W. Jeong, S. Gonzalez, J. L. Hedrick and R. M. Waymouth, *Macromolecules*, 2009, **42**, 1634–1639.
87. J. Raynaud, C. Absalon, Y. Gnanou and D. Taton, *J. Am. Chem. Soc.*, 2009, **131**, 3201–3209.
88. D. Enders and A. Henseler, *Adv. Synth. Catal.*, 2009, **351**, 1749–1752.
89. J. R. Struble and J. W. Bode, *Tetrahedron*, 2008, **64**, 6961–6972.
90. J. R. Struble, J. Kaeobamrung and J. W. Bode, *Org. Lett.*, 2008, **10**, 957–960.

Subject Index